MW00760639

Radio Propagation and Adaptive Antennas for Wireless Communication Networks

WILEY SERIES IN MICROWAVE AND OPTICAL ENGINEERING

KAI CHANG, Editor
Texas A&M University

A complete list of the titles in this series appears at the end of this volume.

Radio Propagation and Adaptive Antennas for Wireless Communication Networks

Terrestrial, Atmospheric, and Ionospheric

Second Edition

NATHAN BLAUNSTEIN

CHRISTOS G. CHRISTODOULOU

Library of Congress Cataloging-in-Publication Data:

Blaunstein, Nathan.
    Radio propagation and adaptive antennas for wireless communication networks : terrestrial, atmospheric, and ionospheric / Nathan Blaunstein, Christos G. Christodoulou. – Second edition.
        pages cm
    Includes bibliographical references and index.
    ISBN 978-1-118-65954-0 (cloth)
    1. Adaptive antennas.   2. Radio wave propagation.   3. Wireless communication systems– Equipment and supplies.   4. Cell phone systems–Equipment and supplies.   I. Christodoulou, Christos G.   II. Title.
    TK7871.67.A33B537 2013
    621.382'4–dc23

                                                                                    2013008675

Printed in the United States of America

10 9 8 7 6 5 4 3 2 1

Contents

Preface

Since the first edition of this book was published, several changes in wireless communications, including new wireless networks, from third generation (3G) to fourth generation (4G), as well as changes in technologies and their corresponding protocols, from WiFi and WiMAX to LTE and other advanced technologies, have taken place. Therefore, this edition has been rearranged to account for several current and modern aspects of wireless networks based on radio propagation phenomena.

This book is intended to appeal to any scientist, practicing engineer, or designer who is concerned with the operation and service of radio links, including personal, mobile, aircraft, and satellite links. It examines different situations in the over-the-terrain, atmospheric, and ionospheric communication channels, including rural, mixed residential, and built-up environments for terrestrial links, atmospheric turbulences, and different kinds of hydrometeors (rain, clouds, snow, etc.). For each channel the main task of the authors of this book is to explain the role of all kinds of obstructions on the corresponding propagation phenomena that influence the transmission of radio signals through such communication channels, both in line-of-sight (LOS) and obstructive non-line-of-sight (NLOS) propagation conditions along the radio path between the transmitter and the receiver antennas. The book also emphasizes how adaptive antennas at the link terminals can be utilized to minimize the deleterious effects of such obstructions.

To introduce the reader to some relevant topics in radio propagation, antennas, and applied aspects of wireless networks design, we decided to divide the book into five parts. Part I, "Fundamentals of Wireless Links and Networks," is comprised of three chapters. Chapter 1 describes briefly the main parameters and characteristics of radio propagation links, as well as the challenges in using adaptive antennas. Multipath phenomena, path loss, large-scale or slow fading, and short-scale or fast fading and all definitions, parameters, and mathematical descriptions are described thoroughly. Chapter 2 introduces the figures of merit and fundamentals of antennas. In Chapter 3 we introduce the reader to the basics of cellular networks, cell patterns, and cell splitting.

Part II, "Fundamentals of Radio Propagation," contains three chapters and describes radio propagation in various media and their applications in smart communication networks, outdoor and indoor, terrestrial, atmospheric, and ionospheric. In Chapter 4, we present the electrodynamics of radio propagation in free space over various terrains based on the Huygens principle and the Fresnel-zone concept. All aspects of terrestrial radio propagation are covered in Chapter 5. First, we start with a description of the characterization of the terrain and propagation scenarios that occur in the terrestrial communication links. Then, a general stochastic approach is used to perform a link budget for different kinds of outdoor communication links—rural, suburban, and urban—based on the physical aspects of the terrain features. Indoor propagation is the subject of Chapter 6, where in the previous edition only some models for practical applications in indoor communications were presented. In this edition, we introduce new stochastic models supported by numerous experiments.

Part III, "Fundamentals of Adaptive Antennas," encompasses three chapters. Chapter 7 describes the main aspects of adaptive (or smart) antenna system technologies, such as antenna phased arrays and digital beamforming, focusing on their special applications in terrestrial, atmospheric, and ionospheric radio propagation for mobile, personal, aircraft, and satellite communication links. In Chapter 8, a general, three-dimensional, stochastic approach is given to predict the joint azimuth-of-arrival (AOA), elevation-of-arrival (EOA), and time-of-arrival (TOA) ray distribution and the corresponding power spectrum distribution in space, angle, time, and frequency domains for different urban environments, using smart antenna technology. Chapter 9 shows, using the theoretical results described earlier in Chapter 8 and some precisely arranged experiments, carried out in different urban sites, how to predict the location of the base station antenna or the subscriber location.

Part IV, "Practical Aspects of Terrestrial Networks Performance: Cellular and Noncellular," consists of two chapters. In Chapter 10, fading phenomena, in land wireless communication links, are described by a proposed stochastic approach. The advantages and disadvantages of the proposed stochastic approach are discussed. Chapter 11 focuses on the cellular and noncellular communication networks design based on radio propagation phenomena.

Finally, Part V, "Atmospheric and Satellite Communication Links and Networks," includes three chapters. In Chapter 12, the effects of the atmosphere and its features (clouds, fog, hydrometeors, rain, turbulences, etc.) on the loss characteristics of any radio signal are described with examples of how to design a link budget for land-atmospheric communication links. In Chapter 13, we give the reader information on how an inhomogeneous ionosphere, containing quasi-regular layers, large-, average- and small-scale sporadic plasma irregularities, affects radio wave propagation. Finally, Chapter 14 describes the different approaches, statistical or physical-statistical, used today in land-satellite communication links, as well as for mega-cell maps performance.

NATHAN BLAUNSTEIN
CHRISTOS G. CHRISTODOULOU

CHAPTER ONE

Wireless Communication Links with Fading

The purpose of this chapter is to familiarize the reader with the basic propagation characteristics that describe various wireless communication channels, such as terrestrial, atmospheric, and ionospheric from VHF to the X-band. Well-known standards in wireless communication [1–10] are introduced for the prediction of path losses and fading effects of any radio signal in various communication links, and finally, new possibilities that can be obtained using smart antennas are discussed.

1.1. RADIO COMMUNICATION LINK

Different radio communication links (land, land-to-air, air-to-air) covering different atmospheric and ionospheric conditions include several components having a plethora of physical principles and processes, with their own independent or correlated working characteristics and operating elements. A simple scheme of such as a radio communication link consists of a transmitter (T), a receiver (R), and a propagation channel. The main output characteristics of such a link depend on the conditions of radio propagation in different kinds of environments, as shown in Figure 1.1. According to Reference 6, there are three main independent electronic and electromagnetic design tasks related to this wireless communication network. The first task is the transmitter antenna operation, including the specification of the electronic equipment that

Radio Propagation and Adaptive Antennas for Wireless Communication Networks: Terrestrial, Atmospheric, and Ionospheric, Second Edition. Nathan Blaunstein and Christos G. Christodoulou.
© 2014 John Wiley & Sons, Inc. Published 2014 by John Wiley & Sons, Inc.

Wireless Propagation Channel

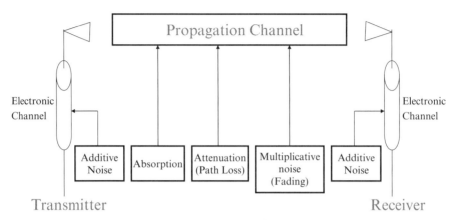

FIGURE 1.1. A wireless communication link scheme.

controls all operations within the transmitter. The second task is to understand, model, and analyze the propagation properties of the channel that connects the transmitting and receiving antennas. The third task concerns the study of all operations related to the receiver.

The propagation channel is influenced by the various obstructions surrounding antennas and the existing environmental conditions. Another important question for a personal receiver (or handheld) antenna is also the influence of the human body on the operating characteristics of the working antenna. The various blocks that comprise a propagation channel are shown in Figure 1.1.

Its main output characteristics depend on the conditions of radiowave propagation in the various operational environments where such wireless communication links are used. Next, we briefly describe the frequency spectrum, used in terrestrial, atmospheric, and ionospheric communications, and we classify some common parameters and characteristics of a radio signal, such as its path loss and fading for various situations which occur in practice.

1.2. FREQUENCY BAND FOR RADIO COMMUNICATIONS

The *frequency band* is a main characteristic for predicting the effectiveness of radio communication links that we consider here. The optimal frequency band for each propagation channel is determined and limited by the technical requirements of each communication system and by the conditions of radio propagation through each channel. First, consider the spectrum of radio frequencies and their practical use in various communication channels [1–5].

Extremely low and *very low frequencies* (ELF and VLF) are frequencies below 3 kHz and from 3 kHz to 30 kHz, respectively. The VLF-band corresponds to waves, which propagate through the wave guide formed by the Earth's surface and the ionosphere, at long distances, with a low degree of attenuation (0.1–0.5 decibel (dB) per 1000 km [1–5]).

Low frequencies (LF) are frequencies from 30 kHz up to 3 MHz. In the 1950s and 1960s, they were used for radio communication with ships and aircraft, but since then they are used mainly with broadcasting stations. Because such radio waves propagate along the ground surface, they are called "surface" waves [1–5]. In terms of wavelength, we call these class of waves the *long* (from 30 kHz to 300 kHz) and *median* (from 300 kHz to 3 MHz) *waves*.

High frequencies (HF) are those which are located in the band from 3 MHz up to 30 MHz. Again, in the wavelength domain, we call these waves the *short waves*. Signals in this spectrum propagate by means of reflections caused by the ionospheric layers and are used for communication with aircrafts and satellites, and for long-distance land communication using broadcasting stations.

Very high frequencies (VHF) (or *short waves* in the wavelength domain) are located in the band from 30 MHz up to 300 MHz. They are usually used for TV communications, in long-range radar systems, and radio-navigation systems.

Ultra high frequencies (UHF) (*ultrashort waves* in the wavelength domain) are those that are located in the band from 300 MHz up to 3 GHz. This frequency band is very effective for wireless microwave links, constructions of cellular systems (fixed and mobile), mobile-satellite communication channels, medium-range radars, and other applications.

In recent decades, radio waves with frequencies higher than 3 GHz (C-, X-, K-bands, up to several hundred gigahertz, which in the literature are referred to as *microwaves* and *millimeter waves*), have begun to be widely used for constructing and performing modern wireless communication channels.

1.3. NOISE IN RADIO COMMUNICATION LINKS

The effectiveness of each radio communication link—land, atmospheric, or ionospheric—depends on such parameters as [9]:

- noise in the transmitter and in the receiver antennas
- noise within the electronic equipment that communicates with both antennas
- background and ambient noise (cosmic, atmospheric, artificial/man-made, etc.).

Now let us briefly consider each type of noise, which exists in a complete communication system. In a wireless channel, specifically, the noise sources can be subdivided into *additive* (or *white*) and *multiplicative* effects, as seen in Figure 1.1 [6, 7, 10].

The *additive noise* arises from noise generated within the receiver itself, such as thermal noise in passive and active elements of the electronic devices, and also from external sources such as atmospheric effects, cosmic radiation, and man-made noise. The clear and simple explanation of the first component of additive noise is that noise is generated within each element of the electronic communication channel due to the random motion of the electrons within the various components of the equipment [5]. According to the theory of thermodynamics, the noise energy can be determined by the average background temperature, T_0, as [1–5]

$$E_N = k_B T_0, \tag{1.1}$$

where

$$k_B = 1.38 \times 10^{-23} W \times s \times K^{-1} \tag{1.2}$$

is Boltzman's constant, and $T_0 = 290$ K $= 17°$C. This energy is uniformly distributed in the frequency band and hence it is called "white noise." The total effective noise power at the receiver input is given by the following expression:

$$N_F = k_B T_0 B_W F, \tag{1.3}$$

where F is the *noise figure* at the receiver and B_w is the bandwidth of the signal. The noise figure represents any additional noise effects related to the corresponding environment, and it is expressed as

$$F = 1 + \frac{T_e}{T_0}. \tag{1.4}$$

Here T_e is the effective temperature, which accounts for all ambient natural (weather, cosmic noise, clouds, rain, etc.) and man-made (industry, plants, power engine, power stations, etc.) effects.

The *multiplicative noise* arises from the various processes inside the propagation channel and depends mostly on the directional characteristics of both terminal antennas, on the reflection, absorption, scattering, and diffraction phenomena caused by various natural and artificial obstructions placed between and around the transmitter and the receiver (see Fig. 1.2). Usually, the multiplicative process in the propagation channel is divided into three types: *path loss*, *large-scale* (or *slow fading*), and *short-scale* (or *fast fading*)

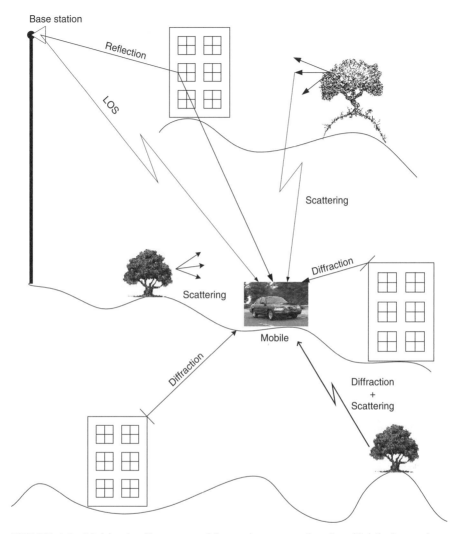

FIGURE 1.2. Multipath effects caused by various natural and artificial obstructions placed between and around the transmitting and the receiving antennas.

[7–10]. We describe these three characteristics of the multiplicative noise separately in the following section.

1.4. MAIN PROPAGATION CHARACTERISTICS

In real communication channels, the field that forms the complicated interference picture of received radio waves arrives via several paths simultaneously, forming a multipath situation. Such waves combine vectorially to give an oscillating resultant signal whose variations depend on the distribution of phases among the incoming total signal components. The signal amplitude variations are known as the *fading* effect [1–4, 6–10]. Fading is basically a spatial phenomenon, but spatial signal variations are experienced, according to the ergodic theorem [11, 12], as temporal variations by a receiver/transmitter moving through the multipath field or due to the motion of scatterers, such as a truck, aircraft, helicopter, satellite, and so on. Thus, we can talk here about space-domain and time-domain variations of EM fields in different radio environments, as well as in the frequency domain. Hence, if we consider mobile, mobile-to-aircraft or mobile-to-satellite communication links, we may observe the effects of random fading in the frequency domain, that is, the complicated interference picture of the received signal caused by receiver/transmitter movements, which is defined as the "Doppler shift" effect [1–7, 10].

Numerous theoretical and experimental investigations in such conditions have shown that the spatial and temporal variations of signal level have a triple nature [1–7, 10]. The first one is the *path loss*, which can be defined as a large-scale smooth decrease in signal strength with distance between two terminals, mainly the transmitter and the receiver. The physical processes which cause these phenomena are the spreading of electromagnetic waves radiated outward in space by the transmitter antenna and the obstructing effects of any natural or man-made objects in the vicinity of the antenna. The spatial and temporal variations of the signal path loss are large and slow, respectively.

Large-scale (in the space domain) or *slow* (in the time domain) fading is the second nature of signal variations and is caused by diffraction from the obstructions placed along the radio link surrounding the terminal antennas. Sometimes this fading phenomenon is called the *shadowing effect* [6, 7, 10].

During shadow fading, the signal's slow random variations follow either a Gaussian distribution or a log-normal distribution if the signal fading is expressed in decibels. The spatial scale of these slow variations depends on the dimensions of the obstructions, that is, from several to several tens of meters. The variations of the total EM field describe its structure within the shadow zones and are called *slow-fading* signals.

The third nature of signal variations is the *short-scale* (in the space domain) or *fast* (in the time domain) signal variations, which are caused by the mutual interference of the wave components in the multiray field. The characteristic scale of such waves in the space domain varies from half wavelength to three wavelengths. Therefore, these signals are usually called *fast-fading* signals.

1.4.1. Path Loss

The path loss is a figure of merit that determines the effectiveness of the propagation channel in different environments. It defines variations of the signal amplitude or field intensity along the propagation trajectory (*path*) from one point to another within the communication channel. In general [1–3, 6–10], the *path loss* is defined as a logarithmic difference between the amplitude or the intensity (called *power*) at any two different points, $\mathbf{r}_1$ (the transmitter point) and $\mathbf{r}_2$ (the receiver point), along the propagation path in the medium. The path loss, which is denoted by L and is measured in decibels, can be evaluated as follows [5]:

for a signal amplitude of $A(\mathbf{r}_j)$ at two points $\mathbf{r}_1$ and $\mathbf{r}_2$ along the propagation path

$$L = 10 \cdot \log \frac{A^2(\mathbf{r}_2)}{A^2(\mathbf{r}_1)} = 10 \cdot \log A^2(\mathbf{r}_2) - 10 \cdot \log A^2(\mathbf{r}_1)$$
$$= 20 \cdot \log A(\mathbf{r}_2) - 20 \cdot \log A(\mathbf{r}_1) \, [\text{dB}]; \tag{1.5}$$

for a signal intensity $J(\mathbf{r}_j)$ at two points $\mathbf{r}_1$ and $\mathbf{r}_2$ along the propagation path

$$L = 10 \cdot \log \frac{J(\mathbf{r}_2)}{J(\mathbf{r}_1)} = 10 \cdot \log J(\mathbf{r}_2) - 10 \cdot \log J(\mathbf{r}_1) \, [\text{dB}]. \tag{1.6}$$

If we assume now $A(\mathbf{r}_1) = 1$ at the transmitter, then

$$L = 20 \cdot \log A(\mathbf{r}) \, [\text{dB}] \tag{1.7a}$$

and

$$L = 10 \cdot \log J(\mathbf{r}) \, [\text{dB}]. \tag{1.7b}$$

For more details about how to measure the path loss, the reader is referred to References 1–3 and 6–10. As any signal passing through the propagation channel passes through the transmitter electronic channel and the receiver electronic channel (see Fig. 1.1), both electronic channels together with the environment introduce additive or white noise into the wireless communication system. Therefore, the second main figure of merit of radio communication channels is the signal-to-noise ratio (SNR or S/N). In decibels this SNR can be written as

$$\text{SNR} = P_R - N_R \, [\text{dB}], \tag{1.8}$$

where P_R is the signal power at the receiver, and N_R is the noise power at the receiver.

1.4.2. Characteristics of Multipath Propagation

Here we start with the general description of *slow* and *fast* fading.

Slow Fading. As was mentioned earlier, the *slow* spatial signal variations (expressed in decibels) tend to have a log-normal distribution or a Gaussian distribution (expressed in watts [W]) [1–4, 6–10]. The *probability density function* (PDF) of the signal variations with the corresponding standard deviation, averaged within some individual small area or over some specific time period, depends on the nature of the terrain, of the atmospheric and ionospheric conditions. This PDF is given by [1–4]

$$PDF(r) = \frac{1}{\sigma_L \sqrt{2\pi}} \exp\left\{ -\frac{(r-\bar{r})^2}{2\sigma_L^2} \right\}. \tag{1.9a}$$

The corresponding *cumulative distributed function* (CDF) (or the total probability itself) is given by [1–4]

$$CDF(Z) \equiv \Pr(r < Z) = \int_0^Z PDF(r)dr. \tag{1.9b}$$

Here $\bar{r} = \langle r \rangle$ is the mean value of the random signal level, r is the value of the received signal strength or voltage envelope, $\sigma_L = \langle r^2 - \bar{r}^2 \rangle$ is the variance or time-average power ($\langle r \rangle$ indicates the averaging operation of a variable r of the received signal envelope), and Z is the slow fade margin giving maximum effect of slow fading on the signal envelope slow variations. Usually, slow fading, described by a Gaussian PDF or CDF, can be presented via the Q-function defined as a *complementary cumulative distribution function* (CCDF) [11, 12]

$$Q(Z) \equiv CCDF(Z) = 1 - CDF(Z) \equiv \Pr(r > Z) \tag{1.10}$$

In References 11 and 12, it was shown that this function is closely related to the so-called error function, erf(w), usually used in description of the stochastic processes in probability theory and in statistical mechanics:

$$Q(w) = \frac{1}{\sqrt{2\pi}} \int_{x=w}^{\infty} \exp\left\{ -\frac{x^2}{2} \right\} dx = \frac{1}{2} \mathrm{erf}\left\{ \frac{w}{\sqrt{2}} \right\}$$

and

$$\mathrm{erf}(w) = \frac{2}{\sqrt{\pi}} \int_0^w \exp(-y^2) dy.$$

Fast Fading. For the case of stationary receiver and transmitter (*static multipath channel*), due to multiple reflections and scattering from various obstructions surrounding the transmitter and receiver, the radio signals travel along different paths of varying lengths, causing such fast deviations of the signal strength (in volts) or power (in watts) at the receiver.

In the case of a *dynamic multipath* situation, either the subscribers' antenna is in movement or the objects surrounding the stationary antennas are moving, so the spatial variations of the resultant signal at the receiver can be seen as temporal variations [11, 12]. The signal received by the mobile at any spatial point may consist of a large number of signals having randomly distributed amplitudes, phases, and angles of arrival, as well as different time delays. All these features change the relative phase shifts as a function of the spatial location and, finally, cause the signal to fade in the space domain. In a dynamic (mobile) multipath situation, the signal fading at the mobile receiver occurs in the time domain. This temporal fading is associated with a shift of frequency radiated by the stationary transmitter. In fact, the time variations, or dynamic changes of the propagation path lengths, are related to the Doppler effect, which is due to relative movements between a stationary base station (BS) and a moving subscriber (MS).

To illustrate the effects of phase change in the time domain due to the Doppler frequency shift (called the Doppler effect [1–4, 6–10]), let us consider a mobile moving at a constant velocity v, along the path XY, as shown in Figure 1.3. The difference in path lengths traveled by a signal from source S to the mobile at points X and Y is $\Delta \ell = \ell \cos\theta = v\Delta t \cos\theta$, where Δt is the time required for the moving receiver to travel from point X to Y along the path, and θ is the angle between the mobile direction along XY and direction to the

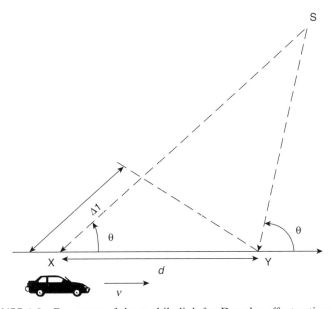

FIGURE 1.3. Geometry of the mobile link for Doppler effect estimation.

source at the current point Y, that is, YS. The phase change of the resultant received signal due to the difference in path lengths is therefore

$$\Delta\Phi = k\Delta\ell = \frac{2\pi}{\lambda}\ell\cos\theta = \frac{2\pi v\Delta t}{\lambda}\cos\theta. \tag{1.11}$$

Hence, the apparent change in frequency radiated, or Doppler shift, is given by f_D, where

$$f_D = \frac{1}{2\pi}\frac{\Delta\Phi}{\Delta t} = \frac{v}{\lambda}\cos\theta. \tag{1.12}$$

It is important to note from Figure 1.3 that the angles θ for points X and Y are the same only when the corresponding lines XS and YS are parallel. Hence, this figure is correct only in the limit when the terminal S is far away from the moving antenna at points X and Y. Many authors have ignored this fact during their geometrical explanation of the Doppler effect [1–4, 10]. Because the Doppler shift is related to the mobile velocity and the spatial angle between the direction of mobile motion and the direction of arrival of the signal, it can be positive or negative depending on whether the mobile receiver is moving toward or away from the transmitter. In fact, from Equation (1.12), if the mobile moves *toward* the direction of arrival of the signal with radiated frequency f_c, then the received frequency is increased; that is, the apparent frequency is $f_C + f_D$. When the mobile moves away from the direction of arrival of the signal, then the received frequency is decreased; that is, the apparent frequency is $f_C - f_D$. The maximum Doppler shift is $f_{Dmax} = v/\lambda$, which in further description will denote simply as f_m.

There are many probability distribution functions that can be used to describe the fast fading effects, such as Rayleigh, Suzuki, Rician, gamma, gamma-gamma, and so on. Because the Rician distribution is more general for description of fast fading effects in terrestrial communication links [1–4, 10], as it includes both line-of-sight (LOS) together with scattering and diffraction with non-line-of-sight (NLOS), we briefly describe it in the following paragraph.

To estimate the contribution of each signal component, at the receiver, due to the dominant (or LOS) and the secondary (or multipath), the Rician parameter K is usually introduced, as a ratio between these components [1–4, 10], that is,

$$K = \frac{\text{LOS} - \text{Component power}}{\text{Multipath} - \text{Component power}} \tag{1.13}$$

The Rician PDF distribution of the signal strength or voltage envelope r can be defined as [1–4, 10]:

$$PDF(r) = \frac{r}{\sigma^2} \exp\left\{-\frac{r^2 + A^2}{2\sigma^2}\right\} I_0\left(\frac{Ar}{\sigma^2}\right), \text{ for } A > 0, r \geq 0, \tag{1.14}$$

where A denotes the peak strength or voltage of the dominant component envelope, σ is the standard deviation of signal envelope, and $I_0(\cdot)$ is the modified Bessel function of the first kind and zero order. According to definition (1.13), we can now rewrite the parameter K, which was defined earlier as the ratio between the *dominant* and the *multipath* component power. It is given by

$$K = \frac{A^2}{2\sigma^2}. \tag{1.15}$$

Using (1.15), we can rewrite (1.14) as a function of K only [1–3, 10]:

$$PDF(r) = \frac{r}{\sigma^2} \exp\left\{-\frac{r^2}{2\sigma^2}\right\} \exp(-K)) I_0\left(\frac{r}{\sigma}\sqrt{2K}\right). \tag{1.16}$$

Using such presentation of Rician PDF, one can easily obtain the mean value and the variance as functions of the parameter K, called also the *fading parameter*. Thus, according to definitions of the mean value and the variance [3, 6], we get

$$\mu_r(K) = \int_0^\infty r \cdot PDF(r) dr = \left[(1+K)I_0\left(\sqrt{2Kr}\right) + KI_1(K/2)\right] \tag{1.17}$$

and

$$\sigma_r^2(K) = \int_0^\infty r^2 \cdot PDF(r) dr = 2 \cdot (1+K) - \mu_r^2. \tag{1.18}$$

Here, $I_1(\cdot)$ is the modified Bessel function of the first kind and first order.

For $K = 0$, that is, the worst case of the fading channel, in expression (1.16), the term $\exp(-K) = 1$ and $I_0(0) = 1$. This worst-case scenario is described by the Rayleigh PDF, when there is NLOS signal only and it is equal to

$$PDF(r) = \frac{r}{\sigma^2} \exp\left\{-\frac{r^2}{2\sigma^2}\right\}. \tag{1.19}$$

Conversely, in a situation of good clearance between two terminals with no multipath components, that is, when $K \to \infty$, the Rician fading approaches a Gaussian one, yielding a "Dirac-delta shaped" PDF described by Equation (1.9a) (see Fig. 1.4). We will use these definitions in Chapter 11 for the link budget design inside a terrestrial radio communication system.

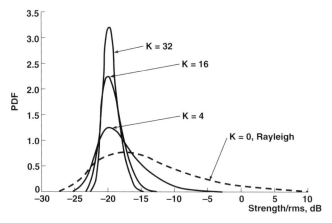

FIGURE 1.4. Rician PDF distribution versus ratio of signal to rms.

Finally, for the practical point of view, we will present the mean and the variance of Rician distribution, respectively. Thus, for $K < 2$, from expression (1.17) it follows, according to Reference 14, that

$$\mu_r(K) = \sqrt{\frac{\pi}{2}} + \sum_{n=1}^{\infty} \frac{2}{\pi} \cdot \frac{(-1)^n}{(2n-1)!} K^n, \qquad (1.20a)$$

whereas for $K \geq 2$,

$$\mu_r(K) = \sqrt{2K}\left(1 - \frac{1}{4K} + \frac{1}{K^2}\right). \qquad (1.20b)$$

The same approximations can be obtained for the variance following derivations of expression (1.18) made in Reference 14. Thus, for $K < 2$,

$$\sigma_r^2(K) = 1 - \left(\frac{\pi}{2} - 1\right)\cdot\exp\left\{-\frac{K}{\sqrt{2}}\right\}, \qquad (1.21a)$$

whereas for $K \geq 2$,

$$\sigma_r^2(K) = 1 - \frac{1}{4K}\left(1 + \frac{1}{K} - \frac{1}{K^2}\right). \qquad (1.21b)$$

Using now relations between the PDF and CDF, we can obtain from expression (1.19) the Rayleigh CDF presentation:

$$CDF(R) \equiv \Pr(r \leq R) = \int_0^R PDF(r)dr = 1 - \exp\left\{-\frac{R^2}{2\sigma_r^2}\right\}. \qquad (1.22)$$

Now, using (1.16) for the Rician PDF, we have a more difficult equation for Rician CDF with respect to Rayleigh CDF due to summation of an infinite number of terms, such as

$$CDF(R) = 1 - \exp\left\{-\left(K + \frac{r^2}{2\sigma_r^2}\right)\right\} \cdot \sum_{m=0}^{\infty} \left(\frac{\sigma_r \sqrt{2K}}{r}\right) \cdot I_m\left(\frac{r \cdot \sqrt{2K}}{\sigma_r}\right). \quad (1.23)$$

Here $I_m(\cdot)$ is the modified Bessel function of the first kind and mth-order. Once more, the Rician CDF depends on one parameter K only and limits to the Rayleigh CDF and Gaussian CDF for $K = 0$ and for $K \to \infty$, respectively. Clearly, the CDF Equation (1.23) is more complicated to evaluate analytically or numerically than the PDF Equation (1.16). However, in practical terms, it is sufficient to use m up to the value where the last term's contribution becomes less than 0.1%. It was shown in Reference 9 that for a Rician CDF with $K = 2$, the 14-dB fading outage probability is about 10^{-2}.

1.4.3. Signal Presentation in Wireless Communication Channels

To understand how to describe mathematically multipath fading in communication channels, we need to understand what kinds of signals we "deal" with in each channel.

Narrowband (CW) Signals. A voice-modulated CW signal occupies a very narrow bandwidth surrounding the carrier frequency f_c of the radio frequency (RF) signal (e.g., the carrier), which can be expressed as

$$x(t) = A(t)\cos[2\pi f_c t + \varphi(t)], \quad (1.24)$$

where $A(t)$ is the signal envelope (i.e., slowly varied amplitude) and $\varphi(t)$ is its signal phase. For example, for a modulated 1-GHz carrier signal by a signal of bandwidth $\Delta f = 2f_m = 8\ kHz$, the fractional bandwidth is very narrow, that is, $8 \times 10^3 \text{Hz}/1 \times 10^9 \text{Hz} = 8 \times 10^{-6}$ or $8 \times 10^{-4}\%$. Since all information in the signal is contained within the phase and envelope-time variations, an alternative form of a bandpass signal $x(t)$ is introduced [1, 2, 6–10]:

$$y(t) = A(t)\exp\{j\varphi(t)\}, \quad (1.25)$$

which is also called the *complex baseband* representation of $x(t)$. By comparing (1.24) and (1.25), we can see that the relation between the *bandpass* (RF) and the *complex baseband* signals are related by:

$$x(t) = \text{Re}[y(t)\exp(j2\pi f_c t)]. \quad (1.26)$$

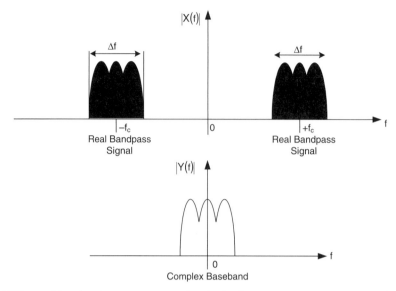

FIGURE 1.5. The signal power presentation in the frequency domain: bandpass (upper figure) and baseband (lower figure).

The relations between these two representations of the narrowband signal in the frequency domain is shown schematically in Figure 1.5. One can see that the complex baseband signal is a frequency-shifted version of the bandpass (RF) signal with the same spectral shape, but centered around a zero frequency instead of the f_c [7]. Here, $X(f)$ and $Y(f)$ are the Fourier transform of $x(t)$ and $y(t)$, respectively, and can be presented in the following manner [1, 2]:

$$Y(f) = \int_{-\infty}^{\infty} y(t)e^{-j2\pi ft}dt = \text{Re}[Y(f)] + j\,\text{Im}[Y(f)] \tag{1.27}$$

and

$$X(f) = \int_{-\infty}^{\infty} x(t)e^{-j2\pi ft}dt = \text{Re}[X(f)] + j\,\text{Im}[X(f)]. \tag{1.28}$$

Substituting for $x(t)$ in integral (1.28) from (1.26) gives

$$X(f) = \int_{-\infty}^{\infty} \text{Re}\left[y(t)e^{j2\pi f_c t}\right]e^{-j2\pi ft}dt. \tag{1.29}$$

Taking into account that the real part of any arbitrary complex variable w can be presented as

$$\text{Re}[w] = \frac{1}{2}[w + w^*]$$

where w^* is the complex conjugate, we can rewrite (1.29) in the following form:

$$X(f) = \frac{1}{2}\int_{-\infty}^{\infty}\left[y(t)e^{j2\pi f_c t} + y^*(t)e^{-j2\pi f_c t}\right] \cdot e^{-j2\pi ft}\,dt. \tag{1.30}$$

After comparing expressions (1.27) and (1.30), we get

$$X(f) = \frac{1}{2}[Y(f - f_c) + Y^*(-f - f_c)]. \tag{1.31}$$

In other words, the spectrum of the real bandpass signal $x(t)$ can be represented by real part of that for the complex baseband signal $y(t)$ with a shift of $\pm f_c$ along the frequency axis. It is clear that the baseband signal has its frequency content centered on the "zero" frequency value.

Now we notice that the mean power of the baseband signal $y(t)$ gives the same result as the mean-square value of the real bandpass (RF) signal $x(t)$, that is,

$$\langle P_y(t)\rangle = \frac{\langle|y(t)|^2\rangle}{2} = \frac{\langle y(t)y^*(t)\rangle}{2} \equiv \langle P_x(t)\rangle. \tag{1.32}$$

The complex envelope $y(t)$ of the received narrowband signal can be expressed according to (1.25), within the multipath wireless channel, as a sum of phases of N baseband individual multiray components arriving at the receiver with their corresponding time delay, $\tau_i, i = 0, 1, 2, \ldots, N - 1$ [6–10]:

$$y(t) = \sum_{i=0}^{N-1} u_i(t) = \sum_{i=0}^{N-1} A_i(t)\exp[j\varphi_i(t, \tau_i)]. \tag{1.33}$$

If we assume that during the subscriber movements through the local area of service the amplitude A_i time variations are small enough, whereas phases φ_i vary greatly due to changes in propagation distance between the BS and the desired subscriber, then there are great random oscillations of the total signal $y(t)$ at the receiver during its movement over a small distance. Since $y(t)$ is the phase sum in (1.33) of the individual multipath components, the instantaneous phases of the multipath components result in large fluctuations, that is, fast fading, in the CW signal. The average received power for such a signal over a local area of service can be presented according to References 1–3 and 6–10 as

$$\langle P_{\mathrm{CW}} \rangle \approx \sum_{i=0}^{N-1} \langle A_i^2 \rangle + 2 \sum_{i=0}^{N-1} \sum_{i,j \neq i} \langle A_i A_j \rangle \langle \cos[\varphi_i - \varphi_j] \rangle. \tag{1.34}$$

Wideband (Pulse) Signals. The typical *wideband* or *impulse* signal passing through the multipath communication channel is shown schematically in Figure 1.6a according to References 1–4. If we divide the time-delay axis into equal segments, usually called bins, then there will be a number of received signals, in the form of vectors or delta functions. Each bin corresponds to a different path whose time of arrival is within the bin duration, as depicted in Figure 1.6b. In this case, the time-varying discrete-time impulse response can be expressed as

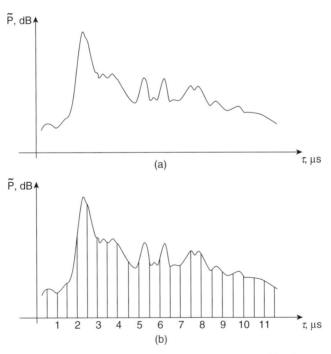

FIGURE 1.6. (a) A typical impulse signal passing through a multipath communication channel according to References 1–4. (b) The use of bins, as vectors, for the impulse signal with spreading.

$$h(t, \tau) = \left\{ \sum_{i=0}^{N-1} A_i(t, \tau) \exp[-j2\pi f_c \tau_i(t)] \delta(\tau - \tau_i(t)) \right\} \exp[-j\varphi(t, \tau)]. \qquad (1.35)$$

If the channel impulse response is assumed to be time invariant, or is at least stationary over a short-time interval or over a small-scale displacement of the receiver/transmitter, then the impulse response (1.35) reduces to

$$h(t, \tau) = \sum_{i=0}^{N-1} A_i(\tau) \exp[-j\theta_i] \delta(\tau - \tau_i), \qquad (1.36)$$

where $\theta_i = 2\pi f_c \tau_i + \varphi(\tau)$. If so, the received power delay profile for a wideband or pulsed signal averaged over a small area can be presented simply as a sum of the powers of the individual multipath components, where each component has a random amplitude and phase at any time, that is,

$$\langle P_{pulse} \rangle = \left\langle \sum_{i=0}^{N-1} \{A_i(\tau) | \exp[-j\theta_i] \|\}^2 \right\rangle \approx \sum_{i=0}^{N-1} \langle A_i^2 \rangle. \qquad (1.37)$$

The received power of the wideband or pulse signal does not fluctuate significantly when the subscriber moves within a local area, because in practice, the amplitudes of the individual multipath components do not change widely in a local area of service.

Comparison between small-scale presentations of the average power of the narrowband (CW) and wideband (pulse) signals, that is, (1.34) and (1.37), shows that when $\langle A_i A_j \rangle = 0$ or/and $\langle \cos[\varphi_i - \varphi_j] \rangle = 0$, the average power for CW signal and that for pulse are equivalent. This can occur when either the path amplitudes are uncorrelated, that is, each multipath component is independent after multiple reflections, diffractions, and scattering from obstructions surrounding both the receiver and the transmitter or the BS and the subscriber antenna. It can also occur when multipath phases are independently and uniformly distributed over the range of $[0. 2\pi]$. This property is correct for UHF/X-wavebands when the multipath components traverse differential radio paths having hundreds of wavelengths [6–10].

1.4.4. Parameters of the Multipath Communication Channel

So the question that remains to be answered is which kind of fading occurs in a given wireless channel.

Time Dispersion Parameters. First, we need to mention some important parameters for wideband (pulse) signals passing through a wireless channel. These parameters are determined for a certain threshold level X (in dB) of the channel under consideration and from the signal power delay profile. These parameters are the mean excess delay, the rms delay spread, and the excess delay spread.

The *mean excess delay* is the first moment of the power delay profile of the pulse signal and is defined as

$$\langle \tau \rangle = \frac{\sum_{i=0}^{N-1} A_i^2 \tau_i}{\sum_{i=0}^{N-1} A_i^2} = \frac{\sum_{i=0}^{N-1} P(\tau_i)\tau_i}{\sum_{i=0}^{N-1} P(\tau_i)}. \tag{1.38}$$

The *rms delay spread* is the square root of the second central moment of the power delay profile and is defined as

$$\sigma_\tau = \sqrt{\langle \tau^2 \rangle - \langle \tau \rangle^2}, \tag{1.39}$$

where

$$\langle \tau^2 \rangle = \frac{\sum_{i=0}^{N-1} A_i^2 \tau_i^2}{\sum_{i=0}^{N-1} A_i^2} = \frac{\sum_{i=0}^{N-1} P(\tau_i)\tau_i^2}{\sum_{i=0}^{N-1} P(\tau_i)}. \tag{1.40}$$

These delays are measured relative to the first detectable signal arriving at the receiver at $\tau_0 = 0$. We must note that these parameters are defined from a single power delay profile, which was obtained after temporal or local (small-scale) spatial averaging of measured impulse response of the channel [1–3, 7–10].

Coherence Bandwidth. The power delay profile in the time domain and the power spectral response in the frequency domain are related through the Fourier transform. Hence, to describe a multipath channel in full, both the delay spread parameters in the time domain and the *coherence bandwidth* in the frequency domain are used. As mentioned earlier, the coherence bandwidth is the statistical measure of the frequency range over which the channel is considered "flat." In other words, this is a frequency range over which two frequency signals are strongly amplitude correlated. This parameter, actually, describes the time-dispersive nature of the channel in a small-scale (local) area. Depending on the degree of amplitude correlation of two frequency separated signals, there are different definitions for this parameter.

The first definition is the *coherence bandwidth*, B_c, which describes a bandwidth over which the frequency correlation function is above 0.9 or 90%, and it is given by

$$B_c \approx 0.02\sigma_\tau^{-1}. \tag{1.41}$$

The second definition is the *coherence bandwidth, B_c*, which describes a bandwidth over which the frequency correlation function is above 0.5 or 50%, or

$$B_c \approx 0.2\sigma_\tau^{-1}. \tag{1.42}$$

There is not any single exact relationship between coherence bandwidth and rms delay spread, and expressions (1.41) and (1.42) are only approximate equations [1–6, 7–10].

Doppler Spread and Coherence Time. To obtain information about the time-varying nature of the channel caused by movements, from either the transmitter/receiver or scatterers located around them, new parameters, such as the *Doppler spread* and the *coherence time*, are usually introduced to describe the time variation phenomena of the channel in a small-scale region. The Doppler spread B_D is defined as a range of frequencies over which the received Doppler spectrum is essentially nonzero. It shows the spectral spreading caused by the time rate of change of the mobile radio channel due to the relative motions of vehicles (and scatterers around them) with respect to the BS. According to References 1–4 and 7–10, the Doppler spread B_D depends on the Doppler shift f_D and on the angle α between the direction of motion of any vehicle and the direction of arrival of the reflected and/or scattered waves (see Fig. 1.3). If we deal with the complex baseband signal presentation, then we can introduce the following criterion: If the baseband signal bandwidth is greater than the Doppler spread B_D, the effects of Doppler shift are negligible at the receiver.

Coherence time T_c is the time domain dual of Doppler spread, and it is used to characterize the time-varying nature of the frequency dispersiveness of the channel in time coordinates. The relationship between these two-channel characteristics is

$$T_c \approx \frac{1}{f_m} = \frac{\lambda}{v}. \tag{1.43}$$

We can also define the coherence time according to References 1–4 and 7–10 as the time duration over which two multipath components of the received signal have a strong potential for amplitude correlation. One can also define the coherence time as the time over which the correlation function of two different signals in the time domain is above 0.5 (or 50%). Then, according to References 7 and 10, we get

$$T_c \approx \frac{9}{16\pi f_m} = \frac{9\lambda}{16\pi v} = 0.18\frac{\lambda}{v}. \tag{1.44}$$

This definition is approximate and can be improved for modern digital communication channels by combining Equations (1.43) and (1.44) to get:

$$T_c \approx \frac{0.423}{f_m} = 0.423\frac{\lambda}{v}. \qquad (1.45)$$

The definition of coherence time implies that two signals arriving at the receiver with a time separation greater than T_c are affected differently by the channel.

1.4.5. Types of Fading in Multipath Communication Channels

Let us now summarize the effects of fading, which may occur in static or dynamic multipath communication channels.

Static Channel. In this case, multipath fading is purely spatial and leads to constructive or destructive interference at various points in space at any given instant in time, depending on the relative phases of the arriving signals. Furthermore, fading in the frequency domain does not change because the two antennas are stationary. The signal parameters of interest, such as the signal bandwidth, B_s, the time of duration, T_s, with respect to the coherent time, T_c, and the coherent bandwidth, B_c, of the channel are shown in Figure 1.7. There are two types of fading that occur in static channels:

A. *Flat Slow Fading* (FSF) (see Fig. 1.8), where the following relations between *signal* parameters of the signal and a channel are valid [7–10]:

$$T_c >> T_S; \quad B_D \cong 0 << B_S; \quad \sigma_\tau < T_S; \quad B_c \sim \frac{0.02}{\sigma_\tau} > B_S \qquad (1.46)$$

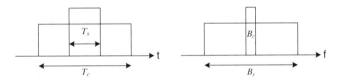

FIGURE 1.7. Comparison between signal and channel parameters.

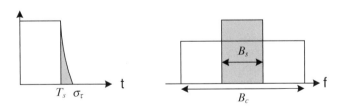

FIGURE 1.8. Relations between parameters for flat slow fading.

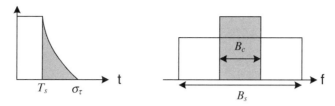

FIGURE 1.9. Relations between parameters for flat fast fading.

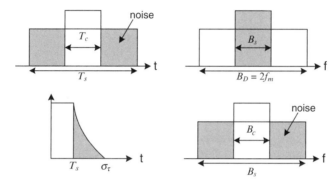

FIGURE 1.10. Relations between parameters for frequency selective fast fading.

Here all harmonics of the total signal are coherent.

B. *Flat Fast Fading* (FFF) (see Fig. 1.9), where the following relations between the parameters of a channel and the signal are valid [7–10]:

$$T_c >> T_S; \quad B_D \cong 0 << B_S; \quad \sigma_\tau > T_S; \quad B_c < B_S \quad\quad (1.47)$$

Dynamic Channel. There are two additional types of fading that occur in a dynamic channel.

A. *Frequency Selective Fast Fading* (FSFF) (see Fig. 1.10), when fast fading depends on the frequency. In this case, following relations between the parameters of a channel and the signal are valid [7–10]:

$$T_c < T_S; \quad B_D > B_S; \quad \sigma_\tau > T_S; \quad B_c < B_S \quad\quad (1.48)$$

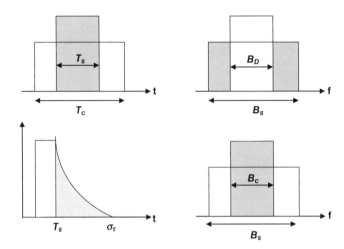

FIGURE 1.11. Relations between parameters for frequency selective slow fading.

B. *Frequency Selective Slow Fading* (FSSF) (see Fig. 1.11), when slow fading depends on the frequency. In this case, following relations between the parameters of a channel and the signal are valid [7–10]:

$$T_c > T_S; \quad B_D < B_S; \quad \sigma_\tau > T_S; \quad B_c < B_S \tag{1.49}$$

Using these relationships between the parameters of the signal and that of a channel, we can define, a priori, the type of fading mechanism which may occur in a wireless communication link (see Fig. 1.12).

1.4.6. Characterization of Multipath Communication Channels with Fading

In previous subsections, we described situations that occur in real communication channels, where natural propagation effects for each specific environment are very actual. Namely, as will be shown in Chapter 13, the ionospheric channel can be considered as a time-varying channel due to scattering from the ionosphere (see Fig. 1.13). Thus, in the ionosphere, due to plasma movements, the parameters of fading have a dispersive character—time dispersive or frequency dispersive. The same effects of the multiplicative noise will be analyzed in Chapter 14 for the land-to-satellite communication channel.

We present another example of terrestrial multipath channel (see Chapters 5 and 8), caused by multipath propagation of rays reflected from building walls (see Fig. 1.14a), diffracted from a hill (see Fig. 1.14b), and scattered from a tree (see Fig. 1.14c). All these channels we define as *time-varying* or *frequency-varying* multipath channels with fading depend on their "reaction" of the propagating radio wave with the environment. Thus, in land communication channels due to multiple scattering and diffraction, the channel becomes fre-

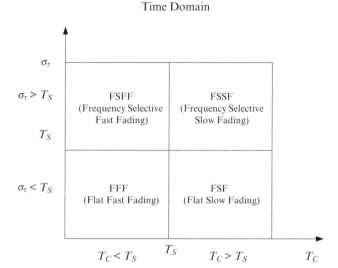

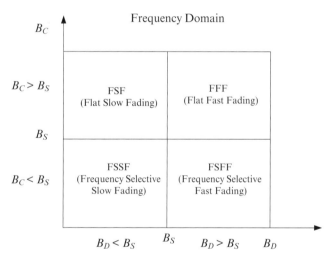

FIGURE 1.12. Common picture of different kinds of fading, depending on relations between the signal and the channel main parameters.

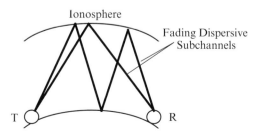

FIGURE 1.13. Multipath phenomena in the land–ionospheric link.

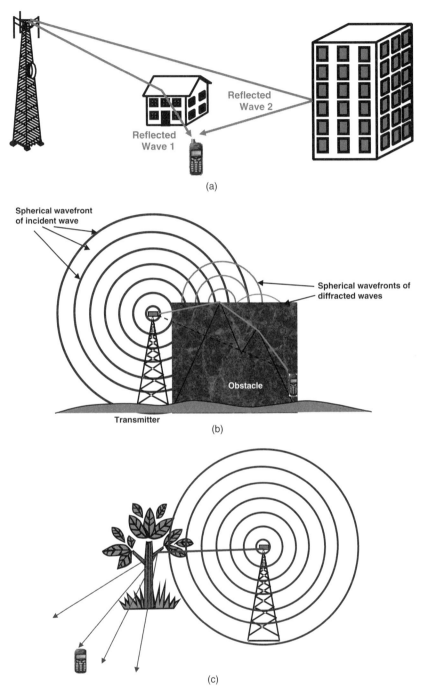

FIGURE 1.14. (a) Specular reflection from walls and roofs. (b) Multiple diffraction from hills. (c) Multiple scattering from tree's leaves.

quency selective. If one of the antennas is moving—mostly the subscriber (MS) antenna—the channel is a *time-dispersive* channel. For the case of a stationary receiver and transmitter (defined earlier as a *static* multipath channel), due to multiple reflections and scattering from various obstructions surrounding the BS and subscriber antennas, we obtain the signal spread with a standard deviation of σ_τ in the time domain as shown in Figure 1.15. As a result, radio signals traveling along different paths of varying lengths cause significant deviations of the signal strength (in volts) or signal power (in watts) at the receiver. This interference picture is not changed in time and can be repeated in each phase of radio communication between the BS and the stationary subscriber (see Fig. 1.16). In the case of a *dynamic* channel, described in the previous subsection, either the subscriber antenna is moving or the objects surrounding the stationary antennas move and the spatial variations of the resultant signal at the receiver can be seen as temporal variations, as the receiver moves

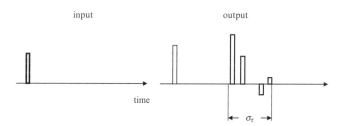

FIGURE 1.15. The multipath delay spread σ_τ in the time-invariant (stationary) channel.

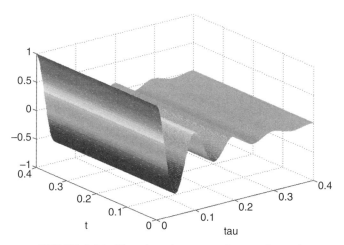

FIGURE 1.16. Time-invariant or stationary channel.

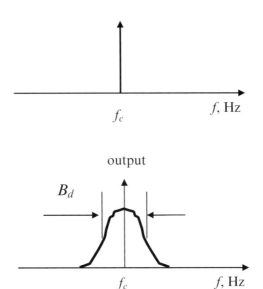

FIGURE 1.17. Doppler spread caused by mobile subscriber.

through the multipath field (i.e., through the interference picture of the field pattern). In such a dynamic multipath channel, signal fading at the mobile receiver occurs in the time domain. This temporal fading relates to a shift of frequency radiated by the stationary transmitter (see Fig. 1.17). In fact, the time variations, or dynamic changes of the propagation path lengths, are related to the Doppler shift, denoted earlier by $f_{d\max}$, which is caused by the relative movements of the stationary BS and the MS. The total bandwidth due to Doppler shift is $B_d = 2f_{d\max}$. In the time-varied or dynamic channel, in any real time t, there is no repetition of the interference picture during crossing of different field patterns by the MS at each discrete time of his movements, as shown in Figure 1.18.

1.5. HIGH-LEVEL FADING STATISTICAL PARAMETERS

In real situations in mobile communications, when conditions in dynamic channels are more realistic (see Fig. 1.12), due to the motion of the receiver or the transmitter, the picture of envelope fading varies. In such realistic scenarios occurring in the urban environment, the fading rate and the signal envelope amplitude are functions of time. If so, for the wireless network designers, it is very important to obtain at the quantitative level a description of the rate, at which fades of any depth occur and of the average duration of a fade below

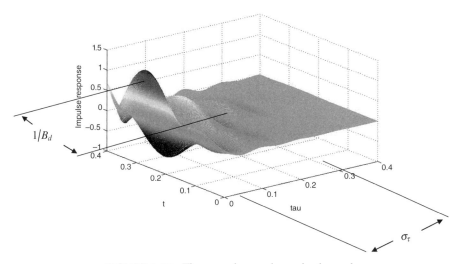

FIGURE 1.18. Time-varying or dynamic channel.

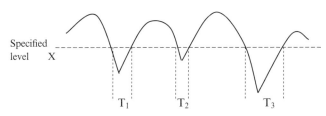

FIGURE 1.19. The illustration of definitions of the signal fading statistical parameters LCR and AFD.

any given depth (usually called the sensitivity or the threshold of the receiver input). Therefore, there are two important high-level statistical parameters of a fading signal, the *level crossing rate* (LCR) and the *average fade duration* (AFD) that are usually introduced in the literature. These parameters are useful for mobile link design and, mostly, for designing various coding protocols in wireless digital networks, where the required information is provided in terms of LCR and AFD.

The manner in which both required parameters, LCR and AFD, can be defined is illustrated in Figure 1.19. As shown in Figure 1.19, the LCR at any specified threshold (i.e., a sensitivity level of the receiver) X is defined as the expected rate at which the received signal envelope crosses that level in a positive-going or negative-going direction. To find this expected rate, we need information about the joint PDF of the specific level X and the slope of the envelope curve $r(t)$, $\dot{r} = dr/dt$, that is, about PDF(X, $\dot{r}$). The same should be done to define AFD, defined as the average period of time for which the receiver signal envelope is below a specific threshold X (see Fig. 1.19).

1.5.1. Level Crossing Rate: A Mathematical Description

Rayleigh Fading Channel. In terms of this joint PDF, the LCR is defined as the expected rate at which the Rayleigh fading envelope, normalized to the local *rms* signal level, crosses a specified level X, let us say in a positive-going direction. The number of level crossings per second, or the LCR, N_x can be obtained, using the following definition [1, 4, 5]:

$$N_X = \int_0^\infty \dot{r} \cdot PDF(X, \dot{r}) d\dot{r} = \sqrt{2\pi} f_m \zeta \exp(-\zeta^2), \qquad (1.50)$$

where, as above, $f_m = \nu/\lambda$ is the maximum Doppler frequency shift and

$$\zeta = \frac{X}{\sqrt{2} \cdot \sigma_r} \equiv \frac{X}{rms} \qquad (1.51)$$

is the value of specific level X, normalized to the local *rms* amplitude of fading envelope (according to Rayleigh statistics $rms = \sqrt{2} \cdot \sigma_r$, see Section 1.4.2). Because f_m is a function of mobile speed v, the value N_x also depends, according to (1.50), on this parameter. For deep Rayleigh fading, there are few crossings at both high and low levels [1–4] with the maximum rate occurring at $\zeta = 1/\sqrt{2}$, that is, at the level 3 dB below the *rms* level.

Rician Fading Channel. In this case, the number of level crossings per second, or the LCR, N_x, can be obtained, using the following result obtained from Reference 13:

$$N_X(X) = \int_0^\infty \dot{r} \cdot PDF(X, \dot{r}) d\dot{r} = \frac{2X\sqrt{2\varsigma}}{\pi^{3/2} \mathrm{K}(0)} e^{-(X^2 + \rho^2)/\mathrm{K}(0)}$$

$$\times \int_0^{\pi/2} \cosh\left(\frac{2X\rho\cos\alpha}{\mathrm{K}(0)}\right) \left[e^{-\xi\rho\sin\alpha} + \sqrt{\pi}\xi\rho\sin\alpha Q(\xi\rho\sin\alpha)\right] d\alpha. \qquad (1.52)$$

Here, as above, X denotes the level of the receiver input; $\rho \equiv |y(t)| = [K/(K+1)]^{1/2}$ is the amplitude of the LOS component of the signal strength; $Q(w)$

is the error function from (1.10); $\varsigma = -\frac{1}{2}K''(0) - \mathrm{Im}\{K'(0)\}^2 / 2K(0)$ and $\xi = [\omega_D \cos\alpha_0 - \mathrm{Im}\{K'(0)\} / K(0)] / \sqrt{2\varsigma}$, where functions $K(0), K'(0)$, and $K''(0)$ are defined in Reference 13 as

$$K(0) = \frac{1}{K+1} \tag{1.53}$$

$$K'(0) = -\frac{i\omega_D}{K+1}\left[\frac{\cos\theta \cdot I_1(\kappa)}{I_0(\kappa)}\right] \tag{1.54}$$

$$K''(0) = \frac{\omega_D^2}{2(K+1)}\left[1 + \frac{\cos 2\theta \cdot I_2(\kappa)}{I_0(\kappa)}\right], \tag{1.55}$$

where $I_n(\kappa)$, $n = 1, 2, 3, \ldots$, are the nth order modified Bessel functions of the first kind, κ determines the beamwidth of arriving waves, and θ denotes the angle between the average scattering direction and the mobile vehicle direction.

Equation (1.52) is a general expression for the envelope LCR and contains (1.50) as a special case of Rayleigh fading, usually used for Doppler effect estimation through the LCR estimates [1, 4, 5].

1.5.2. Average Fade Duration: A Mathematical Description

The AFD, $\langle\tau\rangle$, is defined as the average period of time for which the received signal envelope is below a specific level X (see Fig. 1.19). Its relation with LCR is following

$$\langle\tau\rangle = \frac{1}{N_X}CDF(X). \tag{1.56}$$

Here CDF(X) describes the probability of the event that the received signal envelope $r(t)$ does not exceed a specific level X, that is,

$$CDF(X) \equiv \Pr(r \le X) = \frac{1}{T}\sum_{i=1}^{n}T_i. \tag{1.57}$$

Here T_i is the duration of the fade (see Fig. 1.19) and T is the observation interval of the fading signal.

Rayleigh Fading Statistics. According to the Rayleigh PDF defined by (1.19) and CDF defined by (1.22), the AFD can be expressed, according to (1.56), as a function of ζ and f_m in terms of the *rms* value:

$$\langle\tau\rangle = \frac{\exp(\zeta^2)-1}{\sqrt{2\pi}f_m\zeta}. \tag{1.58}$$

Rician Fading Statistics. Using relation (1.56) between LCR and AFD, as well as equation (1.52) for $N_x(X)$, one can easily investigate the average fade duration using general Rician fading statistics. We will not present these expressions due to their complexity and will refer the reader to the original work in Reference 13.

We also note that it is very important to determine the rate at which the input signal inside the mobile communication link falls below a given level X, and how long it remains below this level. This is useful for relating the SNR (or S/N), during fading, to the instantaneous *bit error rate* (BER). To someone interested in digital systems design, we point out that knowing the average duration of a signal fade helps to determine the most likely number of signal bits that may be lost during this fading. LCR and AFD primarily depend on the speed of the mobile and decrease as the maximum of Doppler shift becomes larger. An example on how to estimate BER using LCR and AFD will be shown in Chapter 11.

1.6. ADAPTIVE ANTENNAS APPLICATION

The main problem with land communication links is estimating the ratio between the coherent and multipath components of the total signal, that is, the Rician parameter K, to predict the effects of multiplicative noise in the channel of each subscriber located in different conditions in the terrestrial environment. This is shown in Figure 1.20 for various subscribers numbered by $i = 1, 2, 3, \ldots$

However, even a detailed prediction of the radio propagation situation for each subscriber cannot completely resolve all issues of effective service and increase the quality of data stream sent to each user. For this purpose, in present and future generations of wireless systems, adaptive or smart antenna systems are employed to reduce interference and decrease the BER. This topic will be covered in detail in Chapters 7, 8, and 11. Here, in Figure 1.21, we present, schematically, the concept of adaptive (smart) antennas, which allows each mobile subscriber to obtain individual service without inter-user interference and by eliminating multipath phenomena caused by multiple rays arriving to his antenna from various directions.

Even with adaptive/smart antennas (see description in Chapter 7), we cannot totally cancel the effects of the environment, especially in urban areas, due to the spread of the narrow antenna beam both in azimuth and elevation domains caused by an array of buildings located at the rough terrain (see Fig. 1.22). Furthermore, if the desired user is located in a shadow zone with respect to the BS antenna, we may expect the so-called masking effect due to guiding effect of crossing straight streets as shown schematically in Figure 1.23. Thus,

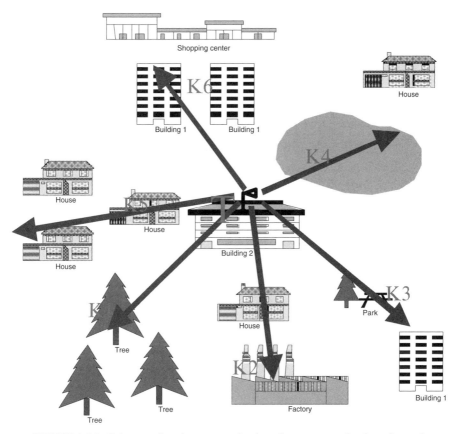

FIGURE 1.20. Scheme of various scenarios in urban communication channel.

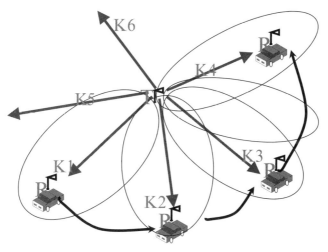

FIGURE 1.21. A scheme for using adaptive antennas for each user located in different conditions in a service area.

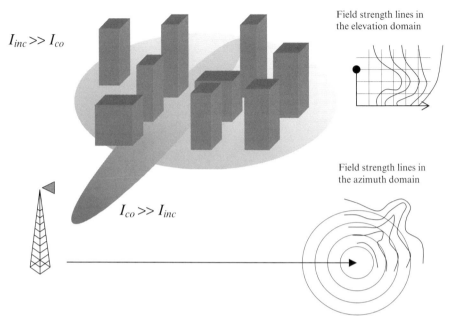

$I_{inc} \gg I_{co}$

Field strength lines in the elevation domain

Field strength lines in the azimuth domain

$I_{co} \gg I_{inc}$

FIGURE 1.22. Effect of built-up area on a narrow-beam adaptive antenna pattern.

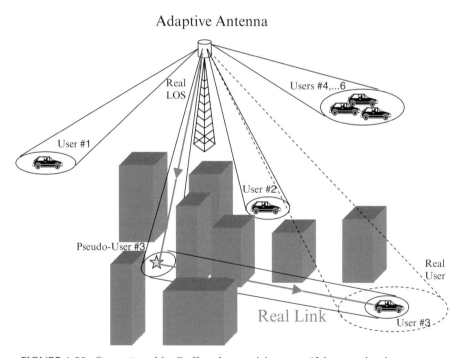

Adaptive Antenna

Real LOS

Users #4,...6

User #1

User #2

Pseudo-User #3

Real User

Real Link

User #3

FIGURE 1.23. Street "masking" effect for servicing user #3 by an adaptive antenna.

instead of the real position of user #3, for the adaptive antenna located at the BS, the "real position" will be at the intersection of two straight crossing streets due to guiding effect and channeling of signal energy transmitted by BS antenna along these two streets. Chapters 5 and 8 will focus on terrain effects where a rigorous analysis of these effects on the wireless systems design will be presented.

REFERENCES

1 Jakes, W.C., *Microwave Mobile Communications*, John Wiley & Sons, New York, 1974.

2 Steele, R., *Mobile Radio Communication*, IEEE Press, New York, 1992.

3 Stuber, G.L., *Principles of Mobile Communications*, Kluwer Academic Publishers, Boston-London, 1996.

4 Lee, W.Y.C., *Mobile Cellular Telecommunications Systems*, McGraw Hill, New York, 1989.

5 Molisch, A.F., *Wireless Communications*, Wiley and Sons, London, 2006.

6 Yacoub, M.D., *Foundations of Mobile Radio Engineering*, CRC Press, New York, 1993.

7 Saunders, S.R., *Antennas and Propagation for Wireless Communication Systems*, John Wiley & Sons, New York, 1999.

8 Bertoni, H.L., *Radio Propagation for Modern Wireless Systems*, Prentice Hall PTR, Upper Saddle River, NJ, 2000.

9 Blaunstein, N., "Wireless Communication Systems," in *Handbook of Engineering Electromagnetics*, R. Bansal, ed., Marcel Dekker, New York, 2004.

10 Rappaport, T.S., *Wireless Communications*, Prentice Hall PTR, New York, 1996.

11 Leon-Garcia, A., *Probability and Random Processes for Electrical Engineering*, Addison-Wesley Publishing Company, New York, 1994.

12 Stark, H. and J.W. Woods, *Probability, Random Processes, and Estimation Theory for Engineers*, Prentice Hall, Englewood Cliffs, NJ, 1994.

13 Tepedelenlioglu, C., et al., "Estimation of Doppler spread and spatial strength in mobile communications with applications to handoff and adaptive transmissions," *J Wireless Communic. Mobile Computing*, Vol. 1, No. 2, 2001, pp. 221–241.

14 Krouk, E. and S. Semionov, eds., *Modulation and Coding Techniques in Wireless Communications*, Wiley & Sons, Chichester, England, 2011.

CHAPTER TWO

Antenna Fundamentals

A radio antenna, transmitting or receiving, is an independent and yet integral component of any wireless communication system. An antenna acts as a transducer that converts the current or voltage generated by the feeding-based circuit, such as a transmission line, a waveguide, or a coaxial cable, into electromagnetic field energy propagating through space and vice versa. In free space, the fields propagate in the form of spherical waves, whose amplitudes are inversely proportional to their distance from the antenna. Each radio signal can be represented as an electromagnetic wave [1] that propagates along a given direction. The wave field strength, its polarization, and the direction of propagation determine the main characteristics of an antenna operation.

Antennas can be divided in different categories, such as wire antennas, aperture antennas, reflector antennas, frequency-independent antennas, horn antennas, printed and conformal antennas, and so forth [2–10]. When applications require radiation characteristics that cannot be met by a single radiating antenna, multiple elements are employed, forming "array antennas." Arrays can produce the desired radiation characteristics by appropriately exciting each individual element with certain amplitudes and phases. The very same antenna array configurations, when combined with signal processing, lead to multiple-beam (switched beam) or adaptive antennas that offer many more degrees of freedom in a wireless system design than using a single antenna [11–14]. The subject antenna arrays, including adaptive arrays, will be studied in detail in Chapter 9. In this chapter, we introduce the basic concepts of

Radio Propagation and Adaptive Antennas for Wireless Communication Networks: Terrestrial, Atmospheric, and Ionospheric, Second Edition. Nathan Blaunstein and Christos G. Christodoulou.

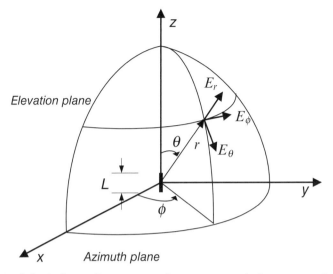

FIGURE 2.1. Spherical coordinate system for antenna analysis purposes. A very short dipole is shown with its nonzero field component directions.

antennas and some fundamental figures of merit, such as radiation patterns, directivity, gain, polarization loss, and so on, that describe the performance of any antenna.

2.1. RADIATION PATTERN

The radiation pattern of any antenna is defined as the relative distribution of electromagnetic energy or power in space. Because antennas are an integral part of all telecommunication systems, the radiation pattern is determined in the far-field region where no change in pattern with distance occurs. Figure 2.1 shows that if we place an antenna at the origin of a spherical coordinate system, the radiation properties of the antenna will depend only on the angles ϕ and θ along a path or surface of constant radius. A trace of the radiated (or received) power at a fixed radius is known as a *power pattern*, while the spatial variation of the electric field along the same radius is called the *amplitude field pattern*.

Although a 3-D visualization of an antenna radiation pattern is helpful, usually, a couple of plots of the pattern as a function of θ, for some particular values of ϕ, plus a couple of plots as a function of ϕ, for some particular values of θ, give sufficient information. For example, Figure 2.2a depicts the 3-D radiation pattern from an ideal or very short dipole. Figure 2.2b shows the xy-plane (azimuthal plane, $\theta = \pi/2$), called the principal E-plane cut, and Figure 2.3c is the xz-plane (elevation plane, $\phi = 0$) or the principal H-plane cut.

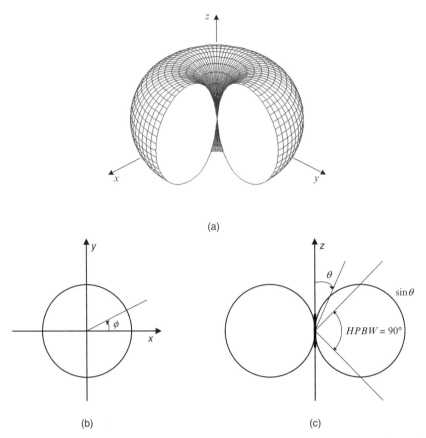

FIGURE 2.2. Radiation field pattern of far field from an ideal or very short dipole: (a) Threedimensional pattern plot; (b) *E*-plane radiation pattern polar plot; and (c) *H*-plane radiation pattern polar plot.

A typical antenna power pattern is shown in Figure 2.3. The upper part depicts a normalized polar radiation pattern in linear, whereas the bottom figure is actually the same pattern but in rectangular coordinates and in decibel scale. The radiation pattern of the antenna consists of various parts, which are known as *lobes*. The *main lobe* (also known as *main beam* or *major lobe*) is the lobe containing the direction of maximum radiation. In the case of Figure 2.3, the main lobe is pointing in the $\theta = 0$ direction. Antennas can have more than one major lobe.

In Chapter 8, we will see that one can create multiple lobes to track several mobile users at the same time from a base station (BS). A *minor lobe* is any lobe other than the main lobe. Minor lobes are usually divided into *side lobes* and *back lobes*. The term side lobe refers to those minor lobes near the main lobe, and by a back lobe we refer to a radiation lobe that is in the opposite

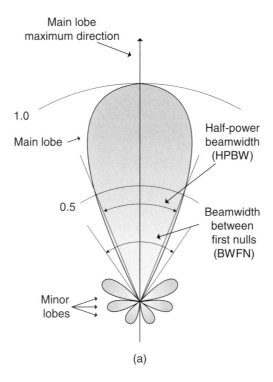

(a)

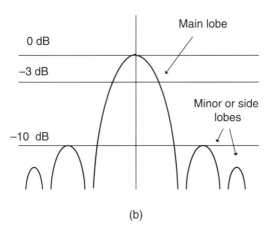

(b)

FIGURE 2.3. Antenna power patterns: (a) A typical polar plot in linear scale and (b) a plot in rectangular coordinates in decibel (logarithmic) scale. The associated lobes and beamwidths are also shown.

direction to that of the main lobe. Minor lobes usually represent radiation in undesired directions that can cause interference in a mobile environment, and they should be minimized. The ratio of levels of the largest side lobe over the major lobe is termed as the *side lobe ratio* or *side lobe level*.

Another term that characterizes a radiation pattern is its *half-power beam-width* (HPBW) in the two principal planes. The HPBW is defined as the angular width of the main lobe within which the radiation intensity is one-half the maximum value of beam (see Fig. 2.3). Sometimes, we also use the *beam-width between the first nulls* (BWFN) around the main beam. The 3-dB beam-width plays a major role in the overall design of an antenna application. As the beamwidth of the radiation pattern increases, the side lobe level decreases, and vice versa. So there is a trade-off between side lobe ratio and beamwidth of an antenna pattern.

Furthermore, the beamwidth of the antenna is also used to describe the resolution capabilities of the antenna to distinguish between two adjacent radiating sources or radar targets. That can play an important role when one uses an antenna to determine the angle of arrival of a radio source.

2.2. FIELD REGIONS OF AN ANTENNA

The space surrounding a transmitting antenna is divided into two main regions: the *near-field* region and the *far-field* region. The near-field region can be further subdivided into two regions: the *reactive near-field* and the *radiating near-field* [1].

Figure 2.4 shows these regions. The first region, which is the closest to the antenna, is called the reactive or induction near-field region. It derives its name

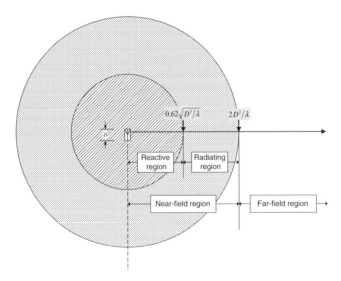

FIGURE 2.4. Field regions of an antenna and some commonly used boundaries.

from the reactive field that lies close to every current-carrying conductor. The reactive field, within that region, dominates over all radiated fields.

For most antennas, the outermost boundary of this region is given by

$$r > 0.62 \sqrt{\frac{D^3}{\lambda}}, \tag{2.1}$$

where r is the distance from the antenna, D is the largest dimension of the antenna, and λ is the wavelength.

Between this reactive near-field region and the far-field region lies the radiating near-field region. Although the radiation fields dominate within this region, the angular field distribution still depends on the distance from the antenna. This region is also called the Fresnel region, a terminology borrowed from the field of optics. The boundaries of this region are

$$0.62 \sqrt{\frac{D^3}{\lambda}} < r < \frac{2D^2}{\lambda}. \tag{2.2}$$

At the outer boundary of the near-field region, the reactive field intensity becomes negligible with respect to the radiated field intensity. The far-field or radiation region, also called the *Fraunhofer region*, begins at $r = 2D^2/\lambda$ and extends outward indefinitely into free space. In this region the angular field distribution of the field of the antenna is not dependent on the distance from the antenna.

2.3. RADIATION INTENSITY

Radiation intensity is a far-field parameter that is used to determine the antenna power pattern as a function of angle:

$$
\begin{aligned}
I(\theta, \phi) &= S_{av} r^2 \\
&= \frac{r^2}{2\eta} |\mathbf{E}(r, \theta, \phi)|^2 \\
&= \frac{r^2}{2\eta} \left[|E_\theta(r, \theta, \phi)|^2 + |E_\varphi(r, \theta, \phi)|^2 \right] \\
&\approx \frac{1}{2\eta} \left[|E_\theta(\theta, \phi)|^2 + |E_\phi(\theta, \phi)|^2 \right],
\end{aligned} \tag{2.3}
$$

where $I(\theta,\phi)$ is the radiation intensity (W/unit solid angle); S_{av} is the Poynting vector (W/m²); $E(r,\theta,\phi)$ is the total transverse electric field (V/m); $H(r,\theta,\phi)$ is the total transverse magnetic field (A/m); r is the distance from antenna to point of measurement (m); η is the intrinsic impedance of medium (Ω per square).

The averaged Poynting vector $\mathbf{S}_{av}$ in Equation (2.3) is derived from:

$$\mathbf{S}_{av} = \frac{1}{2}\text{Re}(\mathbf{E}\times\mathbf{H}^*)\,(\text{W/m}^2), \qquad (2.4)$$

where the notation Re stands for the real part of the complex number and the * denotes the complex conjugate. Note that $\mathbf{E}$ and $\mathbf{H}$ in Equation (2.4) are the expressions for the radiated electric and magnetic fields.

Note that the radiation intensity is independent of distance since in the far field the Poynting vector is entirely radial; that is, the fields are entirely transverse and $\mathbf{E}$ and $\mathbf{H}$ vary as $1/r$. As the radiation intensity is a function of angle, it is related to the power radiated from an antenna per unit solid angle. The measure of a solid angle is the *steradian*, which is defined as the solid angle with its vertex at the center of a sphere of radius r, subtended by a spherical surface area equivalent to that of a square of size r^2 (see Fig. 2.5). But the area of a sphere of radius r is given by $A = 4\pi r^2$, so in a closed sphere there are 4π $r^2/r^2 = 4\pi$ steradian (sr). For a sphere of radius r, an infinitesimal surface area dA can be expressed as

$$dA = r^2 \sin\theta\, d\theta\, d\phi \,(\text{m}^2), \qquad (2.5)$$

and hence the element of solid angle $d\Omega$ of a sphere is given by

$$d\Omega = \frac{dA}{r^2} = \sin\theta\, d\theta\, d\phi \,(\text{sr}). \qquad (2.6)$$

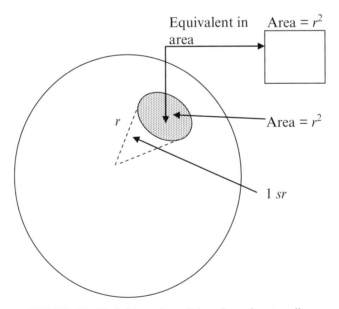

FIGURE 2.5. Definition of a solid angle and a steradian

The total power that can be radiated is given by:

$$P_{rad} = \oiint_\Omega I(\theta, \phi)d\Omega = \int_0^{2\pi} \int_0^\pi I(\theta, \phi)\sin\theta d\theta d\phi. \tag{2.7}$$

Let us consider an *isotropic radiator* as an example. An *isotropic antenna* refers to a hypothetical antenna radiating equally in all directions and its power pattern is uniformly distributed in all directions. That means that the radiation intensity of an isotropic antenna is independent of the angles θ and ϕ, and the total radiated power will be

$$P_{rad} = \oiint_\Omega I_i d\Omega = I_i \int_0^{2\pi} \int_0^\pi \sin\theta d\theta d\phi = I_i \oiint_\Omega d\Omega = 4\pi I_i \tag{2.8}$$

or $I_i = P_{rad}/4\pi$, which is the radiation intensity of an isotropic antenna.

Dividing $I(\theta,\phi)$ by its maximum value I_{max} leads to the *normalized antenna power pattern*, that is,

$$I_n(\theta, \phi) = \frac{I(\theta, \phi)}{I_{max}(\theta, \phi)} \text{ (dimensionless).} \tag{2.9}$$

2.4. DIRECTIVITY AND GAIN

An important parameter which indicates how well radiated power is concentrated into a limited solid angle is *directivity D*. The directivity of an antenna is defined as the ratio of the maximum radiation intensity to the radiation intensity averaged over all directions (i.e., with reference to the isotropic radiator). Thus, the average radiation intensity is found by dividing the total antenna radiated power r by 4π sr, or

$$D = \frac{I_{max}(\theta, \phi)}{I_{av}} = \frac{I_{max}(\theta, \phi)}{I_i} = \frac{I_{max}(\theta, \phi)}{P_{rad}/4\pi} = \frac{4\pi I_{max}(\theta, \phi)}{P_{rad}} \text{ (dimensionless).} \tag{2.10}$$

The narrower the main lobe of the antenna radiation pattern, the larger the directivity of the antenna. Obviously, the directivity of an isotropic antenna is unity. Any other antenna will have a directivity larger than unity (i.e., larger than the isotropic), as shown in Figure 2.6.

Let us consider the directivity of a very short dipole, as an example. The average pointing vector for the dipole is given by [11]

$$S_{av} = \frac{\eta}{2}\left(\frac{I_0 L\beta}{4\pi r}\right)^2 \sin^2\theta \text{ W/m}^2, \tag{2.11}$$

where L is the length of the short dipole, I_0 is the current flowing through the dipole, η is the wave impedance in free space, and $\beta = 2\pi/\lambda$. Using Equation (2.3) we can solve for the radiation intensity, and then we can use Equation

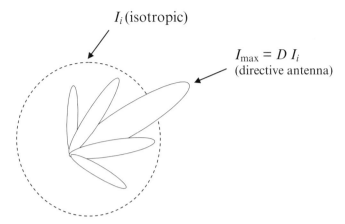

FIGURE 2.6. Directive pattern versus an isotropic one.

(2.10) to obtain a directivity of 1.5. This occurs at the $\theta = 90°$ direction (see Fig. 2.2). Thus, in this direction, a very short dipole can radiate 1.5 times more power than the isotropic radiator. This is often expressed in decibels such that

$$D = 10\log_{10}(d)\,\mathrm{dB} = 10\log_{10}(1.5) = 1.76 \ \mathrm{dB}. \tag{2.12}$$

The *gain* of an antenna is closely associated with directivity, and it is defined as the ratio of the maximum radiation intensity in a given direction to the maximum radiation intensity produced in the same direction from a reference antenna with the same power input. Any convenient type of antenna can be taken as a reference antenna. Usually, the type of reference antenna is determined by the specific application, but the most commonly used one is the isotropic radiator, and thus we can write

$$G = \frac{I_{\max}(\theta, \phi)}{I_i} = \frac{I_{\max}(\theta, \phi)}{P_{in}/4\pi} \ (\text{dimensionless}), \tag{2.13}$$

where the radiation intensity of the isotropic radiator is equal to the input, P_{in}, of the antenna divided by 4π. As the gain of an antenna depends on how efficient it is in converting input power into radiated fields, we need to take into consideration its efficiency before we determine the actual gain. In general, *antenna efficiency* (ε) is defined as the ratio of the power radiated by the antenna to the input power at its terminals:

$$\varepsilon = \frac{P_{rad}}{P_{in}} = \frac{R_r}{R_r + R_{loss}} \ (\text{dimensionless}), \tag{2.14}$$

where R_r is the *radiation resistance* of the antenna; R_r is an equivalent resistance in which the same current flowing at the antenna terminals will produce power equal to that radiated by the antenna. R_{loss} is the *loss resistance* due to any conductive or dielectric losses of the materials used to construct the antenna. So, if we include these losses, a real antenna will have radiation intensity

$$I(\theta, \phi) = \varepsilon I_0(\theta, \phi), \qquad (2.15)$$

where $I_0(\theta,\phi)$ is the radiation intensity of the same antenna with no losses.

Using Equation (2.15) into Equation (2.13) yields the definition of gain in terms of the antenna directivity:

$$G = \frac{I_{\max}(\theta, \phi)}{I_i} = \frac{\varepsilon I_{\max}(\theta, \phi)}{I_i} = \varepsilon D. \qquad (2.16)$$

The values of gain range between zero and infinity, whereas for directivity the values range between unity and infinity. However, though directivity can be found either theoretically or experimentally, the gain of an antenna is almost always determined by a direct comparison of measurement against a reference, usually the standard gain antenna. Gain is expressed also in decibels, that is,

$$G = 10 \log_{10}(g) \, \text{dB} \qquad (2.17)$$

When we use the isotropic antenna as a reference, then we use the dBi notation, which means decibels over isotropic.

2.5. POLARIZATION

2.5.1. Wave and Antenna Polarization

Polarization refers to the direction of the electric field component of an electromagnetic wave. The wave is called *linearly polarized* or *plane polarized*; that is, the locus of oscillation of the electric field vector within a plane perpendicular to the direction of propagation forms a straight line. On the contrary, when the locus of the tip of an electric field vector forms an ellipse or a circle, the wave is called an *elliptically* or *circularly polarized wave,* respectively. There is a tendency to refer to antennas as vertically or horizontally polarized, though it is only their radiations that are polarized. Next, we discuss the mechanics of various polarizations that we encounter in antenna communication systems.

2.5.2. Linear, Circular, and Elliptical Polarization

Consider a plane wave traveling in the positive z-direction, with the electric field component along the x-direction as shown in Figure 2.7a. This wave is linearly polarized, in the x-direction, and its electric field can be expressed as

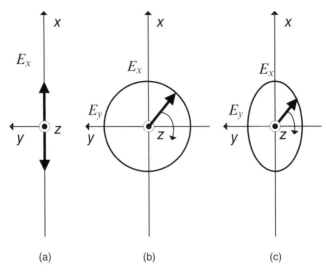

FIGURE 2.7. Polarization of a wave: (a) linear, (b) circular, and (c) elliptical.

$$E_x = E_{x0} \sin(\omega t - \beta z). \tag{2.18}$$

In Figure 2.7b, the wave has both an x and a y electric field component. If the two components E_x and E_y have the same magnitude, then the total (vector) electric field rotates as a function of time, with the tip of the vector forming a circular trace, and the wave is thus called circular polarized. Generally, the wave consists of two electric field components, E_x and E_y, of different amplitude ratios and relative phases that can yield an elliptically polarized wave, as shown in Figure 2.7c. The polarization ellipse may have any orientation, which is determined by its tilt angle, as depicted in Figure 2.8. The ratio of the major to minor axes of the polarization ellipse is called the *axial ratio* (AR).

For any wave traveling in the positive z-direction, the electric field components in the x- and y-directions can be written as

$$E_x = E_{x0} \sin(\omega t - \beta z) \tag{2.19}$$

$$E_y = E_{y0} \sin(\omega t - \beta z + \delta), \tag{2.20}$$

where E_{x0} and E_{y0} are the amplitudes in the x- and y-directions, respectively, and δ is the time-phase angle between them. By manipulating these two components we can show that [1, 11]

$$\frac{E_x^2}{E_1^2} - \frac{2E_x E_y \cos \delta}{E_1 E_2} + \frac{E_y^2}{E_2^2} = \sin^2 \delta. \tag{2.21}$$

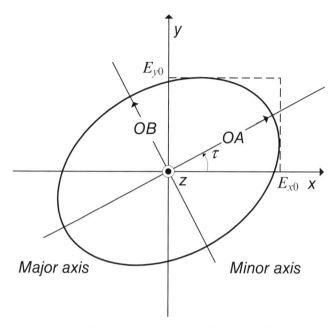

FIGURE 2.8. Polarization ellipse at $z = 0$ of an elliptically polarized electromagnetic wave.

Depending on the values of E_x, E_y, and δ, this equation can be expressed as the equation of an ellipse or of a circle.

The sense of rotation of a circularly or elliptically polarized wave plays an important role in a communication link. It is defined by the direction of rotation of the wave as it propagates toward or away from the observer along the direction of propagation. If, for example, a wave is moving away and its rotation is clockwise, then we say that the wave has a "clockwise" sense of rotation. The most common notation used today is that of the IEEE by which the sense of rotation is always determined observing the field rotation as the wave travels away from the observer. If the rotation is clockwise, the wave is *right-handed* or *clockwise circularly polarized* (RH or CW). If the rotation is counterclockwise, the wave is *left-handed* or *counterclockwise circularly polarized* (LH or CCW). The same applies to elliptically polarized waves.

The polarization state of an antenna is defined as the polarization state of the wave transmitted by the antenna. It is characterized by the sense of rotation and the spatial orientation of the ellipse, if it is elliptically polarized. If the receiving antenna has a polarization that is different from that of the incident wave, a *polarization mismatch* will occur. A polarization mismatch causes the receiving antenna to extract less power from the incident wave.

Polarization loss factor (PLF) is used as a figure of merit to measure the degree of polarization mismatch. It is defined as the square power of the cosine

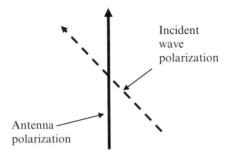

FIGURE 2.9. Definition of PLF.

angle between the polarization states of the antenna in its transmitting mode and the incoming wave (see Fig. 2.9).

$$\text{PLF} = |\cos \gamma|^2. \tag{2.22}$$

Generally, an antenna is designed for a desired polarization. The component of the electric field in the direction of the desired polarization is called the *co-polar component*, whereas the undesired polarization, usually taken in orthogonal direction to the desired one, is known as *cross-polar component*. The latter can be due to a change of polarization characteristics during the propagation or scattering of waves that is known as *polarization rotation*.

An actual antenna does not completely discriminate against a cross-polarized wave due to structural abnormalities of the antenna. The directivity pattern obtained over the entire direction on a representative plane for cross-polarization with respect to the maximum directivity for the desired (co-polar) polarization is called *antenna cross-polarization discrimination* and plays an important factor in determining the antenna performance. Because the elliptical polarization is often used in wireless communication link designs (terrestrial, atmospheric, and ionospheric), we present in the following the main characteristics of the polarized ellipse and the relations between the classical Stokes parameters [15–18], the parameters of the polarized ellipse and the vectors of electromagnetic field components.

2.5.3. Three-Dimensional Characteristics and Vectors of the Polarized Ellipse

Let us consider a spatial coordinate system determined by three orthogonal unit vectors $\{\mathbf{u}_1, \mathbf{u}_2, \mathbf{u}_3\}$. In this general case, any vector of the electrical field strength can be described by the following three components [15]:

$$E(t) = \mathbf{u}_i E_i(t) = \mathbf{u}_i A_i \cos(\omega t + \psi_i), \quad i = 1, 2, 3, \tag{2.23}$$

where A_i and ψ_i are the amplitude and the phase of the component with number i; t is the current time; ω is the angle frequency, $\omega = 2\pi f$. In Chapters 5 and 8, following the notations and parameters introduced there, we will replace the above coordinate system $\{\mathbf{u}_1, \mathbf{u}_2, \mathbf{u}_3\}$ by the well-known Cartesian system, that is, $\mathbf{u}_1 \equiv \mathbf{x}$, $\mathbf{u}_2 \equiv \mathbf{y}$, and $\mathbf{u}_3 \equiv \mathbf{z}$.

The Quadrature Components of the Total Field. The quadrature components of the total field, usually called in the literature the *sine* and the *cosine* components of the total field [16–17], can be defined as

$$S_i = -A_i \sin \psi_i, C_i - A_i \cos \psi_i. \tag{2.24}$$

Using these quadrature components, one can construct the two 3-D vectors, respectively, $\mathbf{S}(S_1, S_2, S_3)$ and $\mathbf{C}(C_1, C_2, C_3)$. In such definitions, we can rewrite Equation (2.23) as

$$\mathbf{E}(t) = \mathbf{S} \sin \omega t + \mathbf{C} \cos \omega t. \tag{2.25}$$

The vectors $\mathbf{S}$ and $\mathbf{C}$ are not time dependent but determine the plane on which lies a trajectory of the vector $\mathbf{E}(t)$ in the time domain. An equation of such a plane can be presented as $\mathbf{C} \times \mathbf{S} = \mathbf{N}$, where $\mathbf{N}$ is the normal to the plane of this trajectory, which is directed in such a way that the field strength vector $\mathbf{E}(t)$ presents a right-hand rotation as shown in Figure 2.10.

Let us define the corresponding elements of the ellipse. The large and the small semi-axes of the ellipse can be determined by the extremes (maximum and minimum) of $|\mathbf{E}(t)|$. According to References 15–18, we can find the extremes of $|\mathbf{E}(t)|^2$, which define each semi-axis of the ellipse:

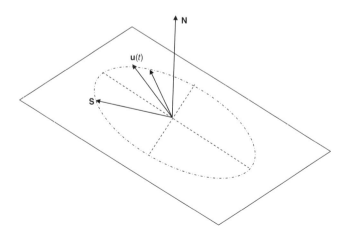

FIGURE 2.10. Geometrical presentation of the hodograph of the vector $\mathbf{E}(t)$.

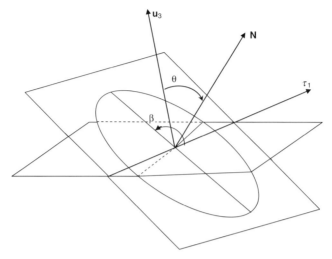

FIGURE 2.11. Geometrical presentation of the parameters of the polarization ellipse.

$$|\mathbf{E}_{min}|^2 = \frac{S^2 + C^2}{2} - \left[\left(\frac{C^2 - S^2}{2}\right)^2 + (\mathbf{C} \cdot \mathbf{S})^2\right]^{1/2} \tag{2.26a}$$

$$|\mathbf{E}_{max}|^2 = \frac{S^2 + C^2}{2} + \left[\left(\frac{C^2 - S^2}{2}\right)^2 + (\mathbf{C} \cdot \mathbf{S})^2\right]^{1/2}. \tag{2.26b}$$

Now, we can define the inclination of the large semi-axis of the ellipse relative to the line of intersection of the plane of the trajectory (described by the vector $\mathbf{E}(t)$ in time) with one of the coordinate plane. In Figure 2.11, the angle of inclination, β, of the large semi-axis can be obtained from the direction determined by the unit vector $\tau_1 = \mathbf{u}_3 \times \mathbf{N}/|\mathbf{u}_3 \times \mathbf{N}|$, oriented along the intersection between the plane of the ellipse and the plane orthogonal to the $\mathbf{u}_3$ (see Fig. 2.11). In References 15–18, the angle of inclination β is determined using the following relations:

$$N^2 \equiv |\mathbf{N}|^2 = C^2 S^2 - (\mathbf{C} \cdot \mathbf{S})^2 \tag{2.27a}$$

$$N_\perp^2 \equiv |\mathbf{u}_3 \times \mathbf{N}|^2 = |CS_3 - SC_3|^2 \tag{2.27b}$$

and

$$\tan 2\beta = \frac{2N\left[(C^2 - S^2) \cdot C_3 S_3 - (C_3^2 - S_3^2)(\mathbf{C} \cdot \mathbf{S})\right]}{(C^2 + S^2) \cdot N_\perp^2 - 2(C_3^2 + S_3^2)N^2}. \tag{2.28}$$

These equations describe relations between the quadrature vectors $\mathbf{S}$ and $\mathbf{C}$ and the parameters N, E_{min}, E_{max}, and β, which fully determine the orientation, the form of the polarized ellipse, and the direction of rotation of the vector $\mathbf{E}(t)$. Since the components of vectors $\mathbf{C}$ and $\mathbf{S}$ cannot be measured directly, they are usually connected with the Stokes parameters, which can be easily measured. We will obtain these parameters in the following section [18–20].

Determining the Parameters of the Polarized Ellipse via Stokes Parameters.

The amplitudes of the high-frequency radio signals can be rewritten as

$$A_i \cos(\omega t + \psi_i) + A_k \cos(\omega t + \psi_k + \delta), \quad i \neq k, \tag{2.29}$$

which can be obtained by summation in pairs of all three components of the total field, $\{E_i\}$, $i = 1,2,3$, by shifting these components' phases by $n\pi/2$, $n = 1,2,3,\ldots$

The values of these amplitudes are fully described in References 18–20. Here, we will introduce only those parameters that have physical meaning and directly related to the Stokes parameters. For this purpose, we will introduce the polar angles, $\varphi \in [0, 2\pi]$ and $\theta \in [0, \pi]$, fixing the position of the normal $\mathbf{N}$ with respect to coordinate axes:

$$\varphi = \tan^{-1} \frac{N_2}{N_1}, \theta = \tan^{-1} \frac{N_3}{N_1}. \tag{2.30}$$

Using References 15–18, instead of Equation (2.26), we can obtain the following expressions for the semi-axes of the ellipse:

$$E_{\max} = \left\{ \frac{I}{2} + \left[\left(\frac{I}{2} \right)^2 - N^2 \right]^{1/2} \right\}^{1/2} \tag{2.31a}$$

$$E_{\min} = \left\{ \frac{I}{2} - \left[\left(\frac{I}{2} \right)^2 - N^2 \right]^{1/2} \right\}^{1/2}. \tag{2.31b}$$

We can also obtain the *elliptical coefficient*, as a ratio of the small semi-axis to the large one, that is,

$$R = \frac{2N}{I + (I^2 - 4N^2)^{1/2}}, \tag{2.32}$$

where $I = \sum_{i=1}^{3} (C_i^2 + S_i^2)$ is the total intensity of the electromagnetic wave.

It is important for our further analysis to show the relationship between the Stokes parameters and the measured amplitudes determined by Equation (2.29). We present only the Stokes parameters that will be important for our future derivations in Chapter 5.

The first of Stokes parameters is simply the full field intensity I, described by (2.33). The other two Stokes parameters can be presented as:

$$Q = I - 2\frac{I_3}{\sin^2\theta}; \quad U = 2\frac{N_1\cos\varphi + N_2\sin\varphi}{\sin\theta}. \tag{2.33}$$

The fourth parameter, V, can be presented as follows:

$$V = 2|\mathbf{C}\times\mathbf{S}| = 2N. \tag{2.34}$$

For further comparison between the Stokes parameters and parameters of the polarization ellipse, we introduce a joint probability density function (PDF) of all six quadrature components of the polarization ellipse in a form of product of six ($i = 1,2,3$) one-dimensional normal density probability functions [15–18]:

$$w(\mathbf{C}, \mathbf{S}) = \prod_{i=1}^{3}\frac{1}{2\pi\sigma_i^2}\exp\left\{-\frac{C_i^2 + S_i^2}{2\sigma_i^2}\right\}. \tag{2.35}$$

Finally, from expression (2.35), a distribution for the following six random parameters can be obtained:

- the three projections, N_1, N_2, N_3, of the vector $\mathbf{N}$ on three coordinate axes
- the full intensity, $I = \sum_{i=1}^{3}(C_i^2 + S_i^2)$, as a first Stokes parameter
- the intensity of the third component, $I_3 = C_i^2 + S_i^2$, of the total field, and its phase, ψ_3.

From the above procedure one can show that the vector $\mathbf{N} = \mathbf{C}\times\mathbf{S}$ and the first Stokes parameter I relate to the six quadrature components of the corresponding vectors $\mathbf{C}$ and $\mathbf{S}$, and to the polarization ellipse parameters such as the elliptical coefficient, as well as the angles of orientations of the field vector trajectory, θ, φ.

Now let us introduce for the above components N_i, $i = 1,2,3$, the spherical variables: N, θ, φ, where $N = |\mathbf{N}|$. Then, we can write

$$N_1 = N\sin\theta\cos\varphi; \; N_2 = N\sin\theta\sin\varphi; \; N_3 = N\cos\theta. \tag{2.36}$$

By introducing now a new variable $n = N/I$, where $n \in (0, 1/2)$, and taking into consideration Equation (2.32), we get

$$R = \frac{2n}{1 + \sqrt{1 - 4n^2}}. \tag{2.37}$$

Sometimes, instead of the elliptic parameter, R, the following parameter is used:

$$p = \frac{1 - R^2}{1 + R^2} \qquad (2.38)$$

Finally, following References 15 and 18, expression (2.35) can be rewritten as:

$$w(I, p, \theta, \varphi, \beta) = \frac{I^2 \cdot p \cdot \sin\theta}{(2\pi)^3} \prod_{i=1}^{3} \frac{\exp(-I_i / 2\sigma_i^2)}{2\sigma_i^2}, \qquad (2.39)$$

where three components ($i = 1, 2, 3$) of the wave intensity equal [18, 15]:

$$I_1 = C_1^2 + S_1^2 = I\left\{\xi(\cos^2\varphi\cos^2\theta - \sin^2\varphi) + \sin^2\varphi - \frac{p\sin 2\beta}{2}\sin 2\varphi\cos\theta\right\}. \qquad (2.40a)$$

$$I_2 = C_2^2 + S_2^2 = I\left\{\xi(\sin^2\varphi\cos^2\theta - \cos^2\varphi) + \cos^2\varphi + \frac{p\sin 2\beta}{2}\sin 2\varphi\cos\theta\right\}. \qquad (2.40b)$$

$$I_3 = C_3^2 + S3_1^2 = I\xi\sin^2\theta. \qquad (2.40c)$$

The computational analysis of the system in (2.39) with three components of the wave intensity, described by (2.40a), (2.40b), and (2.40c) is very complicated. Therefore, the solution can be obtained only for some specific scenarios occurring in wireless communication links.

The above relations and formulas will be used in Section 6.9 to describe the effects of built-up terrain features on the depolarization of propagating waves.

2.6. TERMINAL ANTENNAS IN FREE SPACE

The equations in the previous section allow us to obtain a relation between the power at the transmitter and the power received at the receiver when both antennas are located in free space. This relation is called the *Friis transmission formula* [21]. For two antennas, shown in Figure 2.12, separated by a distance r, large enough so that we are in the far field of both antennas, we get the following:

The receiving antenna received a portion of the incident radiation, that is,

$$P_R = A_R I_T, \qquad (2.41)$$

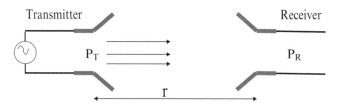

FIGURE 2.12. Transmitter and receiver antennas separated by range r.

where I_T is the radiation intensity of the incident wave and A_R is the effective area of the receiving antenna given by

$$A_R = G_R\left(\frac{\lambda^2}{4\pi}\right). \tag{2.42}$$

Here, G_R is the gain of the receiving antenna, and $\lambda^2/4\pi$ is called the *free-space loss factor*. The received power can now be written as

$$P_R = G_R\left(\frac{\lambda^2}{4\pi}\right)\frac{G_T}{4\pi r^2}P_T, \tag{2.43}$$

with G_T being the gain of the transmitting antenna. If we also include polarization loss, then Equation (2.43) becomes

$$P_R = G_T \cdot G_R\left(\frac{\lambda}{4\pi r}\right)^2 P_T \cdot \text{PLF}. \tag{2.44}$$

This equation is called the *Friis* transmission formula, and it is very essential in designing a communication link between two antennas. Although this particular one is valid for free space only, we will later show how it can be adapted to take into account propagation conditions, other than free space.

2.7. ANTENNA TYPES

There is a large variety of antennas that are used in different branches of wireless communications. The simplest and most commonly used antennas are the *wire* antennas that are used as *dipoles, loops,* or *helical* antennas. Another major antenna category is the *aperture* antennas that appear in the shape of *horns* or *reflectors*. Finally, *array* antennas are used extensively in communications as switched beam antennas or adaptive antennas. For more information on the design and analysis of antennas, we refer the reader to References 1–15, where all types of antennas, mentioned earlier, are fully described. Adaptive and multibeam antennas will be studied in Chapter 8.

REFERENCES

1   Balanis, C.A., *Antenna Theory: Analysis and Design*, 2nd ed., John Wiley & Sons, New York, 1997.

2   Kraus, J.D., *Antennas*, 2nd ed., McGraw-Hill, New York, 1988.

3   Chryssomallis, M. and C.G. Christodoulou, "Antenna radiation patterns" in *John Wiley Encyclopedia of RF and Microwave Engineering*, Wiley, New York, 2005.

4   Drabowitch, S., A. Papiernik, H. Griffiths, J. Encinas, and B.L. Smith, *Modern Antennas*, Chapman & Hall, London, 1998.

5   Kraus J.D. and R. Marhefka, *Antennas*, New York, McGraw-Hill, 2001.

6   Stutzman W.L. and G.A. Thiele, *Antenna Theory and Design*, John Wiley and Sons, Inc., New York, 1981.

7   Weeks, W.L., *Antenna Engineering*, McGraw-Hill, New York, 1968.

8   Milligan, T.A., *Modern Antenna Design*, McGraw-Hill, New York, 1985.

9   Johnson, R.C. (and H. Jasik, editor of first edition), *Antenna Engineering Handbook*, McGraw-Hill, New York, 1993.

10 Lo, Y.T. and S.W. Lee, eds., *Antenna Handbook: Theory, Applications and Design*, Van Nostrand Reinhold, New York, 1988.

11 Siwiak, K., *Radiowave Propagation and Antennas for Personal Communications*, 2nd ed., Artech House, Boston-London, 1998.

12 Vaughan, R. and J. Bach Andersen, *Channels, Propagation, and Antennas for Mobile Communications*, IEEE, London, 2002.

13 Bertoni, H.L., *Radio Propagation for Modern Wireless Systems*, Prentice Hall PTR, Upper Saddle River, NJ, 2000.

14 Saunders, S.R., *Antennas and Propagation for Wireless Communication Systems*, John Wiley & Sons, New York, 1999.

15 Morgan, M. and V. Evans, "Synthesis and analysis of polarized ellipses," in *Antennas of Elliptic Polarization*, Shpuntov, A. I., ed., Foreign Literature, Moscow, 1961, pp. 369–385.

16 Kanare′kin, D.B., N.F. Pavlov, and V.A. Potekhin, *Polarization of Radiolocation Signals*, Sov. Radio, Moscow, 1966 (in Russian).

17 Gusev, K.G., A.D. Filatov, and A.P. Sopolev, *Polarization Modulation*, Sov. Radio, Moscow, 1974 (in Russian).

18 Ponamarev, G.A., A.N. Kulikov, and E.D. Tel'pukhovsky, *Propagation of Ultra-Short Waves in Urban Environments*, Rasko, Tomsk, 1991 (in Russian).

19 Lee, W.C. and R.H. Brandt, "The elevation angle of mobile radio signal arrival," *IEEE Trans. Communic.*, Vol. 21, No. 11, 1973, pp. 1194–1197.

20 Bitler Y.S., "A two-channel vertically spaced UHF diversity antenna system," *Proc. of Microwave Mobile Radio Symp.*, Boulder, CO, 1973, pp. 13–15.

21 Schelknunoff, S.A. and H.T. Friis, *Antenna Theory and Practice*, John Wiley, New York, 1952.

CHAPTER THREE

Fundamentals of Wireless Networks

3.1. CELLULAR NETWORKS CONCEPT

In this chapter we briefly introduce the basics of *cellular* networks, cell patterns and cell splitting. The cellular concept was initially proposed to help designers of land wireless systems to eliminate noise within the propagation channels and to minimize the effects of interference that occur in a multiuser environment within band-limited channels. These phenomena are mostly manifested for moving subscribers, that is, for mobile communication systems due to the Doppler Effect, which causes time-dependent and frequency-dependent fading effects (see Chapter 1).

A cell is the basic geographic unit of a cellular system. The simplest *radio cell* can be constructed using a base station (BS) at the center (or inside) of a cell [1–10] which predicts the coverage area from the BS antenna over small geographic areas having the shape of hexagons as shown in Figure 3.1. The range defines this coverage area where a stable signal from this station can be received. There are several methods to construct the cell pattern, also called the cell map. We start with the regular case based on the hexagon presentation of each cell as shown in Figure 3.1. Each cell size varies depending on the landscape to be covered. Because of constrains imposed in natural terrain and man-made structures, the true shape of cells is not a perfect hexagon, and the cell patterns usually are not regular. These scenarios usually occur in urban and suburban areas and will be covered in Chapter 11.

Radio Propagation and Adaptive Antennas for Wireless Communication Networks: Terrestrial, Atmospheric, and Ionospheric, Second Edition. Nathan Blaunstein and Christos G. Christodoulou.
© 2014 John Wiley & Sons, Inc. Published 2014 by John Wiley & Sons, Inc.

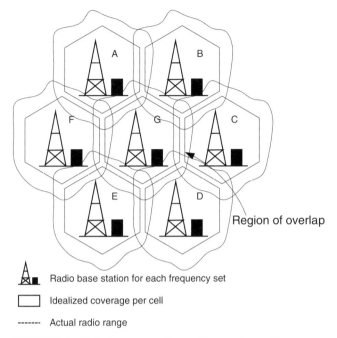

FIGURE 3.1. Cellular map pattern with overlapped hexagons.

It can be seen from Figure 3.1 that some regions overlap with neighboring radio cells, where stable reception from neighboring base stations can be obtained. From this scheme, it is obvious that different frequencies should be used in these cells that surround the central cell. On the other hand, the same frequencies can be used for the cells farthest from the central one. This is the so-called *cell repeating* or *reuse of operating frequencies* principle. At the same time, reuse of the same radio channels and frequencies within the neighboring cells is limited by preplanned *co-channel interference*.

As was mentioned in References 6–10, the *cellular pattern* concept for land wireless communications has been introduced to account for the numerous cells within a small radius, which provide a sufficient signal-to-noise ratio (SNR) and a low level of interference with received signals within the communication channel. According to the literature regarding cellular splitting strategy [1–10], this strategy at early stages of cellular communications has been based on the following principles:

- Use regular hexagon-shaped cells, as the hexagon-shaped cell is more geometrically attractive than the circle-shaped cell, but at the same time hexagon shape is very close to circle shape.
- With an increase of the number of subscribers, the dimensions of the cells must be smaller, usually occurring for the center of cities, where the amount of traffic is bigger and density of buildings is higher.

- Cells are arranged in *clusters*. The cluster size is designated by the letter N and is determined by [6–10]

$$N = i^2 + ij + j^2, \qquad (3.1)$$

where $i, j = 0, 1, 2, \ldots$, that is, only the cluster sizes 3, 4, 7, 9, 12, and so on, are possible. Thus, each hexagonal cell can be packed into clusters "side-to-side" with neighboring cells, as shown in Figure 3.1.

Next, we introduce another main parameter of cellular patterns, called *reuse distance*, D, which defines the distance between two cells utilizing the same limited bandwidth, called *reuse frequencies* or *repeating frequencies*. We will explain this definitions focusing on the popular 7-cell cluster pattern [6–10], which is depicted in Figure 3.2. First, we notice that the allocation of frequencies into seven sets is required. In Figure 3.2, the mean reuse distance is explained in which the cells (denoted by dark color) use the *same frequency bandwidth*. This is a simple way to use the repeat frequency set in the other clusters. Between D and the cell radius R_{cell}, there is a relationship that is called the *reuse ratio*. This parameter, denoted by Q, for a hexagonal cell is a function of cluster size, that is [6–10],

$$Q = \frac{D}{R_{cell}} = \sqrt{3N}. \qquad (3.2)$$

Within other cells in a cluster, interference inside the communication channel can be expected at the same frequencies. Hence, for a 7-cell cluster, there could

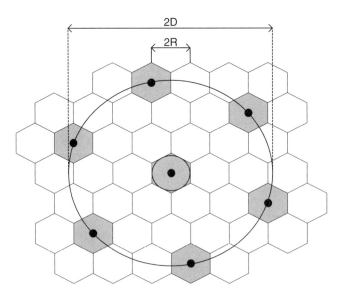

FIGURE 3.2. Frequency reuse plan with reuse factor $N = 7$.

be up to six immediate interferers, as it is shown in Figure 3.2. So, it is apparent that the cellular system concept is closely connected with the so-called *co-channel interference* caused by the frequent reuse of channels within the cellular communication system. To illustrate the concept of co-channel interference, let us consider a pair of cells with radius R, separated by a reuse distance D, as shown in Figure 3.2. As the co-channel site is located far from the transmitter $(D \gg R)$, which is located within the initial cell, its signal at the servicing site will suffer multipath attenuation. To predict the degree of co-channel interference in such a situation with moving subscribers within the cellular system, a new parameter, *carrier-to-interference ratio (C/I)*, is introduced in References 1–10. A co-channel interferer has the same nominal frequency as the desired frequency. It arises from multiple use of the same frequency band. For omnidirectional or isotropic antennas (see definitions in Chapter 2) located inside each site, the theoretical co-channel interference in decibels is given [6–10]:

$$\frac{C}{I} = 10 \log\left[\frac{1}{j}\left(\frac{D}{R}\right)^{\gamma}\right],\tag{3.3}$$

where j is the number of co-channel interferers $(j = 1, 2, \ldots, 6)$, γ is the path-loss slope constant introduced and defined in Chapter 5. For a typical 7-cell cluster $(N = 7)$ with one cell as basic unit (with the transmitter inside it) and with six other interferers $(j = 6,$ as seen from Fig. 3.2) as the co-channel sites, this parameter depends on conditions of the wave propagation within the urban communication channel. As a simple example, presented in Reference 10 for two-ray propagation model above a flat terrain with $\gamma = 4$ (see also Chapter 5), we can rewrite (3.3) as

$$\frac{C}{I} = 10 \log\left[\frac{1}{6}\left(\frac{D}{R}\right)^{4}\right].\tag{3.4}$$

Thus, for $N = 7$, that is, for $D/R_{cell} = \sqrt{3N} = 4.58$, we get $C/I = 18.6$ dB. Using (3.2), we can simplify (3.4) as

$$\frac{C}{I} = 10 \log\left[\frac{1}{6}(3 \cdot N)^{2}\right] = 10 \log(1.5 \cdot N^{2}),\tag{3.5}$$

meaning that the *C/I* ratio is also a function of cluster size N and increases with the increase in the number of cells in each cluster or with decrease of the cell radius R_{cell}.

3.2. SPREAD SPECTRUM MODULATION

Now, we ask a question: what is *spread spectrum modulation*? It is well known that in most current wireless networks, narrowband digital modulation is

usually used. In these cases of narrowband modulation, such as binary amplitude shift keying (BASK), binary frequency shift keying (BFSK), and binary
phase shift keying (BPSK), the signal data (e.g., the baseband signal) have a
useful narrow bandwidth that corresponds to the optimal rate of data stream
passing the channel, corrupted by an AWGN or other noise, to obtain a high
SNR. This narrowband bandwidth is called the *usual* bandwidth in literature
[11–14].

In contrast with this type of narrowband modulation, a different modulation
technique, called spread spectrum (SS) modulation, produces a spectrum for
the transmitted signal with data (bandpass signal) having *much wider* bandwidth than the *usual* bandwidth required to transfer desired information data
stream through a wireless channel.

3.2.1. Main Properties of Spread Spectrum Modulation

Spread spectrum (SS) modulation techniques employ a transmission bandwidth B_w, which is larger than the minimum required bandwidth (BW) of the
information carrying signal B_i (i.e., message with the rate of bits R). Therefore,
the advantage of SS is that many subscribers can simultaneously use the same
bandwidth without significant interference between them. In multiuser interference wireless communication channels, SS modulation becomes very bandwidth efficient.

The second important feature of SS modulation is that SS signals are pseudorandom; that is, they have noise-like properties when compared with the
digital information data. The spreading waveform is controlled by a *pseudo-
noise* (PN) sequence or *pseudo-noise* code (PN code), which is a binary
sequence that appears random but can be reproduced in a deterministic
manner by the corresponding receivers [11–14]. Because each subscriber uses
a unique PN code, which is approximately orthogonal to the codes of other
subscribers located in the area of service, the receivers can separate each user
based on their own unique codes, even though they occupy the same spectral
bandwidth at all times of service.

3.2.2. Main Definitions and Characteristics of Spread Spectrum Modulation

During the spreading process, the original signal power over a broad BW is
defined by the *processing gain* (*PG*), which can be determined as [11–14]

$$PG = \frac{B_t}{B_i}, \tag{3.6}$$

where B_t is the PN-code BW, and B_i is the information data BW (see Fig. 3.3).

Using SS modulation we can overcome the following corrupting effects:

1. Up to a certain number of users, the *interference* between SS signals using
   the same frequency is negligible.

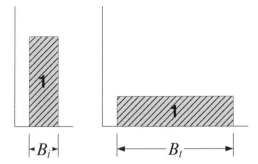

FIGURE 3.3. Schematical presentation of comparison between the information data bandwidth, B_i, and the PN-code bandwidth, B_t.

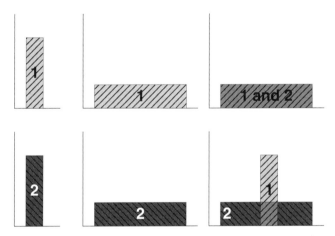

FIGURE 3.4. Information of the desired user #1 can only be recorded at the receiver #1 using his unique PN-code.

Why we can manage to eliminate interference between users using the same frequency band can be seen in Figure 3.4 for the case of two users that do not interfere with one another. In the left and middle parts of the figure one can see that the two users generate an SS signal from their narrowband data signals. In the right part of the figure it is shown that both users transmit their SS signals at the same time (upper picture). But at receiver #1, only the signal of user #1 is coherently summed with its own PN code by the user #1 demodulator and only the user #1 data are recovered. The SS signal of user #2 is contained at the level of the spread signal with low energy over an entire spectrum of frequencies.

2. Another main reason for using SS systems for wireless communication is its resistance to multipath fading that occurs in such channels.

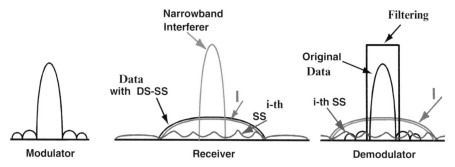

FIGURE 3.5. SS modulation for various kinds of noises and interferers.

Using SS modulation with signals, having uniform energy over a very large bandwidth, at any given time only a small portion of the spectrum will be affected by such kind of fading (see Fig. 3.5). The multipath resistance properties occur due to the fact that the delayed multipath components of the transmitted PN-modulated signal will have close-to-zero correlation with the original PN code, and will thus appear as another uncorrelated user, which is ignored by the receiver. The same situation occurs with any noise or multipath fading or with interference between users (see Fig. 3.5).

3. An SS system is not only resistant to multipath fading, but it can also exploit the multipath components of the total signal to improve the performance of the wireless communication networks. These features of SS modulation we will be discussed in Section 3.2.3.

In summary, SS modulation provides

- Protection against multipath (due to fading) interference
- Multiuser interference rejection
- Antijamming capabilities, especially narrowband jamming
- Low probability of interception (LPI)
- Privacy.

There are a number of SS modulation techniques that generate SS signals:

- *Direct Sequence Spread Spectrum* (DS-SS) *Modulation.* It is based on direct multiplication of the data-bearing signal with a high chip rate spreading PN code.
- *Frequency Hopping* (FH). It based on a jump in carrier frequency, at which the data-bearing signal is transmitted. This jump (change) in frequency occurs according to the corresponding spreading PN code.

- *Time Hopping* (TH). Here, the data-bearing signal is not transmitted continuously. Conversely, it is transmitted in form of short bursts, where the times of the bursts are decided by the spreading PN code, that is, jump in time is noted.
- *Hybrid Modulation* (HM). It based on combination of the above methods to combat with their disadvantages.

3.2.3. Direct Sequence Spread Spectrum Modulation Technique

The block diagram of SS digital communication system is shown in Figure 3.8. Here, the pseudorandom pattern generator, which is connected with the modulator at the transmitter, creates a PN binary-valued sequence, which affects the transmitted signal at the modulator. In other words, the binary data modulate an RF carrier, and then the data carrier is modulated by the PN-code signal which consists of code bits, which can be either "+1" or "–1." A higher BW of PN codes allows spreading. To obtain the desired spreading of the data-bearing signal, the chip rate of the PN-code signal must be much higher that the chip rate of the data-bearing (information) signal. The coded narrowband signal before spreading is modulated using the usual narrowband modulation such as BPSK, DPSK, QPSK, or MSK.

As seen in Figure 3.6, each bit code at the PN sequence is called a *chip* while T_c is called the *chip interval*. So, each pulse in a PN spreading sequence represents a chip, as a rectangular pulse with amplitude equal to +1 or –1, and

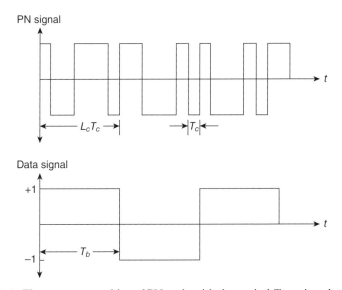

FIGURE 3.6. There are seven chips of PN-code with the period T_c against data bit code with the period $T_b = L_c T_c$; L_c is the length of chips in the one bit code.

duration T_c. Such rectangular pulses are the basic elements in a DS-SS modulation. If we now define a transmitted information bit by its time duration T_b, the bandwidth expansion factor can be expressed as

$$\frac{B_t}{R} = \frac{T_b}{T_c}. \tag{3.7}$$

We now introduce an integer

$$L_c = \frac{T_b}{T_c}, \tag{3.8}$$

which is number of chips per information bit (or the length of chips inside one bit). If so, we can define using this integer, a number of phase shifts that have occurred in the transmitting signal during the bit duration $T_b = 1/R$. Figure 3.6 illustrates the relationships between the PN signal and the information data signal. Here, $T_b/T_c = 7$, that is, $PG = 7$. Using relations between these parameters, one can define the processing gain of DS-SS modulation in the following general form [11–14]:

$$PG = \frac{B_t}{B_i} = \frac{T_b}{T_c} = \frac{E_s}{E_c} = \frac{R_c}{R_s}, \tag{3.9}$$

where E_s is the energy of the data symbol, E_c is the energy of PN code, R_c is the rate of chips in PN sequence, and R_S is the rate of the symbols in the information data sequence.

The received SS signal for a single user can be represented as

$$s_{ss}(t) = \sqrt{\frac{2E_s}{T_s}} m(t) p(t) \cos(2\pi f_c t + \theta), \tag{3.10}$$

where $m(t)$ is a data sequence, $p(t)$ is the PN spreading sequence, f_c is a carrier frequency, and θ is the carrier phase angle at $t = 0$. The data waveform is a time sequence of nonoverlapping rectangular pulses, each of which has amplitude equal to +1 or –1. Each symbol in a data sequence represents a data symbol with duration T_s and energy E_s. If B_t is the bandwidth of SS signal $s_{ss}(t)$ and B_i is the bandwidth of modulated signal $m(t)\cos(2\pi f_c t + \theta)$, the spreading due to $p(t)$ gives $B_t(t) \gg B_i$. If $p(t) = \pm 1$, then $p^2(t) = 1$, and this multiplication yields the dispreading signal $s(t)$ given by

$$s(t) = \sqrt{\frac{2E_s}{T_s}} m(t) \cos(2\pi f_c t + \theta) \tag{3.11}$$

at the input of demodulator. Therefore, the SS demodulation is what we usually call the wideband demodulation, and the data modulation is what we call the narrowband demodulation.

Advantages of DS-SS Technique

- easy code generation
- simple frequency synthesizer (single carrier frequency)
- coherent demodulation of the SS signal is possible
- no synchronization between users is necessary.

Disadvantages of DS-SS Technique

- difficult to acquire and maintain local synchronization between the local code and the received signal
- for correct reception, the locally generated code and the received code sequence must be synchronized in a fraction of chip time; bandwidth is limited to 10–20 MHz.

3.2.4. The Frequency Hopping Technique

In such SS modulation technique, the carrier frequency of the modulated information signal changes periodically, and the hopping pattern is decided by the code signal generated by the PN sequence generator, that is, by a unique PN code for each desired subscriber.

During the time interval T, the carrier frequency remains the same, but after each time interval, the carrier hops to another or possibly the same frequency according to the corresponding PN sequence which is unique for each user. The set of available frequencies the carrier attains is called the *hop-set*.

The frequency occupation of FH-SS modulation technique differs considerably from DS-SS modulation technique. The latter occupies the whole frequency band when information is transmitted, whereas an FH-SS uses only a small part of bandwidth during transmission, but the location of this part differs in time. So, during the entire time of user service, the same average bandwidth finally can be used. This difference is clearly seen in illustrations presented in Figure 3.7. In such an SS modulation technique, the carrier frequency of the modulated information signal changes periodically; the hopping

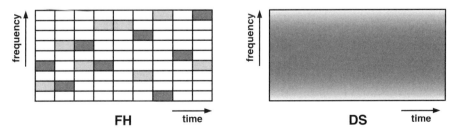

FIGURE 3.7. Comparison between DS-SS and FH-SS modulation techniques.

pattern is decided by the code signal generated by the PN sequence generator. The second difference between DS-SS and FH-SS modulation is in how they reduce the narrowband interference: in DS-SS it is reduced by a factor of PG (*processing gain*), and strong interference may overload the DS receiver. In FH-SS it will cause an error in the system and therefore may be handled by the forward error correction (FEC) [15–22].

In the case of an FH-SS system, several users independently hop their carrier frequency while using BFSK modulation of data signal. If two users are not simultaneously utilizing the same frequency band, the probability of bit error for narrowband BPSK modulation technique can be expressed as [15–22]

$$P_e = \frac{1}{2}\exp\left(-\frac{E_b}{2N_0 B_\omega}\right), \tag{3.12}$$

where E_b is the energy of data bit, N_0 is the power spectral density (PSD) of AWGN, and B_ω is the bandwidth of AWGN.

If now there are $N - 1$ interfering users in the system and M possible hopping channels (called slots), there is a $1/M$ probability that a given interferer will be presented in the desired user's slot and, finally, the overall probability of bit error can be modeled as

$$P_e = \frac{1}{2}\exp\left(-\frac{E_b}{2N_0 B_\omega}\right)\left(1 - \frac{N-1}{M}\right) + \frac{1}{2}\left[\frac{N-1}{M}\right]. \tag{3.13}$$

For $N = 1$, (3.13) reduces to (3.12), the standard probability of error for BPSK data signal modulation. In the general case of $N - 1$ interferers, $E_b/N_0 B_\omega \to \infty$, then the probability of bit error approaches $P_e = 1/2[N - 1/M]$, which illustrates the irreducible error rate due to multiple access interference. Equation (3.13) was obtained by assuming that all users hop their frequencies synchronously.

If the hopping process occurs slower than the data (e.g., symbol or bit) rate, the modulation is considered to be slow frequency hopping (S-FH). Therefore, in such systems, multiple bits or symbols are transmitted at the same frequency.

For S-FH system, where one hop has several symbols or bits, the probability of bit error for $N - 1$ users is more complicated and described by the following formula:

$$P_e = \frac{1}{2}\exp\left(-\frac{E_b}{2N_0 B_\omega}\right)\left[1 - \frac{1}{M}\left(1 + \frac{1}{N_b}\right)\right]^{N-1} + \frac{1}{2}\left\{1 - \left[1 - \frac{1}{M}\left(1 + \frac{1}{N_b}\right)\right]^{N-1}\right\}, \tag{3.14}$$

where N_b is the number of bits per hop.

As it was done earlier for DS-SS modulation, let us show the properties of FH-SS modulation with respect to multiple access capability, multipath interference rejection, narrowband interference rejection, and probability of interception. Thus, the following properties of FH-SS modulation can be emphasized:

- multiple access capability
- multipath interference rejection
- narrowband interference rejection
- LPI.

Next, we will point out advantages and disadvantages of the FH-SS approach.

Advantages of FH-SS Technique

Synchronization is much easier for the FH-SS technique than in the DS-SS technique:

- It has higher SS bandwidths.
- The probability of multiple users transmitting in the same frequency band at the same time is small. A far-located user cannot be interfered by a closely located user, so the near–far performance here is much better than in the DS-SS approach.
- FH-SS offers higher possible reduction of narrowband interference than a DS-SS because of the larger possible bandwidth FH modulation can employ.
- The FH-SS modulation technique has an advantage over the DS-SS technique in that it is not so dependent on the near–far problem, because signals generally do not utilize the same frequency simultaneously, and the relative power levels of signals are not as critical as in the DS-SS system.

Disadvantages of FH-SS Technique

- A highly sophisticated frequency synthesizer is required.
- An abrupt change of the signal during frequency changing bands will lead to an increase in the frequency band occupied. To avoid this, the signal has to be turned off and on when changing frequency.
- Coherent demodulation is difficult because of the problems in maintaining phase relationships during hopping.

3.2.5. Time Hopping Technique

In the time hopping spread spectrum (TH-SS) modulation technique, the data signal is transmitted in rapid bursts at time intervals determined by the code assigned to the user. For this purpose, the time axis is divided to frames, and frames are divided to M time slots (exact definitions will be presented in the next section). During each frame, in the process of data transmission, the user

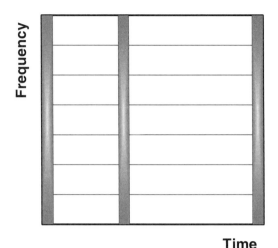

Time

FIGURE 3.8. Time hopping spread spectrum modulation technique plot.

will transmit all its data in *only one* of the M time slots. Therefore, the frequency of data transmission is increased by a factor of M. The time slot is determined by the PN code used for the desired subscriber. Figure 3.8 shows the time-frequency plot of the TH-SS modulation technique. Comparing this figure with Figure 3.7, it can be seen that the TH-SS approach uses the whole wideband spectrum for short periods instead of parts of the spectrum all of the time. The probability of bit error for such technique can be found both for fast TH-SS by using Equation (3.13) and for slow TH-SS by using Equation (3.14), where M now is time-slot channel number and N_b is the number of bits with period T_b inside the one slot-hop time period.

We show the properties of TH-SS modulation with respect to multiple access capability, multipath interference rejection, narrowband interference rejection, and probability of interception. They are

- multiple access capability
- multipath interference rejection
- narrowband interference rejection
- LPI.

Advantages of TH-SS Technique

- Implementation is easier than DS-SS modulation.
- It is useful when the transmitter is average-power limited but not peak-power limited.
- Near–far performance is better than the DS-SS technique.

Disadvantages of TH-SS Technique

- It takes a long time to synchronize the code, and the receiver has a short time to execute it.
- If multiple transmissions occur, a lot of data bits are lost. A good correcting code and data interleaving are necessary.

3.3. MULTIPLE ACCESS TECHNOLOGIES AND NETWORKS

Standard multiple access networks are used to allow many subscribers, stationary or mobile, to share simultaneously a finite amount of wireless frequency spectrum. The sharing of spectrum is required to achieve a high capacity by simultaneously allocating the available bandwidth or available amount of channels to multiusers' access. How this can be done without degradation in the performance of the wireless networks in general is discussed next.

3.3.1. Introduction to Multiple Access Technologies

In a wireless communication link we differentiate the *forward link* or *downlink* from the BS (or radio port) to the desired user (stationary or mobile) and *reverse link* or *uplink* from the desired user to the BS, as shown in Figure 3.9. If this connection is bidirectional, we call it a *full-duplex* link.

Duplexing can be done using frequency or time-domain communication techniques. Thus, *frequency division duplexing* (FDD) provides two distinct bands of frequencies, one for the BS and one for every user connected (see top scheme in Fig. 3.10). *Time division duplexing* (TDD) uses different time instead of frequency to serve both the forward and reverse channels (see bottom scheme in Figure 3.10).

FDD can be performed by simultaneous (time $t = $ const) transmission and reception of information and by careful separation between frequencies for downlink and uplink. The TDD can be performed by time latency due to separate transmitting/receiving of information. If the time split between the uplink and downlink time slot is small, then the transmission and reception of data appears simultaneous to the user. TDD allows communication on a single

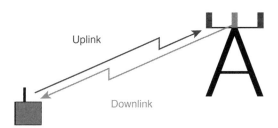

FIGURE 3.9. Uplink and downlink between the BS antenna and the subscriber antenna.

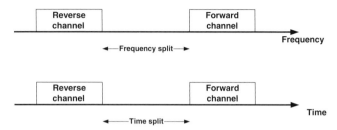

FIGURE 3.10. Frequency division (upper view) and time division duplexing (bottom view).

channel (frequency $f = $ const). Thus, from the beginning, we see several trade-offs between the FDD and TDD approaches.

There are four major access techniques to share the available bandwidth in the wireless communication networks [15–22]:

- frequency division multiple access (FDMA)
- time division multiple access (TDMA)
- code division multiple access (CDMA)
- space division multiple access (SDMA).

These techniques can be grouped into *narrowband* and *wideband* systems.

(a) Narrowband systems

In narrowband systems, the transmission bandwidth of each single channel is smaller than its expected coherence bandwidth. This allows us to divide the available radio spectrum into a large number of narrowband channels. Narrowband channels are usually operated using FDD. If so, the system, called FDMA/FDD, is used, when a subscriber is assigned a particular channel. There is no sharing with other users.

In narrowband TDMA systems (i.e., sharing using TDMA), there are two options to serve users: there are a large number of channels allocated using FDD or TDD. Such systems are called TDMA/FDD and TDMA/TDD, respectively.

(b) Wideband systems

In wideband systems the transmission bandwidth of each single channel is much larger than its expected coherence bandwidth. Here, multipath fading does not greatly affect the received signal within the wideband channel, and frequency-selective fading occurs in only a small fraction of the signal bandwidth. In such a system, the user transmits information in a large part of the radio spectrum. The most widely used techniques are wideband TDMA and CDMA, which can use separately FDD, TDD, or CDD (code division duplexing).

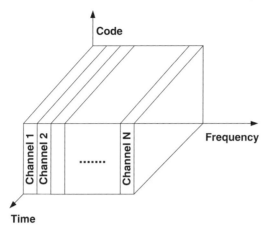

FIGURE 3.11. FDMA schematical presentation regarding other access techniques, CDMA and TDMA.

3.3.2. Frequency Division Multiple Access

The main task of a frequency division multiple access (FDMA) system is to assign individual channels to individual users. Figure 3.11 shows how each user is allocated a unique frequency band or channel. These channels are assigned on demand to users who request service. FDMA is characterized by the following general features:

- one phone circuit at a time per carrier, since channels are narrow (30 kHz)
- at idle states, a channel cannot be used by other users
- after channel assignment, a continuous transmission occurs
- lower framing overheads
- usually is narrowband
- large symbol time compared with the average delay spread caused by multipath fading
- low inter-symbol interference (ISI)
- cost of cell systems is higher than TDMA
- uses duplexers in user unit (mobile or not)
- typical problem is spurious frequencies at BS due to nonlinear filters.

Main Characteristics of FDMA. *Spectral efficiency* is a measure of the efficient use of the frequency spectrum. In FDMA, this measure can be determined by knowing the total number of channels in the FDMA system, N_c, the bandwidth of the system B_s, and that of the channel B_c:

$$\eta_{FDMA} = \frac{B_c N_c}{B_S}. \tag{3.15}$$

The number of channels that can be simultaneously supported in an FDMA system is

$$N_{ch} = \frac{B_S + 2B_{guard}}{B_{ch} + 2B_{guard}}, \tag{3.16}$$

where B_S is the total spectrum of the system, B_{guard} is the guard band at the edge of the allocated spectrum, and B_c is the channel bandwidth.

Throughput efficiency is determined by

$$\eta_{th} = \frac{N_c}{B_S A}, \tag{3.17}$$

where N_c is the total number of channels in the FDMA system, B_s is the bandwidth of the system, and A is the total area of service. So, η_{th} has dimensions of channels/MHz/km^2. Next, we introduce M as the total number of cells in the area of service, $A_c = 2.6R_{cell}^2$ as the area covered by one cell as hexagon in km^2, B_c as the bandwidth of the channel, and N as the reuse factor of the cellular network. The ratio M/N gives the number of clusters in the served area and B_s/B_c gives the number of channels in the cluster. Using these parameters, *throughput* efficiency can be expressed as:

$$\eta_{th} = \frac{(B_S / B_c) \cdot (M / N)}{B_S A_c M} = \frac{1}{B_c A_c N} = \frac{1}{2.6B_c \cdot N \cdot R_{cell}^2}. \tag{3.18}$$

In terms of total traffic carried by the system, η_{th} is measured in Erlangs/MHz/km^2 and is defined as

$$\eta_{th} = \frac{(Total\_Traffic\_Carried\_By\_System)}{B_S A}. \tag{3.19}$$

If the number of channels in the cell, $C = Bs/(B_cN)$, is included, then

$$\eta_{th} = \frac{(Total\_Traffic\_Carried\_By\_C\_Channels)}{B_S A_c}. \tag{3.20}$$

By introducing the trunking efficiency factor, η_t, as a function of the blocking probability (e.g., GOS-probability [5, 18]) and B_S/B_c, we end up with another expression for the spectral efficiency of modulation:

$$\eta_{th} = \frac{\eta_t B_c / (B_c N)}{B_S A_c} = \frac{\eta_t}{B_S A_c N}. \tag{3.21}$$

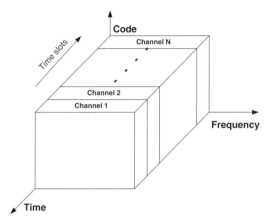

FIGURE 3.12. TDMA schematical presentation regarding other access techniques, CDMA and FDMA.

Finally, the *overall efficiency* of the FDMA system can be found from

$$\eta = \eta_{th}\eta_{FDMA}. \tag{3.22}$$

3.3.3. Time Division Multiple Access

Time division multiple access (TDMA) allows transmission of larger information rates than in an FDMA system. The carrier (radio frequency) is divided to N timeslots (simply, *slots*) and can be shared by N terminals (see Fig. 3.12). Each terminal uses a particular slot different from slots used by the other terminals. For forward and reverse channels in a TDMA/TDD system, different time slots are used: half slots in a frame are for the uplink and half slots for the downlink. In a TDMA/FDD system the same procedure is used, but the carrier frequencies are different for the forward and reverse channels. Thus, a single carrier frequency serves multiples users (see Fig. 3.12).

TDMA includes the following features:

- TDMA shares a single carrier frequency with several users, where each user makes use of nonoverlapping time slots. The number of slots in a frame depends on several factors, such as the modulation technique used, available bandwidth, and so on.
- The TDMA transmission is discontinuous; that is, digital data are transmitted in a buffer-and-burst method. For example, a mobile transmits on *slot 1*, waits during *slot 2*, and receives on *slot 3*, waits again during *slot 4*, and transmits again on *slot 1*. As a consequence of this type of transmission, the gross channel rate is not equal to the communication rate (i.e., the channel throughput seen by the user). The channel throughput must be faster by a factor of at least the number of slots in a frame.

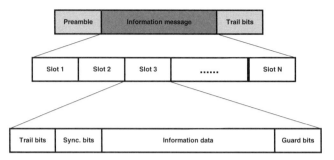

FIGURE 3.13. Frame division on slots; each slot consists of information data, overhead data for using data service, and guard bits to separate each slot and exclude interference between neighboring slots.

- TDMA uses different time slots for transmission and reception, and duplexers are not required.
- In a TDMA frame, the preamble contains the address and synchronization information that both the BS and the subscriber use to identify each other.
- Guard times are utilized to allow synchronization of the receivers between different slots and frames (see Fig. 3.13). The guard times should be minimized and the spectrum broadens.
- High synchronization accuracy is required.
- TDMA has advantage in that it is possible to allocate different time slots per frame to different users (priority).

Main Characteristics of TDMA. Efficiency of TDMA is based on the efficient use of time slots/frames in the system, and it is determined as

$$\eta_{TDMA} = \frac{T_s}{T_f} N_s, \tag{3.23}$$

where T_s is a time of slot duration, T_f is the time of frame duration, and N_s is the number of slots per frame.

Then, the *overall efficiency* of the TDMA system can be found as

$$\eta = \eta_{th}\eta_{TDMA}, \tag{3.24}$$

where η_{th} can be found either by using Equation (3.18), if the strategy of cellular map design with the corresponding parameters of area of service is known, or by using Equation (3.21), if the traffic data and probability of blocking are known for the TDMA network.

The *frame efficiency* is a measure of the percentage of transmission data rate that contains information; that is, it is a percentage of bits per frame, which contain transmission data with respect to overhead data. Efficiency of frame is defined as

$$\eta_f = \left(1 - \frac{b_{OH}}{b_T}\right) \times 100\%. \tag{3.25}$$

Here, the number of overhead bits (e.g., for managing and servicing data) per frame is

$$b_{OH} = N_r b_r + N_t b_p + (N_t + N_r) b_g, \tag{3.26}$$

where N_r is the number of reference bursts per frame, b_r is the number of overhead bits per reference burst, N_t is the number of traffic bursts per frame, b_p is the number of overhead bits per preamble in each slot, and b_g is the number of equivalent bits in each guard time interval. The total number of bits per frame, b_T, is

$$b_T = T_f \cdot R, \tag{3.27}$$

where T_f is the frame duration (see Fig. 3.13), and R is the channel bit rate.

The *number of channels* is the number of TDMA channel slots, which can be found by multiplying the number of TDMA slots per channel by a number of channels available; that is,

$$N = \frac{m(B_t - B_{guard})}{B_{ch}}, \tag{3.28}$$

where m is the maximum number of TDMA users supported on each radio channel.

3.3.4. Code Division Multiple Access

It is a part of spread spectrum multiple access (SSMA), or more strictly speaking, a DS-SS technique, which was described in detail in Section 3.2. What is important to mention here is that in CDMA, each user obtains a unique code (see Fig. 3.14). The narrowband message signal is multiplied by a wideband spreading signal, generated by a noise-like generator, to give a unique PN code to each user.

As previously discussed, with such SS modulation technique, we can resolve some very important problems in wireless communication such as

- data rate (very high in CDMA)
- multiplexing
- use of FDD or TDD
- soft capacity limit, and so on (see Section 3.2).

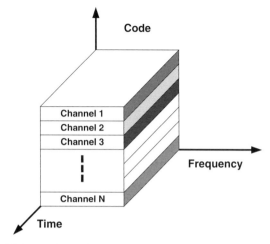

FIGURE 3.14. CDMA schematical presentation regarding other access techniques.

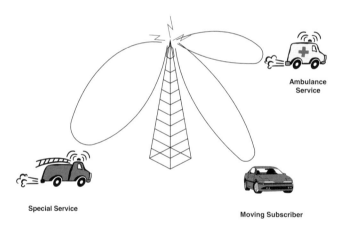

FIGURE 3.15. Spatial division multiple access (SDMA) using antennas with sectors. TDMA and FDMA.

3.3.5. Space Division Multiple Access

The main task of space division multiple access (SDMA) is to control the radiated energy for each user by using the so-called spot-beam antenna (see Fig. 3.15). The main principle of SDMA is that different areas use the same frequency (TDMA, CDMA). Present problems with SDMA are related with issues arising from controlling the reverse channel due to limitations in power control. A detailed discussion on how to use adaptive antennas for SDMA can be found in Chapter 7.

3.4. CAPACITY OF INFORMATION DATA IN MULTIPLE ACCESS NETWORKS

It is interesting to compare FDMA, TDMA, and CDMA in terms of the information rate that each multiple access method achieves in a channel of bandwidth B_w, corrupted by an ideal AGWN noise and the multiplicative noise caused by fading multipath phenomena. Let us compare the capacity of N users, where each user has an average power $P_i = P$, for all $1 \le i \le N$. First, we formulate the capacity term in an ideal band-limited AGWN channel of bandwidth B_w for a single user as

$$C = B_w \log_2\left(1 + \frac{P}{N_0 B_w}\right),\tag{3.29}$$

where $\frac{1}{2}N_0$ is the power spectral density of the additive noise.

In the presence of multiplicative noise with fading, we introduce (as in Chapter 11) in the denominator of Equation (3.29) its spectral density N_m, with its own frequency band, B_Ω. Finally, we can express Equation (3.29), accounting for the fading parameter definition, $K = P/N_m B_\Omega$, as [23]:

$$C = B_w \log_2\left(1 + \frac{P}{N_0 B_w + N_m B_\Omega}\right) = B_w\left[1 + \left(\frac{N_0 B_w}{P} + K^{-1}\right)^{-1}\right].\tag{3.30}$$

Next, we present this general capacity expression for any single subscriber using one of the three multiple access networks, FDMA, TDMA, and CDMA.

In an FDMA network, each user is allocated a bandwidth B_w/N. Hence, the capacity of each user is

$$C_N = \frac{B_w}{N}\log_2\left[1 + \left(\frac{N_0(B_w/N)}{P} + K^{-1}\right)^{-1}\right]\tag{3.31}$$

and the total capacity of the N users is

$$C = N \cdot C_N = B_w \log_2\left[1 + \left(\frac{N_0 B_w}{N \cdot P} + K^{-1}\right)^{-1}\right].\tag{3.32}$$

The total capacity is equivalent to that of a single user (see Equation (3.30)) with average power $P_{av} = N P$. It is interesting to note that for a fixed bandwidth B_w, the total capacity goes to infinity as the number of users increases linearly with N. On the other hand, as the number of users N increases, each user is allocated a smaller bandwidth B_w/N and, consequently, the capacity per user decreases. If we now introduce a normalized capacity per user as C_N/B_w and the energy of signal as a value of the bit energy E_b, we will get the following formula:

$$\frac{C_N}{B_w} = \frac{1}{N}\log_2\left[1+\left(\left(\frac{E_b}{N_0}\right)^{-1}\frac{B_w}{N\cdot C_N}+K^{-1}\right)^{-1}\right].$$ (3.33)

A more compact form of (3.33) can be obtained by defining the normalized total capacity $\tilde{C} = NC_N / B_w$, which is the total bit rate R for all N users per unit of bandwidth. In this case,

$$\tilde{C} = \log_2\left[1+\left(\tilde{C}^{-1}\left(\frac{E_b}{N_0}\right)^{-1}+K^{-1}\right)^{-1}\right]$$ (3.34)

from which we can obtain the ratio E_b/N_0, usually used for practical applications in digital wireless communication systems [23]:

$$\frac{E_b}{N_0} = \frac{K\left(2^{\tilde{C}}-1\right)}{\tilde{C}\left(K+1-2^{\tilde{C}}\right)}.$$ (3.35)

In the absence of multiplicative noise, that is, dealing with a pure AGWN channel only, then $K^{-1}\rightarrow 0(K\rightarrow\infty)$, and Equation (3.35) reduces to the well-known Equation [14–18]

$$\frac{E_b}{N_0} = \frac{2^{\tilde{C}}-1}{\tilde{C}}.$$ (3.36)

In a TDMA system, each user transmits for $1/N$ of the time through the channel of bandwidth B_w, with average power $N\,P$. Therefore, the capacity per user is

$$C_N = \frac{B_w}{N}\log_2\left[1+\left(\frac{N_0 B_w}{N\cdot P}+K^{-1}\right)^{-1}\right],$$ (3.37)

which is identical to the capacity of an FDMA system. However, it must be noted that in TDMA, it may not be possible to transmit the power of $N\,P$ when N is very large. Therefore, there is some threshold in terms of power for the transmitter beyond which it cannot transmit as N exceeds a certain number.

In a CDMA system, each user transmits a pseudorandom signal of a bandwidth B_w and average power $P_{av} = P$. The capacity of the system depends on the level of correlation (cooperation) among the N users. One extreme of uncorrelated CDMA users occurs when the receiver for each user signal does not know the spreading waveforms of other users, or chooses to ignore them in the demodulation process. In that case, the other users appear as interferences at the receiver of each user, and as was shown earlier, they can be rejected. In this case, the multiuser receiver consists of a bank of N single-user receivers. Let us assume that each user's pseudorandom signal waveform is Gaussian and each user signal is corrupted by Gaussian interference power

$(N-1)P$ and additive Gaussian noise $B_w N_0$. Then, the capacity per user, taking into consideration the multiplicative noise too, can be presented in the following form:

$$C_N = B_w \log_2 \left[1 + \left(\frac{N_0 B_w}{P} + (N-1) + K^{-1} \right)^{-1} \right], \qquad (3.38)$$

or, accounting for the bit energy, E_b, the normalized capacity (e.g., spectral efficiency) after straightforward computations can be presented for $N \gg 1$ as:

$$\frac{C_N}{B_w} = \log_2 \left[1 + \frac{C_N}{B_w} \left(\frac{E_b / N_0}{1 + \frac{C_N \cdot N}{B_w} (E_b / N_0) + \frac{C_N}{B_w K} (E_b / N_0)} \right) \right], \qquad (3.39)$$

which, again for a pure AGWN channel (i.e., for $K \to \infty$), reduces to the well-known classical equation [14–16]:

$$\frac{C_N}{B_w} = \log_2 \left[1 + \frac{C_N}{B_w} \left(\frac{E_b / N_0}{1 + (E_b / N_0) \cdot N \cdot C_N / B_w} \right) \right]. \qquad (3.40)$$

For a large number of users, we can use the approximation $\ln(1 + x) < x$, and thus

$$\frac{C_N}{B_w} \leq \frac{C_N}{B_w} \left(\frac{C_N}{B_w} \left(\frac{E_b / N_0}{1 + (E_b / N_0) \cdot N \cdot C_N / B_w + (E_b / N_0) \cdot C_N / B_w K} \right) \right) \log_2 e,$$

or, for the total normalized capacity of all N users ($N \gg 1$), for the fading parameter $K < 1$ (i.e., for the multipath channel with deep fading), and for $E_b/N_0 \gg 1$:

$$\tilde{C} \leq \log_2 e - \frac{1}{E_b / N_0} \leq \frac{1}{\ln 2} \frac{1}{E_b / N_0} - \frac{K}{E_b / N_0} < \frac{1}{\ln 2}. \qquad (3.41)$$

In this case, we observe that the total capacity does not increase with N as in TDMA and FDMA access networks.

We should notice that the above equations for the capacity and/or spectral efficiency of various multiple access systems allow us to estimate the maximum bit rate of information data stream for a given system bandwidth in scenarios occurring not only in the ideal indoor/outdoor channels with AWGN, but also in cases where the multiplicative noise is included. The corresponding algorithms of how to estimate the K-parameter of fading for various outdoor environments are introduced in Chapter 10.

Concluding this chapter, we should emphasize that these basic modulation technologies and multiple access networks have been combined successfully in advanced technologies and modern networks, which we will describe in more detail in Chapter 12 (see also References 24 and 25).

EXERCISES

3.1. Find the parameter reuse ratio, denoted in (3.2) by Q, for the cluster size N changing according to (3.1) for i and j equal from 1 to 3 each.

3.2. Using the definition of the carrier-to-interference ratio (C/I) in decibels indicated by Equation (3.3), plot this cellular characteristic for the following parameters: $j = 1, 2, 6; \gamma = 3, 4; Q = 3, 6$.

3.3. Using Equation (3.9), find the processing gain and R_s, if $T_c = 10ms$, $T_s = 100\mu s$, and $R_c = 200kbps$.

3.4. Using formulas for fast hopping, find the probability of the bit error for 1 user and for 40 users simultaneously under service, if the system bandwidth is 200 MHz and the bandwidth of the hopping channel is 100 KHz. The number of hopping channels $M = 10$.

3.5. Using formulas for slow hopping, find the probability of the bit error for 1 user and for 40 users simultaneously under service, if the system bandwidth is 200 MHz, the bandwidth of the hopping channel is 100 KHz, and the number of bits per one is 3. The number of hopping channels $M = 10$.

3.6. The bandwidth of the FDMA system is 300 MHz and that of a single channel is 150 KHz. The guard frequency interval is 200 Hz. The radius of each cell is 2 km. The cluster size is $N = 7$. Find the number of channels in the FDMA system, the efficiency of FDMA system, the threshold efficiency, and the overall efficiency of the system.

3.7. The bandwidth of the FDMA system is 200 MHz and of the one channel is 100 KHz. The guard frequency interval is 100 Hz. The radius of each cell is 3 km. The total traffic is 70 E per system bandwidth per cell. Find the number of channels in the FDMA system, the efficiency of FDMA, the threshold efficiency, and the overall efficiency of the system.

3.8. Using the information on the threshold efficiency obtained in the previous exercise, find the efficiency of the TDMA system, its overall efficiency, and the number of TDMA channels, if a slot time equals 20 μs, the frame time equals 250 μs, and number of slots per frame is 8.

3.9. Find the efficiency of the frame if the number of the overhead bits equals 405, the frame time is 340 μs, and the information data rate is 5 Mbps.

3.10. Show how one can derive Equation (3.36) from Equation (3.35) and show that Equation (3.35) becomes Equation (3.36) for $K \to \infty$.

REFERENCES

1 Lee, W.Y.C., *Mobile Communication Engineering*, McGraw-Hill, New York, 1985.

2 Jakes, W.C., *Microwave Mobile Communications*, John Wiley and Sons, New York, 1974.

3 Steele, R., *Mobile Radio Communication*, IEEE Press, New York, 1992.

4 Saunders, S.R., *Antennas and Propagation for Wireless Communication Systems*, John Wiley & Sons, New York, 1999.

5 Rappaport, T.S., *Wireless Communications: Principles and Practice*, Prentice Hall, Englewood Cliffs, NJ, 1996.

6 Faruque, S., *Cellular Mobile Systems Engineering*, Artech House, Boston-London, 1994.

7 Feuerstein, M.L. and T.S. Rappaport, *Wireless Personal Communication*, Artech House, Boston-London, 1992.

8 Lee, W.Y.C., *Mobile Cellular Telecommunications Systems*, McGraw Hill Publications, New York, 1989.

9 Linnartz, J.P., *Narrowband Land-Mobile Radio Networks*, Artech House, Boston-London, 1993.

10 Mehrotra, A., *Cellular Radio Performance Engineering*, Artech House, Boston-London, 1994.

11 Simon, M.K., J.K. Omura, R.A. Scholtz, and B.K. Levitt, *Spread Spectrum Communications Handbook*, McGraw-Hill, New York, 1994.

12 Glisic, S. and B. Vucetic, *Spread Spectrum CDMA Systems for Wireless Communications*, Artech House, Boston-London, 1997.

13 Dixon R.C., *Spread Spectrum Systems with Commercial Applications*, John Wiley & Sons, Chichester, 1994.

14 Viterbi, A.J., *CDMA: Principles of Spread Spectrum Communication*, Addison-Wesley Wireless Communications Series, London, 1995.

15 Proakis, J.G., *Digital Communications*, 3rd ed., McGraw-Hill, New York, 1995.

16 Stuber, G.L., *Principles of Mobile Communications*, Kluwer Academic Publishers, Boston-London, 1996.

17 Peterson, R.L., R.E. Ziemer, and D.E. Borth, *Introduction to Spread Spectrum Communications*, Prentice Hall PTR, Upper Saddle River, NJ, 1995.

18 Rappaport, T.S., *Wireless Communications: Principles and Practice*, 2nd ed., Prentice Hall PTR, Upper Saddle River, NJ, 2001.

19 Steele, R. *and L. Hanzo, Mobile Communications*, 2nd ed., John Wiley & Sons, Chichester, 1999.

20 Li, J.S. and L.E. Miller, *CDMA Systems Engineering Handbook*, Artech House, Boston-London, 1998.

21 Molisch, A.F., ed., *Wideband Wireless Digital Communications*, Prentice Hall PTR, Upper Saddle River, NJ, 2000.

22 Paetzold, M., *Mobile Fading Channels: Modeling, Analysis, and Simulation*, John Wiley & Sons, Chichester, 2002.

23   Blaunstein, N. and C. Christodoulou, *Radio Propagation and Adaptive Antennas for Wireless Communication Links*, 1st ed., Wiley & Sons, Hoboken, NJ, 2007.

24   *Cellular Communications*, http://iec.org.

25   Holma H. and Toscala A., *HSDPA/HSUPA for UMTS*, John Wiley & Sons, Chichester, England, 2006.

CHAPTER FOUR

Electromagnetic Aspects of Wave Propagation over Terrain

When both antennas are far from the ground surface, a *free-space propagation* concept is usually used, which is based on the scalar or vector wave equation description, valid for infinite source-free homogeneous media. In Section 4.1, based on Green's theorem and the Huygens's principle, we introduce the Fresnel zone presentation that will be used for the description of terrain and other obstruction effects, such as reflection and diffraction on radio channels. In Section 4.2, we present the main formulas for path loss prediction for a free-space communication link. Next, in Section 4.3, the reflection phenomena due to a flat terrain are described [1–4]. Here, on the basis of Huygens's principle, Fresnel zone concepts, and stationary phase methods, we give main formulas for the resultant reflection coefficients. Section 4.4 deals with the electromagnetic (EM) aspects of radiowave propagation above a rough terrain. Here, the three methods of mathematical derivation of the field strength are presented to obtain the effects of radiowave scattering from ground surfaces with various roughness (large, median, and small) with respect to the wavelength. In Section 4.5, the effect of the ground curvature is considered by using Fock's theory of diffraction. Section 4.6 describes diffraction phenomena caused by a single obstruction placed at the flat ground surface.

4.1. WAVE PROPAGATION IN FREE SPACE

According to References 1–3, the equation of a plane wave, propagating in free space, can be presented as

Radio Propagation and Adaptive Antennas for Wireless Communication Networks: Terrestrial, Atmospheric, and Ionospheric, Second Edition. Nathan Blaunstein and Christos G. Christodoulou.
© 2014 John Wiley & Sons, Inc. Published 2014 by John Wiley & Sons, Inc.

$$\Delta\Psi(\mathbf{r}) - k^2\Psi(\mathbf{r}) = 0. \tag{4.1}$$

Here, Ψ represents each Cartesian component of the electric and magnetic fields of the wave, and $k = 2\pi/\lambda$, where λ is the wavelength.

4.1.1. A Plane, Cylindrical, and Spherical Wave Presentation

The solution of Equation (4.1) is

$$\Psi(\mathbf{r}) = \exp\{i\mathbf{k}\cdot\mathbf{r}\}. \tag{4.2}$$

The waves that satisfy the scalar Equation (4.1) described by the solution (4.2) are called *plane waves*. Wave vector $\mathbf{k}$ denotes the direction of propagation of the plane wave in free space. For any desired direction in the Cartesian coordinate system, the corresponding solution can be immediately obtained from (4.1) and (4.2). For example, if the plane wave propagates along the x-axis, the solution of (4.1) is [1–3]

$$\Psi(x) = A\exp\{ikx\} + B\exp\{-ikx\}. \tag{4.3}$$

This solution describes the waves propagating in the positive direction (with the sign "+") and the negative direction (with sign "–") along the x-axis with phase velocity $v_{ph} = 2\pi f/k = c/\sqrt{\varepsilon_r\mu_r}$, where f is the radiated frequency, $\varepsilon = \varepsilon_0\varepsilon_r$, $\mu = \mu_0\mu_r$, ε_r and μ_r are relative values of the permittivity and permeability of the environment. In free space, the permittivity $\varepsilon \equiv \varepsilon_0 \approx 10^{-9}/36\pi F/m$ ($\varepsilon_r=1$) and permeability $\mu \equiv \mu_0 \approx 4\pi\cdot10^{-7}H/m$ ($\mu_r = 1$). Then, the phase velocity in an ideal free space is simply the speed of light $c = 1/\sqrt{\varepsilon_0\mu_0} = 10^8 m/\sec$.

In the cylindrical coordinate system $\{\rho, \varphi, z\}$, the scalar wave equation, which describes propagation of *cylindrical waves* in free space, can be written as [3]

$$\left(\frac{1}{\rho}\frac{\partial}{\partial\rho}\rho\frac{\partial}{\partial\rho} + \frac{1}{\rho^2}\frac{\partial^2}{\partial\phi^2} + \frac{\partial^2}{\partial z^2}\right)\Psi(\mathbf{r}) = 0. \tag{4.4}$$

This equation has approximate solution, which can be presented in the following exponential form [3, 4]:

$$\Psi(\mathbf{r}) \approx \sqrt{\frac{2}{\pi k_\rho\rho}}\exp\left\{-i\frac{n\pi}{2} - i\frac{\pi}{4}\right\}\exp\left\{i\left[\frac{n}{\rho}(\rho\varphi) + ik_\rho\rho + ik_z z\right]\right\}. \tag{4.5}$$

Here, $\rho\phi$ is the arc length in the ϕ direction, $k_\rho = \sqrt{k^2 - k_z^2}$, and n/ρ can be considered as the component of vector $\mathbf{k}$ if one compares the cylindrical wave presentation as in (4.5) with that for a *plane wave* in (4.2). Consequently, (4.5) looks like a plane wave mainly in the direction $k' = k_z z + k_\rho\rho$, when $\rho\to\infty$.

In the spherical coordinate system $\{r, \theta, \varphi\}$, the scalar wave equation, which describes propagation of *spherical waves* in free space, can be written as in References 1–3:

$$\left(\frac{1}{r^2}\frac{\partial}{\partial r}r^2\frac{\partial}{\partial r}+\frac{1}{r^2\sin^2\theta}\frac{\partial}{\partial\theta}\sin\theta\frac{\partial}{\partial\theta}+\frac{1}{r^2\sin^2\theta}\frac{\partial^2}{\partial\phi^2}+k^2\right)\Psi(\mathbf{r})=0. \quad (4.6)$$

As was shown in References 3 and 4, the spherical wave can be approximated by $\exp\{ikr\}/r$. Thus, one can represent the spherical wave as a plane wave, when $r\rightarrow\infty$.

4.1.2. Green's Function Presentation

Green's function is used in the description of any arbitrary source in an unbounded homogeneous medium, taking into account that each source $s(\mathbf{r})$ can be represented as a linear superposition of point sources. Mathematically this can be expressed as

$$s(\mathbf{r})=\int d\mathbf{r}'s(\mathbf{r}')\delta(\mathbf{r}-\mathbf{r}'). \quad (4.7)$$

The scalar wave equation with the source in the right-hand side can be presented as

$$\nabla^2\Psi(\mathbf{r})-k^2\Psi(\mathbf{r})=s(\mathbf{r}), \quad (4.8)$$

and the corresponding equation for the Green's function in an unbounded homogeneous medium can be presented as

$$\nabla^2G(\mathbf{r},\mathbf{r}')-k^2G(\mathbf{r},\mathbf{r}')=-\delta(\mathbf{r}-\mathbf{r}'). \quad (4.9)$$

The solution of Equation (4.9) is [2–4]

$$G(r)=\frac{1}{4\pi}\frac{\exp\{ikr\}}{r}, \quad (4.10)$$

where $r=|\mathbf{r}-\mathbf{r}'|$. The corresponding solution of (4.8) is

$$\Psi(\mathbf{r})=-\int_V d\mathbf{r}'G(\mathbf{r},\mathbf{r}')s(\mathbf{r}'). \quad (4.11)$$

The geometry of the source $s(\mathbf{r})$ in a space with volume V is shown in Figure 4.1. Using Equation (4.10), one can easily obtain a general solution for the inhomogeneous Equation (4.8) given by

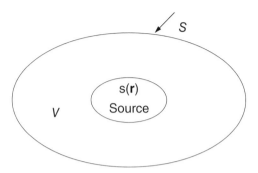

FIGURE 4.1. Geometrical presentation of a source s(r) inside arbitrary volume V bounded by surface S.

$$\Psi(\mathbf{r}) = -\int_V d\mathbf{r}' \frac{\exp\{ik|\mathbf{r}-\mathbf{r}'|\}}{4\pi|\mathbf{r}-\mathbf{r}'|} s(\mathbf{r}'). \tag{4.12}$$

This presentation is valid for any component of an EM wave, propagating in free space, and it satisfies the principle of linear superposition of *point sources* (4.7) for any real source of radiation.

4.1.3. Huygens's Principle

This concept is based on presenting the wave field far from any sources as shown in Figure 4.2. Here, the point of observation A can be either outside the bounded surface S, as shown in Figure 4.2a, or inside, as shown in Figure 4.2b. In other words, according to Huygens's principle, each point at the surface S can be presented as an elementary source of a spherical wave, which can be observed at point A. Mathematically, the Huygens's concept can be explained by the use of Green's functions. First, we multiply the homogeneous Equation (4.1) (without any source) by $G(\mathbf{r},\mathbf{r}')$ and the inhomogeneous Equation (4.9) by $\Psi(\mathbf{r})$. Subtracting the resulting equations from each other and integrating over a volume V containing vector $\mathbf{r}'$ (see Fig. 4.3), yields

$$\Psi(\mathbf{r}') = \int_V d\mathbf{r}\left[G(\mathbf{r},\mathbf{r}')\nabla^2\Psi(\mathbf{r}) - \Psi(\mathbf{r})\nabla^2 G(\mathbf{r},\mathbf{r}')\right] \tag{4.13}$$

from which we obtain *Green's theorem*, or the *second Green formula* [1–4]:

$$\int_V d\mathbf{r}\left[G(\mathbf{r},\mathbf{r}')\nabla^2\Psi(\mathbf{r}) - \Psi(\mathbf{r})\nabla^2 G(\mathbf{r},\mathbf{r}')\right] = \oint_S ds\left[G(\mathbf{r},\mathbf{r}')\frac{\partial\Psi(\mathbf{r})}{\partial\mathbf{n}} - \Psi(\mathbf{r})\frac{\partial G(\mathbf{r},\mathbf{r}')}{\partial\mathbf{n}}\right].$$
$$\tag{4.14}$$

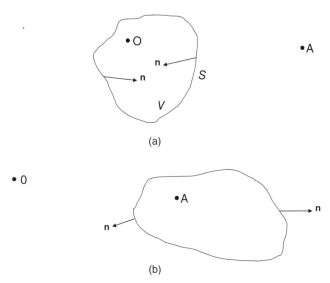

FIGURE 4.2. (a) Geometrical presentation of the Huygens's principle in a bounded surface when the receiver is located at point A outside the bounded surface and the transmitter is located at point O. (b) Geometrical presentation of the Huygens's principle in a bounded surface when the receiver is located at point A inside the bounded surface and the transmitter is located at point O.

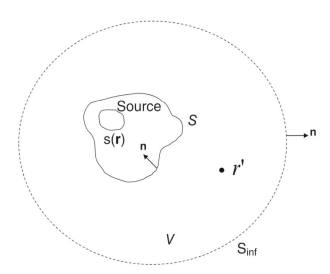

FIGURE 4.3. The geometry for derivation of Green's theorem for the two different boundary conditions at the bounded surface S; Neumann and Dirihlet.

This formula can be simplified using different boundary conditions on surface S. Using the relation between arbitrary scalar functions f and g, fg $\mathbf{n} = f \, \partial g / \partial \mathbf{n}$, we can write

$$\Psi(\mathbf{r}') = \oint_S ds \mathbf{n} \cdot \left[G(\mathbf{r}, \mathbf{r}') \nabla \Psi(\mathbf{r}) - \Psi(\mathbf{r}) \nabla G(\mathbf{r}, \mathbf{r}') \right]. \tag{4.15}$$

Next, if we assume that $\mathbf{n} \, \nabla G(\mathbf{r}, \mathbf{r}') = 0$ at the boundary surface S, defined by radius vector $\mathbf{r}$, Equation (4.15) becomes

$$\Psi(\mathbf{r}') = \oint_S ds \, G(\mathbf{r}, \mathbf{r}') \mathbf{n} \cdot \nabla \Psi(\mathbf{r}). \tag{4.16}$$

If the boundary condition $G(\mathbf{r}, \mathbf{r}') = 0$ is applied at the surface S, then Equation (4.15) becomes

$$\Psi(\mathbf{r}') = -\oint_S ds \, \Psi(\mathbf{r}) \mathbf{n} \cdot \nabla G(\mathbf{r}, \mathbf{r}'). \tag{4.17}$$

Equation (4.15), Equation (4.16), and Equation (4.17) are various forms of Huygens's principle, depending on the definition of Green's function on the bounded surface S. Equations (4.16) and Equation (4.17) state that only $\mathbf{n} \cdot \nabla \Psi(\mathbf{r})$ or $\Psi(\mathbf{r})$ need be known, respectively, on the surface S in order to determine the solution of wave function $\Psi(\mathbf{r}')$ at the observation point $\mathbf{r}'$.

In unbounded homogeneous media, as in free space, Huygens's principle has a clear and physical explanation. Each spherical wave can be presented as a plane wave in the far field. In this case, the elementary spherical waves called wavelets, created by each virtual point source (the dimensions of which are smaller than the wavelength), can be represented by the straight line called the wave fronts, as shown in Figure 4.4. Therefore, the phenomenon of straight-line radiowave propagation is the same as that of the light ray propagation in optics. This is the reason why sometimes in radio propagation the term "ray" is usually used instead of the term "wave."

4.1.4. The Concept of Fresnel Zones for Free Space

The Fresnel zone concept is usually used to describe diffraction phenomena from obstructions in the path of two antennas based on the Huygens's principle. As the latter is useful both for free space and for various finite areas with obstructions, it is important to show both mathematically and physically the meaning of the Fresnel zone concept when both terminal antennas are high enough that the LOS conditions are fully satisfied. In free space, on the basis Huygens's principle, instead of virtual point sources at an arbitrary surface S, as shown in Figure 4.2, Figure 4.3, and Figure 4.4, we introduce virtual sources along some virtual wave front DD', as shown in Figure 4.5. In this figure, points

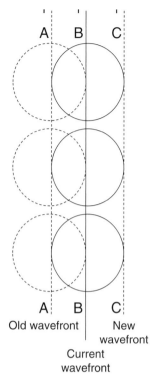

FIGURE 4.4. Geometry of plane wave-ray presentation of Huygens's principle in free space.

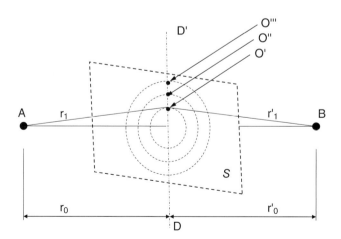

FIGURE 4.5. Presnel-zone concept presentation in free space.

A and *B* denote position of the terminal antennas. The virtual plane *S* is the plane that covers each virtual source located at line DD′ through which the plane *S* is passed. This imaginary plane is normal to the LOS path between two terminals, A and B, and passes across the point *O* at the line AB, as shown in Figure 4.5. For such geometry, the Green's theorem (4.14) can be rewritten for any vector, namely, the Hertz vector in References 1–4, as

$$\Pi(\mathbf{R}) = \int_S ds \frac{\partial \Pi(\mathbf{R}')}{\partial \mathbf{n}} \frac{\exp\{ik|\mathbf{R} - \mathbf{R}'|\}}{|\mathbf{R} - \mathbf{R}'|}, \tag{4.18}$$

where $|\mathbf{R} - \mathbf{R}'| = \mathbf{r}_1'$ is the distance from any point $O^{(i)}$, $i = ', '', ''' \ldots$, at the imaginary plane *S* and the observer at point *B*. If the radiation source located at point A is assumed to be point one with Green's function $G \sim e^{ikr_1}/r_1$, then for any point $O^{(i)}$ (O' in Fig. 4.5), we have, according to

$$\Pi(\mathbf{R}) = \frac{1}{2\pi} \int_S ds \left(\frac{1}{r_1} - ik \right) \frac{r_0}{r_1} \frac{\exp\{ik(r_1 + r_1')\}}{r_1 r_1'} \tag{4.19}$$

as $\partial r_1/\partial n = -r_0/r_1$.

All distances denoted inside the integral are shown in Figure 4.5.

Because the wave in the far field (Fraunhofer zone) between the plane *S* and two terminals *A* and *B*, that is, in the case of $r_1 \gg \lambda$ and $r_1' \gg \lambda$, $1/r_1 \ll ik$, we have one term with fast oscillations $\sim \exp\{ik(r_1 + r_1')\}$, even for small changes of variable *r*, and a second term $\sim ikr_0/r_1$, with very slow variations with change of variable *r*. In this case, the well-known method of stationary phase can be used to derive such an integral, containing both slow and fast terms inside the integrand. Following Reference 4, we can write using

$$\int_{-\infty}^{\infty} \exp\{i\alpha x^2\} dx = \sqrt{i\frac{\pi}{\alpha}} \tag{4.20}$$

and we can rewrite Equation (4.19) as [3, 4]

$$\Pi(B) \approx -\frac{ik}{2\pi r_0 r_0'} \exp\{ik(r_0 + r_0')\} \frac{2i\pi r_0 r_0'}{k(r_0 + r_0')} = \frac{\exp\{ikr\}}{r}, \tag{4.21}$$

where $r = r_0 + r_0'$ is the distance between the source located at point *A* and the observer located at point *B*. So, as follows from Equation (4.21), if the source *A* at the plane *S* creates a field $\sim e^{ikr_1}/r_1$, then the virtual point sources, at the observed point *B* uniformly distributed at *S*, will create a field $\sim e^{ikr}/r$. The same is valid for the direct wave from A to B. This is a main conclusion that results from the Huygens's principle. Moreover, additional analysis of integral in (4.19) shows that the plane *S* can be split into the concentric circles of arbitrary radii. From Figure 4.5, one can see that each wave path through any virtual point $O^{(i)}$ is longer than the direct path AOB, that is, $AO^{(i)}B > AOB$. While passing from one circle to another, the real and the imaginary parts of the integrand in (4.19) change their sign. The boundaries of these circles satisfy the conditions [3, 4]:

$$k\{(r_1 + r_1') - (r_0 + r_0')\} = n\frac{\pi}{2}, \quad n = 1, 2, \ldots. \tag{4.22}$$

These circles are usually called the *Fresnel zones*. Their physical meaning is that only in the first central circle, the virtual sources at the plane S, which lie within the first zone, send to observer at point B radiation with the same phase for each original wave. Sources from two neighboring zones send respective radiation in anti-phase, that is, nulling each other. The radius of the corresponding circle for each Fresnel zone shown at the plane S in Figure 4.5 can be expressed in terms of a zone number n and the distance between points A and B and the imaginary plane S as (see also References 2–4):

$$h_n = \sqrt{\frac{n\lambda r_0 r_0'}{(r_0 + r_0')}} \tag{4.23}$$

from which the radius of the first Fresnel zone is

$$h_1 = \sqrt{\frac{\lambda r_0 r_0'}{(r_0 + r_0')}} \sim \sqrt{\lambda R}, \tag{4.24}$$

where R is the minimal range from each r_0 and r_0'. The width of each circle, Δh, can be easily obtained as [4–6]

$$\Delta h \approx \frac{\pi}{2k}\frac{R}{h} \sim \frac{h_1^2}{h} << h_1, \tag{4.25}$$

where $h = \sqrt{x^2 + y^2}$. From (4.23) and (4.25), the width of the circles decreases with an increase in the zone number n. At the same time, the area of these zones is not dependent on zone number n, that is,

$$2\pi h\Delta h \sim \frac{\pi}{2}\lambda R. \tag{4.26}$$

It is clear that radius of each individual circle depends on the location of the imaginary plane with respect to points A and B, becoming largest at some point midway between A and B. Furthermore, from (4.22) the family of circles has a specific property: The path length from point A to point B via each circle is $n\lambda/2$ longer than the direct path AOB. Thus, for $n = 1$ (first zone) $AO'B -$ AOB $= \lambda/2$, the excess path length for the innermost circle is $\lambda/2$. Other zones will have an excess proportional to $\lambda/2$ with a parameter of proportionality $n = 2, 3, 4, \ldots$. The foci of the points, for which $AO^{(i)}B - AOB = n\lambda/2$, define

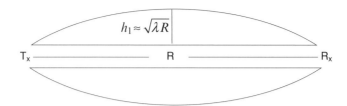

FIGURE 4.6. A free-space pattern of the first Fresnel zone covering both terminals, the transmitter Tx and the receiver Rx.

a family of ellipsoids. The radii of ellipsoids are described by (4.23). Notice that in free space, without any obstructions, only the first ellipsoid is actual and determines the first Fresnel zone with a radius proportional to $\sqrt{\lambda R}$ according to (4.24). This ellipsoid covers an area between two terminal points, the transmitter (T), and the receiver (R) as shown in Figure 4.6. Therefore, despite the fact that in free space both reflection and diffraction phenomena are absent, which causes interference between neighboring zones, the concept of Fresnel zones based on Huygens's principle is very important. It describes the loss characteristics of radio wave passing along a channel of high deviation terminal antennas. This concept allows us to estimate conditions of direct visibility or *clearance of the propagation channel* by using the right-hand term in (4.24) to evaluate the radius of the first Fresnel zone on the basis of the knowledge of the wavelength of the radiated wave and the range between the two terminal antennas. This is very important for link budget design in atmospheric channels where terminal antennas are far from the Earth's surface. This aspect will be discussed in Chapter 12 when we deal with atmospheric communication links.

4.1.5. Polarization of Radio Waves

Despite the fact that in Chapter 2 we gave the main definitions and parameters of wave polarization, to understand this very important aspect of wave propagation in various media, let us introduce some additional aspects of this phenomenon due to various peculiarities of the EM wave propagation. The alignment of the electric field vector **E** of a plane wave relative to the direction of propagation **k** defines the *polarization* of the wave (see Chapter 2). If **E** is transverse to the direction of wave propagation **k**, then the wave is said to be TE wave or *vertically* polarized. Conversely, when **H** is transverse to **k**, the wave is said to be TM wave or *horizontally* polarized. Both of these waves are *linearly polarized*, as the electric field vector **E** has a single direction along the entire propagation axis (vector **k**). If two plane linearly polarized waves of equal amplitude and orthogonal polarization (vertical and horizontal) are combined with a 90° phase difference, the resulting wave will be a *circularly* *polarized* (CP) wave, in which the motion of the electric field vector will describe a circle around the propagation vector.

The field vector will rotate by 360° for every wavelength traveled. Circularly polarized waves are most commonly used in land cellular and satellite communications, as they can be generated and received using antennas which are oriented in any direction around their axis without loss of power [1–3]. They may be generated as either right-hand *circularly polarized* or left-hand *circularly polarized*, depending on the direction of vector **E** rotation (see Fig. 4.7). In the most general case, the components of two waves could be of unequal amplitude, or their phase difference could be other than 90°. The combination

Direction of wave along *z*-axis

Polarization

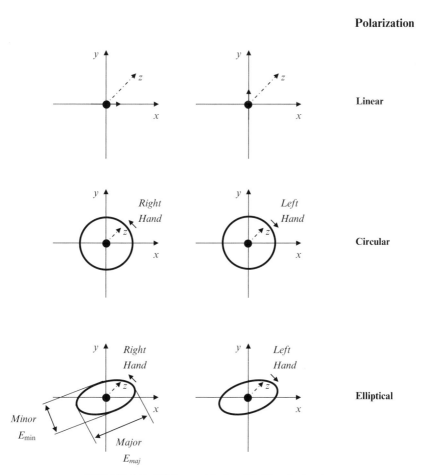

FIGURE 4.7. Different kinds of field polarization.

result is an elliptically polarized wave, where vector $\mathbf{E}$ still rotates at the same rate as for circularly polarized wave, but varies in amplitude with time. In the case of elliptical polarization, the axial ratio, $AR = E_{maj}/E_{min}$, is usually introduced (see Fig. 4.7). AR is defined to be positive for left-hand polarization and negative for right-hand polarization.

Now let us turn our attention to the wave field polarization in the case of a free-space propagation channel. Thus, for both vectors of the EM wave, $\Psi(\mathbf{r}) = \mathbf{E}(\mathbf{r})$ or $\Psi(\mathbf{r}) = \mathbf{H}(\mathbf{r})$, each component of which satisfies the wave Equation (4.1), one can write:

$$\mathbf{E}(\mathbf{r}) = \mathbf{e}_E E_0 \exp(i\mathbf{k} \cdot \mathbf{r}) \tag{4.27a}$$

$$\mathbf{H}(\mathbf{r}) = \mathbf{e}_H H_0 \exp(i\mathbf{k} \cdot \mathbf{r}), \tag{4.27b}$$

where $\mathbf{e}_E$ and $\mathbf{e}_H$ are the constant unit vectors, that is, $|\mathbf{e}_E| = |\mathbf{e}_H| = 1$; E_0 and H_0 are the complex amplitudes, which are constant in space and time. From conditions that satisfy free space propagation without any sources inside, it follows that

$$\mathbf{e}_E \cdot \mathbf{k} = 0, \mathbf{e}_H \cdot \mathbf{k} = 0, \text{ and } \mathbf{e}_H = \frac{\mathbf{k} \times \mathbf{e}_E}{|\mathbf{k}|}. \tag{4.28}$$

The expressions in (4.28) denote, that vectors $\mathbf{E}$ and $\mathbf{H}$ are perpendicular to the direction of wave propagation $\mathbf{k}$, and that vectors $\mathbf{e}_E$, $\mathbf{e}_H$, and $\mathbf{k}$ form a system of orthogonal vectors, where vectors $\mathbf{E}$ and $\mathbf{H}$ oscillate in phase and their ratio is constant.

At the same time, the vector $\mathbf{E}$, described by (4.27a), represents a simple case of a linearly polarized plane wave. To obtain more general cases of wave polarization, we should introduce an additional linearly polarized wave independent of the first one. It can be easily shown that two linearly polarized independent solutions, which satisfy the wave Equation (4.1), can be presented in the following form:

$$\mathbf{E}_1(\mathbf{r}) = \mathbf{e}_1 E_{01} \exp(i\mathbf{k} \cdot \mathbf{r}). \tag{4.29a}$$

$$\mathbf{E}_2(\mathbf{r}) = \mathbf{e}_2 E_{02} \exp(i\mathbf{k} \cdot \mathbf{r}). \tag{4.29b}$$

Consequently, the magnetic fields components of the EM wave satisfy, according to (4.27b) in free space ($\mu_r = 1$), the following relations

$$\mathbf{B}_i = \sqrt{\varepsilon_r} \frac{\mathbf{k} \times \mathbf{E}_i}{|\mathbf{k}|}, \quad \mathbf{B}_i \equiv \mathbf{H}_i, i = 1, 2 \tag{4.30}$$

Here, amplitudes E_{01}, E_{02} and H_{01}, H_{02} are complex values, which enable us to introduce the phase difference between two components, electric and magnetic, of the EM wave. As was mentioned at the beginning of this section, the electric field component fully describes various EM wave polarizations. Thus, the common solution for a plane EM wave, propagating along vector $\mathbf{k}$, can be presented as a linear combination of $\mathbf{E}_1$ and $\mathbf{E}_2$, that is,

$$\mathbf{E}(\mathbf{r}) = \{\mathbf{e}_1 E_{01} + \mathbf{e}_2 E_{02}\}\exp(i\mathbf{k}\cdot\mathbf{r}). \tag{4.31}$$

If now $\mathbf{E}_1$ and $\mathbf{E}_2$ have the same phase, then solution (4.31) describes a *linearly polarized* wave with polarization vector directed at an angle

$$\theta = \tan^{-1}\left(\frac{E_{02}}{E_{01}}\right), \tag{4.32a}$$

with amplitude

$$E = \left(E_{01}^2 + E_{02}^2\right)^{1/2} \tag{4.32b}$$

(see Fig. 4.7, upper part). If $\mathbf{E}_1$ and $\mathbf{E}_2$ have different phases, then the EM wave in Equation (4.31) is *elliptically polarized* (see Fig. 4.7, bottom part). If $\mathbf{E}_1 = \mathbf{E}_2$ and phase difference equals 90°, then the elliptically polarized wave becomes a *circularly polarized* wave. In this case, solution (4.31) becomes:

$$\mathbf{E}(\mathbf{r}) = \{\mathbf{e}_1 \pm \mathbf{e}_2\}\exp(i\mathbf{k}\cdot\mathbf{r}). \tag{4.33}$$

The sign "+" corresponds to anticlockwise rotation, called the wave with *left-hand* circular polarization. The sign "−" corresponds to the wave with right-hand polarization (i.e., with clockwise rotation), as shown in Figure 4.7 (middle part). Let us introduce the orthogonal complex unit vectors:

$$\mathbf{e}_\pm = \frac{1}{\sqrt{2}}(\mathbf{e}_1 \pm j\mathbf{e}_2). \tag{4.34}$$

Then, the common presentation of the polarized wave in (4.31), making use of two linearly polarized waves and two circularly polarized waves described respectively by (4.33) and (4.34), can be rewritten as

$$\mathbf{E}(\mathbf{r}) = \{\mathbf{e}_+ E_+ \pm \mathbf{e}_- E_-\}\exp(i\mathbf{k}\cdot\mathbf{r}), \tag{4.35}$$

where E_+ and E_- are the complex amplitudes of two circularly polarized waves with opposite directions of rotations (see Fig. 4.7). If E_+ and E_- are different, but their phase are equal, then (4.35) describes, as earlier, an elliptically polarized wave with main elliptical axes, large and small, directed along $\mathbf{e}_1$ and $\mathbf{e}_2$, respectively. Introducing now the parameter $q = E_-/E_+$, we can find the ratio of the small and large semi-axes, which equals $(1 - q)/(1 + q)$. For the case $q = \pm 1$, we once more return to the case of a linearly polarized wave.

4.2. PATH LOSS IN FREE SPACE

Let us consider a nonisotropic source placed in free space as a transmitter antenna with P_T watts and with a directivity gain G_T. At an arbitrary large distance r ($r >> \lambda$, where $\lambda = cT = c/f$ is a wavelength) from the source, the radiated power is uniformly distributed over the surface area of a sphere of radius r. If P_R is the power at the receiver antenna, which is located at distance r from the transmitter antenna and has a directivity gain G_R, then the *path loss* in decibels is given by

$$L = 10\log\frac{P_T}{P_R} = 10\log\left[\left(\frac{4\pi r}{\lambda}\right)^2 \Big/ G_T G_R\right] = L_0 + 10\log\left(\frac{1}{G_T G_R}\right). \quad (4.36)$$

Here, L_0 is the path loss for an isotropic point source (with $G_R = G_T = 1$) in free space and can be presented in decibels as

$$L_0 = 10\log\left(\frac{4\pi fr}{c}\right)^2 = 20\log\left(\frac{4\pi fr}{c}\right) = 32.44 + 20\log r + 20\log f, \quad (4.37)$$

where the value 34.44 is obtained from

$$32.44 = 20\log\left(\frac{4\pi \cdot 10^3 (m) \cdot 10^6 (1/s)}{3 \cdot 10^8 (m/s)}\right) = 20\log\left(\frac{40\pi}{3}\right).$$

Notice that all the above formulas are related to the well-known *Friis formula* obtained in Chapter 2. In expression (4.37), the distance r is in kilometers (km), and frequency f is in megahertz (MHz). As a result, the path loss between two directive antennas (receiver and transmitter) finally is given by

$$L_F = 34.44 + 20\log r_{[km]} + 20\log f_{[MHz]} - 10\log G_T - 10\log G_R. \quad (4.38a)$$

It can be presented in a "straight line" form as

$$L_F = L_0 + 10\gamma \log r, \quad (4.38b)$$

where $L_0 = 34.44 + 20\log f - 10\log G_T - 10\log G_R$ and $\gamma = 2$.

4.3. RADIO PROPAGATION ABOVE FLAT TERRAIN

The simplest case of radiowave propagation over a terrain is one where the ground surface can be assumed to be flat and perfectly conductive. The assumption of "flat terrain" is valid for radio links between subscribers up to 10–20 km [4–8]. The second condition of a "perfectly conductive" soil medium can be satisfied only for some special cases, because the combination of conductivity

σ and frequency ω, such as $4\pi\sigma/\omega$, that appears in total formula of permittivity $\varepsilon = \varepsilon_{ro} - i4\pi\sigma/\omega$, play an important role for high frequencies (VHF/L-band, usually used for terrain communication channel design) and finite subsoil conductivity, as well as for small grazing angles of incident waves [1–8]. To introduce the reader to the subject of reflection from the terrain, we start with the simplest case of a perfectly conductive flat terrain.

4.3.1. Boundary Conditions at the Perfectly Conductive Surface

For a perfectly conductive ground surface, the total tangential electric field vector is equal to zero, that is, $\mathbf{E}_\tau = 0$. Consequently, from $\nabla \times \mathbf{E}(\mathbf{r}) = i\omega\mathbf{H}(\mathbf{r})$, the normal component of the magnetic field also vanishes, that is, $\mathbf{H}_n = 0$. At the same time, the tangential component of magnetic field $\mathbf{H}_\tau$ does not vanish because of its compensation by the surface electric current. The normal component of electric field $\mathbf{E}_n$ is also compensated by the electrical charge on the ground surface [1–8]. Thus, for the flat, perfectly conductive ground surface, we have $\mathbf{E}_\tau = 0$ and $\mathbf{H}_n = 0$, which in the Cartesian coordinate system can be rewritten as

$$E_x(x, y, z=0)E_y(x, y, z=0) = H_z(x, y, z=0) = 0. \tag{4.39}$$

Here, $\mathbf{E}_\tau$ is the tangential component of the electric field, and $\mathbf{H}_n$ is the normal component of the EM wave with respect to the ground surface.

4.3.2. Reflection Coefficients

Here, we present the expressions for the complex coefficients of reflection (Γ) for waves with vertical (denoted by index V) and horizontal (denoted by index H) polarization [4–8].

For *horizontal* polarization:

$$\Gamma_H = |\Gamma_H|e^{-j\varphi_H} = \frac{\sin\psi - (\varepsilon_r - \cos\psi)^{1/2}}{\sin\psi + (\varepsilon_r - \cos\psi)^{1/2}}. \tag{4.40a}$$

For *vertical* polarization:

$$\Gamma_V = |\Gamma_V|e^{-j\varphi_V} = \frac{\varepsilon_r \sin\psi - (\varepsilon_r - \cos\psi)^{1/2}}{\varepsilon_r \sin\psi + (\varepsilon_r - \cos\psi)^{1/2}}. \tag{4.40b}$$

Here, $|\Gamma_V|$, $|\Gamma_H|$, and φ_V, φ_H are the magnitude and phase of the reflection coefficient for vertical and horizontal polarization, respectively, $\psi = \pi/2 - \theta_0$ is the grazing angle, and θ_0 is the angle of wave incidence (see Fig. 4.8). The knowledge of reflection coefficient amplitude and phase variations is a very important factor in the prediction of propagation characteristics for different situations in over-the-terrain propagation channels. In practice, for

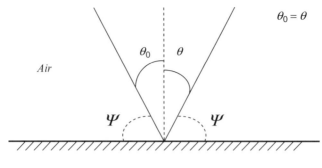

FIGURE 4.8. Specular reflection of the incident ray from a smooth flat terrain.

over-the-terrain wave propagation, the ground properties are determined by the conductivity and the absolute permittivity of the subsoil medium, $\varepsilon = \varepsilon_0 \varepsilon_r$, where ε_0 is the dielectric constant of vacuum, ε_r is the relative permittivity of the ground surface, and $\varepsilon_r = \mathrm{Re}(\varepsilon) + \mathrm{Im}(\varepsilon) = \varepsilon_{\mathrm{Re}} - j60\lambda\sigma$. Here, $\varepsilon_{\mathrm{Re}} \equiv \varepsilon_{r0}$ and $\varepsilon_{\mathrm{Im}} = 60\lambda\sigma$ are the real and imaginary parts of the relative permittivity of the subsoil medium. Because both coefficients presented by (4.40a) and (4.40b) are complex values, the reflected wave will therefore differ in both magnitude and phase from the incident wave. Moreover, both coefficients in (4.40) differ from each other.

Thus, for the case of horizontal polarization, for $\varepsilon_r \to \infty, \sigma \to \infty$ (i.e., for conductive ground surface), the relative phase of the incident and reflected waves, is nearly 180° for all angles of incidence. On the other hand, for very small grazing angles ($\psi << 90°$), as follows from (4.40a), the reflected and incident waves are equal in magnitude but differ by 180° in phase for all actual values of ground permittivity and conductivity, that is, $\Gamma_v = -1$ and $\varphi_H = 180°$.

From (4.40b), the reflection coefficient for a wave with vertical polarization does not change its properties compared with that of horizontal polarization in the case of a real conductive ground surface ($\varepsilon_r \to \infty$, $\sigma \to \infty$), and small grazing angles, that is, for $\psi << 90°$ $\Gamma_v = -1$, and $\varphi_v = 180°$. At the same time, for $\varepsilon_r \to \infty, \sigma \to \infty$ and $0 < \psi < 180°$, we get $\Gamma_v = 1$ (see (4.40b)). However, with an increase of angle ψ, substantial differences appear; that is, both a rapid decrease of magnitude and phase of the reflected wave take place. For $\theta_0 \to \theta_{Br}$ ($\psi \to 90° - \theta_{Br}$), where $\theta_{Br} = \tan^{-1}\sqrt{\varepsilon_r}$ is the Brewster angle, the magnitude $|\Gamma_v|$ becomes a minimum, and the phase φ_v reaches −90°. At values of ψ greater than the Brewster angle, $|\Gamma_v|$ increases again, and the phase φ_v approaches zero, that is, $\Gamma_v \to 1$.

4.4. PROPAGATION ABOVE ROUGH TERRAIN IN LOS CONDITIONS

Now we consider EM wave propagation above a rough terrain. Both terminal antennas, the transmitter and the receiver, are placed above the rough terrain in LOS conditions. Here, multiscattering effects, caused by the terrain roughness, must be taken into account. The total field arrived at the receiving antenna is a superposition of the direct wave, the wave specularly reflected from the quasi-flat ground surface (which together with the direct wave form the coherent part of the signal total intensity, I_{co}), and the waves scattered in all directions from the irregularities of the terrain (which form the incoherent part of the signal total intensity, I_{inc}). In order to predict the propagation loss characteristics of the irregular ground surface and to estimate the role of each kind of waves in the total field, we use the Rayleigh rough-surface criteria and find the influence of each part in the signal total intensity at the receiver [9–14].

Next, we present expressions for both parts of the total signal intensity, the coherent and incoherent, reflected and scattered from the rough ground surface, respectively. These expressions take into consideration the various relations between the dimension of roughness, the wavelength of operation, and the angle of incidence. The interested reader is referred to the original works [15–27] for more details. Here we give recommendations on how to use these expressions for different frequency bands, for different terrain irregularities, and various positions of the receiving and the transmitting antennas.

4.4.1. Scattering from a Rough Ground Surface

A rough terrain can be described, according to References 9–14 and 16, by a "relief function" $z = S(x)$, as shown in Figure 4.9. If the roughness of arbitrary height z is distributed according to Gauss's law, with mean value $\bar{z}$ and a variance of σ^2, then the probability density function (PDF) (see definitions in Chapter 1) of roughness distribution is given by

$$P_z = \frac{1}{\sqrt{2\pi}\sigma}\exp\left\{-\frac{(z-\bar{z})^2}{2\sigma^2}\right\},\tag{4.41}$$

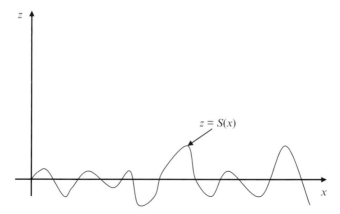

FIGURE 4.9. Relief function presentation for the rough terrain.

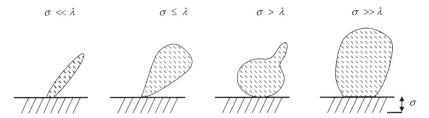

FIGURE 4.10. Different patterns of the scattered wave from a rough terrain.

where the standard deviation of the ground surface roughness around its mean height $\bar{z}$ is $\sigma = \sqrt{<z^2> - \bar{z}^2}$. In Figure 4.10, the criterion of roughness of the terrain is presented schematically for various values of σ for a better understanding of the role of the reflected and scattered waves in the total field pattern. Thus, the case $\sigma = 0$ or $\sigma << \lambda$ (λ is a wavelength) describes the pure reflection from a flat terrain; the case $\sigma \leq \lambda$ describes weak scattering effects from a gently rough surface, where the reflected wave is the dominant contributor to the total field pattern, that is, the coherent component of the wave intensity exceeds its incoherent component, that is, $I_{co} >> I_{inc}$. In the cases $\sigma > \lambda$ and $\sigma >> \lambda$, the terrain is rough and irregular with an increased role in generating a significant scattered wave as a component of the total field pattern. The last two illustrations in Figure 4.10 show the $I_{co} << I_{inc}$ cases.

There are several approximate methods for the total field evaluation in radio propagation channels above a rough terrain. At present, there are three general approaches to solve the wave scattering problem that arises from the rough terrain:

(a) the *perturbation technique* that applies to a surface wtihich is slightly rough and whose surface slope is smaller than unity [4, 9, 21, 25]

(b) the *Kirchhoff approximation* that is applicable to a surface whose radius of curvature is much greater than a wavelength [4, 9, 13–20]

(c) the *Rayleigh approximation* that is applicable to a surface whose curvature is at the same order as the wavelength [9–12].

We describe each approach briefly in the following sections.

4.4.2. The Perturbation Solution

The perturbation method is applicable to a slightly rough surface that will be described herein. Let us consider the height of a rough surface to be given by some function (see Fig. 4.11)

$$z = \varsigma(x, y) \tag{4.42}$$

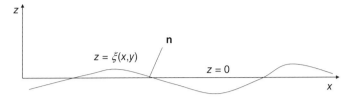

FIGURE 4.11. Geometrical presentation of weaker rough terrain described by perturbation method.

We choose $z = 0$ so that (4.42) represents the deviation from the average height: $\langle \varsigma(x, y) \rangle = 0$. Moreover, the perturbation method is valid when the phase difference due to the height variation is small, that is, when [4, 9]

$$| k \cdot \varsigma(x, y) \cdot \cos \theta_i | << 1$$
$$\left| \frac{\partial \varsigma}{\partial x} \right| << 1, \quad \left| \frac{\partial \varsigma}{\partial y} \right| << 1. \tag{4.43}$$

The boundary condition for the electric field at this surface requires that the tangential components of **E** vanish at the surface $z = \varsigma(x,y)$, that is,

$$\mathbf{E} \times \mathbf{n} = 0, \tag{4.44}$$

where **n** is the vector normal to the surface $z = \varsigma$ at point (x, y). If the surface profile (4.42) and the position of sources are known, then the problem is to determine the field in semi-space $z > 0$, given that the boundary conditions are known [9]. Let us consider the influence of roughness as a small perturbation, that is, the total field is

$$\mathbf{E} = \mathbf{E}^{(0)} + \mathbf{E}^{(1)}, \tag{4.45}$$

where $\mathbf{E}^{(0)}$ is the field that could be derived for the condition $\varsigma = 0$, which a priori is well known using knowledge of specular reflection from smooth terrain obtained from two-ray model. The second term $\mathbf{E}^{(1)}$ that describes the field perturbations can be obtained from wave equation by use of the boundary conditions (4.44). To present the solution of perturbation term, let us consider two special cases, which are practical with regard to over-the-terrain propagation channels.

Let a *vertical dipole* be located at point O as shown in Figure 4.12. Its reflection from flat surface $z = 0$ at the point O_1 according to the reflection theorem must also be directed vertically. By introducing the spherical coordinate systems $\{R, \vartheta, \varphi\}$ and $\{R_1, \vartheta_1, \varphi_1 \equiv \varphi\}$ for each dipole, we can present the components of the nonperturbed field $\mathbf{E}^{(0)}$ as

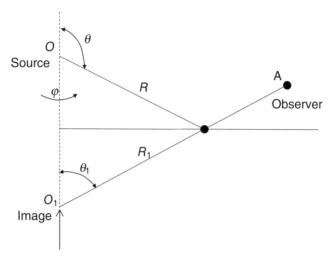

FIGURE 4.12. The geometry of the vertical dipole field scattered from the rough terrain.

$$E_x^{(0)} = \left\{ -k_0^2 p \sin \vartheta \cos \vartheta \frac{e^{i(\omega t - k_0 R)}}{R} - k_0^2 p \sin \vartheta_1 \cos \vartheta_1 \frac{e^{i(\omega t - k_0 R_1)}}{R_1} \right\} \cos \varphi$$

$$E_y^{(0)} = \left\{ -k_0^2 p \sin \vartheta \cos \vartheta \frac{e^{i(\omega t - k_0 R)}}{R} - k_0^2 p \sin \vartheta_1 \cos \vartheta_1 \frac{e^{i(\omega t - k_0 R_1)}}{R_1} \right\} \sin \varphi \qquad (4.46)$$

$$E_z^{(0)} = \left\{ k_0^2 p \sin^2 \vartheta \frac{e^{i(\omega t - k_0 R)}}{R} + k_0^2 p \sin^2 \vartheta_1 \frac{e^{i(\omega t - k_0 R_1)}}{R_1} \right\}.$$

Then, in the plane $z = 0$ $(R = R_1, \vartheta = \pi - \vartheta_1)$,

$$E_x^{(0)} = E_y^{(0)} = 0, \quad E_z^{(0)} = 2k_0^2 p \sin^2 \vartheta \frac{e^{i(\omega t - k_0 R)}}{R}. \qquad (4.47)$$

Here p is the modulus of the momentum of the vertical dipole, which is well known from the literature (see, e.g., References 4, 9, and 11). Because in practical terrain propagation case the source and the observation point are far from the surface $z = 0$, we can present simple formulas for the perturbed part of the total field due to the terrain roughness in the case where the vector of the incident wave lies in the xy-plane (i.e., corresponding to the following conditions: $\varphi = 0$, $pe^{-ik_0 R}/R = qe^{-ik_0 x' \sin \vartheta}$, where q is constant):

$$E_x^{(1)} = \frac{k_0^2}{2\pi} (2q) \int \left\{ ik_0 \cos^2 \vartheta + \sin \vartheta \frac{\partial \varsigma}{\partial x'} \right\} \frac{\partial}{\partial z} \frac{e^{-ik_0(\rho + x' \sin \vartheta)}}{\rho} dx' dy' \qquad (4.48a)$$

$$E_y^{(1)} = \frac{k_0^2}{2\pi} (2q) \int \left\{ \sin^2 \vartheta \frac{\partial \varsigma}{\partial y'} \right\} \frac{\partial}{\partial z} \frac{e^{-ik_0(\rho + x' \sin \vartheta)}}{\rho} dx' dy' \qquad (4.48b)$$

$$E_z^{(1)} = \frac{k_0^2}{2\pi}(2q)\int \left\{ ik_0 \frac{\partial \varsigma}{\partial x'}(\cos^2\vartheta - \sin^2\vartheta) + \left(\frac{\partial^2\varsigma}{\partial x'^2} + \frac{\partial^2\varsigma}{\partial x'^2} \right)\sin\vartheta + \right.$$
$$\left. k_0^2 \cos^2\vartheta\sin\vartheta \right\} \frac{e^{-ik_0(\rho + x'\sin\vartheta)}}{\rho} dx'dy' \tag{4.48c}$$

Here, $\rho = \sqrt{(x-x')^2 + (y-y')^2 + (z-z')^2}$, in which x, y, z are the coordinates of the observed point. For small grazing angles ($\vartheta \to \pi/2$), that is, in the case of slipped incident waves, which is very actual for mobile and personal communication, these formulas can be significantly simplified, for example,

$$E_x^{(1)} \approx -2k_0^2 qe^{-ik_0 x\sin\vartheta}\frac{\partial\varsigma}{\partial x}$$

$$E_y^{(1)} \approx -2k_0^2 qe^{-ik_0 x\sin\vartheta}\frac{\partial\varsigma}{\partial x} \tag{4.49}$$

$$E_z^{(1)} \approx \frac{k_0^2}{2\pi}(2q)\int \left(\frac{\partial^2\varsigma}{\partial x^2} + \frac{\partial^2\varsigma}{\partial x'^2} - ik_0\frac{\partial\varsigma}{\partial x'} \right)\frac{e^{-ik_0(\rho + x'\sin\vartheta)}}{\rho} dx'dy'$$

Since a *horizontal dipole* is located at the point O and oriented along the y-axis, its reflection vector from flat surface $z = 0$ at the point O_1 is oriented in the opposite direction. The same approach, as above, allows us to present the perturbation part of the total field due to the terrain roughness for a horizontal dipole oriented along the y-axis:

$$E_x^{(1)} = 0$$

$$E_y^{(1)} = -\frac{k_0^2}{2\pi}(2q)\int \{ik_0\varsigma\cos\vartheta\}\frac{\partial}{\partial z}\frac{e^{-ik_0(\rho + x'\sin\vartheta)}}{\rho} dx'dy' \tag{4.50a}$$

$$E_z^{(1)} = \frac{k_0^2}{2\pi}(2q)\int \{ik_0\frac{\partial\varsigma}{\partial y'}\cos\vartheta\}\frac{e^{-ik_0(\rho + x'\sin\vartheta)}}{\rho} dx'dy' \tag{4.50b}$$

Then, in the same case of slipped waves (i.e., for small grazing angles ($\vartheta \to \pi/2$)), one can easily obtain from (4.50) very simple formulas for the perturbed part of the total field:

$$E_x^{(1)} = 0$$
$$E_y^{(1)} = -2ik_0^3\varsigma q\cos\vartheta e^{-ik_0 x\sin\vartheta} \approx 0 \tag{4.51}$$
$$E_z^{(1)} = \frac{k_0^3}{2\pi}(2iq)\int \frac{\partial\varsigma}{\partial y'}\cos\vartheta\frac{e^{-ik_0(\rho + x'\sin\vartheta)}}{\rho} dx'dy'$$

Comparison between (4.48) and (4.49), and (4.50) and (4.51), for both kinds of wave field polarizations, shows that the field of the horizontal dipole is less affected by the roughness of the terrain than that of the vertical dipole. The formulas presented here can predict the propagation characteristics over a rough terrain in conditions of direct visibility between the source and the observer if the profile $\varsigma(x,y)$ of the ground surface is known for each situation. Moreover, these formulas allow us to obtain the coherent and incoherent parts of the total field energy. In fact, the coherent power dominates for the case of a smooth surface and is determined by the use of the non-perturbed field $\mathbf{E}^{(0)}$, the components of which are described by (4.46). The incoherent power is determined by the use of the perturbed field $\mathbf{E}^{(1)}$ described by (4.48) and (4.49), and (4.50) and (4.51), for both kinds of field polarization.

The limitation of the perturbation method depends on the requirement of the "smallness" not only for $\nabla\varsigma$ but also for the Earth's surface deviations $\varsigma(x,y)$. But the last condition can be ignored because if we derive the second *perturbation* term $\mathbf{E}^{(2)}$ in expansion (4.45), we obtain for the case of vertical dipole the following condition [9, 11]:

$$(|k_0 \cdot \varsigma \cdot \nabla\varsigma|)^{1/2} << 1 \qquad (4.52)$$

from which, assuming that $|\nabla\varsigma| \approx \varsigma/\ell$, where ℓ is the characteristic length of roughness, we obtain

$$\varsigma << \sqrt{\lambda \cdot \ell}. \qquad (4.53)$$

Therefore, for sufficiently small slope angles, the described perturbation technique is valid even for deviations ς close to or larger than the wavelength λ.

4.4.3. The Kirchhoff Approximation

Now we consider the other limiting case, when the characteristic scales of the Earth's surface roughness significantly exceeds the wavelength size of the radiated field. In this case, the Kirchhoff approximation may be used to obtain a reasonably simple solution. What is very important to note is that this method requires the absence of shadow zones between all roughnesses and/or multireflection and multiscattering between each part of the rough surface at $z = 0$. In other words, we assume that the surface S is slowly varying so that the radius of curvature is much greater than the wavelength (see Fig. 4.13). At each point $\mathbf{r}$ on the quasi-smooth surface S, the wave field is a superposition of the incident field $\mathbf{E}_0$ and the field $\mathbf{E}$ reflected from the plane LL'. This plane is tangential to the surface S at point $\mathbf{r}$, as shown in Figure 4.13. The scattered EM wave at the observation point R can be represented by the values of $\mathbf{E}$ and $\mathbf{H}$ on the surface S and by using the well-known Green's function presentation for the point source, $G = \exp\{ikR\}/R$, that is,

$$\mathbf{E}_i(\mathbf{r} \in S) = \mathbf{E}_0 \frac{\exp\{ikR_1\}}{R_1}, \quad \mathbf{H}_i(\mathbf{r} \in S) = \mathbf{H}_0 \frac{\exp\{ikR_1\}}{R_1}. \qquad (4.54)$$

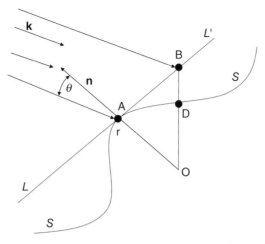

FIGURE 4.13. Geometrical presentation of a quasi-smooth terrain described by Kirchhoff's approximation.

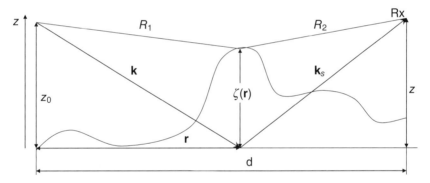

FIGURE 4.14. Reflection from a quasi-smooth terrain.

The final expression for the scattered filed is

$$\mathbf{E}(\mathbf{R}) = \frac{ik}{4\pi} \int_S \frac{e^{ik(R_1+R_2)}}{R_1 R_2} \{[\mathbf{n}\times(\mathbf{H}-\mathbf{H}_0)] + [\mathbf{n}\cdot(\mathbf{E}-\mathbf{E}_0)\cdot\nabla_r R_2]$$
$$- [\nabla_r R_2 \times \mathbf{n}\times(\mathbf{E}-\mathbf{E}_0)]\} \, ds. \tag{4.55}$$

Here, as follows from Figure 4.14, R_1 and R_2 are the distances from the current point $\mathbf{r}(x,y,z=0)$ at the flat surface $z=0$ to the source point O and the observation point R; $\tilde{R}_1$ and $\tilde{R}_2$ are the distances from the current point $\mathbf{r}(x,y,z)$ at the surface S over which the integration in (4.55) takes place; $\varsigma(\mathbf{r})$ is the height of the surface S at the arbitrary point $\mathbf{r}(x,y,z)$. If the source and the observed point are located in the far-field zone relative to surface S, that is, $k\tilde{R}_1 >> 1$ and $k\tilde{R}_2 >> 1$, the integral in (4.55) for the scattered field in direction $\mathbf{k}_s$ can be rewritten as

$$
\mathbf{E}(\mathbf{k}, \mathbf{k}_s) = \frac{ik}{4\pi} \frac{e^{ik(R_{10}+R_{20})}}{R_{10}R_{20}} \int_{S_0} \{[\mathbf{n}\times(\mathbf{H}-\mathbf{H}_0)]+[\mathbf{n}\cdot(\mathbf{E}-\mathbf{E}_0)\cdot\nabla_r R_2]-
$$
$$
-[\nabla_r R_2 \times \mathbf{n} \times (\mathbf{E}-\mathbf{E}_0)]\} \exp\{i[(\mathbf{k}-\mathbf{k}_s)\mathbf{r}+(k_z-k_{sz})\varsigma(\mathbf{r})]\}\frac{d\mathbf{r}}{q_z}. \tag{4.56}
$$

Here, R_{10} and R_{20} are the distances between the arbitrary point $\mathbf{r}(x,y,z=0)$ on the surface S_0, which is the projection of the rough surface S at the plane $z=0$, and the source O and observed point R, respectively. For future analysis of the integral in (4.56), it is convenient to present the distances $\tilde{R}_1$ and $\tilde{R}_2$ through the vector $\mathbf{r}(x,y,z=0)$ that lies on the flat surface $z=0$ and the value of surface height $\varsigma(\mathbf{r})$ at this current point (see Fig. 4.14):

$$
\tilde{R}_1 = \sqrt{r^2+(z_0-\varsigma)^2} \approx R_1 + \alpha_z\varsigma
$$
$$
\tilde{R}_2 = \sqrt{(d-r)^2+(z-\varsigma)^2} \approx R_2 + \beta_z\varsigma \tag{4.57}
$$

where $R_1 = \sqrt{r^2+z^2}$, $R_2 = \sqrt{(d-r)^2+z^2}$; $\alpha_z = -z_0/R_1$ and $\beta_z = z/R_2$ are the z-components of vectors $\mathbf{a}=\nabla_r R_1$ and $\mathbf{b}=-\nabla_r R_2$ (i.e., the projections of these vectors lie at the z-axis). We analyze the expression (4.56) that describes the scattered field for two cases that are useful in practice for over-the-terrain propagation [9–14] by introducing some new variables: $\mathbf{q} = \mathbf{k}_s - \mathbf{k}$, $\mathbf{k} = k\mathbf{a} = k\nabla_r R_1$, $\mathbf{k}_s = k\mathbf{b} = -k\nabla_r R_2$.

In the case of a *perfectly conducting* Earth's surface, the expression (4.56) can be simplified, taking into account that electric and magnetic components of the EM field are mutually perpendicular, $\mathbf{H}_0 = \mathbf{k} \times \mathbf{E}_0/k$, and that we concentrate only on the short-wave approximation, $(q_z\varsigma >> 1)$, that is,

$$
\mathbf{E}(\mathbf{k}, \mathbf{k}_s) \approx \frac{ik}{2\pi} \frac{e^{ik(R_{10}+R_{20})}}{R_{10}R_{20}} \frac{[\mathbf{b}\times(\mathbf{E}_0\times\mathbf{q})]}{q_z} \int_{S_0} \exp\{-i[\mathbf{q}\cdot\mathbf{r}+q_z\varsigma(\mathbf{r})]\}d\mathbf{r}. \tag{4.58}
$$

After statistical averaging of integral (4.58), the average scattered field can be presented as

$$
< \mathbf{E}(\mathbf{k}, \mathbf{k}_s) > = \mathbf{E}^{(0)}(\mathbf{k}, \mathbf{k}_s)\Gamma_f(\psi), \tag{4.59}
$$

where

$$
\mathbf{E}^{(0)}(\mathbf{k}, \mathbf{k}_s) = \frac{ik}{2\pi} \frac{e^{ik(R_{10}+R_{20})}}{R_{10}R_{20}} \frac{[\mathbf{b}\times(\mathbf{E}_0\times\mathbf{q})]}{q_z} \int_{S_0} \exp\{-i\mathbf{q}\cdot\mathbf{r}\}d\mathbf{r} \tag{4.60}
$$

is the field reflected from the area S_0 of the plane $z = 0$, and $\Gamma_f(\psi)$ is the effective reflection coefficient from rough terrain, which for the surface S with a Gaussian distribution, can be presented as [21–27]

$$\Gamma_f(\psi) \approx \exp\{-2k^2\varsigma^2\sin^2\psi\}. \tag{4.61}$$

Here, ψ is the slip angle (see Fig. 4.14). One can see that the effective reflection coefficient decreases exponentially with an increase of roughness height $\varsigma(\mathbf{r})$. Now, by introducing the tensor coefficient of reflections $\Gamma_{j\ell}^E = -\delta_{j\ell} + 2n_j n_\ell$, where the double repeated index ℓ indicates the summation from 1 to 3, and $\delta_{j\ell}$ is the unit tensor (which equal 1, if $j = \ell$, and 0, if $j \neq \ell$), one can finally obtain from (4.58) the solution for the $\mathbf{E}_j$ component of the scattered field [9–14]:

$$\mathbf{E}_{j\ell}(\mathbf{k}, \mathbf{k}_s) \approx \frac{e^{ik(R_{10}+R_{20})}}{4\pi i R_{10} R_{20}} \frac{q^2}{q_z} \Gamma_{j\ell}^E \cdot \mathbf{E}_{0\ell} \int_{S_0} \exp\{-i[\mathbf{q}\cdot\mathbf{r} + q_z\varsigma(\mathbf{r})]\}\, d\mathbf{r}. \tag{4.62}$$

The same result can be obtained for the $\mathbf{H}_j$ component of the scattered field by introducing in (4.62) the following terms: $\mathbf{H}_{j\ell}$, $\mathbf{H}_{01}$, and $\Gamma_{j\ell}^H = -\Gamma_{j\ell}^E$, respectively.

A generalization of the problem for the case of scattering from the impedance rough surface for $k \to \infty$, gives the same result, as expressed in (4.62) for the perfectly conducting ground surface in terms of tensor:

$$\Gamma_{j\ell}^E = \Gamma_V \delta_{j\ell} - \frac{1}{\sin\vartheta}\{(\Gamma_H + \Gamma_V\cos\vartheta)n_j n_\ell + \cos\vartheta(\Gamma_H + \Gamma_V)\alpha_j n_\ell\}$$

$$\Gamma_{j\ell}^H = \Gamma_V \delta_{j\ell} - \frac{1}{\sin\vartheta}\{(\Gamma_V + \Gamma_H\cos\vartheta)n_j n_\ell + \cos\vartheta(\Gamma_H + \Gamma_V)\alpha_j n_\ell\}, \tag{4.63}$$

which are significantly simplified for the case of the perfectly conducting surface and can be presented as

$$\Gamma_{j\ell}^H = -\Gamma_{j\ell}^E = \delta_{j\ell} - 2n_j n_\ell. \tag{4.64}$$

Here, Γ_H and Γ_V are the reflection coefficients presented, previously, by Equation (4.40) in Section 4.3 for the horizontal and vertical polarizations, respectively. Let us note that the expression in (4.62) allows us to treat scattering phenomena from a surface with arbitrary dielectric properties not only for the linearly polarized waves but also for the elliptically polarized waves. Thus, a linearly polarized wave, after undergoing scattering from the impedance rough surface, becomes elliptically polarized. But what is more interesting is that the depolarization phenomenon is not connected with the statistical properties of the rough terrain. It is completely determined by the inclination of the tangential plane LL' to the surface S at the points of specular reflection. The direction of vector $\mathbf{n}_0$ normal to this plane (and, hence, the polarization of reflected field) is related to the direction of wave vector $\mathbf{k}$ of the incident wave and to the direction of the observation point $\mathbf{k}_s$ through the relation, $\mathbf{n}_0 = \mathbf{k}_s - \mathbf{k}/|\mathbf{k}_s - \mathbf{k}|$. Therefore, one can directly use the tensor presentation of the Fresnel reflection coefficients defined in (4.63) with the following conditions, such as $\mathbf{n} = \mathbf{n}_0$ and $\vartheta = \vartheta_0 = \cos^{-1}(\mathbf{n} \cdot \mathbf{b})$, $\mathbf{b} = -\nabla_r R_2$ for the evaluation of the scattered field for different kinds of polarization.

4.4.4. The Rayleigh Approximation

The use of Rayleigh approximation depends not on the dimensions of surface roughness with respect to the wavelength, but mostly on antenna elevation height. For the cases where the coherence length L between two nearby reflected rays is higher than λ, and the roughness is small compared to λ, that is, $\sigma << \lambda$ and $\sigma < \lambda$, the phase difference between field components becomes larger than $\pi/2$. Here the Rayleigh approximation is not as accurate as the Kirchhoff approximation. Sometimes the phase difference is close to $\pi/4$ and $\pi/8$. In that case, a scalar Rayleigh factor in the coherent field can be introduced for such gently rough surfaces, which reduces the energy of the specularly reflected wave. These "above-the-terrain" propagation cases will be examined briefly below by using the effective Kirchhoff reflection coefficients and their corresponding effective permittivity of the rough terrain.

For the high elevated antennas, the roughness is small compared to wavelength ($\sigma < \lambda$). In this case of gentle rough ground surface, the two-ray model that usually applies to smooth terrain and describes the coherent part of the signal can be modified by introducing the reflection coefficient for vertical, Γ_V, and horizontal, Γ_H, polarization, as functions of the effective relative permittivity, ε_{reff} [15–27]:

$$\Gamma_V = \frac{\varepsilon_{reff} \sin \psi - \sqrt{\varepsilon_{reff} - \cos^2 \psi}}{\varepsilon_{reff} \sin \psi + \sqrt{\varepsilon_{reff} - \cos^2 \psi}}. \tag{4.65a}$$

$$\Gamma_H = \frac{\sin \psi - \sqrt{\varepsilon_{reff} - \cos^2 \psi}}{\sin \psi + \sqrt{\varepsilon_{reff} - \cos^2 \psi}}. \tag{4.65b}$$

So, the modified coherent component of the total field intensity is

$$I_{co\,mod} = E_0^2 \left\{ \frac{e^{-jkr_1}}{r_1} + \Gamma_{V,H} \frac{e^{-jkr_2}}{r_2} \right\}^2, \tag{4.66}$$

where ψ is the grazing angle defined earlier, r_1 is the distance of the direct radio path between the antennas, and r_2 is the distance from the transmitter to the point of reflection and from the point of reflection to the receiver, that is, the radio path length of the reflected wave. The equivalent surface impedance is $\eta = \sqrt{\mu_r/\varepsilon_{reff}} \cong 1/\sqrt{\varepsilon_{reff}}$, with the relative permeability $\mu_r \approx 1$, for all nonferromagnetic surfaces.

There are six distinct cases that can be considered here, three for each linear polarization, vertical and horizontal. These three asymptotic cases are valid

(a) for short correlation lengths L and all grazing angles ψ_1
(b) for long correlation scales L and large grazing angles ψ_1
(c) for long correlation scales L and small grazing angles ψ_1.

So, for vertical polarization, the effective surface impedance η has a real part corresponding to a loss of power and an imaginary part corresponding to a reactive, stored energy near the surface. Furthermore, the change in effective surface impedance for vertical polarization is strictly reactive for short correlation lengths ($\lambda >> L$), resistive for large correlation lengths and large grazing angles ($\lambda << L$ and $\psi >> 1/\sqrt{kL}$), and a mixture of both for large correlation lengths and small grazing angles ($\lambda << L$ and $\psi >> 1/\sqrt{kL}$). For the case of horizontally polarized field, the surface impedance and the corresponding effective permittivity can be derived in a similar fashion.

In the case of low antenna elevation with respect to roughness, the criteria of $\sigma \geq \lambda$ and $\sigma >> \lambda$ are generally valid, and the Rayleigh scalar factor can be used as long as the criteria of the Kirchhoff approximation are fulfilled. In this case, instead the specular reflection coefficients, we introduce the following effective reflection coefficients [9–16]:

(a) for vertical polarization

$$\Gamma_V^{ef} = \Gamma_V \exp\left[-2\left(2\pi\frac{\sigma}{\lambda}\sin\psi\right)^2\right], \tag{4.67}$$

where $\exp\left[-2\left(2\pi\frac{\sigma}{\lambda}\sin\psi\right)^2\right]$ is the Rayleigh factor; the coefficient Γ_V is defined by (4.40b) in Section 4.3 and can be reduced to

$$\Gamma_V = -1 + 2\psi\frac{\varepsilon_r}{\sqrt{\varepsilon_r - 1}}. \tag{4.68}$$

(b) for horizontal polarization

$$\Gamma_H^{ef} = \Gamma_H \exp\left[-2\left(2\pi\frac{\sigma}{\lambda}\sin\psi\right)^2\right], \tag{4.69}$$

where Γ_H is defined by (4.40a) in Section 4.3 and can be reduced to

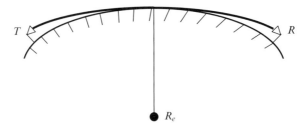

FIGURE 4.15. Geometrical presentation of radio path above a curved terrain.

$$\Gamma_H = -1 + 2\psi \frac{1}{\sqrt{\varepsilon_r - 1}}. \tag{4.70}$$

So, the use of each approximation strongly depends not only on the dimensions of rough structures with respect to the wavelength, but mostly on the terminal antenna elevations as well.

4.5. PROPAGATION ABOVE A SMOOTH CURVED TERRAIN

Let us now consider the case when the terrain is smooth but curved (see Fig. 4.15). In this case, the degree of curvature and diffraction caused by the curved Earth's surface must be taken into account for the evaluation of field characteristics. In practice, for land communications, it is very important to note that the influence of the curvature of the Earth's surface must be taken into account only for radio paths longer than 20–30 km.

4.5.1. Fock's Model

To take into account the terrain curvature and diffraction from the curved terrain, Fock, by introducing the special scales, the *range scale*, $L = \left(\lambda R_e^2 / \pi\right)^{1/3}$, and the *height scale*, $H = 0.5 \cdot \left(\lambda^2 R_e / \pi^2\right)^{1/3}$, respectively, has determined the range of radio path, d, and the heights of both terminal antennas, h_T and h_R, using dimensionless parameters $x = d/L$, $y_1 = h_T/H$, $y_2 = h_R/H$ [28]. The attenuation factor with respect to the flat terrain has a form [28]:

$$F = 2\sqrt{\pi x} \left| \sum_{k=1}^{\infty} \frac{\exp(ixt_k)}{(t_k + p^2)} \frac{A(t_k + y_1)}{A(t_k)} \frac{A(t_k + y_2)}{A(t_k)} \right|, \tag{4.71}$$

where

$$p = i(\pi R_e / \lambda)^{1/3} / \sqrt{\varepsilon_{r0} - i60\lambda\sigma}. \tag{4.72}$$

$R_e = 6375$ km is the actual Earth's radius. By $A(w)$ we denoted a special Airy function, which is related to the special Hankel function of the order 1/3 through

$$A(w) = \sqrt{\pi/3} \exp(-i2\pi/3) w^{1/2} H_{1/3}(2w^{3/2}/3) \qquad (4.73)$$

Here, t_k are the roots of

$$A'(t) - pA(t) = 0. \qquad (4.74)$$

It can be shown that the value of t_k for finite values of p can be estimated as follows:

$$t_k(p) \approx t_k(0) + p/t_k(0), \left| p/\sqrt{t_k} \right| < 1 \qquad (4.75a)$$

and

$$t_k(p) \approx t_k(\infty) + 1/p, \left| p/\sqrt{t_k} \right| > 1. \qquad (4.75b)$$

Let us introduce parameter $\bar{t}_k = t_k \cdot \exp(-2\pi/3)$. Next, we can compute first set of roots $f_i(0)$ and $f_i(\infty)$ of Equation (4.74) as

$$\bar{t}_1(0) = 1.019, \quad \bar{t}_1(\infty) = 2.338$$
$$\bar{t}_2(0) = 3.248, \quad \bar{t}_2(\infty) = 4.088$$
$$\bar{t}_3(0) = 4.820, \quad \bar{t}_3(\infty) = 5.521$$
$$\bar{t}_4(0) = 6.163, \quad \bar{t}_4(\infty) = 6.787$$
$$\bar{t}_5(0) = 7.372, \quad \bar{t}_5(\infty) = 7.994.$$

For the UHF/X-frequency band, in the shadow zones due to ground surface curvature, where $|p| \gg 1$, we can write the attenuation factor as

$$F = U(x)V(y_1)V(y_2). \qquad (4.76)$$

The first term depends on the normalized range x between the antennas

$$U(x) = 2\sqrt{\pi x} \left| \frac{\exp(ixt_1)}{t_1 + p^2} \right|, \qquad (4.77)$$

but the second and the third terms are only functions of the antenna heights (*height parameters*)

$$V(y_{1,2}) = \left| \frac{H(t_1 + y_{1,2})}{H(t_1)} \right|. \qquad (4.78)$$

We must note that according to the definitions mentioned earlier, if both antennas are close to the ground surface (i.e., $y_{1,2} = 0$), then the "height product" $V(y_{1,2}) = 1$. Moreover, for $y_{1,2} < 1$, the "antenna height" factors $V(y_{1,2})$ (in decibels (dB)) are negative; otherwise; they are positive. For $y_{1,2} < 1$, these factors can be approximated as

$$V(y_{1,2}) \cong 20\log(y_{1,2})\,[dB]. \tag{4.79}$$

For $y_{1,2} \geq 1$, some estimates give us

$$V(1) \cong 0\ dB; V(2) \cong 10\ dB; V(4) \cong 20\ dB; V(7) \cong 30\ dB; V(10) \cong 40\ dB.$$

So, the above formulas allow us to compute, with great accuracy, the additional loss due to diffraction at the spherical ground surface both in the geometrical shadow zone and in *zones* of half shadowing. These formulas can be used to predict the diffraction losses of the wave field caused by the Earth's curvature. Once again, the effect of the Earth's curvature must be taken into account only for land radio cases with ranges of more than 10–20 km. At the same time, we must note that for the long radio paths (more than 100 km), the real terrain profile of the path is obviously beyond the capabilities of the Fock model.

4.6. EFFECT OF A SINGLE OBSTACLE PLACED ON A FLAT TERRAIN

Existing obstructions along the radio path lead to additional losses called diffraction losses, which are usually observed in rural areas where some arbitrary obstructions (e.g., hill, mountain) exist. When there is a single obstacle between the transmitter and receiver, which can be modeled by a single "knife edge," losses of the wave energy take place. Such losses in the literature are called *diffraction losses* [29–36]. They can be obtained analytically by using the Fresnel complex integral based on the Huygens's principle discussed in Section 4.2. The total wave field E_{total} after diffraction from the obstruction can be presented in the following form:

$$E_{total} = E_0 \cdot \hat{D} \cdot \exp\{j\Delta\Phi\}, \tag{4.80}$$

where E_0 is the incident wave from the transmitter located in free space; $\hat{D}$ is the diffraction coefficient or matrix [29–36], $\Delta\Phi$ is the phase difference between the diffracted and direct waves mentioned earlier. The main goal of strict diffraction theory is to obtain parameters $\hat{D}$ and $\Delta\Phi$ by using an analytical deterministic approach based on complex Fresnel integral presentation [3]:

$$F(\nu) = \int\limits_0^\nu \exp\left\{-i\frac{\pi}{2}\nu'^2\right\}d\nu' = -F(-\nu). \tag{4.81}$$

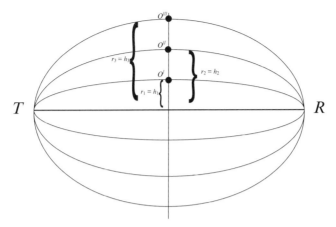

FIGURE 4.16. Geometrical presentation of the Fresnel zones in terms of ellipsoids.

To estimate the effect of diffraction around obstructions, we need a quantitative measure of the required clearance over any terrain obstruction, and, as was shown in Section 4.2, this may be obtained analytically in terms of Fresnel zone ellipsoids drawn around both ends of the radio link, the receiver and the transmitter (see Fig. 4.6). We discussed these zones when we presented free space propagation concept and reflections from a flat terrain. Now, let us introduce the Fresnel zone concept related to diffraction. We show this concept based on the illustration in Figure 4.16, where the cross section radius of any ellipsoid with number n from the family at a distance r_0 and $r_0' = r - r_0$ was presented as a function of the parameters n, r_0 and r_0', which can be written as

$$h_n = \left[\frac{n \lambda r_0 r_0'}{(r_0 + r_0')} \right]^{1/2}. \tag{4.82}$$

The Fresnel integral in (4.81) gives the cumulative effect from several first Fresnel zones covered by the obstruction. In Figure 4.16, the Fresnel (also called *diffraction*) parameter ν in (4.81) is presented by the following formula [29–6]:

$$v_n = h_n \cdot \left[\frac{2(r_0 + r_0')}{\lambda r_0 r_0'} \right]^{1/2} = (2n)^{1/2}. \tag{4.83}$$

From (4.82) and (4.83), one can obtain the physical meaning of the Fresnel–Kirchhoff diffraction parameter ν. Thus, the diffraction parameter v increases with the number n of ellipsoids. All the above formulas are corrected for $h_n \ll r_0, r_0'$, that is, far from the terminal antennas. The volume enclosed by the ellipsoid defined by $n = 1$ is known as the *first Fresnel zone*. The volume between this ellipsoid and the one that is defined by $n = 2$ is the *second Fresnel zone.* The contributions to the total field at the receiving point, from successive Fresnel zones, interfere by giving a very complicated interference picture at the receiver. If a virtual line OO' is placed at the middle of radio path $TO'R$ (i.e., $TO' = O'R$, as shown in Fig. 4.16) then, if the height of the virtual point O' (the virtual source of diffraction) h increases from $h = h_1$ (corresponding to the *first* Fresnel zone) to $h = h_2$ (e.g., to the point O'' defining the limit of the *second* Fresnel zone), then to $h = h_3$ (i.e., to the point O''' defining the limit of the *third* Fresnel zone), and so on, the field at the receiver R will oscillate. The amplitude of oscillations would essentially decrease as a smaller amount of wave energy penetrates into the outer zone relative to the inner zone.

If, for example, some obstacles that we may model by the simple knife edge (with height above the line-of-sight line TOR, h, denoted in Figure 4.17 as $OO^{(n)}$) lies between the receiver and the transmitter at distances r_0 and r_0', respectively, the Fresnel parameter can be presented as [1–8]

$$v = h\left[\frac{2(r_0 + r_0')}{\lambda r_0 r_0'}\right]^{1/2} = 2\left[\frac{\Delta r}{\lambda}\right]^{1/2}, \qquad (4.84)$$

and the phase difference $\Delta\Phi$ between the direct ray from the source placed at the point O (denoted TOR) and the diffracted ray from the point $O^{(n)}$ (denoted $TO^{(n)}R$) can be obtained in the standard manner by use of a simple presentation of the path difference, Δr, and the phase difference, $\Delta\Phi$, between these rays. From the geometry of the problem, shown in Figure 4.17, and using relationship (4.84) between Δr and v, the phase difference, $\Delta\Phi$, can be presented as

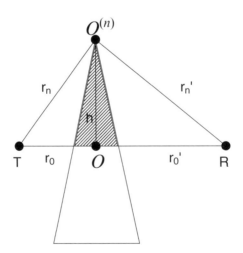

FIGURE 4.17. Geometrical presentation of the knife-edge diffraction.

$$\Delta\Phi = \frac{2\pi}{\lambda}\Delta r = \frac{\pi}{2}v^2. \tag{4.85}$$

From the previous discussions, it is clear that any radio path in obstructive conditions requires a certain amount of a clearance around the central ray if free-space propagation is to occur. This effect can be understood by using the principle of *Fresnel clearance*, which is important in design of point-to-point radio links, where communication is required along a single radio path. This clearance can be explained in terms of Fresnel zones. Thus, the first Fresnel zone (for $n = 1$) encloses all radio paths for which the additional path length Δr, defined in (4.84), does not exceed $\lambda/2$, and, according to (4.85), a phase change is $\Delta\Phi_1 = \pi$. The second Fresnel zone (for $n = 2$) encloses all paths for which the additional path length Δr does not exceed $2\cdot\lambda/2 = \lambda$, and correspondingly, $\Delta\Phi_2 = 2\pi$, and so on. The corresponding radius of the first Fresnel zone h_1 can be derived by setting $\Delta r = \lambda/2$ in (4.82). As a result,

$$h_1 = \left(\frac{\lambda r_0 r_0'}{r_0 + r_0'}\right)^{1/2} = \left(\frac{300 \cdot r_0 r_0'}{fr}\right)^{1/2}, \tag{4.86}$$

where f is measured in gigahertz and $r = r_0 + r_0'$ in kilometers. The shape of the first Fresnel zone and the effect of the obstruction on the clearance are clearly illustrated by Figure 4.18. The clearance due to the diffraction effect from the obstruction is about 60% of the first Fresnel zone, which normally in practice is considered an adequate value for the land rural point-to-point radio links. To finish this analysis, we must mention that the Fresnel zone principle, as well as the clearance explanation, is correct for the case where $r > r_0$, $r_0' >> h_n$, which is adequate for the most practical cases of land radio link designs.

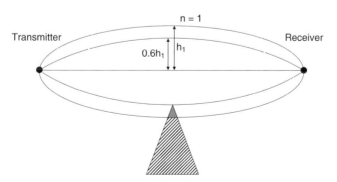

FIGURE 4.18. The clearance effect in the presence of knife-edge obstruction.

EXERCISES

4.1. Using the condition $\mathbf{n} \cdot \nabla G(\mathbf{r}, \mathbf{r}') = 0$, show that Equation (4.16) can be derived from Equation (4.15).

4.2. Using the condition $\nabla G(\mathbf{r}, \mathbf{r}') = 0$ show that Equation (4.17) can be derived from Equation (4.15).

4.3. Using the free space model, find the path loss (in decibels) between the transmitter antenna with gain $G_T = 8dBm$ and the receiver antenna with gain $G_T = 5dBm$, separated at a distance of 3.5 km. The radiated frequency is 1.8 GHz.

4.4. Based on Equation (4.40a) and (4.40b), find the coefficients of specular reflection from a flat terrain for the vertical and horizontal polarization, respectively. Use the following parameters: $\varepsilon_r = 6$ and $f = 2.4$ GHz. Plot both coefficients versus the grazing angle ψ that changes from 0 to π in steps of $\pi/12$.

4.5. Based on Equation (4.67), Equation (4.68), Equation (4.69), and Equation (4.70), find coefficients of diffusive reflection for both the vertical and horizontal polarizations, based on the Rayleigh principle for the following parameters: $\sigma = 3.8m$, $\varepsilon_r = 80$, and $f = 3.3$ GHz. Plot both coefficients versus the grazing angle ψ that changes from 0 to π in steps of $\pi/12$.

4.6. Using the relations between the virtual height of the obstruction and the Fresnel coefficient, find both of these parameters for the four Fresnel zones, that is for $n = 1, 2, 3, 4$. The distance from the transmitter and the virtual obstruction is $r_0 = 1.1$ km and between the obstruction and the receiver is $r_0' = 1.5$ km. The radiated frequency is 2.2 GHz.

REFERENCES

1   Jackson, J.D., *Classical Electrodynamics*, John Wiley & Sons, New York, 1962.

2   Kong, J.A., *Electromagnetic Wave Theory*, John Wiley & Sons, New York, 1986.

3   Dudley, D.G., *Mathematical Foundations for Electromagnetic Theory*, IEEE Press, New York, 1994.

4   Al'pert, Ia. L., V.L. Ginsburg, and E.L. Feinberg, *Radio Wave Propagation*, State Printing House for Technical-Theoretical Literature, Moscow, 1953.

5   Jakes, W.C., *Microwave Mobile Communications*, John Wiley and Sons, New York, 1974.

6   Lee, W.Y.C., *Mobile Cellular Telecommunications Systems*, McGraw Hill Publications, New York, 1989.

7   Saunders, S.R., *Antennas and Propagation for Wireless Communication Systems*, John Wiley & Sons, New York, 1999.

8   Steele, R., Mobile *Radio Communication*, IEEE Press, New York, 1992.

9   Bass, F.G. and I.M. Fuks, *Wave Scattering from Statistically Rough Surfaces*, Pergamon Press, Oxford, 1979.

10 Tatarskii, V.I. and M.I. Charnotskii, "On the universal behavior of scattering from a rough surface from small grazing angles," *IEEE Trans. Anten. Propagat.*, Vol. 46, No. 1, 1995, pp. 67–72.

11 Voronovich, A.G., *Wave Scattering from Rough Surfaces*. Springer-Verlag, Berlin, 1994.

12 Backmann, P. and A. Spizzichino, *The Scattering of Electromagnetic Waves from Rough Surfaces*, Artech House, Boston-London, 1963.

13 Felsen, L. and N. Marcuvitz, *Radiation and Scattering of Waves*, Prentice Hall, Englewood Cliffs, NJ, 1973.

14 Ishimaru, A., *Electromagnetic Wave Propagation, Radiation, and Scattering*, Prentice-Hall, Englewood Cliffs, NJ, 1991.

15 Miller, A.R., R.M. Brown, and E. Vegh, "New derivation for the rough-surface reflection coefficient and for the distribution of the sea-wave elevations," *IEE Proc.*, Vol. 131, Pt. H, No. 2, 1984, pp. 114–116.

16 Ogilvy, J.A., *Theory of Wave Scattering from Random Rough Surfaces*, IOP, Bristol, England, 1991.

17 Rice, S.O., "Reflection of electromagnetic waves from slightly rough surfaces," *Comm. Pure Appl. Math.*, Vol. 4, No. 3, 1951, pp. 351–378.

18 Barrick, D.E. and W.H. Peake, "A review of scattering from surfaces with different roughness scales," *Radio Sci.*, Vol. 3, No. 7, 1968, pp. 865–868.

19 Barrick, D.E., "Theory of HF and VHF propagation across the rough sea—Parts I and II," *Radio Sci.*, Vol. 6, No. 3, 1971, pp. 517–533.

20 Barrick, D.E., "First order theory and analysis of MF/HF/VHF scatter from the sea," *IEEE Trans. Anten. Propagat.*, Vol. 20, No. 1, 1972, pp. 2–10.

21 Wait, J.R., "Perturbation analysis for reflection from two-dimensional periodic sea waves," *Radio Sci.*, Vol. 6, No. 3, 1971, pp. 387–391.

22 Valenzuela, G.R., "Scattering of electromagnetic waves from a slightly rough surface moving with uniform velocity," *Radio Sci.*, Vol. 3, No. 1, 1968, pp. 12–21.

23 Valenzuela, G.R., "Scattering of electromagnetic waves from a tilted slightly rough surface," *Radio Sci.*, Vol. 3, No. 6, 1968, pp. 1057–1066.

24 Valenzuela, G.R., "The effective reflection coefficients in forward scatter from a dielectric slightly rough surface," *Proc. IEEE*, Vol. 58, No. 12, 1970, pp. 1279–1285.

25 Krishen, K., "Scattering of electromagnetic waves from a layer with rough front and plane back (small perturbation method by Rice)," *IEEE Trans. Anten. Propagat.*, Vol. 19, No. 4, 1970, pp. 573–576.

26 Davies, H., "The reflection of electromagnetic waves from a rough surface," *Proc. IEEE*, Vol. 101, No. 2, 1954, pp. 209–214.

27 Bullington, K., "Reflection coefficient of irregular terrain," *Proc. IRE*, Vol. 42, No. 11, 1954, pp. 1258–1262.

28 Fock, V.A., *Electromagnetic Diffraction and Propagation Problems*, Pergamon Press, Oxford, 1965.

29 Keller, J.B., "Diffraction by an aperture," *J. Appl. Phys.*, Vol. 28, 1957, pp. 857–893.

30   Keller, J.B., "Geometrical theory of diffraction," *J. Opt. Soc. Amer.*, Vol. 52, No. 1, 1962, pp. 116–131.

31   James, G.L., *Geometrical Theory of Diffraction for Electromagnetic Waves*, 3rd ed., Peter Peregrines, London, UK, 1986.

32   Honl, H., A.W. Maue, and K. Westpfahl, *Theory of Diffraction*, Springer-Verlag, Berlin, 1961.

33   Kouyoumjian, R.G. and P.H. Pathak, "A uniform geometrical theory of diffraction for an edge in a perfectly conducting surface," *Proc. IEEE*, Vol. 62, No. 9, 1974, pp. 1448–1469.

34   Russel, S.T.A., C.W. Boston, and T.S. Rappaport, "A deterministic approach to predicting microwave diffraction by buildings for micro cellular systems," *IEEE Trans. Anten. and Propag.*, Vol. 41, No. 12, 1993, pp. 1640–1649.

35   Dougherty, H.T. and L.J. Maloney, "Application of diffraction by convex surfaces to irregular terrain situations," *Radio Phone*, Vol. 68B, 1964, p. 239.

36   Anderson, L.J. and L.G. Trolese, "Simplified method for computing knife edge diffraction in the shadow region," *IRE Trans. Anten. Propagat.*, Vol. AP-6, 1958, pp. 281–286.

CHAPTER FIVE

Terrestrial Radio Communications

In this chapter, we consider wave propagation in various terrain environments based on the description of propagation characteristics such as the propagation (or path) loss, L, and the slope parameter γ that describes signal decay. These main parameters are very crucial in predicting land communication channels. First, in Sections 5.1 and 5.2, we introduce the reader to a brief description of the terrain features and various propagation situations in terrestrial communications related to the terminal antenna positions with respect to building rooftops. In Section 5.3, we continue the description of the propagation channel when two antennas are placed on a flat terrain and under line-of-sight (LOS) conditions, when a free-space propagation concept can be used and is described by a two-ray model. In Section 5.4, we consider radio propagation in a "hilly terrain," where we replace the hill by a "knife edge" and introduce Lee's empirical model. Next, in Section 5.5, we present a unified approach on how to predict radio losses in rural forest area links based on a stochastic model that describes multiscattering effects from trees. This model is compared with standard empirical, analytical, and statistical models. Section 5.6 describes radio propagation in mixed residential areas based on the same stochastic approach, but taking into consideration only a single scattering from houses and trees. Section 5.7 introduces the reader to the problems of radio propagation in urban and suburban areas, where we consider two typical situations in the urban environment: (a) urban grid-plan buildings distribution with straight crossing rows of streets and (b) urban areas with randomly distributed buildings placed on a rough terrain. Here, we present the unified

Radio Propagation and Adaptive Antennas for Wireless Communication Networks: Terrestrial, Atmospheric, and Ionospheric, Second Edition. Nathan Blaunstein and Christos G. Christodoulou.
© 2014 John Wiley & Sons, Inc. Published 2014 by John Wiley & Sons, Inc.

stochastic approach which generalizes stochastic models presented in Sections 5.5 and 5.6 by taking into account the building overlay profile and effects of diffraction from buildings' roofs. Then, we compare this general model with those which are mostly used for predicting loss characteristics in such terrestrial communication links. Finally, in Section 5.8, we describe effects of depolarization of radio waves in built-up areas based on results obtained in Section 5.7.

5.1. CHARACTERIZATION OF THE TERRAIN

The process of classifying *terrain configurations* is a very important stage in the construction of propagation models above the ground surface and in predicting the signal/wave attenuation (or "path loss," defined in Chapter 1) within each specific propagation channel.

These terrain configurations can be categorized as [1–14]

- flat ground surface
- curved, but smooth terrain
- hilly terrain
- mountains.

The *built-up areas* can also be simply classified as [1–4]

- rural areas
- mixed residential areas
- suburban areas
- urban areas.

Many experiments carried out in different built-up areas have shown that there are many specific factors that must be taken into account to describe specific propagation phenomena. Recently, a new standard for terrain classification has been introduced for the analysis of urban topographic maps [1–4]. This standard is based on the following terrain characteristics:

1. position and distribution of buildings regarding the observer
2. dimensions of buildings or useful built-up area
3. number of buildings in the area under test
4. height of ground surface and its degree of "roughness" [15–27]
5. presence of vegetation.

Using these specific characteristics and parameters of the terrain, we can easily classify various kinds of terrain by examining the topographic maps for each deployment of radio communication systems.

5.2. PROPAGATION SCENARIOS IN TERRESTRIAL COMMUNICATION LINKS

As remarked earlier, a very important characteristic of the propagation channel is the location/position of both antennas with respect to the obstacles placed around them. Usually, there are three possible situations:

(a) Both antennas, receiver and transmitter, are placed higher than any obstacles (in a built-up area this means that they are above rooftop level) (Fig. 5.1a).
(b) One of the antennas is higher than obstacles' height (namely, the roofs), but the second one is lower (Fig. 5.1b).
(c) Both antennas are below the obstacles' level (Fig. 5.1c).

In the first situation they are in *direct visibility* or LOS conditions. In the last two situations, one or both antennas are in *clutter* or obstructive conditions. In all these cases the profile of the terrain surface is very important and may vary

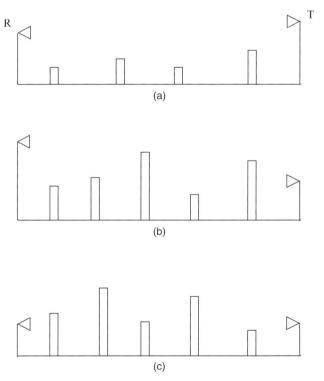

FIGURE 5.1. The three possible locations of the terminal antennas with respect to building height profile.

from flat and smooth, to a curved surface, and finally to a rough and hilly terrain.

5.3. PROPAGATION OVER A FLAT TERRAIN IN LOS CONDITIONS

Instead of using the complicated formulas in Section 4.2 to describe radiowave propagation above the flat terrain, the "two-ray" model is usually used. Let us briefly describe this situation that widely occurs in land communication channels.

5.3.1. Two-Ray Model

The *two-ray* or *two-slope* model was first proposed in the early 1960s for describing the process of radiowave propagation over a flat terrain [1–6]. Let us briefly consider the two-ray model, which is based on the superposition of a direct ray from the source and a ray reflected from the flat ground surface, as shown in Figure 5.2. Earlier, in Chapter 2, the *Friis formula* for the direct wave in free space was presented. We will rewrite it here in the following form:

$$E = \sqrt{30 G_T G_R P_T} \,/\, r_1, \tag{5.1}$$

where r_1 is the radio path of the direct wave as presented in Figure 5.2. The total field at the receiver is the sum of direct and received waves [7], that is,

$$E_R = E_T \left(1 + \frac{d}{d_1} |\Gamma| e^{-jk\Delta d - j\Phi}\right), \tag{5.2}$$

where $\Gamma(\psi)$ is the reflection coefficient described by Equation (4.40a) and Equation (4.40b) in Chapter 4 for horizontal and vertical polarization, respectively; $\Delta r = r_2 - r_1$ (see Fig. 5.2) is the difference in the radio paths of the two waves. $\Delta\varphi = k \cdot \Delta r$ is the phase difference between the reflected and direct waves, which can be presented according to the geometry in Figure 5.2, as

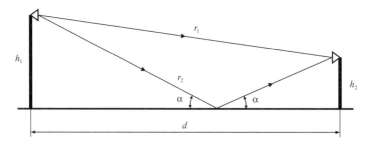

FIGURE 5.2. Geometrical presentation of the two-ray model.

$$\Delta\varphi = k\Delta d = \frac{2\pi}{\lambda}r\left[\left[1+\left(\frac{H_R+H_T}{r}\right)^2\right]-\left[1+\left(\frac{H_R-H_T}{r}\right)^2\right]\right], \quad (5.3)$$

where h_R and h_T are the receiver and transmitter antenna heights, respectively, and r is the distance between them. For $r_1 \gg (h_T \pm h_R)$ and $r_2 \gg (h_T \pm h_R)$, using the assumption that $r_1 \approx r_2 \approx r$, the phase difference in (5.3) can be written as

$$\Delta\varphi = \frac{4\pi h_R h_T}{\lambda \cdot r}. \quad (5.4)$$

If we now assume that $G_R \approx G_T = 1$ (valid for isotropic or omnidirectional antennas, see Chapter 2) and that $\Gamma(\psi) \approx -1$ for the farthest ranges from transmitter (i.e., small grazing angles), we can obtain the magnitude of the signal power at the receiver as [11]

$$|P_R| = |P_T|\left(\frac{\lambda}{4\pi d}\right)^2 |1+\cos^2 k\Delta d - 2\cos k\Delta d + \sin^2 k\Delta d|$$
$$= |P_T|\left(\frac{\lambda}{4\pi d}\right)^2 \sin^2 \frac{k\Delta d}{2} \quad (5.5)$$

From Equation (5.5) one can determine the distance between a receiver and a transmitter for which maximum power is received, taking into account the following conditions:

$$\frac{k\Delta r}{2} \approx \frac{\pi}{2}, \quad \sin\frac{k\Delta r}{2} \approx 1. \quad (5.6)$$

This distance is called the *critical* or *break point range*, denoted by r_b, and is given approximately by the following equation:

$$r_b \approx \frac{4h_R h_T}{\lambda}. \quad (5.7)$$

Then, following definition of the *path loss* introduced in Chapter 1, and using

$$L = 20\log|E_i| + 20\log|1+|\Gamma|e^{-jk\Delta d - j\Phi}|, \quad (5.8)$$

we can easily obtain the path loss over a flat terrain by using the definition of break point range, r_b, and the mathematical description of the "straight line" (as in Reference 7)

$$\text{for } r \leq r_b \text{ } L = L_b + 10\gamma\log\left(\frac{r}{r_b}\right), \quad \gamma = 2 \quad (5.9a)$$

$$\text{for } r \leq r_b \text{ } L = L_b + 10\gamma \log\left(\frac{r}{r_b}\right), \quad \gamma = 4, \quad\quad (5.9b)$$

where L_b is the path loss in free space at the distance that equals the critical range, that is, $r = r_b$, which can be calculated from the following formula [7]:

$$L_b = 32.44 + 20\log r_{b[km]} + 20\log f_{[MHz]}. \quad\quad (5.10)$$

From equations (5.9a) and (5.9b), there are two modes of field intensity decay. One is $\sim r^{-q}$, $q = 2$ for $r \leq r_b$, and second is $\sim r^{-q}$, $q = 4$ for $r > r_b$. From the free space model, the range dependence between the terminal antennas is $\sim r^{-2}$.

Also, for large distances between antennas, that is, $r > r_b$, we get from Equation (5.5) $\sin^2 k\Delta r/2 \approx (k\Delta r/2)^2$ with $\Delta r = 2h_R h_T/r$ (see Equation 5.3 and Equation 5.4). After some straightforward manipulations, we obtain the formula that describes the signal equation decay $\sim r^{-4}$, and usually called the *flat terrain* (FT) *model* for $G_T = G_R = 1$ [1, 2]:

$$L_{FT} = 40\log r_{[m]} - 20\log h_T - 20\log h_R. \quad\quad (5.11)$$

From the two-ray model, the break point is within the range of $r_b = 150–300$ m from the source; at that point, the $\sim r^{-2}$ mode transforms into the $\sim r^{-4}$ mode. This effect depends, according to Equation (5.7), on both antennas' heights and the wavelength of operation. Hence, the two-ray model covers both the free-space propagation model in close proximity to the source and flat terrain propagation model at ranges far from the source. Most equations above have been obtained for isotropic or omnidirectional antennas. For more directive antennas, their gain has to be introduced, as was done in Chapter 2 for Friis formula.

5.4. PROPAGATION OVER A HILLY TERRAIN IN NLOS CONDITIONS

Equation (4.80) describes the effect of a hill, as a "knife-edge" obstruction, which was introduced in Section 4.5 to estimate the effect of diffraction losses. To accurately obtain the diffraction coefficient and the diffraction losses, the formula with the complex Fresnel integral presentation (4.81) should be used [28–35]. However, it is a very time-consuming computational task [1–6], and so, *empirical* models are commonly used instead. We present below Lee's empirical model, on which most empirical and semiempirical models are based [36–40].

5.4.1. Lee's Model

A frequently used empirical model, developed by Lee [6], gives the following expressions for the knife-edge diffraction losses in decibel:

$$L(v) = L_{\Gamma}^{(0)} = 0 (dB), v \le -1, \tag{5.12a}$$

$$L(v) = L_{\Gamma}^{(1)} = 20\log(0.5 - 0.62 \cdot v)(dB), -0.8 < v < 0, \tag{5.12b}$$

$$L(v) = L_{\Gamma}^{(2)} = 20\log\{0.5 \cdot \exp(-0.95 \cdot v)\}(dB), 0 < v < 1, \tag{5.12c}$$

$$L(v) = L_{\Gamma}^{(3)} = 20\log[0.4 - (0.1184 - (0.38 - 0.1 \cdot v)^2)^{1/2}](dB), 1 < v < 2.4, \tag{5.12d}$$

$$L(v) = L_{\Gamma}^{(4)} = 20\log\left(\frac{0.225}{v}\right)(dB), v > 2.4. \tag{5.12e}$$

These equations are used for the cases where several knife edges are placed along the radio path between the two terminal antennas. In this case, a simple summation of the loss from each individual edge is obtained, according to Equation (5.12). This approach gives a sufficiently correct result (see results of corresponding empirical models described in References 41–44).

5.5. PROPAGATION IN RURAL FOREST ENVIRONMENTS

Vegetation presents another significant effect on radiowave propagation, such as scattering and absorption by trees with their irregular structure of branches and leaves. Predictions of signal decay in the case of irregular terrain at frequencies less than 500 MHz have been made by a number of authors [45–48] during the 1950s and 1960s. During the 1970s, vegetation and foliage losses have been reported [49–51] at frequencies up to 3 GHz but for relatively few paths. For forest-type environments [52–56], trees exhibit mainly absorbing and scattering effects and very little diffraction effects.

5.5.1. A Model of Multiple Scattering in a Forest Area

In References 57–60, a stochastic approach was proposed to investigate the absorbing and multiple scattering effects that accompany the process of radiowave propagation through forested areas. This is a combination of probabilistic and deterministic approaches, which describe the random media scattering phenomenon. The geometrical optics approximation is used to account for propagation over a series of trees modeled as absorbing amplitude/phase screens with rough surfaces. This stochastic approach allows the designer to obtain the absorption effects from trees using their real physical parameters, such as permittivity and conductivity, as well as the random distribution of their branches and leaves.

The Forest Terrain Description. Let us consider an array of trees as cylinders with randomly distributed surfaces, all placed on a flat terrain (Fig. 5.3). Also assume that the reflecting properties of the trees are randomly and independently distributed, but they are statistically the same [61, 62]. The values of the

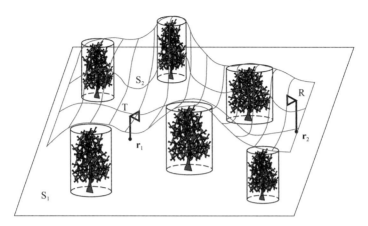

FIGURE 5.3. The profile of a forested nonregular area.

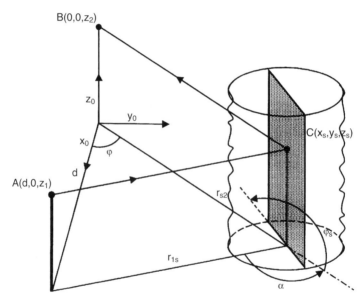

FIGURE 5.4. Scattering from a cylinder with a rough surface which is used to model a tree.

reflection coefficient are complex with uniformly distributed phase in the range $[0, 2\pi]$. Thus, the average value of the reflection coefficients is zero, that is, $\langle \Gamma(\varphi_s, r_s) \rangle = 0$. The geometry of the problem is shown in Figure 5.4, where $A(\mathbf{r}_1)$ denotes the location of the transmitting antenna at height z_1, $B(\mathbf{r}_2)$ is the location of the receiving antenna at height z_2. Let us derive an average measure of field intensity for waves passing through the layer of trees after multiple scattering. In this case, we consider each tree as a phase-amplitude cylindrical screen. Figure 5.3 shows an array of these (screens) placed at $z = 0$ (a flat surface S_1). The trees have average height $\bar{h}$ and width $\bar{d}$ [60]. These trees are randomly and independently distributed, and they are oriented in arbitrary directions at the plane $z = 0$ with equal probability and with average density ν (per square kilometers). In the case where both antennas are placed within the forest environment and are lower than the average tree height $\bar{h}$, that is, $0 < z_2, z_1 < \bar{h}$, then the multiscattering effects are predominant and must be taken into account. In this case we can present the range of direct visibility (LOS under conditions) between the two terminal antennas $\bar{\rho} = \gamma_0^{-1}$, where γ_0 is tree density:

$$\gamma_0 = 2\bar{d}\,\nu/\pi \tag{5.13}$$

The "roughness" of a tree's surface is described by introducing a correlation function for the reflection coefficient Γ. We define the correlation function for the reflection coefficient Γ for $\ell_h, \ell_v << \bar{\rho}, \bar{d}, \bar{h}$ as [59, 60]

$$K(\mathbf{r}_{2S}, \mathbf{r}_{1S}) = <\Gamma_{2S} \cdot \Gamma_{1S}^*> = \Gamma \cdot \exp\left\{-\frac{|\rho_{2S} - \rho_{1S}|}{\ell_h} - \frac{|z_{2S} - z_{1S}|}{\ell_v}\right\}, \tag{5.14}$$

where $\mathbf{r}_{2s}$ and $\mathbf{r}_{1s}$ are the points at the surface of an arbitrary tree (see Fig. 5.6); Γ is the absolute value of the reflection coefficient, given by (4.40a–b) in Chapter 4 for two kinds of field polarization: $\Gamma_{2S} = \Gamma(\mathbf{r}_{2S})$ and $\Gamma_{1S} = \Gamma(\mathbf{r}_{1S})$.

The Average Field Intensity. Taking into account the wave field presentation and the Green's theorem for our problem introduced in Chapter 4, we can present the field over the rough terrain using Green theorem in integral form [57, 58]:

$$U(\mathbf{r}_2) = U_i(\mathbf{r}_2) + \int_S \left\{U(\mathbf{r}_s)\frac{\partial G(\mathbf{r}_2, \mathbf{r}_s)}{\partial \mathbf{n}_s} - G(\mathbf{r}_2, \mathbf{r}_s)\frac{\partial U(\mathbf{r}_s)}{\partial \mathbf{n}_s}\right\}dS, \tag{5.15}$$

where $U_i(\mathbf{r}_2)$ is the incident wave field, $\mathbf{n}_s$ is the vector normal to the terrain surface S at the scattering point $\mathbf{r}_s$, $G(\mathbf{r}_2,\mathbf{r}_s)$ is the Green's function of the semi-space defined in Chapter 4, which we rewrite as [57–59]:

$$G(\mathbf{r}_2, \mathbf{r}_1) = \frac{1}{4\pi}\left[\frac{\exp[ik|\mathbf{r}_2 - \mathbf{r}_1|]}{|\mathbf{r}_2 - \mathbf{r}_1|} \pm \frac{\exp[ik|\mathbf{r}_2 - \mathbf{r}_1'|]}{|\mathbf{r}_2 - \mathbf{r}_1'|}\right]. \tag{5.16}$$

Here, $\mathbf{r}_1'$ is the point symmetrical to $\mathbf{r}_1$ relative to the Earth's surface S_1; $k = 2\pi/\lambda$, λ is the wavelength. In integral (5.15), the random surface S (relief of the terrain with obstructions) is treated as the superposition of an ideal flat ground surface S_1 ($z = 0$) and the rough surface S_2 is created by the tops of the obstructions (see Fig. 5.3). We construct the Green's function in the form of (5.16) to satisfy a general electrodynamic approach; that is, to describe both vertical (sign "+" in (5.16)) and horizontal (sign "−" in (5.16)) polarizations with their corresponding boundary conditions. In fact, by introducing the Green's function (5.16) with the "+" sign in (5.15), we satisfy the Dirihlet boundary conditions at the flat (nondisturbed) Earth's surface $S_1(z = 0)$. That means, $G_{z=0} = 2$ and $\partial u/\partial \mathbf{n}_s = 0$ (the same conditions were stated in Chapter 4).

At the same time, using the sign "−" we satisfy the Neumann boundary conditions at the plane $z = 0$: $G_{z=0} = 0$ and $u = 0$ (the same conditions were stated in Chapter 4). Hence, if the source is described by Equation (5.16), we can exclude the integration over the nondisturbed surface S_1, assuming the surface S_1 as perfectly reflecting. Next, by using the well-known Kirchhoff's approximation described earlier, we can determine the scattered field $U_r(\mathbf{r}_s)$ from the forest layer as a superposition of an incident wave $U_i(\mathbf{r}_2)$, the reflection coefficient $\Gamma(\varphi_s, \mathbf{r}_s)$, and the shadow function $Z(\mathbf{r}_2, \mathbf{r}_1)$. The shadow function equals one, if the scattered point $\mathbf{r}_s$ inside the forested layer can be observed from both points $\mathbf{r}_1$ and $\mathbf{r}_2$ of the transmitter and receiver locations (as shown in Fig. 5.4), and equals zero in all other cases. Taking into account all these assumptions, Equation (5.15) can be rewritten, according to References 57–60, as

$$U(\mathbf{r}_2) = Z(\mathbf{r}_2, \mathbf{r}_1)\tilde{G}(\mathbf{r}_2, \mathbf{r}_1)$$
$$+ 2\int_{S_2} \left\{ Z(\mathbf{r}_2, \mathbf{r}_s, \mathbf{r}_1)\Gamma(\varphi_s, \mathbf{r}_s)\tilde{G}(\mathbf{r}_s, \mathbf{r}_1) \cdot (\mathbf{n}_s \cdot \nabla_s)\tilde{G}(\mathbf{r}_2, \mathbf{r}_s) \right\} dS, \qquad (5.17)$$

where $\nabla_s = (\partial/\partial x_s, \partial/\partial y_s, \partial/\partial z_s)$, $\varphi_s = \sin^{-1}(\mathbf{n}_s \cdot \mathbf{r}_s - \mathbf{r}_1/|\mathbf{r}_s - \mathbf{r}_1|)$ (see Fig. 5.10), and $\tilde{G}(\mathbf{r}_2, \mathbf{r}_1)$ is the normalized Green's function.

To solve (5.17), we apply Twersky's approximation [63] to (5.17), which does not take into account mutual multiple scattering effects. Twersky's approximation states that the contributions of multiple scattered waves are additive and independent. This approximation together with that of $\langle \Gamma \rangle = 0$ makes it possible to obtain the coherent part of the total field by averaging (5.17) over the reflecting properties of each tree and over all tree positions:

$$<U_2> = <Z_{21}> \tilde{G}_{21} \qquad (5.18)$$

Z_{12} is the "shadowing" function that describes the probability of existence of obstructions in the radio path of the terminal antennas. Following now References 57–60, we, after some straightforward computations, after averaging over all tree (screen) positions and over the screen orientations for each scattered point, will obtain the coherent part of the total field by accounting (5.38) [57–60]:

$$<I_{co}> = \frac{1}{(4\pi)^2} \frac{\exp(-\gamma_0 r)}{r^2} \left| 2\sin\frac{kz_1 z_2}{r} \right|^2. \qquad (5.19)$$

The incoherent part of total field intensity can be found after straightforward computations following References 57–60 as

$$<I_{inc}> \approx \frac{\gamma_0 \Gamma}{(4\pi)^2}\left[\frac{\Gamma^3}{4(8)^3}\frac{\exp(-\gamma_0 r)}{r} + \frac{\Gamma}{32}\left(\frac{\pi}{2\gamma_0}\right)^{1/2}\frac{\exp(-\gamma_0 r)}{r^{3/2}} + \frac{1}{2\gamma_0}\frac{\exp(-\gamma_0 r)}{r^2}\right].$$

(5.20)

Since

$$<I_{total}> = <I_{co}> + <I_{inc}>,$$ (5.21)

we can evaluate the total path loss

$$L_{total} = 10\log[\lambda^2(<I_{co}> + <I_{inc}>)].$$ (5.22)

The first two terms in (5.20) are important only at long distances $r > r_{kr}$ ($r_{kr} < 8^3/\pi\gamma_0\Gamma^2$) from the transmitter.

5.5.2. Comparison with Other Models

There are many models, empirical, semiempirical, and deterministic, which have been curried during the last three decades [45–56, 64–77] to predict propagation characteristics in forest environments. It was found that there are a lot of factors that affect radio propagation above and inside a forest area. Therefore, according to the Recommendations of ITU-RP.833.3 [72] and based on previous experimental and theoretical studies [45–56, 64–77], we introduce *the radio path coefficient of attenuation* (or *loss*), $\alpha(f)$ parameter measured in decibels per meter, in the forested areas, which depends on all the effects of forest influence that can be taken into account, such as

- damping due to energy loss at tree branches and tree leaves
- diffraction at tree elements leading to deviations of amplitude and phase of the incident radio wave
- complicated interference structure of the total field scattered by leaves
- depolarization of the incident radio wave
- reflection of the radio wave from the top boundary of the forest massive, called the *lateral wave*, and so on.

This coefficient strongly depends on the radio frequency of the transmitted signals. Here we will summarize the main results from previous research as follows: it was found that the forest mass can be transparent for radio waves at frequencies $f < 10$MHz with $\alpha(f) < 10^{-3} - 10^{-4}$dB/m; semitransparent for $10 < f < 100$ MHz, where the loss coefficient lies in the range of $10^{-3} < \alpha(f) <$

10^{-2} dB/m; strongly absorbed at frequencies $100 < f < 2000$ MHz, where $10^{-2} < \alpha(f) < 10^{-1}$ dB/m; and finally, fully nontransparent at frequencies $f > 2$GHz, where $\alpha(f) < 10^{-1}$ dB/m.

We will compare each model with results obtained from the stochastic model. An attenuation law for radio waves passing through the forest mass was described by Tamir in his deterministic model [48–50]. This deterministic model accounts for diffraction effects of vegetation hidden within the forest layer and describes the array of trees as a homogeneous dielectric slab. However, as was shown experimentally, this model is a good approximation for a forest up to frequencies of 200–500 MHz, because the "gaps" between neighboring trees in such a model of a "continuous dielectric slab" should be less than a wavelength. There are two scenarios that have been examined: (1) one of the terminal antennas (mainly the Tx antenna) is over the "forest slab layer" and the other (mainly the Rx antenna) is hidden into such a dielectric slab, and (2) both antennas are lower than the "forest dielectric slab" overlay profile. In both cases, the role of the surface wave, called lateral, was studied and was found that this wave, caused by diffraction from the "forest slab tops" and propagated along the "forest overlay profile," forms the additional losses to the LOS component between the terminal antennas. Due to this additional term, the total field intensity for the range $0 < r < 1$–2 km in References 48–50 is proportional to $r^{-2.5} - r^{-3}$, instead of r^{-2}, as in the case of LOS conditions (see Section 5.1).

A more realistic deterministic model for the forest mass was proposed and analyzed by Popov and colleagues in References 73–77. They proposed two models: (1) one is the same as the one proposed by Tamir, that is, presenting the forest mass as vertically oriented in a space-layered structure (called Model I), and (2) the second assumes rows of trees having quasi-spherical top structure, with "gaps" between them (called Model II). Both models were compared with existing experiments and were shown that for high frequencies, $900 < f < 5000$ MHz (interesting for modern cellular and noncellular networks), the path loss according to Model I changes from –16.2 dB to –90 dB at the length of the radio path inside forest of 100 m. At the same time, Model II predicts the path loss at 100 m length at the range of –34 to –189.0 dB. Experiments mentioned in References 73–77 have been carried out for some specific frequencies, such as $f = 1200$ MHz and $f = 3200$ MHz, giving, at the distance of 100 m inside forest mass, a path loss of –35 dB and –50 dB, respectively. Theoretical predictions yield –21.6 dB (Model I) and –45.4 dB (Model II) for 1200 MHz, and –57.6 dB (Model I) and –121 dB (Model II) for 3200 MHz. At lower frequencies, $100 < f < 900$ MHz, the proposed models predict the attenuation coefficient from –0.018 dB/m to –0.0378 dB/m (for 100 MHz) and –0.162 dB/m to –0.34 dB/m (for 900 MHz). At the same time, they predict the wave attenuation law $\sim r^{-\gamma}$, with $\gamma = 3$–4, that is, much stronger attenuation effects compared with those predicted by the Tamir model [48–50].

From Equation (5.20) it follows that the stochastic model presented above fully overlaps with both predicted laws with the parameter γ varying from 2.5

to 4.0. In fact, for zones far from the transmitter, the two last terms in (5.48) are the dominant terms, and they predict a field intensity attenuation from $\exp(-\gamma_0 r)/r$ to $\exp(-\gamma_0 r)/r^{3/2}$ that can be expanded from $r^{-2.5}$ to $r^{-4.5}$ [60].

A good agreement was also obtained by comparing the results of the stochastic model with those obtained by Weissberger, who created the empirical model based on statistical analysis of numerous experiments carried out in forest environments [64]. According to his empirical model, the total path loss varies according to law $\sim r^{-3.0}$ in the close zone from the transmitting antenna ($1 \leq r \leq 14$ m) and $\sim r^{-2.588}$ far from the transmitter ($15 \leq r \leq 400$ m) within the vegetation and the trees' mass.

Comparison with the statistical model described in References 65–67, where single scattering with signal decay law of $r^{-2} - r^{-2.5}$ was obtained using the Born's approximation, also shows that the stochastic model described in Section 5.7 is more general and covers all cases of multiple diffraction (in close zones) and multiple scattering (in far zone).

5.6. PROPAGATION IN MIXED RESIDENTIAL AREAS

Let us consider an array of houses and trees as blocks and cylinders with randomly distributed surfaces which are placed on a flat terrain (see Fig. 5.5). Such obstructions are mainly present in mixed residential areas. The characterization of the propagation properties of such environments has been thoroughly investigated in Reference 78.

5.6.1. Statistical Description of Mixed Residential Area

We assume that the reflecting properties of the houses and trees are randomly and independently distributed, but they are statistically the same. The values

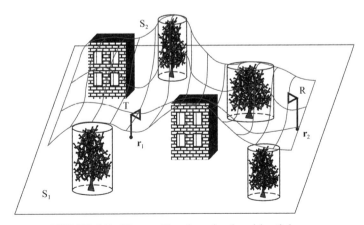

FIGURE 5.5. The profile of a mixed residential area.

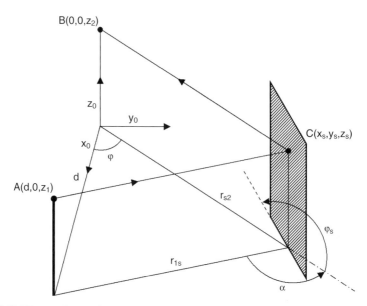

FIGURE 5.6. Scattering from a nontransparent screen that models a house.

of the reflection coefficient are complex with a uniformly distributed phase in the range $[0, 2\pi]$. Thus, we consider each house or tree as a phase-amplitude screen (see Fig. 5.6). The reflection properties of these screens are described by the complex reflection coefficient with a uniformly distributed phase in the range $[0, 2\pi]$ and with the correlation scales in horizontal, ℓ_h, and vertical, ℓ_v, directions, respectively. Both scales characterize the correlation function of the reflection coefficient, which describes reflection properties of residential houses and trees and can be presented as [78]

$$K_\Gamma(\mathbf{r}_S, \mathbf{r}_S') = \Gamma(\varphi_S) \cdot \exp\left\{-\frac{|\xi|}{\ell_v} - \frac{|\eta|}{\ell_h}\right\}, \tag{5.23}$$

where $\Gamma(\varphi_S)$ is the amplitude distribution of the reflection coefficient over angles φ_S. The absolute value of $\Gamma(\varphi_S)$ is defined by classical Equation (4.40) in Chapter 4 both for a vertical and horizontal field polarization. In (5.23), to obtain $K_\Gamma(\mathbf{r}_S, \mathbf{r}_S')$, we introduce a new variable $\xi = |\mathbf{r}_S' - \mathbf{r}_S|$ and construct, at the surface of reflected rough screen (see Fig. 5.6), the local coordinate system $\{\xi, \eta\}$ with the origin at the point $\mathbf{r}_S$. The geometry of the problem is shown in Figure 5.5, where $A(\mathbf{r}_1)$ is the point of the transmitting antenna location at the height z_1 and $B(\mathbf{r}_2)$ is the point of receiving antenna location at the height z_2. As was shown in Reference 36, to derive an average measure of field intensity for waves passing through the mixed layer of houses and trees, one needs to use the single-scattering approach and take into account diffraction from building roofs. In this case, we consider one of the antennas to be higher than the average mixed-layer height, $\bar{h}$, that is, $z_2 > \bar{h} > z_1$. The field component which passes through such a layer, after multiple scattering, is smaller than that of the single-scattering case. Thus, only a single-scattering problem with diffraction from the mixed layer tops should be considered here. Moreover, because in residential areas the height of trees and houses are at the same level (i.e., uniformly distributed in the vertical plane), we can exclude the influence of the terrain profile on propagation effects within such a channel, as it was done previously for the forest environment.

5.6.2. The Average Field Intensity

Taking into account basic description of signal strength [78], we determine the correlation function of the total field for the approximation of single scattering, $K(\mathbf{r}_2, \mathbf{r}_2') = \langle U(\mathbf{r}_2)U^*(\mathbf{r}_2') \rangle$, in the following form:

$$K(\mathbf{r}_2, \mathbf{r}_2') = 4k^2 \int_{S_2} dS_2 \int_{S_2} dS_2' \cdot \langle Z(\mathbf{r}_2, \mathbf{r}_S, \mathbf{r}_1) \cdot Z(\mathbf{r}_2', \mathbf{r}_S', \mathbf{r}_1) \cdot K_\Gamma(\mathbf{r}_S, \mathbf{r}_S')$$

$$\times \sin \psi_S \cdot \sin \psi_S' \cdot G(\mathbf{r}_2, \mathbf{r}_S) \cdot G(\mathbf{r}_S, \mathbf{r}_1) \cdot G^*(\mathbf{r}_2', \mathbf{r}_S') \cdot G^*(\mathbf{r}_S'\mathbf{r}_1) \rangle. \tag{5.24}$$

To derive the correlation function, we must average expression (5.24) over the positions of the reflecting surfaces of the obstructions (houses and trees) and over their number and their reflecting properties. First, let us average Equation (5.24) over the reflection coefficient of each obstruction as a random screen over the phase interval $[0, 2\pi]$, and denote this result by $K_\Gamma(\mathbf{r}_2, \mathbf{r}_2')$. Assuming that the correlation scales introduced earlier are smaller than the obstructions sizes and the average distances between obstructions, that is, ℓ_h and $\ell_v \ll \bar{h}, \bar{d}, \bar{L}$, but $k\ell_h \gg 1$, $k\ell_v \gg 1$, we integrated (5.24) over the variables ξ and η and taking into account (5.23), we get after averaging over the ensemble of obstructions that are randomly distributed at the ground surface for $k\ell_h \gg 1, k\ell_v \gg 1$:

$$< I(\mathbf{r}_2) > \equiv K(\mathbf{r}_2, \mathbf{r}_2) = 4\gamma_0 \int_V (d\mathbf{r}) \exp\left\{ -\gamma_0 \left[r + \tilde{r} \frac{\bar{h} - z}{z_2 - z} \right] \right\} \Gamma\left(\frac{\alpha}{2}\right) \sin^2 \frac{\alpha}{2}$$

$$\times \frac{4k\ell_h |G(\mathbf{r}_2, \mathbf{r})|^2}{1 + (k\ell_h)^2 (\cos\psi_s - \cos\varphi_s)^2} \frac{4k\ell_v |G(\mathbf{r}, \mathbf{r}_1)|^2}{1 + (k\ell_v)^2 (\cos\theta_2 - \cos\theta_1)^2}, \tag{5.25}$$

where $\ell = |\mathbf{r}_2' - \mathbf{r}_2|$; all angles, which are shown in Figure 5.7, can be defined as

$$\cos\varphi = \left(\frac{\mathbf{r}_2 - \mathbf{r}_1}{|\mathbf{r}_2 - \mathbf{r}_1|} \cdot \frac{\mathbf{r}_2 - \mathbf{r}_S}{|\mathbf{r}_2 - \mathbf{r}_S|} \right), \cos\varphi' = \left(\frac{\mathbf{r}_2 - \mathbf{r}_1}{|\mathbf{r}_2 - \mathbf{r}_1|} \cdot \frac{\mathbf{r}_2 - \mathbf{r}'}{|\mathbf{r}_2 - \mathbf{r}_2'|} \right) \tag{5.26a}$$

$$\sin\theta_1 = (z_S - z_1)/ |\mathbf{r}_S - \mathbf{r}_1|, \sin\theta_2 = (z_2 - z_S)/ |\mathbf{r}_2 - \mathbf{r}_S|. \tag{5.26b}$$

All parameters and functions presented in (5.26) are described earlier. Here, the integration is over the layer volume $V \equiv \{ x, y \in (-\infty, +\infty); z \in (0, \bar{h}) \}$, and

$$r = \sqrt{(x - x_1)^2 + (y - y_1)^2}, \tilde{r} = \sqrt{(x_2 - x)^2 + (y_2 - y)^2}. \tag{5.27}$$

In Equation (5.25), the Green's functions were obtained according to geometrical optics approximation, for $\ell_h, \ell_v << \bar{h}, \bar{d}, \bar{L}$, and $k\ell_h >> 1, k\ell_v >> 1$:

$$|G(\mathbf{r}_2, \mathbf{r})|^2 \approx \frac{1}{16\pi^2} \frac{1}{|\mathbf{r}_2 - \mathbf{r}_1|} \tag{5.28a}$$

$$|G(\mathbf{r}, \mathbf{r}_1)|^2 \approx \frac{1}{4\pi^2} \frac{1}{|\mathbf{r} - \mathbf{r}_1|} \sin^2 \frac{kzz_1}{|\mathbf{r} - \mathbf{r}_1|}. \tag{5.28b}$$

Finally, using (5.28a) and (5.28b) in (5.25), for $(z_2 - \bar{h})/\bar{h} >> \gamma_0 d \cdot e^{-\gamma_0 d}$, where $d = |\mathbf{r}_2 - \mathbf{r}_1|$, yields the following expression for the incoherent part of the total field intensity (for single scattering from each obstacle, see Fig. 5.7):

$$\langle I_{inc} \rangle = \frac{\Gamma}{8\pi} \cdot \frac{\lambda \cdot \ell_h}{\lambda^2 + \left[2\pi \ell_h \bar{L} \gamma_0 \right]^2} \cdot \frac{\lambda \cdot \ell_v}{\lambda^2 + \left[2\pi \ell_v \gamma_0 (\bar{h} - z_1) \right]^2} \frac{(z_2 - \bar{h})}{d^3}. \tag{5.29}$$

This formula is more general than the ones obtained in References 57 and 58 because it accounts for the limit dimensions of obstructions in both the vertical and horizontal directions according to (5.23).

The average intensity of the field through the mixed layer is the sum of the intensity of the scattering wave defined in Equation (5.29) (incoherent part) and of the intensity of the coherent part $<I_{co}>$ created by the wave coming from the source. The straightforward evaluations of (5.25) allow us to obtain $<I_{co}>$ as [68]:

$$< I_{co} >= \exp \left\{ -\gamma_0 d \frac{\bar{h} - z_1}{z_2 - z_1} \right\} \left[\frac{\sin(kz_1 z_2 / d)}{2\pi d} \right]^2. \tag{5.30}$$

Finally, the corresponding path loss can be obtained by substituting expressions (5.29) and (5.30) in Equation (5.21) and Equation (5.22), respectively.

5.7. PROPAGATION IN URBAN ENVIRONMENTS

Here, we consider two specific urban propagation environments observed from the topographic maps of most cities [69–83]:

(a) regularly distributed rows of buildings and streets, and
(b) nonregularly distributed buildings placed on a rough terrain with various orientations relative to the transmitting and receiving antennas.

In Section 5.7.1, we start with the multislit street waveguide model, which was found to be in good agreement with experimental data of wave propagation in

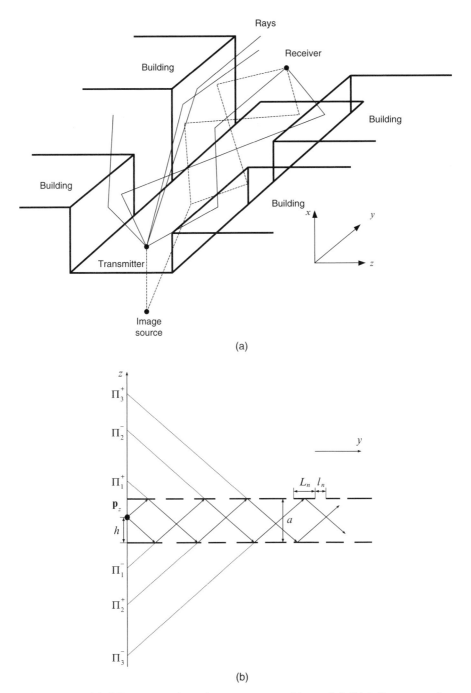

FIGURE 5.7. (a) 3-D presentation of a street waveguide model. (b) 2-D presentation of the multislit street model with the width *a*.

urban areas with regular crossing-street grid layouts [79–81]. In Section 5.7.2, we will discuss the situation when an array of buildings is randomly distributed at a rough terrain surface, and we present the 3-D stochastic multiparametric model obtained in References 86 and 87 accounting for buildings' overlay profile. Then, in Section 5.7.3, we compare the stochastic model with the frequently used empirical, semiempirical, and deterministic analytical models for predicting loss in various built-up areas [82–85, 88–93].

5.7.1. Propagation in Urban Areas with Regularly Distributed Rows of Buildings

Here we consider several urban propagation environments. Earlier, we started to describe the simplest case of EM wave propagation in the urban environment, when both antennas are placed above a flat ground surface in conditions of LOS and below the rooftop level. As was shown in Section 5.3, in LOS conditions, the corresponding propagation characteristics, as path loss and radio coverage, can be determined using the well-known two-ray model. We will briefly discuss later the multislit waveguide model (for LOS propagation) that describes specific scenario, when both antennas are located along the street in LOS conditions and their heights are lower than the building rooftops. It was found in References 79–81 that they are in good agreement with experimental data of wave propagation in urban areas with a regular crossing-street grid.

Street Waveguide Model. As mentioned in References 59 and 79–81, the conditions of LOS propagation along a straight street, in which a base station (BS) is located, is of great importance in defining the coverage area for antennas located below because of the low path loss as compared to propagation over the rooftops. At the same time, a "multislit waveguide" model has been introduced recently for describing the propagation of EM waves in a city environment with regularly planned streets, that is, a model of straight streets with buildings lined up on the sides [79–81]. The street is seen as a planar multislit waveguide with a Poisson distribution of screens (building walls) and slits (intervals between buildings). The dielectric properties of the building walls are taken into account by introducing the electrical impedance as a function of their surface permittivity and conductivity. In Figure 5.7a, a 3-D waveguide model of a city region with regularly planned buildings, and with a receiver and transmitter, is shown.

We notice that the condition $h_T, h_R < h_b$ is a main condition for the validity of the proposed street waveguide model [79–81]. Here, h_b is the height of the buildings lining up the street, and h_T and h_R are the transmitter and receiver antenna heights, respectively. The reflection from the ground surface is also considered using an imaginary source. The projection of such waveguide on the zy-plane presents the 2-D impedance parallel multislit waveguide with randomly distributed screens and can be considered as a model of a city street (see Fig. 5.7b). One waveguide plane is placed at the waveguide (street) side $z = 0$, and the second one at $z = a$, where a is a street width (see Fig. 5.7b). The screen length L_n and slit length l_n are distributed according to the Poisson distribution with the average values of $<L> = L$ and $<l> = l$, respectively [59, 79–81]:

$$f(L_n) = L^{-1} \exp\left\{-\frac{L_n}{L}\right\}, f(l_n) = l^{-1} \exp\left\{-\frac{l_n}{l}\right\}. \tag{5.31}$$

The dielectric properties of buildings' walls are usually described by the surface impedance:

$$Z_{EM} \sim \varepsilon_r^{-\frac{1}{2}}, \varepsilon_r = \varepsilon_{r_0} - j60\lambda\sigma. \tag{5.32}$$

In a real city scenario, the screen and slit lengths are much greater than the radiation wavelength λ, that is, $L_n >> \lambda, l_n >> \lambda$. In this case, we can make use of approximations provided to us from the geometrical theory of diffraction (GTD). According to the GTD, the reflected and diffracted waves have the same nature, and the total field can be presented as a superposition of direct (incident) fields from the source and waves reflected and diffracted fields from the screens.

Average Field Intensity in the Multislit Street Waveguide. Reflection from screens and their corners is taken into account by introducing the special "telegraph signal" functions $f_1(y)$ and $f_2(y)$ defined for the first and the second waveguide walls, respectively, as [81]

$$f_{1,2}(y) = \begin{cases} 1, \text{on the screen} \\ 0, \text{on the slit} \end{cases}. \tag{5.33}$$

Next, we introduce the image sources as presented in Figure 5.7b and denote them for the first reflection from surface $z = a$ by the symbol "+" and for the first reflection from surface $z = 0$ by the symbol "−".

Multiple Reflected Modes inside the Multislit Waveguide. In the first stage, we construct the reflected wave fields (called reflected waveguide modes) from the first, second, and so on, reflections taking place from the waveguide wall $z = a$, and then from the wall $z = 0$. Following the procedure described in Reference 81, using the induction method, we obtain for the n-time reflected wave field (the first reflection taking place from the plane $z - a$) the following expressions [81]:

for even $n = 2m, m = 1, 2, 3 \ldots$

$$\Pi_{z_n}^+ = \frac{e^{ikr_n}}{r_n} f_1\left[\frac{(a-h)y}{na-h+z}\right] f_2\left[\frac{(2a-h)y}{na-h+z}\right] \times \ldots \times f_1\left[\frac{((n-1)a-h)y}{na-h+z}\right] f_2\left[\frac{(na-h)y}{na-h+z}\right]$$

$$\tag{5.34a}$$

for odd $n = 2m + 1, m = 1, 2, 3, \ldots$

$$\Pi_{z_n}^+ = \frac{e^{ikr_n}}{r_n} f_1 \left[\frac{(a-h)y}{(n+1)a-h-z} \right] f_2 \left[\frac{(2a-h)y}{(n+1)a-h-z} \right] \times \ldots$$
$$\times f_2 \left[\frac{((n-1)a-h)y}{(n+1)a-h-z} \right] f_1 \left[\frac{(na-h)y}{(n+1)a-h-z} \right]. \tag{5.34b}$$

The same procedure can be used for the first reflection taking place from the second waveguide wall $z = 0$, that is, for the image sources Π_n^- (see Fig. 5.7b). After similar geometric consideration, we obtain the following expressions [81]:for even $n = 2m, m = 1, 2, 3, \ldots$

$$\Pi_{z_n}^- = \frac{e^{ikr_n'}}{r_n'} f_2 \left[\frac{hy}{na+h-z} \right] f_1 \left[\frac{(a+h)y}{na+h-z} \right]$$
$$\times \ldots \times f_2 \left[\frac{((n-2)a+h)y}{na+h-z} \right] f_1 \left[\frac{((n-1)a+h)y}{na+h-z} \right] \tag{5.35a}$$

for odd $n = 2m + 1, m = 1, 2, 3, \ldots$

$$\Pi_{z_n}^- = \frac{e^{ikr_n'}}{r_n'} f_2 \left[\frac{hy}{(n-1)a+h+z} \right] f_1 \left[\frac{(a+h)y}{(n-1)a+h+z} \right] \times \ldots$$
$$\times f_1 \left[\frac{((n-2)a+h)y}{(n-1)a+h+z} \right] f_2 \left[\frac{((n-1)a+h)y}{(n-1)a+h+z} \right]. \tag{5.35b}$$

Average Field in the Impedance Multislit Street Waveguide. The statistical moments of the reflected field inside the multislit street waveguide relate to the statistical moments of "telegraph signal" functions $f_1(y)$ and $f_2(y)$ through the parameter of slit density (e.g., parameter of discontinuity) $x = L/L + l$ as [81]:

$$\langle f_i(y) \rangle = x = \frac{L}{L+l}, i = 1, 2 \tag{5.36a}$$

$$\langle f_i(y_1)f_i(y_2) \rangle = x^2 K(y_1 - y_2) \tag{5.36b}$$

$$\langle f_i(y_1)f_i(y_2)f_i(y_3) \rangle = x^3 K(y_1 - y_2)K(y_2 - y_3) \tag{5.36c}$$

and, finally,

$$\langle f_i(y_1)f_i(y_2) \ldots f_i(y_2) \rangle = x^n \prod_{\nu=1}^{n-1} K(y_{\nu-1} - y_\nu) \tag{5.36d}$$

where $K(w)$ is the correlation function of the "telegraph" signal functions

$$K(w) = \chi^2 \left\{ 1 + \frac{l}{L} \exp\left[-\left(\frac{1}{L} + \frac{1}{l} \right) |w| \right] \right\}. \tag{5.37}$$

Taking into account the fact that the slit and screen distributions in the street waveguide are statistically independent, that is,

$$\langle f_1(y) \cdot f_2(y) \rangle = \langle f_1(y) \rangle \cdot \langle f_2(y) \rangle = \chi^2, \tag{5.38}$$

and using the relationships described by (5.36), we derive for the *n*-times reflected fields the expression as a sum of two terms. The first one describes the average reflected field inside the waveguide for a first reflection from the wall at $z = a$:

$$\langle \Pi_{zn}^+ \rangle = \frac{e^{ikr_n}}{r_n} \chi^n R^n \prod(\alpha, \beta) e^{iK((n+1)a - z - h)} \times K^{n-2} \left(\frac{2ay}{(n+1)a - z - h} \right), n = 2m + 1 \tag{5.39a}$$

$$\langle \Pi_{zn}^+ \rangle = \frac{e^{ikr_n}}{r_n} \chi^n R^n \prod(\alpha, \beta) e^{iK(na + z - h)} \times K^{n-2} \left(\frac{2ay}{na + z - h} \right), n = 2m, m = 1, 2, 3 \ldots \tag{5.39b}$$

The second term describes the average reflected field inside the waveguide for a first reflection from the wall at $z = 0$:

$$\langle \Pi_{zn}^- \rangle = \frac{e^{ikr_n'}}{r_n'} \chi^n R^n \prod(\alpha, \beta) e^{iK((n-1)a + z + h)} \times K^{n-2} \left(\frac{2ay}{(n-1)a + z + h} \right), n = 2m + 1 \tag{5.40a}$$

$$\langle \Pi_{zn}^- \rangle = \frac{e^{ikr_n'}}{r_n'} \chi^n R^n \prod(\alpha, \beta) e^{iK(na - z + h)} \times K^{n-2} \left(\frac{2ay}{na - z + h} \right), n = 2m, m = 1, 2, 3 \ldots \tag{5.40b}$$

Here, $R = (K - kZ_{EM})/(K + kZ_{EM})$ is the coefficient of reflections from the impedance walls. As was shown in Reference 81, in (5.39) and (5.40), we can assume that the correlation functions $K^{n-2}(w) \approx 1$.

Using this fact and after some straightforward calculations, according to a specific procedure of integration of (5.40) [81] over the pole points determined by

$$\rho_n = (k^2 - K_n^2)^{\frac{1}{2}}; K_n = \pm \frac{\pi n}{a} + i \frac{\ln|\chi|}{a} - \frac{\varphi_n}{a} = \operatorname{Re} K_n + \operatorname{Im} K_n, n = 1, 2, 3 \ldots, \tag{5.41}$$

we finally obtain for the *discrete spectrum* for the case $z > h$

$$\Pi_z(x,y,z) = \frac{D}{2a} \frac{H_0^{(1)}(\rho_n r)}{\left[R_n - 2\dfrac{kZ_{EM}}{(K_n + kZ_{EM})^2} \right]} \times$$

$$\times \left\{ 2R_n \cos[K_n(z-h)] + \frac{1}{\chi}\left[e^{-iK_n(z+h)} + e^{iK_n(z+h-2a)} \right] \right\} \tag{5.42a}$$

and for the case $z < h$

$$\Pi_z(x,y,z) = \frac{D}{2a} \frac{H_0^{(1)}(\rho_n r)}{\left[R_n - 2\dfrac{kZ_{EM}}{(K_n + kZ_{EM})} \right]} \times$$

$$\times \left\{ R_n e^{iK_n(z-h-a)} + e^{-iK_n(z+h+a)} + e^{iK_n(z+h-3a)} + \frac{1}{R_n} e^{-iK_n(z-h+3a)} \right\}. \tag{5.42b}$$

Each index n in the pole points ρ_n from (5.41) corresponds to a waveguide mode of an average reflected field n-time reflected from the street waveguide.

The coefficient of reflection of normal modes in the impedance ($Z_{EM} \neq 0$) multislit waveguide, $R_n = K_n - kZ_{EM}/K_n + kZ_{EM}$, can be described by using the presentation of its phase, φ_n, and its magnitude, $|R_n|$:

$$|R_n| = \frac{\sqrt{\left[(\operatorname{Re} K_n)^2 + (\operatorname{Im} K_n)^2 - (kZ_{EM})^2 \right]^2 + 4(\operatorname{Im} K_n) Z_{EM}^2}}{(\operatorname{Re} K_n + kZ_{EM})^2 + (\operatorname{Im} K_n)^2} \tag{5.43a}$$

$$\varphi_n = \tan^{-1} \frac{2\operatorname{Im} K_n kZ_{EM}}{(\operatorname{Re} K_n)^2 + (\operatorname{Im} K_n)^2 - (kZ_{EM})^2}. \tag{5.43b}$$

It is easy to show that for $r/a \gg 1$ (i.e., far from the transmitting antenna), this discrete waveguide mode spectrum can be significantly simplified. Thus, for the case $z > h$, we obtain [81]:

$$\Pi^d \approx \frac{C}{\sqrt{r}} \exp(i\rho_n^{(0)} r) \exp\left[-\frac{|\ln \chi| R_n|}{\rho_n^{(0)} a}\left(\frac{\pi n - \varphi_n}{a} \right) r \right], \tag{5.44}$$

where $\rho_n^{(0)} = \sqrt{k^2 - (n\pi/a)^2}$, C is a constant that determines intrinsic parameters of the electric dipole as a source of radiation.

For the case of an ideal conductive multislit waveguide model, when $Z_{EM} = 0$ and $|R_n| = 1$, $\varphi_n = 0, 180°$, we obtain from (5.42a) and (5.44), respectively, for $z > h$:

$$\Pi_z(x,y,z) = \frac{D}{2a} H_0^{(1)}\left(\sqrt{k^2 - K_n^2}\, r \right) \times$$

$$\times \left\{ 2\cos[K_n(z-h)]\frac{1}{\chi}\left[e^{-iK_n(z+h)} + e^{iK_n(z+h-2a)} \right] \right\} \tag{5.45}$$

and

$$\Pi^d \approx \frac{C}{\sqrt{r}} \exp\left(i\rho_n^{(0)}r\right) \exp\left[-\frac{|\ln \chi|}{\rho_n^{(0)}a}\left(\frac{\pi n}{a}\right)r\right]. \qquad (5.46)$$

In both waveguides, impedance and ideal conductive, the normal modes of the discrete spectrum attenuate according to (5.42a) and (5.44), (5.45), and (5.46) exponentially inside the discontinuous multislit waveguide, as cylindrical waves $\sim 1/\sqrt{r}$.

The extinction lengths of the normal modes decay are given by the following formula:

$$\zeta_n = \frac{\rho_n^{(0)}a}{\left(\dfrac{\pi n - \varphi_n}{a}\right)\ln\|\chi \mid R_n\|}. \qquad (5.47)$$

The extinction lengths depend on the number of reflections "n," on the waveguide (street) width "a," on the parameter of discontinuity χ, and on the parameter of wall's surface dielectric properties (defined by R_n). In the case of impedance waveguide ($|R_n| \neq 1$), the character of the reflected mode attenuation depends on the real values of the electrical impedance Z_{EM}. With increasing Z_{EM} ($Z_{EM} > 0$), the extinction lengths become smaller, and the normal waves in the impedance multislit waveguide attenuate faster than in the case of the ideal conductive multislit waveguide. The same tendency is observed with an increase in the number of reflections n: the normal reflected modes in multislit waveguide with numbers $n \geq 5$ attenuate very quickly, and the corresponding extinction length ζ_n decreases. On the other hand, an increase in the slit density χ (i.e., decrease of gaps between buildings) leads to a decrease of the reflected wave attenuation factor. In the limit of a continuous waveguide ($\chi = 1$), the normal waves with numbers $n < 5$ (called the "main reflected modes" [14]) also propagate without appreciable attenuation at large distances the effect, which depends on the wall's dielectric properties (i.e., on the parameter Z_{EM}).

The *continuous spectrum* of the total field can be presented for $r/a \gg 1$ as [81]:

$$\Pi^c \approx \sqrt{2}De^{i\frac{3\pi}{4}}\frac{1-\chi|R_n|}{1+\chi|R_n|}\frac{e^{ikr}}{4\pi r}. \qquad (5.48)$$

As can be seen from Equation (5.48) in the discontinuous street waveguide with impedance walls, the continuous part of the total field propagates as a spherical wave $\sim e^{ikr}/r$ and reduces to the continuous waveguide case in the limit $\chi = 1$. But, if in the ideal conductive discontinuous waveguide, when $|R_n| = 1$ and $Z_{EM} = 0$, $\Pi^c = 0$ for the large distances ($r \gg a$) [14], in the impedance ideal waveguide with continuous walls the Π^c of the total field does not vanish because for the case $\chi = 1$ and $|R_n| \neq 1$, Π^c, as can be seen from Equation (5.48) differs from zero. This is a new principal result which is absent in the case of an ideal conductive waveguide with continuous walls (i.e., continuous) [14].

Path Loss along the Straight Streets. Following the previously constructed Equation (5.42) and Equation (5.46) for $r \gg a$, we can, through the definition of the average field intensity inside the impedance multislit waveguide, $\langle I \rangle = \langle (\Pi^d + \Pi^c)\cdot(\Pi^d + \Pi^c)^* \rangle$, and the relation between the path loss and intensity, $L = 10\log\langle I \rangle$, finally obtain the expression for the path loss of radio-wave intensity:

$$L \approx 32.1 + 20\log_{10} f_0 - 20\log_{10}\left[\frac{(1-\chi|R_n|)^2}{(1+\chi|R_n|)^2}\right] + 17.8\log_{10} r - 40\log_{10}|\Gamma_g| +$$

$$+ 8.6\left\{|\ln\chi|R_n|\left|\frac{\pi n - \varphi_n}{a}\right|\frac{r}{\rho_n^{(0)}a}\right\}.$$

(5.49a)

The expression of the ground reflection coefficient Γg is presented by (4.40) in Chapter 4. For a perfectly conductive discontinuous multislit street waveguide, $Z_{EM} = 0$, $|R_n| = 1$, $\varphi_n = 0$ and

$$L \approx 32.1 + 20\log_{10} f_0 - 20\log_{10}\left[\frac{(1-\chi)^2}{(1+\chi)^2}\right] + 17.8\log_{10} r + 8.6\left\{|\ln\chi|\left[\frac{\pi n}{a}\right]\frac{r}{\rho_n^{(0)}a}\right\}.$$

(5.49b)

Like classical waveguides, most of the energy is conveyed by the first mode. Hence, taking $n = 1$ gives an accurate estimation of the resulting path loss along the street in LOS conditions. Using the two-ray model (see Section 5.3), we can also obtain from the proposed waveguide model the break point r_b, which determines the attenuation of the path loss with distance as r^{-2}, for $r < r_b$, and varying as r^{-4}, for $r > r_b$ [69–71]. The formula for the break point not only covers both these cases, but also shows the changes of position of break point r_b, which depends on the geometry of the streets and their structures, is given by

$$r_b = \frac{4h_T h_R}{\lambda}\frac{[(1+\chi|R_n|)/(1-\chi|R_n|)]\cdot[1-h_b/a+h_T h_R/a^2]}{|R_n|^2}.$$

(5.50)

In all above formulas we assumed that the absolute values of diffraction coefficients Dmn from the buildings' corners are close to unity. Analyzing Equation (5.50), one can see that for *wide avenues*, when $a > h_b > h_T, h_R$, and $\chi \to 0$, for $|R_n| \approx 1$, the break point is $r_b = 4h_T h_R/\lambda$, that is, the same formula obtained from the two-ray model. Beyond the break point, the field intensity attenuates exponentially [69–71]. This law of attenuation, obtained experimentally, states that the attenuation mode of field intensity beyond the break point is $\sim r^{-q}$, $q = 5 - 7$. This result does not follow from the two-ray model but can be explained using the waveguide model.

In the case of *narrow streets*, when $a < h_T$, $h_R < h_b$, and $\chi \rightarrow 0$, the range of the break point tends to infinity for the observed wavelength bandwidth $\lambda = 00.1 - 0.3$ m used in wireless communications [79–81]. So, in the case of narrow streets, the two-ray model cannot describe the absence of the break point and of two-mode field intensity decay. In the case of narrow streets, the multislit waveguide model predicts the exponential attenuation of the total field at the street level and compares well with numerous experimental data obtained in microcellular propagation channels (see detailed discussions in References 79–81). Hence, the waveguide model is more general than the two-ray model and covers all situations occurring in a street scenario.

Radio Propagation along and across the Street-Grid Waveguides. From the proposed multislit waveguide model presented above, we can analyze qualitatively the distribution of the antenna pattern located within a street with gaps (slits) between buildings, as nontransparent screens, lining this street. Figure 5.8a is a simple sketch that indicates the way in which the radiowave field strength for the omnidirectional antenna (with pattern angle in the horizontal plane $\varphi = \pi$) may vary because of channeling street orientation. As it is clearly seen, for a continuous and perfectly conductive waveguide [69, 70], the antenna pattern is not changed when the wave travels along the street waveguide (see Fig. 5.8a, top panel).

In the case of a real street with randomly distributed buildings (screens) and gaps (slits), the angle φ is smaller than π because there exists losses of wave energy through the slits (see Fig. 5.8a, bottom panel). In the case of a multislit street waveguide with the impedance building walls, the angle φ can be derived from the results described earlier according to the 2-D waveguide model presented in Figure 5.7b. Thus, the angle φ inside the multislit waveguide at the half level of field intensity can be presented for $y/a \gg 1$ as

$$\varphi = \tan^{-1} \frac{a \cdot \log 2}{y \cdot \|\log \chi \, | \, R_n \, \|}. \tag{5.51}$$

Here, y is the distance from the transmitting antenna along the waveguide; all other parameters are defined above. From (5.51), it follows that in the case of $y/a \gg 1$ and $\chi \approx 0 (L \ll l)$, we get propagation above a flat terrain described previously and $\varphi \approx 0$. For the case $\chi \approx 1 \ (L \gg l)$, we get $\varphi \approx \pi$; that is, we limit ourselves in this case to an ideal continuous waveguide depicted at the top segment in Figure 5.8a.

Let us ask a question: what will be new in a situation of rectangular-crossed street plan of a city and why is this question so important for prediction of signal intensity distribution along the radial and crossing side streets? Numerous

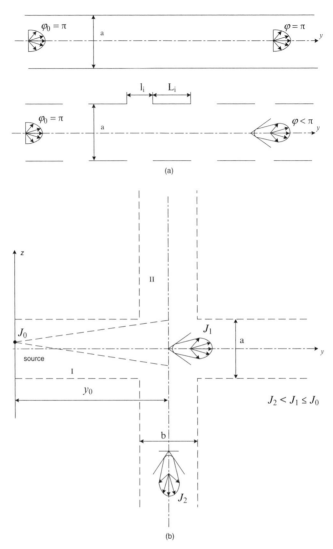

FIGURE 5.8. (a) The simplified scheme of a continuous street waveguide with omm-antenua inside (top view), and of a broken waveguide (bottom view), (b) The simplified scheme of the rectangular crossing-street waveguides.

measurements carried out in cities with regularly distributed crossing-street pattern have shown that the redistribution of the signal strength is very complicated, showing the channeling phenomenon both along radial and side streets that is very difficult to predict using a strict theoretical treatment [82–95].

Thus, measurements described in References 82 and 83 and carried out in New York City indicated that in an urban environment with regularly distributed intersecting buildings, as screens, it has been observed that buildings lining the streets work as waveguides, affecting the propagation direction of the radio waves. Moreover, it has been found that the subscribers at the street level moving radially from the BS, or on the streets parallel to these, may receive a signal 10–20 dB higher than that received when moving on the perpendicular streets. It was found that this effect is more significant in the microcell areas—up to 1–2 km away from the BS, becoming negligible at distances above 5–10 km, that is, in the macrocell areas [83–85]. As was mentioned in References 84 and 87, the existing simple approach, dealing with the relative field strength presentation by a density of arrows along the various streets, can only indicate the way in which field strength may vary in an urban area following to streets' orientation but cannot predict the real field strength distribution between streets and their intersections. Using the above approach, the total path loss at the crossing-street level is a simple arithmetic summation of path loss at the radial street where the BS is located, and of path loss from the intersection to the side street, that is, $L_{total} = L_{radial} + L_{side}$ [86].

At the same time, experiments carried out in the crossing-street areas of Central London at 900 MHz and 1.7–1.8 GHz have shown a complicated 2-D shape of microcell coverage, similar to a Christmas tree with the BS near the "foot of the tree" [88, 89]. Such complicated redistribution of field energy among the rectangular crossing streets cannot be understood using a simple geometric optic model, even taking into account diffraction from the building corners [91–95].

In References 79–81 and 96, it was shown that for the case of both terminal antennas that are below the rooftop level, propagation of waves among the "Manhattan-shaped" crossing-street grid can be successfully described by a 2-D crossing-waveguides model, which is based on results obtained from the 2-D multislit waveguide model presented previously. Such scenario can be presented as a portrait of the two rectangular-crossed discontinuous street waveguides sketched in Figure 5.8b. In this case, the redistribution of field intensity from the transmitting antenna depends on the relation between the antenna pattern angle φ and angle φ_0 which defines the area of observation of the antenna as the intersection of waveguides (see Fig. 5.8b). For $y_0 >> a$, we can estimate this angle as $\varphi_0 = a/y_0$. If the antenna pattern angle $\varphi < \varphi_0$, the effect of the street discontinuity is not significant, and only a small part of the source energy penetrates the side strcct waveguide (noted by II in Fig. 5.8b). This means that the intensity loss is small and at the distance y from the source inside the radial waveguide (noted by I in Fig. 5.8b), the wave intensity J is simply $J \approx J_0$. In the case when $\varphi \approx \varphi_0$, some field energy from the transmitting antenna penetrates into the side street waveguide II from the main radial street I. The amount of energy loss depends on the parameter of discontinuity in the main radial waveguide and on the distance from the source. In this case, the intensity in the waveguide I equals $J \approx J_0 \tan[\varphi(\chi, y_0)] < J_0$, where $\varphi(\chi, y_0)$ is presented by (5.51) with $y = y_0$. For the case $\varphi > \varphi_0$, the waveguide modes, propagating along the radial waveguide I, easily penetrate into the side waveguide II. The field intensity loss now depends on the gap distribution between buildings lining both waveguides, the main and the side, and equals $J \approx J_0 \tan[\varphi(\chi_1, y_0)] \tan[\varphi(\chi_2, |z|)] \ll J_0$, where $\varphi(\chi_2, |z|)$ is presented by (5.51) too, but for the side waveguide by replacing y with $|z|$.

As was shown in References 79 and 80, the low-order wave modes in the radial waveguide I generate high-order wave modes ion the side waveguide II and, conversely, high-order wave modes in a radial waveguide generate low-order modes in a side waveguide. This is why the total field intensity in the main waveguide cannot be simply divided into two equal parts at the intersection of two rectangular crossing waveguides, as was done in References 94 and 95. The redistribution of wave energy near each intersection inside the rectangular grid of crossing streets depends on the processes inside each waveguide and on the parameters of discontinuity, that is, on building distribution in each radial and rectangular crossing street inside the grid.

5.7.2. Propagation above Urban Irregular Terrain

In Sections 5.5 to 5.7, we dealt with propagation models, which describe radio propagation above the irregular terrain, typical for rural environments containing obstructions such as hills, mountains, and trees. Some of these models adequately describe the situation in the urban environment, mostly in the suburban areas, where effects of foliage, usually negligible in city centers, can be quite important. At the same time, the effects of trees are similar to those of buildings, introducing additional path losses and producing spatial signal variations.

In Section 5.7.1, we considered the case when both communicating antennas were located in LOS conditions, but assumed that streets and buildings were uniformly distributed on a flat terrain. Now we will consider the situation where the buildings are randomly distributed over an irregular terrain, as is the main case of a city topography, and will present the 3-D stochastic multiparametric model based on the same approach proposed for forested and mixed residential areas.

Statistical Description of Urban Terrain. Let us consider an array of buildings randomly distributed in a terrain. Using the approach in References 57–59, the coordinate system $\{x, y, z\}$ is placed at the plane $z = 0$ on the ground surface. The heights of the rough ground surface are described by the generalized function $Z(x, y)$ according to Shwartz [57–59] (see Fig. 5.9). The shadow function $Z(\mathbf{r}_2, \mathbf{r}_1)$, presented in integrals (5.17) and (5.24) for forested ad mixed areas, will also be used for the urban environments. However, the situation in built-up areas is more complex as we must also take into account the buildings' overlay profile and other specific features of built-up terrain. In this case, the shadow function is a product of different probability functions which we will briefly present later following the approach in References 96 and 97.

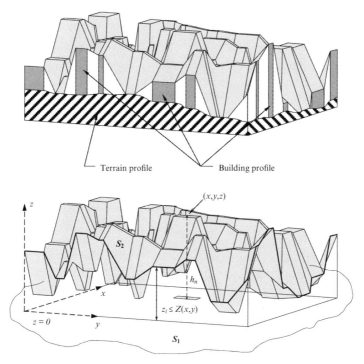

FIGURE 5.9. The nonregular buildings' overlay profile $z = Z(x, y)$.

Probability of LOS between Subscribers. The next formula determines the probability that there is a direct visible link between two arbitrary observers inside the layer of city buildings. Thus, if $<L>$ is the average length of screens (buildings) surrounding the points $A(\mathbf{r_1})$ and $B(\mathbf{r_2})$ (see Fig. 5.10), then the probability that there is no intersection of the line AB with any of the building screens is equal to [57–59]

$$P(\mathbf{r_1}, \mathbf{r_2}) \equiv P_{12} = \exp\{-2 < L > v r_{12} / \pi\} \qquad (5.52)$$

from which we can easily define the one-dimension building density parameter γ_0 (in km^{-1}) as

$$\gamma_0 = 2 < L > v/\pi. \qquad (5.53)$$

Here, v is the density of buildings in the investigated area of 1 km^2. Parameter γ_0 determines the average minimal horizontal distance of LOS, $\bar{\rho}$ as $\bar{\rho} = \gamma_0^{-1}$ [57–59].

Influence of City Building Profile. The probability that the arbitrary subscriber antenna location is inside a built-up layer is described by the complimentary

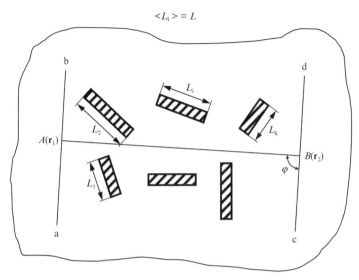

FIGURE 5.10. Line-of-sight conditions between terminal antennas in the urban scene.

cumulative distribution function (CCDF) $P_h(z)$ [CCDF = 1-CDF], which was introduced in References 59 and 97 as the probability that a point z is located below the buildings' roofs level.

$$P_h(z) = \int_z^\infty w(h_n)dh_n. \qquad (5.54)$$

Here, $w(h_n)$ is the probability density function (PDF) which determines the probability that each subscriber antenna, stationary or mobile, with a vertical coordinate z, is located inside the built-up layer; that is, $z < h_n$, where h_n is the height of building with number n (see Fig. 5.9). Let us now consider the influence of the city buildings profile on the average field intensity. Here we use definitions introduced in References 59, 86, and 87 to obtain a more general description of built-up relief functions. Taking into account the fact that the real profiles of urban environment are randomly distributed, as shown in Figure 5.10, we can present, according to References 57, 59, 96, and 97, CCDF defined in the following form:

$$P_h(z) = H(h_1 - z) + H(z - h_1)H(h_2 - z)\left[\frac{(h_2 - z)}{(h_2 - h_1)}\right]^n, \quad n > 0, 0 < z < h_2, \qquad (5.55)$$

where the function $H(x)$ is the Heaviside step function, which equals 1 for $x > 0$, and 0 for $x < 0$. Using this, we can now introduce the built-up layer

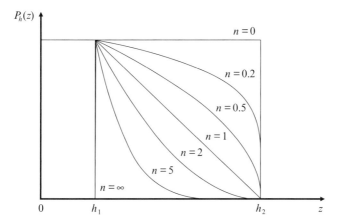

FIGURE 5.11. Buildings height distribution functions P_h, (z) versus the current height z for various parameters n of built-up profile.

profile "between the two terminal antennas" that is described by the following function [97]:

$$F(z_1, z_2) = \int_{z_1}^{z_2} P_h(z)dz. \tag{5.56}$$

To understand the influence of a built-up area relief on the signal intensity, let us first examine the height distribution function $P_h(z)$. The graph of this function versus height z of a built-up overlay is presented in Figure 5.11. For $n \gg 1$ $P_h(z)$ describes the case where the buildings are higher than h_1 (this is a very rare case as most buildings are at the level of a minimal height h_1). The case when all buildings have heights close to h_2 (i.e., most buildings are tall) is given by $n \ll 1$. For n close to zero, or n approaching infinity, most buildings have approximately the same level which equal h_2 or h_1, respectively. For $n = 1$, we have the case of building heights uniformly distributed in the range h_1 to h_2.

The same result follows from an analysis of the built-up layer profile $F(z_1, z_2)$. For the case when the minimum antenna height is above the rooftop level, that is, $z_2 > h_2 > h_1$, then, according to Reference 97,

$$F(z_1, z_2) = H(h_1 - z_1)\left[(h_1 - z_1) + \frac{(h_2 - h_1)}{(n+1)}\right]$$
$$+ H(z_1 - h_1)H(h_2 - z_1)\frac{(h_2 - z_1)^{n+1}}{(n+1)(h_2 - h_1)^n}, \tag{5.57a}$$

and for the case the minimum antenna height is below the rooftop level, that is, $z_2 < h_2$, we derived [97]

$$F(z_1, z_2) = H(h_1 - z_1) \left[(h_1 - z_1) + \frac{(h_2 - h_1)^{n+1} - (h_2 - z_2)^{n+1}}{(n+1)(h_2 - h_1)^n} \right]$$
$$+ H(z_1 - h_1) H(h_2 - z_1) \frac{(h_2 - h_1)^{n+1} - (h_2 - z_2)^{n+1}}{(n+1)(h_2 - h_1)^n}. \tag{5.57b}$$

From Equations (5.57a) and (5.57b) we can determine the average building height as

$$\bar{h} = h_2 - n(h_2 - h_1)/(n+1), \tag{5.58}$$

which reduces to

$$\bar{h} = (h_1 + h_2)/2 \tag{5.59}$$

for the case $n = 1$ of a uniformly distribution profile investigated in References 57 and 58.

As there are many geometrical factors in the built-up layer profile, the antenna heights z_1 and z_2, the minimum and maximum building heights h_1 and h_2, and the building relief that appear in Equation (5.57a) and Equation (5.57b), we consider their effects on function $F(z_1, z_2)$ separately.

In Figure 5.12, $F(h_T, h_R, n) = F(z_1, z_2)$ given by expressions (5.57a) or (5.57b), for $z_2 > h_2 > h_1$ or $z_2 < h_2$, respectively, is depicted as a family of curves versus the receiver antenna height [97]. The discrete parameters are denoted by ",". The parameters are the transmitter antenna height, ranging between 10 m (bottom curve) and 100 m (top curve), and n. The minimum and maximum heights of the buildings' overlay profile are indicated by dotted vertical lines. We have chosen $n = 1$ which corresponds to a uniform distribution of building heights. By inspection of the displayed curves, it is obvious that for a constant transmitter antenna height, as the receiver antenna height increases, the value of $F(z_1, z_2)$ becomes smaller, and the effect of the building layer on the path loss is reduced. Thus, for a transmitter antenna height $h_T = 40$ m, as the receiver antenna height h_R increases from 40 m (at the bottom level of rooftops) to 50 m (some intermediate building height value), $F(h_R)$ decreases sharply from 15 m to 0. A more gradual decrease of $F(h_R)$ is evidenced for higher transmitter antennas, (e.g., see curve for $h_T = 100$ m).

In Figure 5.13, we examined the role of the parameter n on $F(h_T, h_R, n)$ at a constant transmitter antenna height $h_T = 100$ m. The values chosen were $n = 0.1, 1, 10$ that describe predominantly tall buildings, uniformly distributed heights, and predominantly low building heights, respectively. This provides a transition of the built-up area from that of a typically residential area with predominantly small buildings (the bottom curve in Fig. 5.13 corresponding to $n = 10$), to that of a dense city center with predominantly tall buildings (the top curve in Fig. 5.13 corresponding to $n = 0.1$).

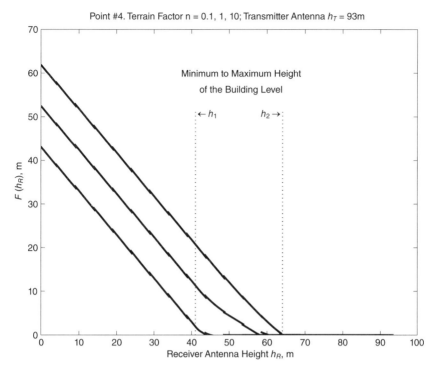

FIGURE 5.12. Distribution of $F(h_R)$ versus the receiver antenna height h_R for various heights of the transmitter antenna: $h_T = 10$. 40, and 100 m. and h_1 and h_2 are the minimum and maximum of built-up profile.

We therefore can state that the proposed method of characterizing the terrain and its associated building overlay provide good information regarding the nature of the profiles *vis-à-vis* the pertinent evaluation of terminal antennas, the transmitter, and the receiver.

Dimensions of the Reflected Surface Sections. Let us consider the case when LOS visibility exists between two points $\mathbf{r}_1$ and $\mathbf{r}_2$ (Fig. 5.10). Let us now determine the probability that given a point $A(\mathbf{r}_1)$, the horizontal segment inside the building (as a nontransparent screen) can be observed (see Fig. 5.14). If a horizontal segment with length l could be seen from point $\mathbf{r}_1$, a vertical segment with width l can be seen from this point as well. The vertical screen forms an angle Ψ with line AB. After some straightforward calculations, we can obtain the probability of direct visibility of segment cd with the length l, from point A at a range r_{12} as [57, 58]:

$$P_{cd} = \exp(-\gamma_0 \gamma_{12} r_{12} - \nu \varepsilon_{12} r_{12} | \sin \Psi |), \qquad (5.60)$$

where

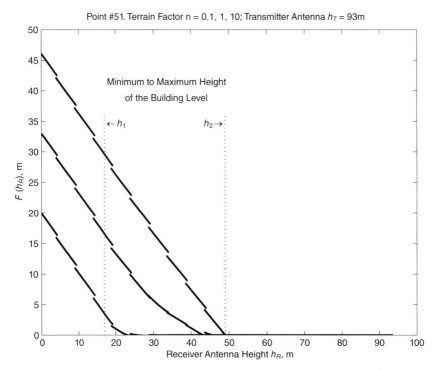

FIGURE 5.13. Distribution of $F(h_R)$ versus the receiver antenna height h_R (for a transmitter antenna $h_T = 93$ m) for various parameters $n = 0.1, 1, 10$.

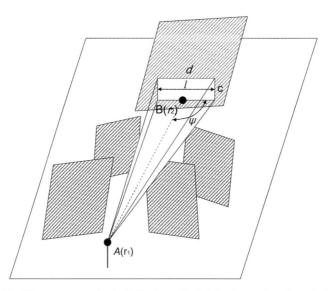

FIGURE 5.14. The screen at the building's wall with horizontal and vertical segments of length l. illuminated under the angle ψ by the source located at the point $A(r_1)$.

$$\varepsilon_{12} = (z_2 - z_1)^{-1} \int_{z_1}^{z_2} (z - z_1)(z_2 - z_1)^{-1} P_h(z)[1 - XP_h(z)]^{-1} dz. \qquad (5.61)$$

Here, the multiplier X in the integrand of (5.77) determines the probability of the event when the projection of the point $\mathbf{r}(x, y, z)$ on the plane $z = 0$ hits inside an arbitrary building (as shown in Fig. 5.14). When $X = 1$ and $z > z_1, z_2$, Equation (5.61) becomes

$$\varepsilon_{12} = (z_2 - z_1)^{-1} \int_{z_1}^{z_2} P_h(z)(z - z_1)(z_2 - z_1)^{-1} dz. \qquad (5.62)$$

The Spatial Distribution of Scattering Points. The role of the single-scattering case is very important when one of the antennas (mainly, the BS antenna) is above the roof level, and the other one is below it. This case is presented in Figure 5.15, where the scattered point C is inside the building contour with the average height $\bar{h}$. The building orientation is determined by the angle Ψ_s to the vector $(\mathbf{r}_s - \mathbf{r}_1)$. The receiver (or transmitter) is placed at point $B(\mathbf{r}_2)$, and the transmitter (or receiver) is placed at point $A(\mathbf{r}_1)$. If we now introduce the polar

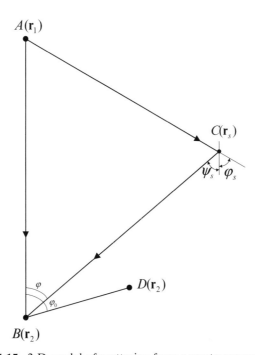

FIGURE 5.15. 2-D model of scattering from a nontransparent screen.

coordinate system (r, φ) with point B as a base point on the plane $z = 0$ (see Fig. 5.15), then for the discrete distributed scatterers (i.e. scattered points), the projection of the density of scattered points distribution at the plane $z = 0$ (i.e., their 1-D density) can be presented as follows [57–59]:

(a) for $z_1, z_2 < \bar{h}$

$$\tilde{\mu}(r, \varphi) = \frac{\nu \cdot \gamma_0}{2} \sin^2(\alpha/2) \cdot r \cdot (r + \tilde{r}) \exp\{-\gamma_0(r + \tilde{r})\} [\text{km}^{-1}] \qquad (5.63)$$

(b) for $z_1 < \bar{h}, z_2 < \bar{h}$

$$\tilde{\mu}(r, \varphi) = \left(\frac{\nu \gamma_0 r}{2\bar{h}}\right) \sin^2\left(\frac{\alpha}{2}\right) \exp\{-\gamma_0(r + \tilde{r})\} \int_0^{\bar{h}} \left[(r + \tilde{r}) - r\left((z_2 - \bar{h})/(z_2 - z)\right)^2\right] \times$$

$$\times \exp\left[\gamma_0 r\left(\frac{z_2 - \bar{h}}{z_2 - z}\right)\right] dz \quad [\text{km}^{-1}]$$

$$(5.64a)$$

or after integration over the built-up layer with height $\bar{h}$, we get finally for $\gamma_0 r \gg 1$

$$\tilde{\mu}(r, \varphi) = \frac{\nu}{2} \cdot \sin^2\left(\frac{\alpha}{2}\right) \left\{\frac{\gamma_0 \bar{h} r(r + \tilde{r})}{z_2} \exp\left[-\gamma_0\left(\tilde{r} + \frac{\bar{h}r}{z_2}\right)\right]\right\} +$$

$$+ \frac{\nu}{2} \cdot \sin^2\left(\frac{\alpha}{2}\right) \left\{\frac{(z_2 - \bar{h})\tilde{r}}{\bar{h}} \exp[-\gamma_0 r]\right\} = \tilde{\mu}_1(r, \varphi) + \tilde{\mu}_2(r, \varphi) \quad [\text{km}^{-1}]$$

$$(5.64b)$$

where $\tilde{r} = (d^2 + r^2 - 2rd \cos\varphi)^{1/2}$; and $\bar{h}$ is the average building height. Comparing Equation (5.63) and Equation (5.64b), one can see that the first summand in (5.64b), $\tilde{\mu}_1(r, \varphi)$, is the same as that described in expression (5.63) for the case of $z_2 = \bar{h}$. Both of these expressions describe *rare scatterers* which are distributed over a large area of a city, far from the receiver. The addition of significant changes in the scatterer distribution, for the case $z_2 > h$, yields the second term in (5.64b), given by $\tilde{\mu}_2(r, \varphi)$. For $z_2 = \bar{h}$, its value is zero, but even some small increase of z_2 above $\bar{h}$ (i.e., when $z_2 > \bar{h}$), yields a significant influence on the total scatterer distribution according to (5.64b). It describes the "illumination" of a small area near the upper boundary of a building layer, in the $\bar{P}$-region of a moving transmitter.

The Distribution of Reflected Points. In built-up areas, reflections are the most interesting single-scattering events described by geometrical optics. We can present the probability of distribution of the reflection points within a building layer as [57, 58]

$$\mu(\tau, \varphi) = \frac{\nu\gamma_0 d^3}{4} \frac{(\tau^2 - 1)}{(\tau - \cos\varphi)} P_h(z_c) \exp\{-\gamma_0\gamma_{12}\tau d\}, \qquad (5.65)$$

where $P_h(z_c)$ is described by Equation (5.54) with variable

$$z_c = z_2 - \frac{(\tau^2 - 1)}{2(\tau - \cos\varphi)} \frac{(z_2 - z_1)}{\tau}. \qquad (5.66)$$

Here, τ ($\tau = (r + \tilde{r})/d$) is the relative time of single-scattered waves propagating from the transmitter to the receiver through the built-up region using the function presentation (5.64b). We also assume that the height of point B (receiver) is higher than that of point A (transmitter) (i.e., $z_2 > z_1$), and is higher than the average building height $\bar{h}$, that is, $z_2 > \bar{h}$. The contribution of each level in the building layer, described by (5.65), is different from zero only for those values of τ and φ for which the coordinate z_c lies inside the building layer (i.e., $0 < z_c < h$).

To analyze the distribution of reflection points $\mu(\tau, \varphi)$ in the angle-of-arrival (AOA) domain and in the time-of-arrival (TOA) domain, assuming a uniform building layer, we first integrate (6.65) over τ and then over φ. This also assumes that within this layer, the distribution of building heights is uniform (i.e., $h_i = \bar{h} = $ constant). Next, we introduce the nondimensional parameter $\varsigma = (z_2 - \bar{h})/(z_2 - z_1)$, which describes effects of the difference between the terminal antennas compared with that for BS antenna with respect to average building height.

Let us now examine qualitatively how the distribution of these reflection points is changed at the plane (x, y) (at the real terrain surface). We construct the regions G at which approximately 90% of reflected points is located. The boundaries of such a region consist of the arcs of ellipses with $\tau = \tau_{0.9}$, where $\tau_{0.9}$ is determined from the following relation:

$$\int_1^{\tau_{0.9}} d\tau \int_0^{2\pi} d\varphi \mu(\tau, \varphi) = 0.9 \int_1^{\infty} d\tau \int_0^{2\pi} d\varphi \mu(\tau, \varphi), \qquad (5.67)$$

and from the arcs of circles, the equation of which is

$$\tau^2 - 1 = 2\varsigma\tau(\tau - \cos\varphi). \qquad (5.68)$$

Equation (5.68) can be presented by using nondimensional coordinates $\xi = x/d$ and $\eta = y/d$ in the following form:

$$\left(\xi - \frac{\varsigma^2}{2\varsigma - 1}\right)^2 + \eta^2 = \left[\frac{\varsigma(1 - \varsigma)}{2\varsigma - 1}\right], z_2 \geq h > z_1. \qquad (5.69)$$

From Figure 5.16, we can see how the region G and its boundaries are changed with changes in the height factor ς from 0 to 1. In Figure 5.16, the region G and its boundaries (arcs of ellipses) are presented by the dotted curves, and the arcs of circles are presented by the continuous curves. These curves were constructed for the ranges between terminals $d \approx 500–600$ m (between terminals), which is close to conditions of the most experiments carried out in built-up areas (see Chapter 11).

Estimations show that the region G is limited by a single ellipse with two foci, A and B, for $z_2 = h$ (i.e., $\varsigma \approx 0$). The distribution of the reflection points is maximum near these points (see Fig. 5.16). The distribution of the reflection points does not equal zero at the segment $[AB]$, because there are some intersections of the segment with one of the arbitrary buildings (screens) that crosses the path AB (see Fig. 5.10). When the height of point B increases with respect to the rooftop height (i.e., when $z_2 > h$), the region G, where these reflections are observed, is formed mainly around the transmitting point A. That means there is no reflection in the neighborhood of the receiving point B. Also, for $\varsigma = 0.2 (\bar{h} = 20$ m, $z_2 = 25$ m$)$, the region that is "prohibited" for reflections has the shape of a circle, the center and the boundary of which are

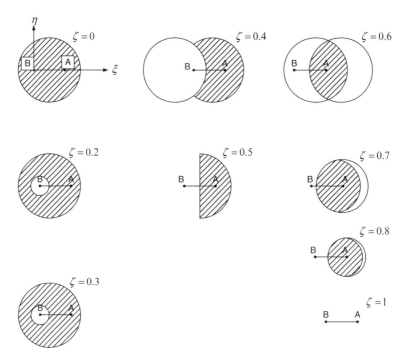

FIGURE 5.16. The region G and its boundaries change with changes in height factor ς from 0 to 1.

determined by (5.68). Moreover, an increase in the height factor ς $(z_2 > \bar{h})$ spreads this region (for $\varsigma = 0.4, h = 20$ m, $z_2 = 33$ m) to occupy the entire left half-plane (for $\varsigma = 0.5, h = 20$ m, $z_2 = 40$ m). Any further increase in ς $(\varsigma \rightarrow 1)$ limits the reflections to the neighborhood of point A (see in Fig. 5.16 the circles and arcs for $\varsigma = 0.6, z_2 = 50$ m and $\varsigma = 0.8, z_2 = 100$ m).

Effects of Multiple Scattering form Obstructions. To analyze the multiple scattering phenomena caused by the buildings (e.g., nontransparent screens), we assume, as in References 57, 96, and 97, that the distribution of all obstructions placed above the rough terrain is satisfied by Poisson's distribution law. Consequently, the probability of the event for at least one ray being received after n-time scattering from the randomly distributed screens is

$$P_n = 1 - \exp\{-<N_n(r)>\}. \tag{5.70}$$

Here, the average amount of n-time scattered rays from the screens can be obtained from the probability of the scattered points distribution $\mu_n(\mathbf{r}_0|\mathbf{r}_1, \mathbf{r}_2, \dots, \mathbf{r}_n)$:

$$\langle N_n(\mathbf{r}, \mathbf{r}_0)\rangle = \int \cdots \int \mu_1(\mathbf{r}_0|\mathbf{r}_1, \mathbf{r}_2, \dots, \mathbf{r}_n) \dots \mu_n(\mathbf{r}_0|\mathbf{r}_1, \mathbf{r}_2, \dots, \mathbf{r}_n) d\mathbf{r}_n d\mathbf{r}_{n-1} \dots d\mathbf{r}_1, \tag{5.71}$$

where

$$\mu_i(\mathbf{r}_0|\mathbf{r}_1, \mathbf{r}_2, \dots, \mathbf{r}_n) = \exp\left\{-\gamma_0 \sum_{i=0}^{n}|\mathbf{r}_{i+1} - \mathbf{r}_i|\right\} \prod_{i=1}^{n} \frac{\gamma_0\nu}{2}\left\{|\mathbf{r}_{i+1} - \mathbf{r}_i| + |\mathbf{r}_i - \mathbf{r}_{i-1}|\sin^2\left(\frac{\alpha_i}{2}\right)\right\}.$$

Here, the angle α_i is an angle between vectors $(\mathbf{r}_{i+1} - \mathbf{r}_i)$ and $(\mathbf{r}_i - \mathbf{r}_{i-1})$ for all $i = 1, 2, \dots, n$; $(\mathbf{r}_0, \mathbf{r}_1, \mathbf{r}_2, \dots, \mathbf{r}_n)$ are the radius-vectors of points $A, C_1, C_2, \dots, C_n, B$, respectively (see Fig. 5.17). The examples of average values of once-, twice- and three-times-scattered rays from the randomly distributed buildings can be presented by using the Macdonald functions $K_n(w)$ of the order $n = 1$; 2; 3, respectively:

$$\langle N_1(\mathbf{r})\rangle = \frac{\pi\nu r^2}{4} K_2(\gamma_0 r) \tag{5.72a}$$

$$\langle N_2(\mathbf{r})\rangle = 9(\pi\nu r^2)^2 \left[\frac{K_4(\gamma_0 r)}{8!} + \left(\frac{\pi\gamma_0 r}{2}\right)^{-1/2} \frac{K_{7/2}(\gamma_0 r)}{7!}\right] \tag{5.72b}$$

$$\langle N_3(\mathbf{r})\rangle = 8(\pi\nu r^2)^3 \left\{(\gamma_0 r)^{-1}\frac{K_5(\gamma_0 r)}{10!} + \left(\frac{\pi\gamma_0 r}{2}\right)^{-1/2} \frac{K_{11/2}(\gamma_0 r)}{11!}\right\}. \tag{5.72c}$$

The probability of occurrence of a single-scattered wave (curve 1), a doubly scattered wave (curve 2), and a three-time scattered wave (curve 3), calculated

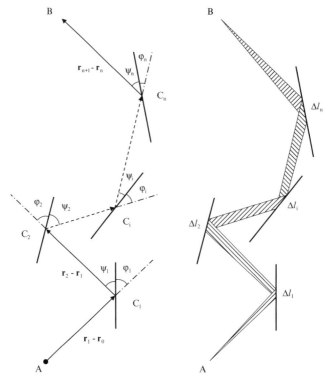

FIGURE 5.17. Geometry of multiple scattering by n randomly distributed buildings, as nontransparent screens.

according to (5.71) and (5.72) and observed at the range of 1–2 km from the source, is presented in Figure 5.18. In microcellular conditions ($r < 1$–2 km), the probability of observing these rays at the receiver for single-to-three-time scattered waves is equal to the unity. For short ranges from the transmitter, only single-scattered waves can be observed. On the other hand, in the far field, the effect of the multiscattering becomes stronger than the single-scattering effect. All of the above-mentioned probability formulas were substituted in the corresponding integral (5.24) instead of the shadow functions $Z_{\alpha\beta}$ for signal field intensity evaluation.

3-D Stochastic Model. The analysis used in Section 5.7 can be adopted to evaluate the average signal intensity distribution in the space domain in an urban communication channel. Here, all functions and parameters that describe statistical properties of the rough built-up terrain and buildings, as scatterers (see Fig. 5.9), will be taken into account to derive Equation (5.24).

There is a difference between the mixed residential areas and the built-up areas with buildings larger than the wavelength and a corresponding correlation scale ℓ_h. It turns out that we can exclude the influence of the reflecting

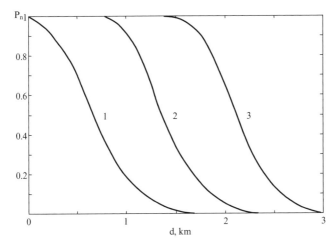

FIGURE 5.18. Probability of single, double, and triple scattering versus distance d between the transmitting and the receiving antennas.

properties of these buildings' walls, in the horizontal directions, and describe the screens' reflecting properties only in the vertical plane (1-D case, analyzed in References 57 and 58 by rewriting (5.23) proportional to $\exp\{-|\xi|/\ell_v\}$). Also, in the case of the built-up terrain, as was mentioned earlier, we need to take into account the building layer profiles according to Equation (5.56) and Equation (5.57). From Figure 5.12 and Figure 5.13, one can see that this factor plays a significant role in signal power decay.

The theory of the average field intensity has been derived for 3-D model in References 96 and 97, for the case of $\gamma_0 r_{12} = \gamma_0 d \gg 1$ and for the quasi-homogeneous built-up profiles. Similar to our treatment of the mixed residential areas, the expression for the incoherent part of the total field intensity can be presented, taking into account single and double diffracted waves shown in Figure 5.19a and b, respectively. Let us briefly examine the influence of diffraction phenomena caused by buildings' rooftops on the field intensity attenuation. To account for this effect, we use the Huygens–Kirchhoff approximation, described earlier. For the derivation of the diffraction field, we introduce, according to References 57 and 96, the surface S_B of virtual sources that is normal to the building layer S and the surface of infinite semi-sphere S_R that contains the source of radiation inside it, as shown in Figure 5.19a. The effect of all virtual sources placed inside the semi-sphere S_R is negligible, because it is limited to zero when the radius of this semi-sphere goes to infinity. Thus, the field at the receiver, presented by Green's theorem (see Chapter 4) can be rewritten as

$$U(\mathbf{r}_2) = 2ik \int_{S_B} \{U(\mathbf{r}_{S_B}) \cdot G(\mathbf{r}_2, \mathbf{r}_{S_B}) \cdot \cos\psi_{S_B}\} dS_B, \tag{5.73}$$

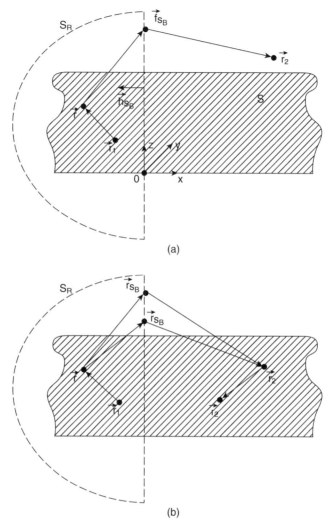

FIGURE 5.19. (a) Geometry of single scattering and diffraction over a built-up layer (b) Geometry of double scattering and diffraction over a built-up layer.

where $U(\mathbf{r}_{S_B})$ is the field at the surface S_B obtained by use of approximation (5.73) for single diffraction (see Fig. 5.19a); $\cos\psi_{S_B} = (\mathbf{n}_{S_B} \cdot (\mathbf{r}_2 - \mathbf{r}_{S_B})/|\mathbf{r}_2 - \mathbf{r}_{S_B}|)$, $\mathbf{n}_{S_B}$ is the unit vector normal to surface. The average intensity of received field $<I(r2)> = \langle U(\mathbf{r}_2) \cdot U^*(\mathbf{r}_2) \rangle$ can be presented according (5.73) as:

$$< I(\mathbf{r}_2) >= 4k^2 \int\limits_{S_B} dS_B \int\limits_{S_B} dS'_B \cdot K(\mathbf{r}_{S_B}, \mathbf{r}'_{S_B}) \cdot G(\mathbf{r}_2, \mathbf{r}'_{S_B}) \cdot G(\mathbf{r}_2, \mathbf{r}_{S_B}) \cos\psi_{S_B} \cos\psi'_{S_B},$$

(5.74)

where $K(\mathbf{r}_{S_B}, \mathbf{r}'_{S_B})$ is the correlation function of the total field at points $\mathbf{r}_{S_B}$ and $\mathbf{r}'_{S_B}$ located at surface S_B when the source is located at the point $\mathbf{r}_1$:

$$K(\mathbf{r}_{S_B}, \mathbf{r}'_2) = 4k^2 < \int_{S_B} dS_B \int_{S_B} dS'_B \cdot Z(\mathbf{r}_2, \mathbf{r}_{S_B}, \mathbf{r}_1) \cdot Z(\mathbf{r}'_2, \mathbf{r}'_{S_B}, \mathbf{r}_1) \cdot \Gamma(\varphi_{S_B}, \mathbf{r}_{S_B}) \times$$

$$\Gamma^*(\varphi'_{S_B}, \mathbf{r}'_{S_B}) \cdot \sin \psi_{S_B} \cdot \sin \psi'_{S_B} \cdot G(\mathbf{r}_2, \mathbf{r}_{S_B}) \cdot G(\mathbf{r}_{S_B}, \mathbf{r}_1) \cdot G^*(\mathbf{r}'_2, \mathbf{r}'_{S_B}) \cdot G^*(\mathbf{r}'_{S_B}, \mathbf{r}_1) > . \tag{5.75}$$

Here, the reflection coefficient $\Gamma(\varphi_S, \mathbf{r}_S) = \Gamma \exp\{-\xi/l_v\}$ and the shadow function $Z(\mathbf{r}_2, \mathbf{r}_1)$ is a superposition of all probability functions defined earlier. By averaging (5.75) over the spatial distribution of the nontransparent screens, over their number, and over the reflection properties of these screens, we obtain the following formula for the single-scattered field (single diffraction from the building rooftops) [96, 97]:

$$\langle I(\mathbf{r}_2) \rangle = \frac{\Gamma \lambda l_v}{8\pi \left[\lambda^2 + (2\pi \ell_v \gamma_0 F(z_1, z_2))^2 \right] d^3} \left[(\lambda d / 4\pi^3) + (z_2 - h)^2 \right]^{1/2}. \tag{5.76}$$

The same result can be obtained for the twice-diffracted waves as shown in Figure 5.19b. Using the same presentation of average intensity of total field, as (5.24), we get

$$< I(\mathbf{r}_2) > = 4k^2 \int_{S_B} dS_B \int_{S_B} dS'_B \cdot K(\mathbf{r}_{S_B}, \mathbf{r}'_{S_B} \mid \mathbf{r}, \mathbf{r}_1) \cdot K(\mathbf{r}_{S_B}, \mathbf{r}'_{S_B} \mid \mathbf{r}', \mathbf{r}_1) \cos \psi_{S_B} \cos \psi, \tag{5.77}$$

where $K(\mathbf{r}_{S_B}, \mathbf{r}'_{S_B} \mid \mathbf{r}, \mathbf{r}_1)$ and $K(\mathbf{r}_{S_B}, \mathbf{r}'_{S_B} \mid \mathbf{r}', \mathbf{r}_1)$ are the correlation function of total field at the surface of virtual sources of diffraction, which is determined by (5.75). Next, by averaging (5.77), over the distribution of the nontransparent screens, over their number, and over the reflection properties of screens, we can derive the formula for the double-diffracted and double-scattered field as:

$$\langle I_{inc}(\mathbf{r}_2) \rangle = \frac{\Gamma^2 \lambda^3 l_v^2}{24\pi^2 \left\{ \lambda^2 + [2\pi l_v \gamma_0 F(z_1, z_2)]^2 \right\}^2 d^3} \left[\frac{\lambda d}{4\pi^3} + (z_2 - h)^2 \right]. \tag{5.78}$$

Using now Equation (5.54) for the distribution function $P_h(z)$, the coherent part of the total field intensity can be obtained as [59, 87]:

$$\langle I_{co}(\mathbf{r}_2) \rangle = \exp \left\{ -\gamma_0 d \frac{F(z_1, z_2)}{(z_2 - z_1)} \right\} \frac{\sin^2 (kz_1 z_2 / d)}{4\pi^2 d^2}. \tag{5.79}$$

The difference between Equation (5.76) and Equation (5.78), and those obtained for the forest and mixed residential areas, presented by (5.20) and (5.29), respectively, is that here we introduce single and double diffraction effects as well as a new relief function $F(z_1, z_2)$, given by Equation (5.57). This relief function is better suited to handle more realistic and general cases of terrain and buildings overlay, as well as for different configurations of transmitter and receiver antennas. Comparison of formulas obtained in References 57–59 and Equation (5.76) and Equation (5.78) shows that in References 57–59, the restricted case $n = 1$ of a uniform distribution profile was assumed, while Equation (5.76) and Equation (5.78) give more latitude in describing more general distributions. Finally, the total average field intensity is written as

$$\langle I_{total} \rangle = \langle I_{inc} \rangle + \langle I_{co} \rangle. \tag{5.80}$$

Hence, the path loss is presented as [76, 77]

$$L_{total} = 10 \log \{ \lambda^2 (\langle I_{inc} \rangle + \langle I_{co} \rangle) \}. \tag{5.81}$$

Finally, the signal average intensity decay obtained in Sections 5.6 and 5.7 for various environments is valid only for the case of an irregular but not curved terrain, and hence, it is only valid for radio links shorter than 5–10 km.

5.7.3. Comparison with Existing Models

Let us compare results obtained from the stochastic, multiparametric model and those obtained from other well-known and frequently used models.

Comparison with Empirical Models. The well-known empirical Okumura–Hata model discussed in References 98 and 99 is based on numerous measurements of the average power within the communication channel carried out in and around Tokyo, gives the path loss attenuation for urban environments, as a function of distance between "mobile-base station," d, which can be presented, as in Section 5.3, by using the straight line mathematical description

$$L(urban) = L_q + 10\gamma \log d. \tag{5.82}$$

Here,

$$L_q = 69.55 + 26.16 \log f - 13.82 \log h_T - a(h_R)$$
$$\gamma = (44.9 - 6.55 \log h_T) / 10, \tag{5.83}$$

where, f is the radiated frequency, h_R and h_T are the heights of the receiving and transmitting antenna, respectively. Parameter $a(h_R)$ changes for various terrain types [98, 99]. Thus,

- for medium-size cities,

$$a(h_R) = (1.1 \log f - 0.7) h_R - (1.56 f - 0.8) \tag{5.84a}$$

- for large cities,

$$a(h_R) = 8.29 \cdot (\log 1.54 h_R)^2 - 1.1, \quad f_0 \leq 200\ MHz$$
$$a(h_R) = 3.2 \cdot (\log 11.75 h_R)^2 - 4.97, \quad f_0 > 200\ MHz \tag{5.84b}$$

Then, for suburban areas [98, 99],

$$L = L(urban) - 2\left[\log\left(\frac{f_0}{28}\right)\right]^2 - 5.4 \quad (dB), \tag{5.85}$$

and for open and rural areas,

$$L = L(urban) - 4.78[\log f]^2 - 18.33 \log f - 40.94 \quad (dB). \tag{5.86}$$

By analyzing the parameter of signal attenuation γ from (5.83) for various heights of the transmitting antenna, it was found that the signal power attenuation varies as

$$L(W) \propto d^{-\gamma}, \gamma = 3.0 - 3.8. \tag{5.87}$$

Comparison with the Standard COST-231 Model. The Walfisch and Ikegami semiempirical model first discussed in Reference 100 gives a good path loss prediction for dense built-up areas of medium and large size cities and therefore was taken as the standard model. The model is based on an analytical approach, developed by Bertoni and his colleagues in References 5 and 101–103.

It was shown that the diffraction from the roofs and corners of buildings plays a significant role, and the total field depends not only on the reflected waves, but mainly on the diffracted waves. Following the work in References 5 and 101–103, the semiempirical model developed by Walfisch and Ikegami considered two options for the BS locations: one above and one below the rooftops, in an environment with regularly distributed nontransparent buildings with various heights and different separation distances between them. Moreover, the semiempirical model takes into account important urban parameters such as building density, average building height, and street width. The antenna height is generally lower than the average buildings' height, so that the waves are guided along the street. We present again the path loss dependence on the range d between the terminal antennas in urban environments taken from References 5 and 100–103 by using the straight line mathematical description:

$$L = L_q + (20 + K_d) \cdot \log d = L_q + 10 \gamma \log d. \tag{5.88}$$

Here

$$L_q = 32.4 + (30 + K_f) \cdot \log f + 20 \log \Delta h_R + L(0) - 16.9 - 10 \log a \quad (5.89a)$$

$$\gamma = (20 + K_d)/10, \; K_d = 18 - 15(\Delta h_T / h_b) \quad (5.89b)$$

$$K_f = -4 + 0.7[(f/925) - 1], \quad \text{for urban areas,} \quad (5.89c)$$

where $\Delta h_R = h_b - h_R$, $\Delta h_T = h_T - h_b$, $h_T > h_b > h_R$, h_b are the average height of building roofs, and a is the average street width. Finally, it follows from (5.88) that the signal power dependence can be presented as

$$L(W) \propto d^{-\gamma}, \quad \gamma = 2.6 - 3.8. \quad (5.90)$$

At the same time, the stochastic multiparametric model discussed in previous sections and the corresponding Equation (5.76) and Equation (5.78) presented there, give the signal intensity decay law versus the range between terminal antennas d as

$$L(W) \propto d^{-\gamma}, \quad \gamma = 2.5 - 3.0. \quad (5.91)$$

This result is very close to those predicted by the two other models mentioned earlier.

Comparison with the Ray Tracing Model. Ray tracing is a technique based on geometrical optics, and an easily applied approximate method for estimating a high frequency electromagnetic field. In the ray-tracing method, rays launched from a BS transmitter in all directions (in azimuth and elevation angles) undergo reflection, transmission, and diffraction at various components on the way to the receiver. All rays reaching the receiver are summed up to determine the total field strength.

The electric field E_i of the ith ray arriving at the receiver is represented by using the reflection coefficient R_j for the jth reflector, transmission coefficient T_k for the kth transmission, diffraction coefficient D_l for the lth diffracting wedge, as well as the transmitter and receiver gain G_T and G_R, and path length d. The total electric field E at the receiver is given in terms of the addition of various rays,

$$E = \sum_i E_i = \sum_i \left(\frac{K \cdot P_T \cdot G_T \cdot G_R}{d^2} \prod_j R_j \cdot \prod_k T_k \cdot \prod_l D_l \right), \quad (5.92)$$

where P_T is the transmitter power, and K is a constant determined by such factors as wavelength.

Below, we show the ray-tracing results for four specific radio paths, Rx1 (Fig. 5.20a) to Rx4 (Fig. 5.20d) presented in Reference 59 during measurements carried out in Tokyo University (see details of this series of experiment in Chapters 10 and 11). In these computations of the ray tracing and the

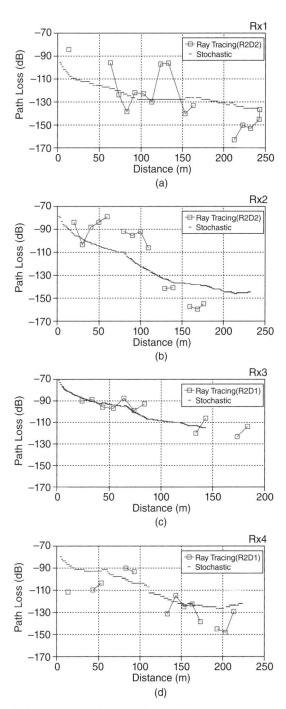

FIGURE 5.20. (a) Comparison of stochastic model (continuous curve) with the ray tracing model (rombs connected by segments) for the route of the subscriber Rx1. (b) The same as in panel a, but for the subscriber Rx2. (c) The same as in panel a, but for the subscriber Rx3. (d) The same as in panel a, but for the subscriber Rx4.

stochastic models, the frequency of 5 GHz ($\lambda = 0.06$ m) was used and it was assumed that all of the buildings are composed by concrete walls with the following electric parameters: $\varepsilon_{r=6.76}, \sigma = 2.3 \cdot 10^{-3} (S/m)$. For the ground surface, the following electric parameters were taken: $\varepsilon_r = 3.0, \sigma = 1.0 \cdot 10^{-4} (S/m)$. The electric property of the concrete buildings was assumed to be consistent with the assumption of the reflection coefficient of $\Gamma = 0.85$. In Fig. 5.20a–d, the ray-tracing result was calculated every 10 m; the same was done for the stochastic model.

The ray-tracing results in Fig. 5.20a–d are mainly based on two-time reflection (R2) and two-time diffraction (D2) rays, which seem to be consistent with the work given in Reference 59. The two-time reflection (R2) and one-time diffraction (D1) for the paths of Rx3 and Rx4 were taken into account here. Comparison between the results by the stochastic modeling and ray-tracing result is performed in Figure 5.20a–d for the paths Rx1–Rx4, respectively.

As follows from the presented illustrations, the agreement between the stochastic and ray-tracing models is good, and the stochastic result is considered to reflect the general trend obtained by the ray-tracing estimation. At shorter distances ($d < 50$ m), we notice a rather weaker agreement between the stochastic and ray-tracing results as suggested by a discrepancy between both models, whereas at distances larger than 50–70 m, they are closer to each other. This can be explained by the fact that far from the BS antenna, stochastic effects of the rough built-up terrain become dominant and a cumulative "multiray effect" described by the deterministic ray-tracing model is close to those obtained from the stochastic multiparametric model.

5.8. DISTRIBUTION OF POLARIZED PARAMETERS IN URBAN ENVIRONMENTS WITH MULTIPATH

In Chapter 2, we introduced the main polarization characteristics of an arbitrary electromagnetic wave and introduced the main parameters of the polarization ellipse.

As was shown above, in dense urban environments, the multipath phenomena are usually observed at the receiver. At close ranges with high-elevated BS antenna and low-elevated moving subscriber (MS) antennas [5, 57–59, 86–93], the distribution of angle of arrivals in the azimuth domain is very wide. This is due to the multiray phenomena at the observation point. The random distribution of the number of rays and the direction of each arriving ray in the azimuth (horizontal) domain lead to a broadening pattern in the azimuth domain.

As for the elevation (e.g., vertical) plane, the energy spectrum of the total wave [5, 59, 87], is concentrated mainly within a small angle range around the direction between the BS and MS. So, in our analysis we can assume that the energy spectrum in the elevation domain (in the vertical plane, for the component with $i = 3$) is contained within a narrow angular range.

5.8.1. Distribution of Wave Polarization Characteristics in Urban Environments

Let us continue the analysis of Equation (2.39) introduced in Chapter 2, but now for the specific case of a dense built-up terrain [104–106]. Results described in Chapter 2 allow us to state that differences between the statistical characteristics of waves, σ_1^2 and σ_2^2, in the horizontal plane, defined by field components with $i = 1$ (along **x**-axis) and $i = 2$ (along **y**-axis), weakly affect the distributions of the polarized parameters, such as R, θ, and β. Therefore, we can differentiate in our analysis the vertical direction **z**, defined by the depolarization parameter $\sigma_3^2 \equiv \sigma_\parallel^2$, from directions **x** and **y**, placed at the horizontal (ground) plane, and defined by depolarization parameters $\sigma_1^2 \equiv \sigma_{1\perp}^2$ and $\sigma_{2\perp}^2$. All definitions and the corresponding geometry of the problem were presented in Chapter 2 (in Fig. 2.11).

According to the above assumptions, we can, with a great degree of accuracy, put $\sigma_{1\perp} = \sigma_{2\perp} = \sigma_\perp$, as the cross-polarized energy deviation parameters, and $\sigma_3 \neq \sigma_\perp$. This assumption allows us to rewrite Equation (2.39) from Chapter 2 to describe the total probability function distribution over all polarization parameters as

$$w(I, p, \theta, \varphi, \beta) = \frac{I^2 \cdot p \cdot \sin\theta}{8(2\pi)^2 \sigma_\perp^4 \sigma_\parallel^3} \exp\left(-\frac{I}{2}\left[\frac{1 - \xi\sin^2\theta}{\sigma_\perp^2} - \frac{\xi\sin^2\theta}{\sigma_\parallel^2}\right]\right). \qquad (5.93)$$

It can be noted that for the case under consideration, the probability density is independent of the azimuth angle ϕ. Integration of (5.93) over parameters $I \in (0, \infty)$ and $\varphi \in (0, 2\pi)$ gives

$$w(p, \theta, \beta) = \frac{1}{\pi}\left(\frac{\sigma_\perp}{\sigma_\parallel}\right)^2 \frac{p \cdot \sin\theta}{\left\{1 - \xi\left[1 - \left(\frac{\sigma_\perp}{\sigma_\parallel}\right)^2 \sin^2\theta\right]\right\}^3}. \qquad (5.94)$$

Integration of (6.93) over $p \in (0, 1)$, and then over β gives the 1-D power density distribution:

$$w(\theta) = \left(\frac{\sigma_\perp}{\sigma_\parallel}\right)^2 \frac{p \cdot \sin\theta}{2\left[\cos^2\theta - \left(\frac{\sigma_\perp}{\sigma_3}\right)^2 \sin^2\theta\right]^{3/2}}. \qquad (5.95)$$

The same procedure, but over angle θ, gives the 1-D power density distribution of β:

$$w(\beta) = \frac{1}{\pi} \left(\frac{\sigma_\perp}{\sigma_\parallel}\right)^2 \frac{(1+\eta)^2(1+\eta_0)}{(\eta-\eta_0)^2} \times$$

$$\times \left\{\frac{\eta-3\eta_0}{2}\sqrt{\eta}\tan^{-1}\sqrt{\eta} + \eta_0\sqrt{\eta_0}\tan^{-1}\sqrt{\eta_0} + \frac{\eta}{2}\frac{\eta-\eta_0}{1+\eta}\right\},$$

(5.96)

where $\eta = \eta_0 \dfrac{1-\cos 2\beta}{1+\eta_0\cos 2\beta}$, $\eta_0 = \dfrac{\sigma_\parallel^2 - \sigma_\perp^2}{\sigma_\parallel^2 + \sigma_\perp^2}$.

The 1-D power density of the distribution of parameter p can be obtained from (5.93) by integrating over θ and then over angle β, to get:

$$w(p) = 2Cp\left\{\frac{1}{1-\eta_0^2 p^2} + \frac{\eta_0^2}{2(1-\eta_0^2 p^2)^2}\right\}$$

(5.97)

where $C = 2(1-\eta_0^2)\left[\eta_0^2 - 2\ln(1-\eta_0^2)\dfrac{1-\eta_0^2}{\eta_0^2}\right]^{-1}$.

In the two limiting cases, (1) $\sigma_\perp = \sigma_\parallel$ (stochastic processes are equal both in the vertical and the horizontal plane), and (2) $\sigma_\perp << \sigma_\parallel$ (stochastic processes in the horizontal plane are not significant with respect to those in the vertical plane), we get, respectively,

$$\sigma_\perp = \sigma_\parallel : w(p) = 2p$$

(5.98a)

$$\frac{\sigma_\perp}{\sigma_\parallel} \to 0 : w(p) \to \delta(p-1),$$

(5.98b)

where $\delta(p-1)$ is the general Dirac function, equaling 0, when $p \neq 1$, and ∞, when $p = 1$. The expression (5.98b) describes the case of linearly polarized field (for which $\sigma_\perp/\sigma_{\parallel=0}$), concentrated at the vertical plane and has a unimodal form close to the delta-function distribution (see also Reference 58).

Taking into consideration the relations between parameter p and the elliptic coefficient, R, (see Chapter 2),

$$R = \left(\frac{1-p}{1+p}\right)^{1/2}$$

(5.99)

we can obtain from (5.97) the 1-D power density function for the elliptic parameter R as

$$w(R) = \frac{8CR(1-R^2)}{(1+R^2)^3}\left[\frac{1}{\left[1-\eta_0^2\left(\dfrac{1-R^2}{1+R^2}\right)^2\right]^2} + \frac{\eta_0^2}{2\left[1-\eta_0^2\left(\dfrac{1-R^2}{1+R^2}\right)^2\right]^2}\right],$$

(5.100)

which for the same two limiting cases mentioned earlier gives

$$\sigma_\perp = \sigma_\| : w(R) = 8R(1-R^2)/(1+R^2)^3 \tag{5.101a}$$

$$\sigma_\perp / \sigma_\| \to 0 : w(R) \to \delta(R). \tag{5.101b}$$

So, for the case of a dominant linearly polarized component of the field in the vertical plane, we have again a unimodal distribution of signal energy proportional to the δ-function (or Dirac function). These results are in a good agreement with those presented in References 57 and 94–96, both theoretically and experimentally.

The Line of Sight (LOS) between the BS and MS Antennas. As was mentioned earlier such a scenario is valid when the stationary BS antenna is at a higher level than the MS antenna. In this case we have both LOS and multipath components of the wave arriving at the MS antenna. As was shown in References 5, 57–59, and 87, only the third component of the total field is predominant in such a scenario, and should be accounted for at the vertical plane (e.g., in elevation domain).

In this case, we can present this component, as a linearly polarized wave in the vertical plane, in the following form:

$$\mathbf{E}_{03}(t) = \mathbf{u}_3 A_{03} \cos(\omega t + \psi_{03}) = \mathbf{u}_3 E_{03}(t) \tag{5.102}$$

The rest of multipath waves are distributed according to the normal (e.g., Gaussian) law both in the horizontal and vertical planes. This statement, as was mentioned earlier, was verified by numerous experiments carried out in various outdoor links.

Let us now introduce a Cartesian coordinate system $\{x, y, z\}$. In this case, $\sigma_{1\perp} \equiv \sigma_{x\perp}$, $\sigma_{2\perp} \equiv \sigma_{y\perp}$ and $\sigma_{3\|} = \sigma_{z\|}$. The existence of a non-zero deterministic component $E_{03} \equiv E_{0z}$ does not influence the azimuthally distributed multipath waves, which are distributed in such a manner that the difference between $\sigma_{x\perp}$ and $\sigma_{y\perp}$ is small and do not affect the polarization parameters in the horizontal $x0y$-plane. Therefore, again we can use the assumption that $\sigma_{x\perp} = \sigma_{y\perp} = \sigma_\perp$, which corresponds to the real situation of polarized signals observed in urban areas. For further simplicity we set $\sigma_{z\|} = \sigma_\|$ and $E_{0z} \equiv E_{0\|}$.

The joint power density distribution function of these five parameters, defined earlier, through the phase of the LOS component, ψ_3, is :

$$w(I, p, \theta, \beta, \varphi) = \frac{I^2 p \cdot \sin\theta}{32\pi^2 \sigma_\perp^4 \sigma_\|^2} \exp\left\{ -\frac{I}{2}\left[\frac{1-\xi\sin^2\theta}{\sigma_\perp^2} + \frac{\xi\sin^2\theta}{\sigma_\|^2} \right] \right\} \times$$
$$\times \left(\frac{A_{0\|}\sqrt{I\xi}}{\sigma_\|^2}\sin\theta \cdot \cos\psi_3 - \frac{A_{0\|}^2}{2\sigma_\|^2} \right) \tag{5.103}$$

Here $\xi = 0.5 \cdot (1-p \cos 2\beta)$, $\xi \in [0, 1]$, $p = (1-R^2)/(1 + R^2)$, R is the elliptic parameter. After integration over $\psi_3 \in [0, 2\pi]$, we finally get:

$$w(I, p, \theta, \beta, \varphi) = \frac{I^2 p \cdot \sin \theta}{16\pi^2 \sigma_\perp^4 \sigma_\parallel^2} \exp\left\{-\frac{A_{0\parallel}^2}{2\sigma_\parallel^2}\right\} I_0\left(\frac{A_{0\parallel}^2 \sqrt{I\xi}}{\sigma_\parallel^2}\right) \times$$
$$\times \exp\left\{-\frac{I}{2}\left[\frac{1-\xi \sin^2 \theta}{\sigma_\perp^2} + \frac{\xi \sin^2 \theta}{\sigma_\parallel^2}\right]\right\}$$
(5.104)

Now, using well-known formula based on Rician distribution taken from Reference 57,

$$\int_0^\infty x^2 e^{-x} I_0(\sqrt{Kx}) dx = \left[2 + K + \frac{1}{16} K^2\right] \cdot e^{K/4}$$
(5.105)

and equation (5.104) we can obtain the probability density function $w(p,\theta,\beta)$:

$$w(p, \theta, \beta) = \frac{1}{2\pi}\left(\frac{\sigma_\perp^2}{\sigma_\parallel^2} \cdot \frac{p \sin \theta}{X^3}\right) \cdot \exp\left\{-\frac{\sigma_\parallel^2}{\sigma_\perp^2} q + \frac{q}{2}\right\} \times$$
$$\times \cdot \left[\left(1 - 2q + \frac{q^2}{2}\right) + \frac{2q - q^2}{X} + \frac{q^2}{2X^2}\right]$$
(5.106)

In equation (5.105) $K = \frac{A_{0\parallel}^2}{\sigma_\parallel^2} \equiv \frac{LOS - component}{Multipath\ component}$ is the Rician (e.g., fading) parameter. Also,

$$q = \frac{1}{2} \frac{A_{0\parallel}^2}{\sigma_\parallel^2} \frac{\sigma_\perp^2}{\sigma_\parallel^2 - \sigma_\perp^2}, X = 1 - \frac{\sigma_\parallel^2 - \sigma_\perp^2}{2\sigma_\parallel^2}(1 - p\cos 2\beta) \sin^2 \theta.$$
(5.107)

Next, under LOS conditions (e.g., clearance exceeding 75–80%) the linearly polarized component in the vertical plane is the dominant component and for the multipath component we can assume that $\sigma_\parallel^2 \gg \sigma_\perp^2$. Experiments carried both in outdoor and outdoor environments [5, 57, 59], have shown that for the vertically polarized direct wave the condition $\sigma_\parallel^2 > \sigma_\perp^2$ was observed in more than 90% of the measurements, both for open and quasi-open radio ranges. In such situations the distribution of signal power in the vertical plane, $W(\theta)$, and that in the horizontal plane, $W(\varphi)$, had pure unimodal distribution with characteristic maxima at the points $\theta = \pi/2$ and $\beta = \pi/2$ [57, 58]. In other words, some concentration of "Gaussian" profiles of signal energy around the direction of LOS in the azimuth and elevation domains are usually observed.

We can now derive the probability density function of the signal energy distribution versus the elliptic parameter p, defined earlier. Thus, we return to Equation (5.93), which for our previous assumptions, $\sigma_\parallel^2 > \sigma_\perp^2$ and $A_{0\parallel}^2 \gg \sigma_\parallel^2 + \sigma_\perp^2$, can be derived as follows: First, we introduce new variables,

$y = \cos\theta$ and $\varsigma = \pi - 2\beta$, and present the integrand in (5.93) as a set of functions of y and ς. By eliminating all terms of order $\sim y^2, y^3, \ldots$ and of $\sim \varsigma^2, \varsigma^3, \ldots$. And integrating the integral in (5.93) over y and ς, we obtain $w(p)$ as:

$$w(p) \approx \left(\frac{\sigma_\perp}{\sigma_\|}\right)^6 \left(\frac{p}{1+p}\right)^{1/2} \frac{(1+p)^2}{\left[(1+p)+(\sigma_\perp/\sigma_\|)^2(1+p)\right]^3} \times$$

$$\times \exp\left\{-\frac{1}{2}\left(\frac{A_{0\|}}{\sigma_\|}\right)^2 \frac{(1-p)}{(1-p)(\sigma_\perp/\sigma_\|)^2(1+p)}\right\} \qquad (5.108)$$

Equation (5.108) satisfies a well-known property of the probability of the total energy distribution, i.e.,

$$\int_0^1 w(p)dp = 1 \qquad (5.109)$$

The Non Line of Sight (NLOS) between the BS and MS Antennas. In the case where both terminal antennas, the BS and the MS, are below the average rooftop level, the distribution of signal power can be multi-modal both in the vertical and horizontal planes with a non-symmetrical form in the signal energy pattern [57–59]. However, as in the first case mentioned in the previous section, there is no any correlation between the vertical (e.g., "elevation") component and the horizontal (e.g., "azimuth") components of the total field. We can now state that $\sigma_{x\perp} \neq \sigma_{y\perp}$, and both of them are differ from $\sigma_z \equiv \sigma_\|$. In this case, integration of the common Equation (5.93), over the normalized parameters I and p, with limits $I \in (0, 1)$ and $p \in (0, 1)$, gives:

$$w(\theta, \beta, \varphi) = \frac{4\sin\theta}{2\pi^2\sigma_{x\perp}^2\sigma_{y\perp}^2 C_2(C_1\cos 2\beta + C_2\sin 2\beta + C_3)} \qquad (5.110)$$

Here we used notations introduced in Section 5.4. In this case parameters C_i, $i = 1, 2, 3$ can be written according to as:

$$C_1 = \frac{\cos^2\theta\cos^2\varphi - \sin^2\varphi}{\sigma_{x\perp}^2} + \frac{\cos^2\theta\sin^2\varphi - \cos^2\varphi}{\sigma_{y\perp}^2} + \frac{\sin^2\theta}{\sigma_\|^2},$$

$$C_2 = \left(\frac{1}{\sigma_{y\perp}^2} - \frac{1}{\sigma_{x\perp}^2}\right)\sin\theta\sin 2\varphi, \qquad (5.111)$$

$$C_3 = \frac{\cos^2\theta\cos^2\varphi + \sin^2\varphi}{\sigma_{x\perp}^2} + \frac{\cos^2\theta\sin^2\varphi + \cos^2\varphi}{\sigma_{y\perp}^2} + \frac{\sin^2\theta}{\sigma_\|^2}$$

Integrating (5.110) over $\beta \in (-\pi/2, \pi/2)$ in the complex plane, by introducing a new variable $z = e^{i2\beta}$, leads to a contour integral along a circle with radius $|z| = 1$, which gives us a solution in the pole of the second order at the point

$$z_p = \frac{-C_3 \pm \sqrt{C_3^2 - C_2^2 - C_1^2}}{C_1 - iC_2} \tag{5.112}$$

Using Cauchy's theorem for such kinds of poles, we finally get a simple formula for the joint power density function distribution in the azimuth and elevation domains:

$$w(\theta, \varphi) = \frac{2\sin\theta}{\pi\sigma_{x\perp}^2 \sigma_{y\perp}^2 \sigma_{\parallel}^2 (C_3^2 - C_2^2 - C_1^2)^{3/2}} \tag{5.113}$$

Introducing parameters C_1^2, C_2^2, C_3^2 from (5.111), we finally get:

$$w(\theta, \varphi) = \frac{\sigma_{x\perp}\sigma_{y\perp}\sigma_{\parallel}}{4\pi} \frac{\sin\theta}{\left[\sigma_{\parallel}^2 \cos^2\theta + \sin^2\theta(\sigma_{y\perp}^2 \sin^2\varphi - \sigma_{x\perp}^2 \cos^2\varphi)\right]^{3/2}} \tag{5.114}$$

Integrating equation (5.114) over $\theta \in (0, \pi/2)$ gives the ordinary power density distribution in the azimuth domain

$$w(\varphi) = \frac{\sigma_{x\perp}\sigma_{y\perp}}{2\pi} \frac{1}{(\sigma_{y\perp}^2 \sin^2\varphi - \sigma_{x\perp}^2 \cos^2\varphi)} \tag{5.115}$$

All the above-mentioned formulas show that the distribution of the corresponding density functions depend on the ratio between the corresponding deviations of signal power in the vertical and horizontal planes, that is, on $\sigma_{x\perp}/\sigma_{y\perp}$, $\sigma_{\perp}/\sigma_{\parallel}$, etc.

5.8.2. Influence of the Built-Up Terrain Characteristics on Wave Depolarization

In References 57–59, 86, and 87, the 2-D and 3-D multiparametric stochastic models were presented to investigate the influence of the buildings density and their overlay profile on the signal intensity distribution in the space domain. The main results of these investigations were presented earlier in Sections 5.7 and 5.8. We now present results when the distribution of signal field strength in the vertical and horizontal plane are uncorrelated, that is, $\langle U_{\parallel}(\mathbf{r}_2)U_{\perp(\mathbf{r}_2)}\rangle = 0$. Then, using definitions for the *rms* of signal intensity depolarization with zero-mean Gaussian PDFs (see discussions presented in the previous section) in the vertical plane $\sigma_{\parallel}^2 = \langle U_{\parallel}(r_2)U_{\parallel}^*(r_2)\rangle = \langle I_{inc}\rangle_{\parallel}$ and in the horizontal plane $\sigma_{\perp}^2 = \langle U_{\perp}(r_2)U_{\perp}^*(r_2)\rangle = \langle I_{inc}\rangle_{\perp}$, respectively, we can derive:

$$\sigma_{\parallel}^2 = I_0 \frac{\Gamma \cdot \lambda \cdot \ell_v}{\lambda^2 + [2\pi\ell_v\gamma_0 F(h_R, h_T)]^2} \frac{(h_T - \bar{h})}{d} \tag{5.116a}$$

and

$$\sigma_\perp^2 = I_0 \frac{\Gamma}{8\pi} \cdot \frac{\lambda \cdot \ell_h}{\lambda^2 + [2\pi\ell_h < L > \gamma_0]^2} \qquad (5.116b)$$

Here, ℓ_v and ℓ_h are the correlation scales of the reflection coefficients in the vertical and horizontal planes, respectively, I_0 is the intensity of the transmitted signal, and $\bar{h}$ is the average building height. All other parameters are defined earlier in previous sections.

As was shown in References 59 and 77, by investigating numerous urban overlay profiles, the case of $n = 1$ is the most realistic profile that occurs in the urban environment, for which simply $F(h_R, h_T) \to (\bar{h} - h_R)$. Then, we can simplify (6.156a) by

$$\sigma_\parallel^2 = I_0 \frac{\Gamma \cdot \lambda \cdot \ell_v}{\lambda^2 + [2\pi\ell_v\gamma_0(\bar{h} - h_R)]^2} \frac{(h_T - \bar{h})}{d} \qquad (5.117)$$

Below, we will analyze the above formulas for two scenarios, LOS and NLOS. First, we will analyze the ratio $\sigma_\parallel^2 / \sigma_\perp^2$ for various building profiles and density of buildings that are typical for many small and large cities. From equation (6.116) and (5.117), this ratio depends mostly on the density of buildings, γ_0, and on the elevations of both terminal antennas, h_R and h_T, with respect to the average building height $\bar{h}$, that is,

$$\frac{\sigma_\parallel^2}{\sigma_\perp^2} = 8\pi \left(\frac{\ell_v}{\ell_h}\right) \frac{\left(1 + \dfrac{[2\pi\ell_h\gamma_0\langle L\rangle]^2}{\lambda^2}\right)}{\left(1 + \dfrac{[2\pi\ell_v\gamma_0 F(h_R, h_T)]^2}{\lambda^2}\right)} \frac{(h_T - \bar{h})}{d} \qquad (5.118)$$

From the above equation, for a high building density located at the ground surface and a "smooth" building profile in the vertical plane (i.e., when $(\bar{h} - h_R)$ and $(h_T - \bar{h})$ are close), for distances more than 100 m, and for $(\ell_v/\ell_h) \approx 1$, we get $\sigma_\parallel^2 / \sigma_\perp^2 < 1$. In this case, the degree of depolarization due to multi-ray phenomena in the horizontal plane, caused by multiple scattering, diffraction and reflection, becomes stronger with respect to that in the vertical plane. Conversely, for the "sharp shaped" building profile and small building density located at the ground surface for the same conditions, as above, we get $\sigma_\perp^2 / \sigma_\parallel^2 < 1$. Finally, when signal energy, due to multipath phenomena, becomes random in both the horizontal and vertical plane (large city scenario), we get $\sigma_\perp^2 / \sigma_\parallel^2 \approx 1$. This qualitative analysis allows us now to investigate the parameters of the polarization ellipse of the incident radio wave for various scenarios occurring in the urban environment with different ratios $\sigma_\perp^2 / \sigma_\parallel^2$.

As mentioned previously, the main parameters of the radiowave depolarization is the elliptic parameter R determined by the density distribution function (PDF) of the signal $w(R)$ defined by (5.100) for different ratios of $\sigma_\perp / \sigma_\parallel$. Therefore, it is important to analyze the effects of changes in this characteristic of

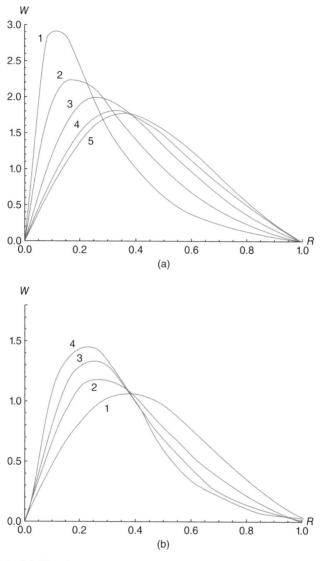

FIGURE 5.21. (a) Signal power versus the parameter of ellipse R for $K = 7, 5, 3, 1, 0$ (curves 1, 2, 3, 4, 6) for two ratios $\sigma_\perp = 0.5 \cdot \sigma_\parallel$ and $\sigma_\perp = \sigma_\parallel$. (b) The same, as in panel a, but for inverse notations, when $K = 0, 1, 3, 5, 7$ corresponds to curves denoted by 1, 2, 3, 4, 5, respectively.

the signal elliptic polarization for specific situations occurring in the urban environment with strong fading.

Results of computations for $K = 7, 5, 3, 1, 0$ (curves 1, 2, 3, 4, 6) for two ratios $\sigma_\perp = 0.5 \cdot \sigma_\parallel$ and $\sigma_\perp = \sigma_\parallel$ are shown in Figure 5.21a. In Figure 5.21b, these curves correspond to the inverse notations, that is, $K = 0, 1, 3, 5, 7$ (curves 1, 2, 3, 4,

5), respectively, using similar notations for curves. One can see that for the case of strong depolarization of the signal (i.e., deformation of the polarized ellipse), caused by multipath phenomena, both in the vertical plane (due to effects of the building profile) and in the horizontal plane (due to an increase of building density at the ground surface), $w(R)$ is close to the a pure Rayleigh distribution (with $K = 0$).

REFERENCES

1 Jakes, W. C., *Microwave Mobile Communications*, New York: John Wiley and Son, 1974.

2 Lee, W. Y. C., *Mobile cellular Telecommunications Systems*, New York: McGraw Hill Publications, 1989.

3 Saunders, S. R., *Antennas and Propagation for Wireless Communication Systems*, New York: John Wiley & Sons, 1999.

4 Steele, R., *Mobile Radio Communication*, New York: IEEE Press, 1992.

5 Bertoni, H. L., *Radio Propagation for Modern Wireless Systems*, New Jersey: Prentice Hall PTR, 2000.

6 Rappaport, T. S., *Wireless Communications*, New York: Prentice Hall PTR, 1996.

7 Milstein, L. B., D. L. Schilling, R. L. Pickholtz et al., "On the feasibility of a CDMA overlay for personal communications networks," *IEEE Select. Areas in Commun.*, vol. 10, no. 4, 1992, pp. 665–668.

8 Fock, V. A., *Electromagnetic Diffraction and Propagation Problems*, Oxford: Pergamon Press, 1965.

9 Al'pert, Ia. L., V. L. Ginsburg, and E. L. Feinberg, *Radio Wave Propagation*, Moscow: State Printing House for Technical-Theoretical Literature, 1953.

10 Bass, F. G., and I. M. Fuks, *Wave Scattering from Statistically Rough Surfaces*, Oxford: Pergamon Press, 1979.

11 Voronovich, A. G., *Wave Scattering from Rough Surfaces*. Berlin: Springer-Verlag, 1994.

12 Backmann, P. and A. Spizzichino, *The Scattering of Electromagnetic Waves from Rough Surfaces*, Boston-London: Artech House, 1963.

13 Ishimaru, A., *Electromagnetic Wave Propagation, Radiation, and Scattering*, Englewood Cliffs, New Jersey: Prentice-Hall, 1991.

14 Felsen, L. B.m and N. Marcuvitz, *Radiation and Scattering of Waves*, New York: IEEE Press, 1994.

15 Miller, A. R., R. M. Brown, and E. Vegh, "New derivation for the rough-surface reflection coefficient and for the distribution of the sea-wave elevations," *IEE Proc.*, vol. 131, Pt. H, no. 2, 1984, pp. 114–116.

16 Ogilvy, J. A., *Theory of Wave Scattering from Random Rough Surfaces*, London, 1987.

17 Rice, S. O., "Reflection of electromagnetic waves from slightly rough surfaces," *Comm. Pure Appl. Math.*, vol. 4, no. 3, 1951, pp. 351–378.

18   Barrick, D. E., and W. H. Peake, "A review of scattering from surfaces with different roughness scales," *Radio Sci.*, vol. 3, no. 7, 1968, pp. 865–868.

19   Barrick, D. E., "Theory of HF and VHF propagation across the rough sea-Parts I and II," *Radio Sci.*, vol. 6, no. 3, 1971, pp. 517–533.

20   Barrick, D. E., "First order theory and analysis of MF/HF/VHF scatter from the sea," *IEEE Trans. Anten. Propagat.*, vol. 20, no. 1, 1972, pp. 2–10.

21   Wait, J. R., "Perturbation analysis for reflection from two-dimensional periodic sea waves," *Radio Sci.*, vol. 6, no. 3, 1971, pp. 387–391.

22   Valenzuela, G. R., "Scattering of electromagnetic waves from a slightly rough surface moving with uniform velocity," *Radio Sci.*, vol. 3, no. 1, 1968, pp. 12–21.

23   Valenzuela, G. R., "Scattering of electromagnetic waves from a tilted slightly rough surface," *Radio Sci.*, vol. 3, no. 6, 1968, pp. 1057–1066.

24   Valenzuela, G. R., "The effective reflection coefficients in forward scatter from a dielectric slightly rough surface," *Proc. IEEE*, vol. 58, no. 12, 1970, pp. 1279–1285.

25   Krishen, K., "Scattering of electromagnetic waves from a layer with rough front and plane back (small perturbation method by Rice)," *IEEE Trans. Anten. Propagat.*, vol. 19, no. 4, 1970, pp. 573–576.

26   Davies, H., "The reflection of electromagnetic waves from a rough surface," *Proc. IEEE*, vol. 101, no. 2, 1954, pp. 209–214.

27   Bullington, K., "Reflection coefficient of irregular terrain," *Proc. IRE*, vol. 42, no. 11, 1954, pp. 1258–1262.

28   Keller, J. B., "Diffraction by an Aperture," *J. Appl. Phys.*, vol. 28, 1957, pp. 857–893.

29   Keller, J. B., "Geometrical theory of diffraction," *J. Opt. Soc. Amer.*, vol. 52, no. 1, 1962, pp. 116–131

30   James, G. L., *Geometrical Theory of Diffraction for Electromagnetic Waves*, 3-nd ed., London, UK: Peter Peregrines, 1986.

31   Honl, H., A. W. Maue, and K. Westpfahl, *Theory of Diffraction*, Berlin: Springer-Verlag, 1961.

32   Kouyoumjian, R. G., and P. H. Pathak, "An uniform geometrical theory of diffraction for an edge in a perfectly conducting surface," *Proc. IEEE,* vol. 62, no. 9, 1974, pp. 1448–1469.

33   Russel, S. T. A., C. W. Boston and T. S. Rappaport, "A deterministic approach to predicting microwave diffraction by buildings for micro cellular systems," *IEEE Trans. Anten. and Propag.*, vol. 41, no. 12, 1993, pp. 1640–1649.

34   Dougherty, H. T., and L. J. Maloney, "Application of diffraction by convex surfaces to irregular terrain situations," *Radio Phone*, vol. 68B, 1964, p. 239–241.

35   Anderson, L. J., and L. G. Trolese, "Simplified method for computing knife edge diffraction in the shadow region," *IRE Trans. Anten. Propagat.*, vol. AP-6, 1958, pp. 281–286.

36   Vogler, L. E., "The attenuation of electromagnetic waves by multiple knife-edge diffraction," *NTIA Report*, 1981, pp. 81–86.

37   Vogler, L.E., "An attenuation function for multiple knife-edge diffraction," *Radio Sci.*, vol. 17, no. 9, 1982, pp. 1541–1546.

38 Andersen, J. B., "UTD multiple-edge transition zone diffraction," *IEEE Trans. Anten. Propagat.,* Vol. 45, No. 7, 1997, pp. 1093–1097.

39 Lee, S. W., "Path integrals for solving some electromagnetic edge diffraction problems," *J. Math. Phys.,* vol. 19, no. 10, 1978, pp. 1414–1422.

40 Bertoni, H. L., and J. Walfisch, "A theoretical model of UHF propagation in urban environment," *IEEE Trans. Anten. Propagat.,* vol. 36, no. 12, 1988, pp. 1788–1796.

41 Bullington, K., "Radio propagation at frequencies about 30 Mc," *Proc. IRE,* vol. 35, no. 10, 1947, pp. 1122–1136.

42 Epstein, J., and D. W. Peterson, "An experimental study of wave propagation at 850 Mc," *Proc. IRE,* vol. 41, no. 5, 1953, pp. 595–611.

43 *Atlas of Radio Wave Propagation Curves for Frequencies Between 30 and 10,000 Mc/s,* Radio Research Lab., Ministry of Postal Services, Tokyo, Japan, 1957, pp. 172–179.

44 Deygout, J., "Multiple knife-edge diffraction of microwaves," *IEEE Trans. Anten. Propagat.,* Vol. 14, No. 4, 1966, pp. 947–949.

45 LaGrone, A. H., and C. W. Chapman, "Some propagation characteristics of high UHF signals in the immediate vicinity of trees," *IRE Trans. Anten. Propagat,* vol. AP-9, 1961, pp. 957–963.

46 Sachs, D. L., and P. J. Wyatt, "A conducting-slab model for electromagnetic propagation of lateral waves in an inhomogeneous jungle," *Radio Science,* vol. 3, 1968, pp. 125–134.

47 Reudink, D. O., and M. F. Wazowicz, "Some propagation experiments relating foliage loss and diffraction loss at X-band and UHF frequencies," *IEEE Trans. on Commun.,* vol. COM-21, 1973, pp. 1198–1206.

48 Dence, D., and T. Tamir, "Radio loss of lateral waves in forest environments," *Radio Science,* vol. 4, 1969, pp. 307–318.

49 Tamir, T., "On radio-wave propagation in forest environments," *IEEE Trans. Anten. Propagat.,* vol. AP-15, 1967, pp. 806–817.

50 Tamir, T., "Radio wave propagation along mixed paths in forest environments," *IEEE Trans. Anten. Propagat.,* vol. AP-25, 1977, pp. 471–477.

51 Swarup, S., and R. K. Tewari, "Depolarization of radio waves in jungle environment," *IEEE Trans. Anten. Propagat.,* Vol. AP-27, 1979, pp. 113–116.

52 Vogel, W. J., and J. Goldhirsch, "Tree attenuation at 869 MHz derived from remotely piloted aircraft measurements," *IEEE Trans. Anten. and Propagat.,* vol. AP-34, 1986, pp. 1460–1464.

53 Vogel, W. J., and J. Goldhirsch, "Tree attenuation at 869 MHz derived from remotely piloted aircraft measurements," *IEEE Trans. Anten. and Propagat.,* vol. AP-34, 1986, pp. 1460–1464.

54 Lebherz, M., W. Weisbeck, and K. Krank, "A versatile wave propagation model for the VHF/UHF range considering three dimensional terrain," *IEEE Trans. Anten. Propagat.,* vol. AP-40, 1992, pp. 1121–1131.

55 Matzler, C., "Microwave (1–100 GHz) dielectric model of leaves," *IEEE Trans. Geoscience and Remote Sensing,* vol. 32, 1994, pp. 947–949.

56 Seker, S., "Multicomponents discrete propagation model of forest," *IEE Proc. Microwave Anten. Propag.,* Vol. 142, No. 3, 1995, pp. 357–363.

57 Ponomarev, G. A., A. N. Kulikov, and E. D. Telpukhovsky, *Propagation of Ultra-Short Waves in Urban Environments*, Tomsk, USSR: Rasko, 1991.

58 Blaunstein, N., "Distribution of angle–of–arrival and delay from array of building placed on rough terrain for various elevation of base station antenna," *Journal of Communic. and Networks*, vol. 2, no. 4, 2000, pp. 305–316.

59 D. Katz, D., N. Blaunstein, M. Hayakawa, and Y. Sanoh Kishiki, "Radio maps design in Tokyo City based on stochastic multi-parametric and deterministic ray tracing approaches," *Antennas and Propagation Magazine*, Vol. 51, No. 6, 2009, pp. 200–208.

60 Blaunstein, N., I. Z. Kovacs, Y. Ben-Shimol, J. Bach Andersen et al., "Prediction of UHF path loss for forest environments," *Radio Sci.*, Vol. 38, No. 3, 2003, pp. 251–267.

61 Furutsu, K., "On the statistical theory of electromagnetic waves in a fluctuating medium (I)," *J. Res. NBS*, Vol. 67D (Radio Propagation), No. 3, 1963, pp. 303–323.

62 Furutsu, K., *On the Statistical Theory of Electromagnetic Waves in a Fluctuating Medium* (II), National Bureau of Standards Monograph 79, Boulder, Co., USA, 1964.

63 Twersky, V., "Multiple scattering of electromagnetic waves by arbitrary configurations," *J. Math. Phys.*, vol. 8, 1967, pp. 569–610.

64 Weissberger, M. A., "An initial critical summary of models for predicting the attenuation of radio waves by trees," *ESD-TR-81-101*, EMC Analysis Center, Annapolis, MD, USA, 1982.

65 Torrico, S. A., and R. H. Lang, "Bistatic scattering effects from a tree in a vegetated residential environment," *Proc. of National URSI Meeting*, Boulder, Colorado, 5-8 January, 1998, pp. 24–25.

66 Torrico, S. A., H. L. Bertoni, and R. H. Lang, "Modeling tree effect on path loss in a residental environment," *IEEE Trans. Anten. Propagat.*, vol. 46, 1998, pp. 107–119.

67 Lang, R. H., "Electromagnetic backscattering from a sparse distribution of lossy dielectric scatterers," *Radio Sci.*, vol. 16, pp. 15–30, 1981.

68 McPetrie, J. S., and L. M. Ford, "Experiments on propagation of 9.2 cm wavelength, especially on the effects of obstacles," *J. Inst. Electrical Engineers*, London, vol. 93, no. 3-A, 1946, pp. 531–543.

69 Labrone, A. M., P. E. Martin, and C. W. Chapman, "High gain measurement of VHF and UHF behind a grove of treks," *IRE Trans. Anten. Propagat.*, vol. 9, 1961, pp. 487–491.

70 Murray, O. M., "Attenuation due trees in the VHF/UHF bands," *Marconi Rev.*, vol. 37, no. 192, 1974, pp. 41–50.

71 Herbstrei, J. W., and W. Q. Crichlow, "Measurement of the attenuation of radio signals by jungles," *Radio Sci.*, vol. 68D, no. 8, 1964, pp. 903–911.

72 *Recommendation in Vegetation*. Recommendation of ITU-RP.833.3 (1992-1994-1999-2001).

73 Popov, V. I., "Propagation of radio waves in the forest," *Report N 3731*, L'vov Polytechnical Institute (LPI), L'vov, Ukraine, 1981/1983 (in Russian).

74 Popov V., A. Chaiko, V. Homicky, and N. Mogorit, "Effective complex dielectric permittivity of forest media for radio waves," *Latvian Journal of Physics and Technical Sciences*, No. 2, 2001, pp. 46–50.

75 Popov, V. "UHF radio wave propagation through woodlands in cellular mobile communication systems," *Proc. of 44th Int. Sci. Conf.*, Riga, Latvia, October 11–13 2003, Seria 6, 2004, 12 pages.

76 Popov, V. "VHF radio wave propagation through woodlands in cellular mobile communication systems," *Proc. of 44th Int. Sci. Conf.*, Riga, Latvia, October 11–13 2003, Seria 6, 2004, 12 pages.

77 Popov V. I., *Fundamentals of Cellular Communication Standard GSM*, Moscow: ECOTRENDS, 2005 (in Russian).

78 Blaunstein, N., D. Censor, D. Katz, A. Freedman, I. Matityahu, "Radio propagation in rural residential areas with vegetation," *J. Electromagnetic Waves and Applications*, Vol. 17, No. 7, 2002, pp. 1039–1041; *Progress In Electromagnetic Research*, PIER 40, 2003, pp. 131–153.

79 Blaunstein, N., and M. Levin, "VHF/UHF wave attenuation in a city with regularly spaced buildings," *Radio Sci.*, vol. 31, no. 2, 1996, pp. 313–323.

80 Blaunstein, N., and M. Levin, "Propagation loss prediction in the urban environment with rectangular grid-plan streets," *Radio Sci.*, vol. 32, no. 2, pp. 453–467, 1997.

81 Blaunstein, N., "Average field attenuation in the nonregular impedance street waveguide," *IEEE Trans. Anten. and Propagat.*, vol. 46, no. 12, 1998, pp. 1782–1789.

82 Black, D. M., and D.O. Reudink, "Some characteristics of radio propagation at 800 MHz in the Philadelphia area," *IEEE Trans. Veh. Technol.*, vol. 21, no. 1, 1972, pp. 45–51.

83 Reudink, D. O., "Comparison of radio transmission at X-band frequencies in suburban and urban areas," *IEEE Trans. Anten. Propagat.*, Vol. 20, No. 4, 1972, pp. 400–405.

84 Chan, G. K., "Propagation and coverage prediction for cellular radio systems," *IEEE Trans. Veh. Technol.*, vol. 40, no. 5, 1991, pp. 665–670.

85 Harley, P., "Short distance attenuation measurements at 900 MHz and 1.8 GHz using low antenna heights for microcells," *IEEE J. Select. Areas Communic.*, Vol. 7, No. 1, 1989, pp. 5–11.

86 Stewart, K., and D. Schaeffer, "The microcellular propagation environments," *Proc. of Symp. on Microcellular Technology*, II., USA, March 1992, pp. 19–26.

87 Crosskopf, R., "Prediction of urbam propagation loss," *IEEE Trans. Anten. Propagat.*, vol. 42, no. 5, 1994, pp. 658–665.

88 Chia, S. T. S., "Eadiowave propagation and handover criteria for microcells," *British Telecom Tech. J.*, vol. 8, no.1, 1990, pp. 50–61.

89 Steele, R., "The cellular environment of lightweight hand-held portable," *IEEE Communic. Magazine*, no. 1, 1989, pp. 20–29.

90 Dersch, U., and E. Zollinger, "Propagation mechanisms in microcell and indoor environments," *IEEE Trans. Veh. Technol.*, vol. 43, no. 10, 1994, pp. 1058–1066.

91 Xia, H. H., H. L. Bertoni, L. R. Maciel et al., "Radio propagation characteristics for line-of-sight microcellular and personal communications," *IEEE Trans. Anten. Propagat.*, vol. 41, no. 10, 1993, pp. 1439–1447.

92 Rustako, A. J., Jr., N. Amitay, M.J. Owens, and R.S. Roman, "Radio propagation at microwave frequencies for line-of-sight microcellular mobile and

personal communications," *IEEE Trans. Veh. Technol.*, vol. 40, no. 2, 1991, pp. 203–210.

93 Maciel, I. R., H. L. Bertoni, and H. H. Xia, "Unified approach to prediction of propagation over buildings for all ranges of base station antenna height," *IEEE Trans. Veh. Technol.*, vol. 42, no. 1, 1993, pp. 41–45.

94 Tan , S.Y., and H. S. Tan, "UTD propagation model in an urban street scene for microcellular communications," *IEEE Trans. Electromag. Compat.*, vol. 35, no. 4, 1993, pp. 423–428.

95 Tan, S. Y., and H.S. Tan, "A theory of propagation path loss characteristics in a city street-grid scene," *IEEE Trans. Electromagn. Compat.*, Vol. 37, 1995, pp. 333–342.

96 Blaunstein, N. "Prediction of cellular characteristics for various urban environments," *IEEE Anten. Propagat. Magazine*, vol. 41, No. 6, 1999, pp. 135–145.

97 Blaunstein, N., D. Katz, D. Censor, A. Freedman, I. Matityahu, and I. Gur-Arie, "Prediction of loss characteristics in built-up areas with various buildings' overlay profiles," *IEEE Anten. Propagat. Magazine*, vol. 43, no. 6, 2001, pp. 181–191.

98 Okumura, Y., E. Ohmori, T. Kawano, and K. Fukuda, "Field strength and its variability in the VHF anf UHF land mobile radio service," *Review Elec. Commun. Lab.*, vol. 16, 1968, pp. 825–843.

99 Hata, M., "Empirical formula for propagation loss in land mobile radio services," *IEEE Trans. Veh. Technol.*, vol. VT-29, 1980, pp. 317–325.

100 Saleh Faruque, *Cellular Mobile Systems Engineering*, Artech House, Boston-London, 1994.

101 Walfisch, J. and H. L. Bertoni, "A theoretical model of UHF propagation in urban environments," *IEEE Trans. Anten. and Propagat.*, vol. AP-38, 1988, pp. 1788–1796.

102 Xia, H. H. and H. L. Bertoni, "Diffraction of cylindrical and plane waves by an array of absorbing half screens," *IEEE Trans. Anten. and Propagat.*, vol. 40, 1992, pp. 170–177.

103 Bertoni, H. L., W. Honcharenko, L.R. Maciel, and H.H. Xia, "UHF propagation prediction for wireless personal communications," *Proc. IEEE*, Vol. 82, No. 9, 1994, pp. 1333–1359.

104 Morgan, M, and V. Evans, "Synthesis and analysis of polarized ellipses," in *Antennas of Elliptic Polarization*, NY, 1961, 385 p.

105 Lee, W. C., and R. H. Brandt, "The elevation angle of mobile radio signal arrival," *IEEE Trans. Communic.*, vol. 21, No. 11., 1973, pp. 1194–1197.

106 Bitler Y. S., "A two-channel vertically spaced UHF diversity antenna system," *Proc. of Microwave Mobile Radio Symp.*, Boulder, CO, 1973, pp. 13–15.

CHAPTER SIX

Indoor Radio Propagation

Indoor use of wireless systems poses one of the biggest design challenges, as indoor radio propagation is essentially a Black Art. Personal communications systems (PCS), wireless local area networks (WLANs), wireless private branch exchanges (WPBXs), and Home Phoneline Network Alliance (HomePNA, IEEE 802.11x, etc.) are the services that are being deployed in indoor areas on an increasing scale. The latter application of indoor wireless networks is proving to have a large market as it will be integrated to the emerging Digital Subscriber WLAN technologies. The present deployment of WLAN services is reaching out to offices, schools, hospitals, and factories. The increasing demand for indoor radio applications, such as wireless LAN, "Smart house," and so on, develops a need to design and analyze those systems wisely and efficiently. An important consideration in successful implementation of the PCS is indoor radio communication. In the design process of those systems, the designer is required to place the picocell antennas (at ranges not more than 100 m) in a way that will provide an optimal coverage of the building area [1–4]. Recent standardization of femto and pico base stations (BS), also called the femto access points (FAPs) and pico access points (PAPs), deployed in indoor environments, availability of the 4-G mobile broadband technologies for indoor environments, incorporation of the self-organized networks (SONs) algorithms into the optimization techniques of the 4-G mobile services, growing demand for mobile data services, and a constantly increasing number of indoor mobile subscribers motivate significant research investments in areas of high data rate communications over the wireless broadband MIMO media and

Radio Propagation and Adaptive Antennas for Wireless Communication Networks: Terrestrial, Atmospheric, and Ionospheric, Second Edition. Nathan Blaunstein and Christos G. Christodoulou.
© 2014 John Wiley & Sons, Inc. Published 2014 by John Wiley & Sons, Inc.

location-based services (LBS). Implementation of both high-data rate MIMO communications and the LBS require accurate modeling of the indoor propagation conditions. Furthermore, decisions on the evolution of wireless technologies are made based on the fundamental understanding of the radio frequency (RF) propagation phenomena.

Indoor radio communication covers a wide variety of situations ranging from communication with individuals walking in residential or office buildings, supermarkets, or shopping malls, to fixed stations sending messages to robots in motion in assembly lines and factory environments of the future. The indoor radio propagation modeling efforts can be divided in two categories. In the first category, transmission occurs between a unit located outside a building and a unit inside [5–7]. Expansion of current cellular mobile services to indoor application of the two types of services has been the main thrust behind most of the measurements in this category.

In the second category the transmitter and the receiver are both located inside the building [8–11]. Establishment of specialized indoor communication systems has motivated most of the researchers in this category. Although the impulse response approach is compatible with both, it has been mainly used for measurements and modeling effort reported in the second category.

There is large variety of different models developed in recent decades to describe the propagation of signals in indoor environments [12–23]. Their ability to predict the behavior of signals in indoor communication channels is crucial, and the confusion and lack of correlation between these models diminish their usefulness. The thorough understanding of these models and their unification to a more applicable one will allow a better behavioral prediction and better capabilities in the design of indoor communication networks. The indoor radio propagation environment is very complex and has many specific features and characteristics [1, 3, 4]. Adding all these variables together produces a very complex problem that has to be dealt with efficiently and elegantly. Every indoor communication system, as well as wireless outdoor system (see Chapter 1), has a different structure and requirements due to their various applications. Therefore, giving an accurate answer to each indoor communication system using the same models is complex. Path loss is difficult to calculate for an indoor environment. Because of the variety of physical barriers and materials within the indoor structure, the signal does not predictably lose energy. Walls, ceilings, and other obstacles usually block the path between receiver and transmitter. Depending on the building construction and layout, the signal usually propagates along corridors and into other open areas. In some cases, transmitted signals may have a direct path (line-of-sight [LOS]) to the receiver. LOS examples of indoor spaces are warehouses, factory floors, auditoriums, and enclosed stadiums. In most cases, the signal path is obstructed (non-line-of-sight [NLOS]). Finally, those who are involved in the wireless discipline whether as a designer or as a user must be aware of the different construction materials used for the interior and exterior walls, and of the location of a building for the best position of WLAN radio equipment. For optimal performance, the user should also consider work activities. Ultimately, the WLAN user needs to understand the relationship between indoor propagation effects and how WLAN performance is affected.

The indoor and the outdoor channels are similar in their basic features: they both experience multipath dispersions caused by a large number of reflectors and scatterers. As illustrated further in this book, they can be described using the same mathematical models. However, there are also major differences, which we want to describe here briefly.

The conventional outdoor mobile channel (with an elevated base antenna and low-level mobile antennas) is stationary in time and nonstationary in space. The temporal stationary picture is observed due to the fact that the signal dispersion is usually caused by large fixed objects (such as buildings). In comparison, the effects caused by people and vehicles in motion are negligible. The indoor channel, on the contrary, is stationary neither in space nor in time. Temporal variations in the indoor channel statistics are due to the motion of people and equipment around the low-level portable antennas.

The indoor channel is characterized by higher path losses and sharper changes in the mean signal level, as compared to the mobile channel [3, 24, 25]. Furthermore, applicability of a simple negative-exponent distance-dependent path loss model, well established for the outdoor channels (see Chapter 5), is not universally accepted for the indoor channel. Rapid motion and high velocities, typical of mobile users, are absent in an indoor environment. The Doppler shift effects, that is, the frequency-selective *fast fading* effects (see Chapter 1) in the indoor channel, are therefore negligible.

Maximum excess delay for the mobile channel is typically several microseconds if only the local environment of the mobile is considered, and more than 100 μs without distant reflectors, and 10–20 μs with distant reflectors. The indoor channel, on the contrary, is characterized by excess delays of less than 1 μs, and an *rms* delay spread in the range of several tens to several hundreds of nanoseconds (most often less than 100 ns [3, 26]).

As a result, for the same level of intersymbol interference (ISI), transmission rates can be much higher and the bit-error-rate (BER) can be much lower in indoor environments [27]. Also, the relatively large outdoor mobile transceivers are powered by the vehicle's battery with an antenna located away from the mobile user. This is in contrast with lightweight portables normally operated close to the user's body. As a result, much higher transmitted powers are feasible in a mobile environment.

Finally, the indoor radio channel differs from the outdoor mobile or personal radio channel in two principal aspects: the distances covered, which are much smaller, and the variability of the environment, which is much greater for smaller transmitter–receiver separation distances. The effects occurring in the time delay (TD) or time-of-arrival (TOA) and angle-of-arrival (AOA) domains will be discussed in Chapter 8. Here, we only will show the importance to account for the signal distribution not only in path loss-distance domain, but also in joint TOA and AOA domains for positioning and service of each subscriber located in indoor environment of service.

6.1. MAIN PROPAGATION PROCESSES AND CHARACTERISTICS

The propagated electromagnetic signal in the indoor environment can undergo three primary physical mechanisms. These are reflection, diffraction, and scattering. The following definitions assume small signal wavelength, large distances (relative to wavelength), and sharp edges.

Reflection occurs when the radio wave impinges on an obstacle whose dimensions are considerably larger than the wavelength of the incident wave. A reflected wave can either decrease or increase the signal level at the reception point. Reflections occur from the ground surface and from buildings and walls. In practice, not only metallic materials, but also dielectrics (or electrical insulators) cause reflections. Other materials will reflect part of the incident energy and transmit the rest. The exact amount of transmission and reflection is also dependent on the angle of incidence, material thickness, and dielectric properties. The actual signal levels reflected from insulators depends, in a very complicated way, on many characteristics such as geometry, different materials' characteristics, and so on. Major contributors to reflection are walls, floors, ceilings, and furniture.

Diffraction occurs when direct visibility between the transmitter and the receiver can be obstructed by sharp obstacles (edges, wedges, etc.), the dimensions of which are considerably larger than the signal wavelength. The secondary waves resulting from the obstructing surface are present throughout the space and even behind the obstacle, giving rise to a bending of waves around the obstacle, even when a LOS path does not exist between the transmitter and receiver. At high frequencies of UHF/X-bands, diffraction, like reflection, depends on the geometry of the object as well as the amplitude, phase, and polarization of the incident wave at the point of diffraction. It is a deterministic process where the cumulative effect of rays arriving at the receiver can be described by the Fresnel integrals introduced in Chapter 4.

Scattering occurs when the medium through which the wave travels contains the obstacles whose dimensions are smaller than or comparable with the wavelength and where the number of obstacles per unit volume is large. Scattered waves are produced by rough surfaces, small objects, or by other irregularities in the channel. The nature of this phenomenon is not similar to the reflection and diffraction because radio waves are scattered in a greater number of directions with random phase and amplitude deviations, and at the receiver, a random cumulative effect is observed. From all the above-mentioned effects, scattering is most difficult to predict.

Multipath phenomena of multiple reflection, diffraction, and scattering give rise to additional radio propagation paths beyond the direct LOS path between the radio transmitter and receiver. Figure 6.1 shows how a transmitted radio

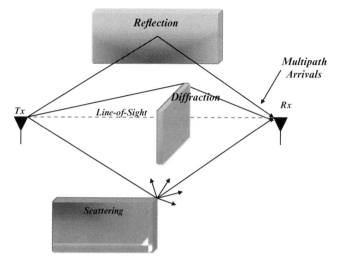

FIGURE 6.1. Multipath effects.

wave, in the indoor environment, reaches the receiving antenna in more than one path. It is clear that in the indoor propagation situation, it is very difficult to design an "RF friendly" building that is free from multipath reflections, diffraction around sharp corners, or scattering from wall, ceiling, or floor surfaces. To describe all these phenomena, the following characteristics of the channel, the same as for outdoor propagation, are usually used, for example, the attenuation or *path loss*, the *fast* and *slow fading* described in detail in Chapter 1. Regarding the indoor testing, fading effects are also caused by human activities inside buildings and are usually defined as slow variations of the total signal. Sometimes oscillating metal-bladed fans can cause rapid fading effects, which can be described using multipath time delay spreading. As the signal can take many paths before reaching the receiver antenna, the signals will experience different arrival times. Thus, a spreading in time (as well as frequency) can occur. Typical values for indoor spreading are less than 100 ns. Different arrival times ultimately create further degradation of the signal.

At the same time, the indoor radio channel differs from the traditional outdoor radio channel in two aspects: the distances covered, which are much smaller, and the variability of the environment, which is much greater for a much smaller range of T-R separation distance. It has been observed that propagation within buildings is strongly influenced by specific features such as the construction materials of the building and the building type. As explained previously, indoor radio propagation is dominated by reflection, diffraction, and scattering. However, the conditions are much more variable than in outdoor environments. For example, signal levels vary greatly depending on whether the interior doors are open or closed inside a building. The place where antennas are mounted also impacts large-scale propagation. Antennas mounted at the desk level in a partitioned office receive vastly different signals than those mounted on the ceiling.

Also, the smaller propagation distances make it more difficult to insure far-field radiation for all receiver locations and all types of antennas. One main reason for indoor signal losses is the partitions. Partitions losses can be divided into two kinds:

1. *Partition losses at the same floor.* Buildings have a wide variety of partitions and obstacles, which form the internal and external structures. Houses typically use a wood frame partition with plastic board to form internal walls and have also wood or nonreinforced concrete between floors. Office buildings, on the contrary, often have large, open areas (open plane), which are constructed by using movable office partitions so that the space may be reconfigured easily, and use metal-reinforced concrete between floors. Partitions that are formed as part of the building structure are called hard partitions, and partitions that may be moved and which do not span to the ceiling are called soft partitions. Partitions vary widely in their physical and electrical characteristics, making it difficult to apply general models to specific indoor installation.

2. *Partition losses between floors.* The losses between floors of a building are determined by its external dimensions and wall material, as well as by the type of construction used to create the floors and the external surroundings. Even the number of windows in a building and the presence of tinting (with attenuated radio energy) can impact losses between floors.

6.2. MODELING OF LOSS CHARACTERISTICS IN VARIOUS INDOOR ENVIRONMENTS

This section outlines models for path loss within buildings. As mentioned earlier, there is not a single theoretical model for path loss and fading effects prediction in indoor communications. For each separate situation (i.e., propagation along the corridor, inside the room, between floors and walls), a corresponding model is employed. Here we focus the reader's attention to the most widely used propagation models in today's practical applications.

6.2.1. Numerical Ray-Tracing UTD Model

Ray tracing and the unified theory of diffraction (UTD) have been used successfully in predicting the behavior of indoor communication channels [1, 28–30]. Here we present a UTD model for the analysis of complex indoor radio environments, in which microwave WLAN systems operate. The model employs a heuristic UTD diffraction coefficient capable of taking into account not only the effects of building walls, floors, and corners, but also the presence of metallic and penetrable furniture. A numerical tool based on an enhanced 3-D beam-tracing algorithm, which includes diffraction phenomena, has been developed to compute the field distribution, providing description of the scattered field and a physical insight into the mechanisms responsible for the multipath phenomenon. Numerical results show that the electromagnetic field distribution and the channel performance are significantly influenced by the diffraction processes arising from the presence of furniture.

The Field Prediction. The electromagnetic field is represented in terms of diffracted and ray-optics fields. The various elements of the environment are modeled as junctions of thin flat multilayered lossy or lossless structures. The geometric optics (GO) field is computed by means of reflection **R** and transmission **T** matrices, while the diffracted field is evaluated by means of a suitable UTD heuristics diffraction coefficient **D**. The adopted diffraction coefficient accurately models the field interaction with furniture edges and junctions between thin flat plates of different materials so that all significant field processes, which take place in the indoor environment, are rigorously modeled. As in indoor environments, the field contributions arising from double diffraction are small [1, 28–30]; only a single diffraction process is considered in this model. The field prediction procedure described previously is outlined in Figure 6.2a extracted from Reference 30. The structure which is illuminated by an incident electric field E^i can be modeled by a thin flat penetrable plate located near a partially reflecting plane. To simplify the graphical representation, only the rays that have experienced up to two interactions are taken into account. Finally, the direct, reflected, diffracted, and diffracted-reflected ray fields are taken into consideration [30].

Based on such an assumption, the broadcast beam-tracing algorithm was proposed in Reference 30 for the computation of the electromagnetic field and the characteristics of the radio channel.

The Beam-Tracing Algorithm. This numerical algorithm consists of two parts. The *first* determines the ray optical paths, while the *second* evaluates the electromagnetic field distribution. The field radiated from the antenna is modeled by means of beams shooting from the antenna location toward all space directions, independently of the observation point. During the propagation, the beam can impinge, totally or partially, on a surface describing the environment (see Fig. 6.2), it can capture the observation point, or finally, it may not intercept any of the environment elements. In the first case, using Snell's law, the transmitted and the reflected beams are evaluated. If the beam partially impinges on the surface, it is split in new beams in a way that they totally intercept, or not, the surface under construction. Then, the ray optical paths of the diffracted field are determined. To this end, a subdivision of the diffracted ray tube, identified by the two Keller's cones whose tips coincide with

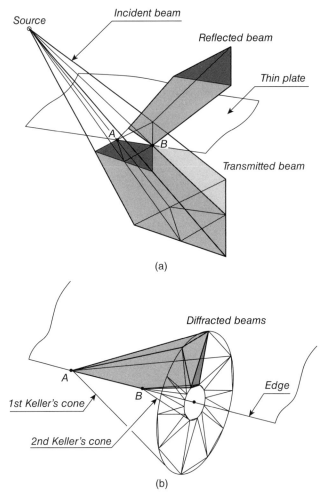

FIGURE 6.2. (a) The beam impinges on a surface (extracted from Reference 30). (b) A subdivision of the diffracted ray tube (extracted from Reference 30).

the extremes of the segment excited by the incident ray beam, is performed (see Fig. 6.2b [30]). If the beam does not intercept any obstructions, it does not produce any secondary beams, and, consequently, it is removed from the field computation procedure. For each observation point lighted by the beam, the exact ray path is computed by means of the image method.

In the second part of the beam-tracing algorithm, the GO and diffracted fields are evaluated using the reflection, transmission, and diffraction matrices at the points where the incident field impinges. In the numerical procedure, only the edge diffraction processes excited by the GO field are taken into account.

To increase the numerical accuracy of the computational analysis, one can take into account the GO field contribution that has experienced up to five reflections/transmissions. The diffracted field arising from any scattering object is considered to be excited either by the LOS GO field, or by the GO contributions that have experienced up to three reflections/transmissions. Finally, the diffracted field contribution is taken into account whether it reaches the observation point directly or after three reflections/transmissions processes.

The complete analysis was carried out in References 1 and 28–30 using ray-tracing approach. On the basis of experimental data and numerical results of the UTD ray model, it was shown that the presence of furniture in the LOS region gives rise to greater field diffusion and additional attenuation of the received signal. This effect decreases efficiency of the channel performance in wireless indoor communication systems. Because of its numerical accuracy and limited computational requirements, the UTD ray-tracing models can be successfully employed to estimate the channel performance and the total field distribution (radio coverage) directly during the design phase of indoor wireless communication.

6.2.2. Physical Waveguide Model of Radio Propagation inside a Building Corridor

This model is an analytic model of radiowave propagation along an impedance corridor as a waveguide. This model, which differs from other models [14, 18], allows us to analyze the electromagnetic field distribution inside a building corridor to obtain an expression for the attenuation (extinction) length and the path loss.

The Geometry of the Problem. In what follows, we briefly present the guiding effects of the corridor based on the same theoretical approach that was followed for the outdoor street environment and described in detail in Chapter 5, Section 5.7. In other words, we model the corridor by a 2-D impedance parallel waveguide (Fig. 6.3a).

As $d >> \lambda$, where d is the corridor's width and λ is the wavelength, we can use the approximation of geometrical theory of diffraction (GTD). This approximation is valid as long as the first Fresnel zone, $\sim(\lambda x)^{1/2}$, equals or does not exceed the width of corridor d. In this case, $x \leq$ 30–50 m; $\lambda =$ 3–10 cm (L/X-band); $d =$ 2–3 m; $(\lambda x)^{1/2} \leq d$. The electrical properties of walls are defined by the surface impedance $Z_{TE} \sim \varepsilon^{-1/2}$, $\varepsilon = \varepsilon_0 - j(4\pi\sigma/\omega)$, where ε is the dielectric permittivity of the wall's surface, ε_0 is the dielectric constant of the vacuum, σ is the conductivity, and $\omega = 2\pi f$ is the angular frequency of the radiated wave.

We consider the 2-D problem of wave reflection without taking into account the reflection from the corridor's floor and ceiling because the corridor's height H and the position of the transmitter/receiver $h =$ 2–3 m are usually larger values than λ. Let us also assume, according to the 3-D geometry

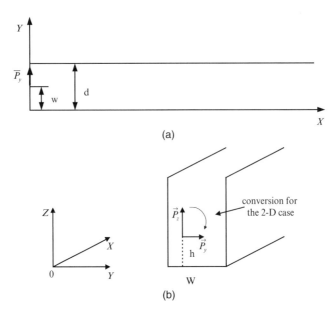

FIGURE 6.3. (a) The waveguide model of corridor; a view from the top. (b) The corridor in the 2-D case.

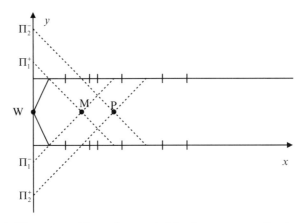

FIGURE 6.4. The waveguide modes created by the corresponding image sources.

presented in Figure 6.3b, that a vertical electric dipole is placed at the point $(0, w, h)$ at the (y, z)-plane, as shown in Figure 6.3b. In Figure 6.4, we return to the 2-D case to differentiate the slits (gaps due to windows, open doors, etc.) with the average length $\bar{l}$, and the screens (nontransparent room walls lining the corridor) with average length $\bar{L}$ are ranged by vertical segments.

To convert the problem to a 2-D case, we must consider the dipole to be oriented along the y-axis, that is, the horizontal dipole with respect to the (x, y)-plane, which corresponds to the well-known electromagnetic field equation described by the Hertz's potential vector $\Pi_y^i(x, y)$ [2]:

$$\nabla^2\Pi_y^i(x, y) - k^2\Pi_y^i(x, y) = \frac{4\pi i}{\omega}\mathbf{p}_y\delta(x)\delta(y - w). \tag{6.1}$$

The solution of such an equation can be presented using the Green's function presentation [2]:

$$\Pi_y^i(x, y) = \frac{4\pi i}{\omega}\mathbf{p}_y\frac{e^{ik\rho}}{\rho}. \tag{6.2}$$

Here, $\mathbf{p}_y$ is the electric momentum of a point horizontal electric dipole, $\rho = \sqrt{x^2 + y^2}$ is the distance from the source.

Total Field in 2-D Continuous Impedance Waveguide. The reflected field in a continuous waveguide (solid waveguide with no slits) can be determined according to results obtained in Section 5.7 as the sum of reflected modes replaced by the image sources (as shown in Fig. 6.4).

The normal mode expression inside the impedance wave guide (called the *discrete spectrum* of the total field) is given by (see Chapter 5):

$$\Pi_n(x, y) = D_1 e^{i\rho_n^{(0)}x}\exp\left\{-\frac{\|\ln\chi|R_n\|}{\rho_n^{(0)}d}\left(\frac{\pi n}{d}\right)x\right\}, \tag{6.3}$$

where $\rho_n^{(0)} = \sqrt{k^2 - K_n^2} = \sqrt{k^2 - (n\pi/d)^2}$ is a wave number of the propagating waveguide modes and $R_n = K_n - kZ_{EM}/K_n + kZ_{EM}$ is the complex reflection coefficient of waveguide modes; $D_1 = 2DR_n/i\rho_n^{(0)}d$; and $K_n = \pi n/d + i\ln(\chi|R_n|)/d - \varphi_n/d$ is the wave number of normal modes of number n that propagate along the waveguide with width d; φ_n is the phase of the reflection coefficient of normal modes of number n; $k = 2\pi/\lambda$; $\chi = \bar{L}/\bar{L}+\bar{l}$ is the parameter of discontinuity of the multislit waveguide, and D is the parameter of electrical dipole including its momentum $\mathbf{p}_y$ [2].

Following Reference 2 (see also Chapter 5), we can present the continuous spectrum of the total field for $x/d \gg 1$ as

$$\Pi_c \approx \sqrt{2}De^{i\left(\frac{3\pi}{4}\right)}\frac{1 - \chi|R_n|}{1 + \chi|R_n|}\frac{e^{ikx}}{x} \tag{6.4}$$

For the case of a perfectly conductive continuous waveguide, when $\chi \to 1$, $|R_n| = 1$, $Z_{EM} = 0$, we obtain $\Pi_c = 0$; that is, in the case of the ideal conductive waveguide, discussed in Section 5.8, the continuous part Π_c of the total field vanishes, and only the discrete spectrum of the normal waves propagates along the ideal waveguide without attenuation, according to (6.3). The above ideal scenario of the continuous waveguide corresponds to that where there are no slits, for example, when windows and doors are closed. Conversely, in the case of $\chi \to 0$, the discrete spectrum of guiding waves, described by (6.3), vanishes and continuous spectrum of the total field (6.4) converts to the classical case of spherical wave propagation in free space [2]. Indeed, in this scenario, the waveguiding effect disappears since there is no actual corridor environment and the considered case turns to be an open environment. So, the above-mentioned two limiting cases, the classical continuous waveguide and the free space propagation case, allow us to prove the validity of the corridor discontinuous waveguide model.

In real indoor situations, the maximal guiding effect of waves (i.e., waves accumulating effect) can be achieved when $0 < \chi < 1$, and some portion (small or significant) of the transmitted energy is dissipated outside the corridor, leading to a lower wave guiding effect.

The majority of reflected fields are waves created by specular reflection, assuming low wall roughness. Therefore, the total radiowave intensity can be derived as a function of a distance between the transmitter (T_x) and the receiver (R_x) by substituting (6.3) and (6.4) into the approximate expression, summarizing the intensity of modes with multiple reflections in the space domain. Finally, the intensity of the total field can be approximately obtained as

$$I(x) \approx \sum_{n=1}^{n_{\max}} [(\Pi_n + \Pi_c) \cdot (\Pi_n + \Pi_c)^*], \qquad (6.5)$$

where $(\Pi_n + \Pi_c)^*$ is the complex conjugate of $(\Pi_n + \Pi_c)$.

The path loss of the radio wave can be presented as [2]:

$$
\begin{aligned}
L &\approx 32.1 - 20\log_{10} f_0 - 20\log_{10}\left[\frac{1 - \chi[R_n]^2}{1 + \chi|R_n|^2}\right] \\
&\quad + 17.8\log_{10} x + 8.6\left\{-[\ln \chi|R_n|]\left(\frac{\pi n}{d}\right)\frac{x}{\rho_n^{(0)}d}\right\},
\end{aligned}
\qquad (6.6)
$$

where x is the distance between two terminals, receiver and transmitter, along the corridor.

Equation (6.6) can be rewritten [31] to describe path loss in the joint TOA and AOA domains, and differentiate the free space propagation loss from the guiding wave propagation loss along the corridor, that is,

$$L_t(t, \varphi) = 10\log_{10}(I_{total}(\tau, \varphi)) = L_o + L_{su}(t, \varphi), \qquad (6.7)$$

where L_o is a free space path-loss, which represents the attenuation along the direct path between the transmitter and the receiver, and L_{su} is a model of the path-loss induced by the wave guiding effect of the cylindrical waves propagation, or

$$
\begin{aligned}
L_{su}(t, \varphi) = 20\log_{10} \sum_{n=1}^{n_{max}} & \left\{ \left[\frac{(1 - \chi |R_n(t, \varphi)|)^2}{(1 + \chi |R_n(t, \varphi)|)^2} \right] \right. \\
& \left. + 8.6 \left[-|\ln(\chi \cdot |R_n(t, \varphi)|)| \frac{d(\pi n(t, \varphi) - \varphi_n)}{a^2 \cdot \rho_n^{(0)}} \right] \right\}.
\end{aligned}
\tag{6.8}
$$

The first term in (6.8) represents the joint influence of the reflection coefficient and the waveguide discontinuity parameter on the waveguide-induced path loss. Note that in scenarios where the wave number n increases, the first term in (6.8) approaches zero. The second term in (6.8) represents the influence of the waveguide geometry on the path loss. In scenarios with large wave number n, the second term in (6.8) increases, and therefore decreases the probability of the waveguide mode accumulation in the waveguide. Note that the path loss function along the corridor is obtained in Reference 31 by taking into account the spatial and temporal signal interaction with environment in the joint AOA-TOA domain.

Figure 6.5 explains the phenomenon described previously by presenting the wavemode number n, of the discrete spectrum accumulated at the total received field as a function of $r \equiv d/n$. It demonstrates the geometrical relation between the relative TOAs and AOAs as a function of propagating modes. The expressions for the relative TOA, τ, and AOA, φ, for the first three propagation modes are [31]:

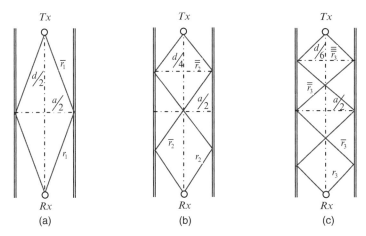

FIGURE 6.5. Wave mode number (ray reflections number) n geometrical presentation: (a) $n = 1$, (b) $n = 2$, (c) $n = 3$.

$$n = 1: \tau_1 = \frac{r_1 + \overline{r_1}}{d} = \frac{2\sqrt{(d/2)^2 + (a/2)^2}}{d}, \varphi_1 = \tan^{-1}\left(\frac{a}{d}\right) \quad (6.9a)$$

$$n = 2: \tau_2 = \frac{r_2 + \overline{r_2} + \overline{\overline{r_2}}}{d} = \frac{2 \cdot 2\sqrt{(d/(2 \cdot 2))^2 + (a/2)^2}}{d}, \varphi_2 = \tan^{-1}\left(\frac{2 \cdot a}{d}\right) \quad (6.9b)$$

$$n = 3: \tau_3 = \frac{r_3 + \overline{r_3} + \overline{\overline{r_3}} + \overline{\overline{\overline{r_3}}}}{d} = \frac{2 \cdot 3\sqrt{(d/(2 \cdot 3))^2 + (a/2)^2}}{d}, \varphi_3 = \tan^{-1}\left(\frac{3 \cdot a}{d}\right). \quad (6.9b)$$

Following similar arguments, the relative TOA and AOA of the n wave mode (n reflections of the propagating rays in the corridor) can be derived as

$$n: \tau_n = \frac{2 \cdot n \cdot \sqrt{\left(\frac{d}{2 \cdot n}\right)^2 + \left(\frac{a}{2}\right)^2}}{d} \quad \varphi_n = \tan^{-1}\left(\frac{a \cdot n}{d}\right). \quad (6.10)$$

Analysis of the Waveguide Corridor Model. Let us present some examples of simulation of the total path loss L in decibels (dB) in the space domain (i.e., along the radio path between the transmitter and the receiver) according to (6.6) versus distance between the transmitter and receiver. For our numerical computation, we considered the following parameters: the width of the corridor $d = 3$ m, the conductivity of walls $\sigma = 0.0133$ S/m, and the signal frequency $f = 900$ MHz [32]. The results of these path loss computations, according to (6.6), are shown in Figure 6.6a for the guiding modes with the number n varying from 1 to 10. For $n \geq 3$, the effect of these modes is negligible at ranges beyond 20 m, and we just have to subtract the attenuation from the first two main modes of the original signal power in order to get the total power of a signal (in decibels) for each distance d between the transmitter and the receiver located along the corridor waveguide. This effect was also shown in Reference 32, where it was experimentally obtained that only one to two main modes are important in the range of 10 and more meters from the transmitter. Therefore, in Figure 6.6b, we present the total field attenuation as a sum of the first two waveguide modes that fully describes the total path loss inside the corridor as a guiding structure versus the distance from the transmitter. We will compare this theoretical prediction of the path loss with the real experiment carried out in Reference 32 along the corridor.

Next, we analyze the total path loss along the corridor for both in AOA and TOA domains according to (6.8). Thus, Figure 6.7 shows the discrete spectrum of the accumulated average path loss at each normal waveguide mode n at the receiver in the AOA-TOA domain. The discrete nature of this spectrum is dictated by the considered specular reflection, which induces propagation of the discrete waveguide modes in the indoor corridor. The filled circles represent the signal strength level at each discrete propagating waveguide mode. Figure 6.7 shows that a variation in the discontinuity parameter χ causes a variation in the spread of the received wave intensity in the AOA domain.

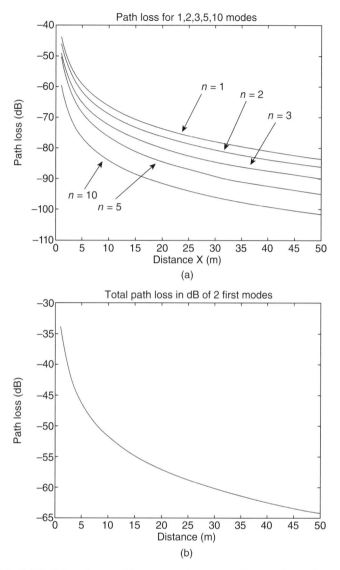

FIGURE 6.6. (a) Path loss for $n = 10$ wave modes versus distance from the transmitter. (b) Path loss for two first modes versus distance from the transmitter.

More detailed information on how the parameter of discontinuity affects signal power distribution in the joint AOA-distance and AOA-TOA domains will be discussed in Chapters 7 and 8 on the basis of recent experiments carried out in indoor environments [31, 33–34].

Note that the intensity distribution in the AOA domain can be explained by the dependence of the reflection coefficient R_n on the angle of incidence.

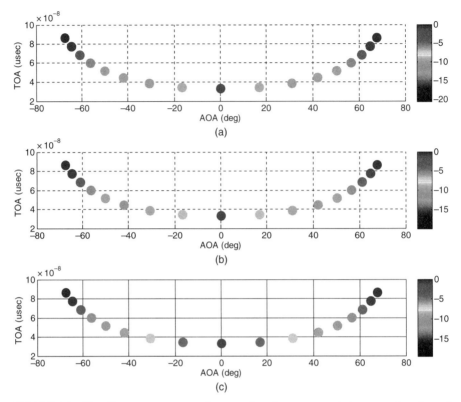

FIGURE 6.7. The discrete spectrum of accumulated average path loss as a function of the brokenness parameter at the receiver in AOA-TOA domain considering the LOS scenario along the corridor with R_x-T_x distance of 10 m and corridor width of 3 m with brokenness parameter (a) $\chi = 0.5$, (b) $\chi = 0.7$, (c) $\chi = 0.9$.

Thus, the reflection coefficient approaches $R_n \to 1$ when the angle of incidence approaches 180° [31, 33]. This observation agrees with Equation (6.8) and its geometrical interpretation in Figure 6.5 and Figure 6.7, where it is shown that the incidence angle θ decreases with an increase in the waveguide mode n.

6.2.3. Physical Model of Radio Propagation between Floors and Walls

Bertoni et al. [1, 21, 22], developed a theoretical model, based on the GTD, which explains the propagation between a transmitter and a receiver located on different floors of a building. Depending on the structure of the building and the location of the antennas, either direct ray propagation through floors or diffraction outside the building will determine the propagation character-istics and the range dependence of the signal. There are two paths over which propagation can take place:

1. paths that involve transmission through the floors
2. paths having segments outside the building and involving diffraction at window frames.

The paths through the floors include the direct ray, the multiple-reflected rays, and the rays that are transmitted through semitransparent walls and floors. These rays are contained entirely within the building perimeter. The diffracted ray paths involve transmission outside the building through windows and diffraction into paths that run alongside the face of the building and then reenter through another window at a different floor. For propagation of the direct ray through semitransparent floors, as indicated by path T in Figure 6.8, extracted according to Reference 1, the electromagnetic field strength in general reaching a receiving site is given by [21, 22]

$$|\mathbf{E}|^2 = \frac{Z_0 P_e}{4\pi L^2} \prod_m T_{floor,m}^2 \prod_n T_{wall,n}^2. \tag{6.11}$$

Here $Z_0 = 120\,\pi\,\Omega = 377\,\Omega$ is the free-space wave impedance, P_e is the effectively transmitted power and L is the direct distance between the transmitter (Tx) and the receiver (Rx) antennas. T_{floor} and T_{wall} are the loss coefficients of each floor and wall, respectively, passed by the direct ray. Such a direct ray, passing through three floors and two interior walls, is indicated in Figure 6.8. If one knows the absolute value of the reflection coefficient Γ of each wall and floor, then T_{floor} or T_{wall} can be calculated [21, 22]:

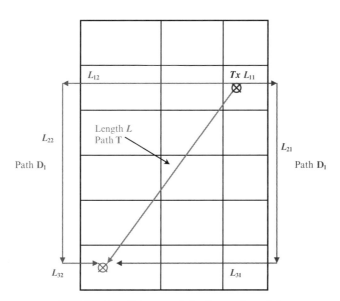

FIGURE 6.8. Scheme of the Bertoni model.

$$T = \sqrt{X(1 - |\Gamma|^2)}, \tag{6.12}$$

where X is a constant, obtaining from the concrete experiment. The signal can also reach other floors via paths that involve diffraction. Referring to paths D_1 and D_2 in Figure 6.8 (in general denoted as D_i), the field reaching the receiver via one such diffracted path is given by [21, 22]

$$|\mathbf{E}|^2 = \frac{Z_0 P_e}{4\pi} \frac{\prod_i D^2(\alpha_i) \prod_j T_{glass,j}^2 \prod_k T_{wall,k}^2}{\prod_m \sum_n L_{nm}}, \tag{6.13}$$

where L_{nm} is the length of D_i diffracted path. In the geometry of the concrete experiment carried out in the hotel schematically presented in Figure 6.8 according to References 21 and 22, L_{nm} is the length of D_1 and D_2, where $\prod_m \sum_n L_{nm} = (L_{11} + L_{21} + L_{31})(L_{12} + L_{22} + L_{32})$ and $T_{glass(m)}$ and $T_{wall(n)}$ are the transmission coefficients through glass and through interior walls crossed by path segments. In (6.13), $D(\alpha_i)$ is the diffraction coefficient for a propagating ray bending through angle α_i.

Depending on the construction of the building and its window frames, different choices may be made for the diffraction coefficient. For simplicity in investigating the relative strength of the total field associated with the direct ray and the diffracted ray, the coefficient for an absorbing wedge, obtained by Keller's diffraction theory [1], was used:

$$D(\alpha_i) = \frac{1}{2\pi k} \left[\frac{1}{2\pi + \alpha_i} - \frac{1}{\alpha_i} \right], \tag{6.14}$$

where $k = 2\pi/\lambda$ is the wave number. Thus, when propagation takes place through the floors, the signal will decrease rapidly with the number of floors separating the transmitter and the receiver.

On the contrary, if propagation occurs via diffracted paths, the signal will be small even for separation by a single floor but will decrease a bit slower with increased separation. For testing the model, an experiment was made according to Figure 6.8, in the frequency of 852 MHz, where, according to presented geometry, the angle $\alpha_i = \pi/2$. Measurements have shown that in each floor, the attenuation was about 12–13 dB. From various experiments, the coefficients for the walls, windows, and floors are

$$T_{wall} = 2.2 \, dB; \quad T_{glass} = 0.25 \, dB; \quad T_{floor} = 13.0 \, dB. \tag{6.15}$$

The total received power in decibels at the R_x position can be calculated according to (6.9) for the direct path through floors and walls as

$$P_{rDirect} = 10 \log_{10} \left[\lambda^2 |E|_{direct}^2 / (Z_0 \cdot 4\pi) \right] [dB], \tag{6.16}$$

where $P_{r\,Direct}$ is the power gain from direct propagation wave, $|E|^2$ is calculated according to (6.11) and $\lambda = c/f$ is the wavelength, $Z_0 = 120\pi\ [ohm]$ is the impedance in free space, and according to (6.11)

$$P_{rDiff} = 10\log_{10}\left[\lambda^2 |E|^2_{diff} / (Z_0 \cdot 4\pi)\right] [\text{dB}], \tag{6.17}$$

where $P_{r\,Diff}$ is the power gained from diffracted propagation wave and $|E|^2$ is calculated according to (6.11). Then the total received power will be

$$P_{rtotal} = P_{rDirect} + P_{rDiff}\ [\text{dB}]. \tag{6.18}$$

Additional numerical analysis of Bertoni's model and comparison with numerous experiments carried out by other researchers (see Section 6.3 later) have shown that despite the fact that this model offers very precise physical calculations that are suitable for different kinds of buildings, the attenuation effects due to shadowing caused by diffraction from internal obstructions are not taken into consideration. This type of attenuation must be accounted for because it can decrease the total strength of radio signal that reaches the receiver by 10–15 dB. However, the shadow effect is actual only at the upper floors, that is, when the difference between antenna locations is more than two to three floors.

Also another difficulty with the implementation of Bertoni's model is related to the fact that it requires a priori knowledge of the precise building architecture and the establishment of various propagation paths which by all means is a very difficult task to achieve. The estimation of path loss through walls and floors according to (6.11) is more precise compared with other existing empirical models (see Equation (6.20) and Equation (6.21)). Therefore, we will use (6.11) in future budget link design of indoor communication links, taking into account shadow effects caused by the internal obstructions located within the radio path between the two terminal antennas, following Reference 32 or the receipt proposed in Chapter 5 for link budget design.

6.2.4. Empirical Models

Such models are based mostly on numerous experiments carried out in various indoor environments as the best-fit prediction to the corresponding measured data. We start with a very simple model that modifies the well-known dual-slop model, usually used in outdoor environments (so-called two-ray model, see Chapter 5), and then we introduce the most applicable empirical model that is currently used for loss characteristics prediction in indoor communication links.

Modified Dual-Slop Model. The challenging problem in applying the well-known dual-slop models from the outdoor environment to the indoor environment is that we need to account for the wall and floor factors. In Reference

35, to characterize indoor path loss, a fixed path loss exponent $\gamma = 2$, just as in free space (see Chapter 5), was used, plus additional attenuation factors (in decibels) per floor, α_f, and per wall, α_w, timing on the number of floors, N_f, and walls, N_w, respectively, that is,

$$L = L_0 + 20\log r + N_f\alpha_f + N_w\alpha_w, \tag{6.19}$$

where r is the straight-line distance between the terminal antennas and L_0 is a free-space path loss at the referenced range of 1 m.

As no values for the wall and floor factors were reported in Reference 35, an improved model was developed, called the ITU-R model [36]. According to this dual-slop approach, only the floor loss is accounted for explicitly. The loss between points located at the same floor is accounted by changing the path loss exponent γ. The frequency effect is accounted for in the same manner as in free space (see Chapter 5), producing the following total path loss (in decibels):

$$L = 10\gamma\log r + 20\log f + L_f(N_f) - 28, \tag{6.20}$$

where γ is shown in Table 6.1 [36], and $L_f(N_f)$ is the floor attenuation factor, which varies with the number of penetrated floors N_f, as shown by Table 6.2 [36].

TABLE 6.1. Path loss exponent γ for the ITU-R model [36]

Frequency [GHz]	Residential	Environment	
		Office	Commercial
0.9	–	3.3	2.0
1.2–1.3	–	3.2	2.2
1.8–2.0	2.8	3.0	2.2
4.0	–	2.8	2.2

TABLE 6.2. Floor attenuation factor $L_f(N_f)$ in decibels for the ITU-R model [36]

Frequency [GHz]	Residential	Environment	
		Office	Commercial
0.9	–	9 (1 floor) 19 (2 foors) 24 (3 floors)	–
1.8–2.0	$4N_f$	$15 + 4(N_f - 1)$	$6 + 3(N_f - 1)$

Rappaport's Path Loss Prediction Model. Rappaport and his associates [3, 11, 13, 20] made a lot of experiments in various indoor environments in different locations and sites. The main goal of these experiments was to achieve unique parameters of attenuation and loss prediction on different kind of multifloor buildings.

Distance-Dependent Path Loss Model. In References 3 and 20, it was assumed that the mean path loss $\bar{L}$ is an exponential function of distance d with the power n:

$$\bar{L}(d) \propto \left(\frac{d}{d_0}\right)^{\gamma},\tag{6.21}$$

where $\bar{L}(d)$ is mean path loss, γ is the mean path loss exponent which indicates how fast path loss increases with distance; d_0 is a reference distance, usually chosen equal to 1 m in indoor communication links; and d is the transmitter–receiver separation distance. Absolute mean path loss, in decibels, is defined as the path loss from the transmitter to the reference distance d_0, plus the additional path loss [3], that is,

$$\bar{L}(d) = L(d_0) + 10 \cdot \gamma \cdot \log\left(\frac{d}{d_0}\right) [\text{dB}].\tag{6.22}$$

For these data, $L(d_0)$ is the reference path loss due to free space propagation from the transmitter to a 1-m reference distance, and calculated by

$$L(d_0) = 20\log\left(\frac{4\pi d_0}{\lambda}\right) [\text{dB}].\tag{6.23}$$

This empirical model takes into account the effects of shadowing by introducing in (6.22) a term X_σ, which describes the statistical character of slow fading within the indoor link and, as a random variable, satisfies the lognormal distribution with a standard deviation of σ in decibels (see definitions in Chapter 1). Then the total path loss within building equals, in decibels, [3]

$$L(d) = L(d_0) + 10 \cdot \gamma \cdot \log\left(\frac{d}{d_0}\right) + X_\sigma [\text{dB}].\tag{6.24}$$

For this model, the exponent γ and standard deviation σ were determined as parameters that are functions of building type, building wing, and number of floors between Tx and Rx. Thus, a model to predict the path loss for a given environment is given by [3]

$$L(d) = \bar{L}(d) + X_\sigma [dB],\tag{6.25}$$

TABLE 6.3. Path loss exponent and standard deviation for various types of buildings based on [3] measurements at a carrier frequency of 914 MHz

Place	γ	σ (dB)	Number of Locations
All Buildings			
All locations	3.14	16.3	634
Same floor	2.76	12.9	501
Through one floor	4.19	5.1	73
Through two floors	5.04	6.5	30
Through three floors	5.22	6.7	30
Grocery store	1.81	5.2	89
Retail store	2.18	8.7	137
Office Building 1			
Entire building	3.54	12.8	320
Same floor	3.27	11.2	238
West wing 5th floor	2.68	8.1	104
Central wing 5th floor	4.01	4.3	118
West wing 4th floor	3.18	4.4	120
Office Building 2			
Entire building	4.33	13.3	100
Same floor	3.25	5.2	37

where X_σ is a zero mean lognormal distributed random variable with standard deviation σ and accounts for attenuation due to diffraction from the environment. Table 6.3 summarizes the mean path loss exponent γ, standard deviation σ about the mean $\bar{L}$ for different indoor environments, and the number of measurement locations used to compute the statistics for each building. From Table 6.3, it can be seen that the parameters for path loss prediction for all antenna locations are $\gamma = 3.14$ and $\sigma = 16.3$ dB. This large value of σ is typical for data collected from different building types and indicates that only 68% of actual measurements will be within ±16.3 dB of the predicted mean path loss. As stated in References 3 and 20, these parameters may be used in modeling the first-order prediction of mean signal strength when only the Tx–Rx separation is known but no specifics about the building. In multifloor environments, (6.22) is used to describe the mean path loss as a function of distance. Equation (6.22) emphasizes that the mean path loss exponent is a function of the number of floors between Tx and Rx. The values of γ (*multifloor*) are given in Table 6.3 for use in (6.22), and this equation can be rewritten as [3]

$$\bar{L}(d) = L(d_0) + 10 \cdot \gamma(multifloor) \cdot \log\left(\frac{d}{d_0}\right). \tag{6.26}$$

TABLE 6.4. Average floor attenuation factor in decibels for one, two, three, and four floors in two office buildings [3]

Building	FAF	σ	Number of Locations
Office Building 1			
Through one floor	12.9	7.0	52
Through two floors	18.7	2.8	9
Through three floors	24.4	1.7	9
Through four floors	27.0	1.5	9
Office Building 2			
Through one floor	16.2	2.9	21
Through two floors	27.5	5.4	21
Through three floors	31.6	7.2	21

Floor Attenuation Factor (FAF). In (6.26), γ (*multifloor*) is a function of the number of floors between *Tx* and *Rx*. Alternatively, a constant floor attenuation factor (FAF) (in decibels), which is a function of the number of floors and building type, was added in References 3 and 20 to the mean path loss described by Equation (6.22), which uses the "same floor" path loss exponent for a particular building type:

$$\bar{L}(d) = L(d_0) + 10 \cdot n(same\ floor) \cdot \log\left(\frac{d}{d_0}\right) + FAF\ [dB], \qquad (6.27)$$

where *d* is in meters and $L(d_0)$ is the free space path loss determined by (6.23). Table 6.4 [3], gives the floor attenuation factors FAF (in decibels) and the standard deviation σ (in decibels) between the measured and predicted path loss and the number of discrete measurement locations used to compute the statistics.

Soft Partition and Concrete Wall Attenuation Factor. The above formulas include the effects of *Tx–Rx* separation, building type, and the number of floors between the *Tx* and *Rx*, and the first step for including site information to improve propagation predictions. There are often obstructions between the transmitter and receiver even when the terminals are on the same floor.

The model considers the path loss effects of *soft partition* and *concrete walls* between the *Tx* and *Rx*. The model assumes that path loss increases with distance as in *free space* ($\gamma = 2$), as long as there are no obstructions between the *Tx* and *Rx*. Then, attenuation factors for each soft partition and concrete walls that lie directly between *Tx* and *Rx* are included. Let *p* be the number of soft partitions, and q is the number of concrete walls in the direct path between Tx and Rx. The mean path loss predicted by this model is

$$\bar{L}(d) = 20 \cdot \log_{10}\left(\frac{4\pi d}{\lambda}\right) + p \cdot \text{AF(soft partition)} + q \cdot \text{AF(concrete wall)} \, [dB],$$

(6.28)

where AF(soft partition) is the attenuation factor per soft partition and AF(concrete wall) is the attenuation factor per concrete wall. Typical values for AF are 1.4 dB for soft partition and 2.4 dB for concrete wall.

Numerical Simulations of Rappaport's Model. To compare different approaches described by Equation (6.26) and Equation (6.27), let us introduce some typical parameters obtained experimentally by Rappaport et al., as presented in Table 6.3 and Table 6.4. Thus, for simulation purposes, we used the following parameters: $f = 915$ MHz ($\lambda = 0.32$ m); γ for the same floor $= 3.27$; $\gamma = 4.19$ for the 1st floor; $\gamma = 5$ for the 2nd floor; $\gamma = 5.22$ for the 3rd floor; $\gamma = 5.35$ for the 4th floor; $\gamma = 5.45$ for the 5th floor; FAF for the 1st floor $= 12.9$ dB; FAF for the 2nd floor $= 18.7$ dB; FAF for the 3rd floor $= 24.4$ dB; FAF for the 4th floor $= 26$ dB; FAF for the 5th floor $= 27$ dB.

In Figure 6.9, the path loss versus the number of floors is presented according to (6.26) with γ(multifloor), and in Figure 6.10 it is according to the FAF model (6.27) with γ(same floor). From Figure 6.9, the attenuation of radio wave penetrating through the first three floors increases linearly and then according to the square root curve dependence, which is in a good agreement with Bertoni's model, taking into account the diffraction path loss described by Equation (6.13).

Additional analysis of Rappaport's model has shown that there are some difficulties to using this model in practical situations for indoor environments. Rappaport's model relies heavily on experimental data to determine the required parameters that can be used for modeling purposes.

Let us now compare two approaches, Bertoni's and Rappaport's, through the prism of experimental data. Rappaport's model is based on experimental factors (FAF and γ), and it does not provide the radiowave propagation characteristic, such as the attenuation inside buildings, in the case when the transmitter and the receiver are located at different floors.

Bertoni's model states that when there is a several-floor separation between the terminal antennas, the additional loss occurs due to the diffraction path through the frames of the windows according to (6.13). The slow fading effect, called shadowing (see definitions in Chapter 1), can easily be represented in the Bertoni model as the additional effect of the wave field that comes from the diffraction paths. In the Rappaport model, the shadowing effect is accounted by the FAF factor and by the path loss exponent γ. Obviously, the slow fading factor must be added to Rappaport's model during the measurement phase, where parameters FAF and γ are usually determined. In general, Rappaport's

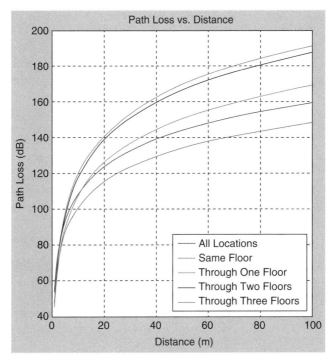

FIGURE 6.9. Path loss versus distance from the transmitter for various scenarios inside building.

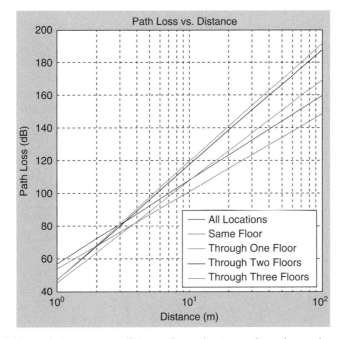

FIGURE 6.10. Path loss versus distance from the transmitter for various scenarios inside building.

model is suitable for buildings with a low number of floors (up to 4–5). The results are very similar to Bertoni's model in the lower floors when radio propagation mostly occurs through the direct path as described by (6.11). When taller buildings are tested, Bertoni's model is a more appropriate model to use. In the following paragraph, we indicate additional improvements to Bertoni's model by accounting for the shadow effects.

Suggested Model. As mentioned above, the models that predict link budgets for the indoor environment are complex, sometimes using parameters without any physical meaning and explanation. Here we suggest a model, proposed in References 31–33, for the radio propagation between floors that takes into consideration the physical media and parameters of the total path loss obtained from experiments. In general, the suggested model will follow the formula

$$L_{total} = \bar{L} + X_\sigma \ [dB], \tag{6.29}$$

which is similar to (6.25) as shown in the Rappaport model, but now $\bar{L}$ is the loss achieved from a direct propagated ray with NLOS features and X_σ is a zero mean of lognormally distributed random variable with standard deviation σ in decibels and accounts for attenuation from diffracted propagated waves. The parameter X_σ can be easily obtained from experiments made in different building environments, and $\bar{L}$ must be calculated according to the direct NLOS attenuation described by (6.11) following the Bertoni model. In other words, model (6.29) is a combination of Bertoni's physical model of direct propagation through floors (6.11) and Rappaport's empirical model by estimation of the parameter X_σ from the approach presented in Chapter 11 (Section 11.2) for link budget design or from experimental data. The suggested path loss model is based on two essential aspects. First, it uses Bertoni's prediction to obtain the received power signal along the radio path of rays penetrated through floors and described by (6.11) and second, it uses Rappaport's statistical measured σ and the method of fade margin estimation according to the procedure of link budget design described in Chapter 11 (see also Reference 32).

6.3. LINK BUDGET DESIGN VERIFICATION BY EXPERIMENTAL DATA

In order to investigate the accuracy of the well-known models and the suggested above models of radio propagation along the corridor (6.8) and between floors and walls described by (6.29) with the help of (6.11) and (6.16) for the average path loss evaluation, we carried out some special experiments within several four-storied buildings.

Path Loss along the Corridor

The system consists of two main parts: the first is a wireless access point (BreezeCom AP10) connected to a power supply, and the second is a laptop with a wireless LAN card (BreezeCom SA-PCR) (see Figure 6.11a,b according to References 32 and 33). The laptop was located on a portable surface in order to separate it from the floor. Table 6.5 presents several important technical specifications of the system. The signal was measured at different locations 10 times in terms of 2–3 minutes. This assures that local interferences such as electrostatic waves, cellular communication, and moving objects can be

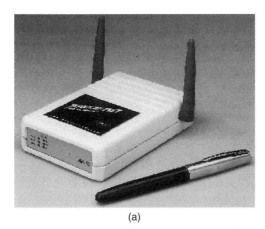

(a)

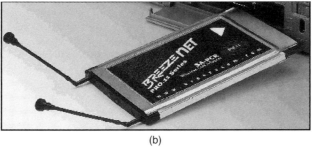

(b)

FIGURE 6.11. (a) The transmitter setup presentation. (b) The receiver setup presentation.

TABLE 6.5. Technical specifications of the system

Transmission Method	Spreading in spectrum and skipping in frequency
Frequency Spectrum	2.4–2.4835 GHz
Brooding Time	32, 64, 128 ms
Transmission Power	Up to 100 mW (20 dBm)
Sensitivity	
@1 Mbps	–81 dBm
@2 Mbps	–75 dBm
@3 Mbps	–67 dBm
Antenna Division	2 Antenna

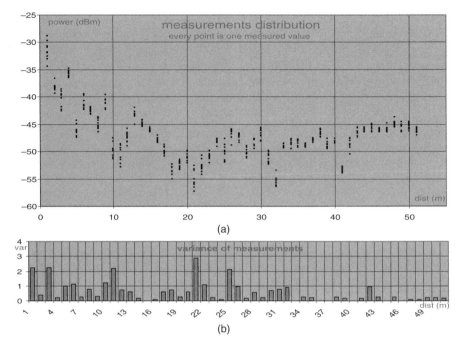

FIGURE 6.12. (a) Measurements along the corridor; each point represents one measured value of the signal path loss (in dBm). (b) Variance of measured data presented in panel a.

eliminated. In the building used for measurement, there was a 51-m corridor with glass and metal doors at the edges. The transmitting and receiving stations were placed on a portable laptop surface. The transmitting access points were placed in the beginning and in the middle of the corridor. The results are presented in Figure 6.12a with the variance of the measured data shown in Figure 6.12b, from which it follows that with an increase in distance between the transmitter and receiver (more than 20–25 m), the saturation of the effect of attenuation is observed. The standard deviation of the measured data does not exceed 2–3 dB. There are few spikes that were probably caused by the variance in the architecture characteristics of the walls and by some local obstructions such as people walking along corridor and interference from the additional cellular communication networks.

The situation is changed significantly when experiment was carried out with the terminal antennas separated by one floor. Thus, the variance of measured data obtained for the scenario where the terminal antennas were separated by one floor, shown in Figure 6.13a, increases twice compared with the previous case depicted in Figure 6.12a and can achieve 7–8 dB (see Fig. 6.13b). Comparison of these results with those obtained, theoretically, using the corridor waveguide model performed for the same floor, are presented in Figure 6.14 by the continuous curve. It is clearly seen that the mean difference between

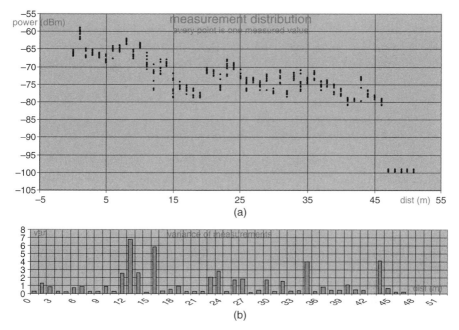

FIGURE 6.13. (a) Measurements along the corridor when the transmitting and the receiving antennas are separated by one floor; each point represents one measured value of the signal path loss (in dBm). (b) Variance of measured data presented in panel a.

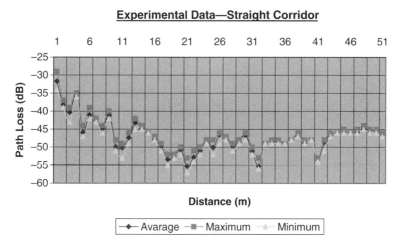

FIGURE 6.14. Experimental data measured along the corridor.

the theoretical prediction and experimental data does not exceed 2–3 dB at the beginning of corridor, becomes 4–5 dB in the middle sites, and reaches maximum difference of 9.794 dB at the end of the corridor, where an intersection with another crossing corridor exists. So, the corridor waveguide model is a good predictor of radio coverage inside the straight corridor except for some intersections with other crossing corridors within the tested building.

Link Budget for Indoor Links between the Floors and Walls

All experiments have been carried out in different campuses of the Ben-Gurion University, Israel, each of which is a typical three-floor university campus, comprising long hallways and contiguous enclosed classrooms with windows. All outside and inside walls are constructed of concrete. There are large windows along the corridors (north wing) and inside every classroom (south wing). Each classroom is furnished with chairs and tables having the same size and height and made of metal and wood. During the experiments, all windows were closed in each floor (both along the corridor and inside each classroom). The receiver (Rx) and the transmitter (Tx) were separated with obstructions between them, that is, having both NLOS and LOS conditions. The transmitter was located in a fixed position at the first floor. The receiver was moved from one location to another within the measurement area from the third floor to the second floor, both along the corridor and inside rooms lining each corridor. Thus, Figure 6.15 shows the radio coverage of the signal path loss (i.e., the relative power attenuation with respect to the transmitted power, see definition in Chapter 1). It is clearly seen that the signal power has complicated spatial distribution inside the room when both terminal antennas are located along the corridor outside the concrete room. The total signal path

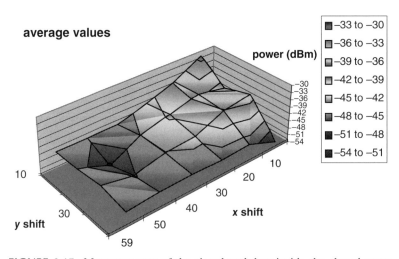

FIGURE 6.15. Measurements of the signal path loss inside the closed room.

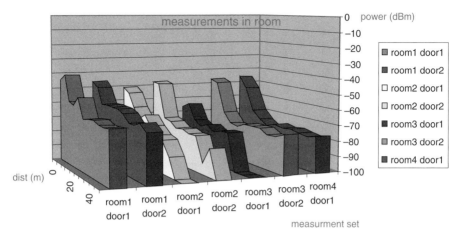

FIGURE 6.16. Measured signal path loss through rooms located along the corridor — effects of walls attenuation.

loss varies from –30 dB to –50 dB with regard to the transmitter located along the corridor. Full analysis of effect of walls for one set of measurements carried out along the corridor for rooms located at the same floor and that lie along the radio path between the terminal antennas is presented in Figure 6.16. It is clearly seen that each wall attenuates the signal power approximately in 3–5 dB, with complicated spatial distribution of the signal power inside each room.

On the basis of numerous experimental data and measurement analysis, a preliminary suggestion was done that the proposed model, $L_{total}[dB] = \bar{L}[dB] + X_\sigma[dB]$, based on the combination of Bertoni's formula of direct penetration through floors and walls (6.11) and the additional attenuation X_σ, which accounts the *lognormal* shadowing effects caused by internal structures and obstructions, predicts the path loss measurements with the smallest deviation from experiment results (see Fig. 6.17 and Fig. 6.18). According to receipt described in Chapter 11 or in Reference 32, the probability for shadowing in the selected area can also be found. Figure 6.17 presents the simulation according to Bertoni's model (6.18) for conditions of the experiment described earlier, and the same simulation according to the suggested model (6.29) is shown by Figure 6.18.

Below we present the distinct difference between the suggested model for predicting path loss between floors (6.29) and Bertoni's path loss prediction model (6.18), which takes into account diffraction by window corners for the receiver at the third floor (see Fig. 6.19) and then the receiver at the second floor (see Fig. 6.20).

From Figure 6.19a,b, the suggested model achieved better agreement with measurements, with an average error of 4.76 dB, than with Bertoni's model where the average error exceeds 10 dB. On the third floor where Rx was

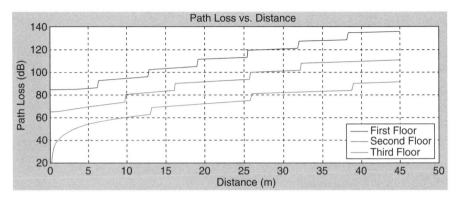

FIGURE 6.17. Path loss versus distance for different scenarios inside the building.

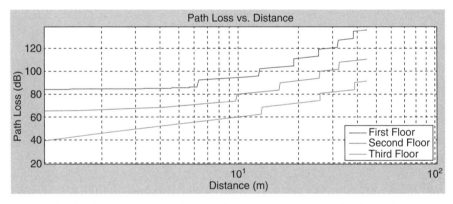

FIGURE 6.18. Path loss versus distance for different scenarios inside the building.

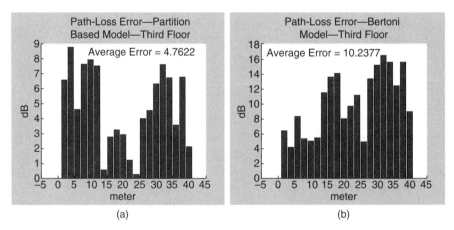

FIGURE 6.19. (a) Error of suggested model compared with measurements. (b) Error of Bertoni's 3rd floor simulation to measurements.

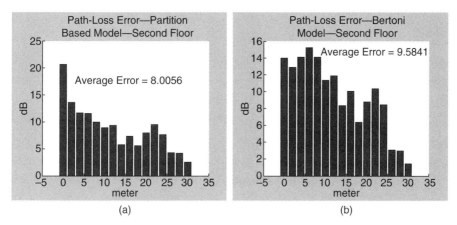

FIGURE 6.20. (a) Error of suggested model compared with measurements. (b) Error of Bertoni's 2nd floor simulation compared with measurements.

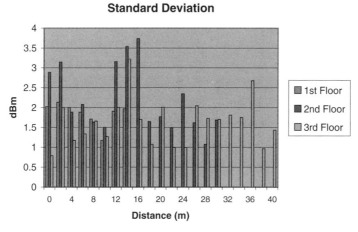

FIGURE 6.21. Standard deviation obtained from the statistical analysis of all scenarios.

located, the shadowing term was evaluated to be $X_\sigma = 12.9$ dB. On the second floor (see Fig. 6.20a,b), again, the suggested model achieved better results with error of 8.00 dB compared with 9.58 dB obtained from Bertoni's model. On the second floor, the term of shadowing was $X_\sigma = 8.1$ dB. In computations and comparison with Bertoni's model, the wall attenuation factor of 4 dB (for concrete wall) and the floor attenuation factor of 13 dB (for mixed concrete walls) were accounted [32].

The cumulative effect of deviation from the theoretical prediction, based on Equation (6.29) and measured data, for different three-story buildings is shown in Figure 6.21.

According to these results, we can conclude that the suggested model is very simple in terms of calculation and that it takes into account the slow fading (statistical approach) that other models omit. At the same time, the proposed model and the corresponding simulation results do not take into account objects such as furniture, people, and their movements. Therefore, some deviation error between simulation and actual link test results should be expected.

Finally, we present some experimental results in the form of a straight-line model as a best fit to the measured data to follow the same procedure used in outdoor communication links (see Chapter 11). Thus, in Figure 6.22, Figure 6.23, and Figure 6.24, the total path loss versus distance, according to experimental data, are shown for the third, second, and first floors, respectively. The path loss exponent for each floor has been obtained from the approximate equations of straight lines. In Figure 6.22, both antennas are at the same, third,

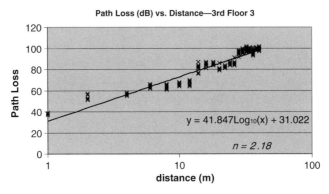

FIGURE 6.22. Best fit obtained for experiment carried out at the same, third, floors.

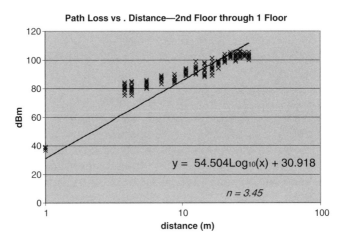

FIGURE 6.23. Best fit obtained for experiment carried out through one floor between antennas.

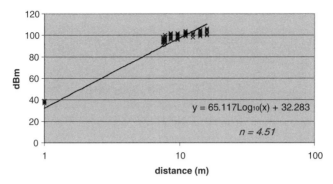

FIGURE 6.24. Best fit obtained for experiment carried out through two floors between antennas.

floor, that is, in LOS conditions. As was found for the LOS conditions, between the transmitter and receiver, the attenuation parameter equals $\gamma = 2.18$. For the receiver located at the second floor, that is, one floor below the transmitter, we found (see Fig. 6.23) that $\gamma = 3.45$. Finally, for the receiver located on the first floor having two-floor difference with the transmitter location, the attenuation parameter is $\gamma = 4.51$ as a best fit to the experimental data (see Fig. 6.24).

These results are very close to those obtained by Rappaport in his numerous experiments for different kinds of buildings (see Table 6.3). Thus, we can summarize that the suggested model (6.29) can be successfully used for predicting the total path loss inside buildings, for different antenna positions and different floors if we take into account Bertoni's formulas strictly (6.11), and (6.18) for the direct path loss between floors and walls and the shadow margin, which can be obtained either from experimental data or by using the method of shadow effect estimation described in Chapter 11 for the link budget design.

REFERENCES

1 Bertoni, H.L., *Radio Propagation for Modern Wireless Systems*, Prentice Hall PTR, Upper Saddle River, NJ, 2000.

2 Blaunstein, N., "Average field attenuation in the nonregular impedance street waveguide," *IEEE Trans. Anten. and Propagat.*, Vol. 46, No. 12, 1998, pp. 1782–1789.

3 Rappaport T.S., *Wireless Communications*, Prentice Hall PTR, New York, 1996.

4 Saunders, S.R., *Antennas and Propagation for Wireless Communication Systems*, J. Wiley & Sons, New York, 1999.

5 Cox, D.C., R.R. Murray, and A.W. Norris, "Measurements of 800 MHz radio transmission into buildings with metallic walls," *AT&T Bell Lab. Tech. J.*, Vol. 62, 1983, pp. 2695–2717.

6   Davidson, A. and C. Hill, "Measurement of building penetration into medium building at 900 and 1500 MHz," *IEEE Trans. Veh. Technol.*, Vol. 46, 1997, pp. 161–167.

7   Turkmani, A.M.D. and A.F. de Toledo, "Modeling of radio transmission into and within multistory buildings at 900, 1800, and 2300 MHz, *IEE Proc.*–1, Vol. 40, 1993, pp. 462–470.

8   Alexander, S.E., "Radio propagation within buildings at 900 MHz," *Electronics Letters*, Vol. 18, No. 21, 1982, pp. 913–914.

9   Hashemi, H., "The indoor radio propagation channel," *Proc. IEEE*, vol. 81, No. 7, 1993, pp. 943–968.

10 Lemieux, J.F., M. Tanany, and H.M. Hafez, "Experimental evaluation of space/frequency/polarization diversity in the indoor wireless channel," *IEEE Trans. Veh. Technol.*, Vol. 40, No. 3, 1991, pp. 569–574.

11 Rappaport, T.S., "Characterization of UHF multipath radio channels in factory buildings," *IEEE Trans. Antennas Propagat.*, vol. 37, No. 8, 1989, pp. 1058–1069.

12 Devasirvatham, D.M., M.J. Krain, and T.S. Rappaport, "Radio propagation measurements at 850 MHz, 1.7 GHz, and 4.0 GHz inside two dissimilar office buildings," *Electronics Letters*, Vol. 26, No. 7, 1990, pp. 445–447.

13 Rappaport, T.S. and D.A. Hawbaker, "Wide-band microwave propagation parameters using cellular and linear polarized antennas for indoor wireless channels," *IEEE Trans. on Communications*, Vol. 40, No. 2, 1992, pp. 231–242.

14 Tarng, J.H., W.R. Chang, and B.J. Hsu, "Three-dimensional modeling of 900-MHz and 2.44-GHz radio propagation in corridors," *IEEE Trans. Veh. Technol.*, Vol. 46, 1997, pp. 519–526.

15 Gibson, T.B. and D.C. Jenn, "Prediction and measurements of wall intersection loss," *IEEE Trans. Antennas Propagat.*, Vol. 47, 1999, pp. 55–57.

16 Lafortune, J.F. and M. Lecours, "Measurement and modeling of propagation losses in a building at 900 MHz," *IEEE Trans. Veh. Technol.*, Vol. 39, 1990, pp. 101–108.

17 Arnod, H.W., R.R. Murray, and D.C. Cox, "815 MHz radio attenuation measured within two commercial buildings," *IEEE Trans. Antennas Propagat.*, Vol. 37, 1989, pp. 1335–1339.

18 Whitman, G.M., K.S. Kim, and E. Niver, "A theoretical model for radio signal attenuation inside buildings," *IEEE Trans. Veh. Technol.*, Vol. 44, 1995, pp. 621–629.

19 Seidel, S.Y. and T.S. Rappaport, "Site-specific propagation prediction for wireless in-building personal communication system design," *IEEE Trans. Veh. Technol.*, Vol. 43, 1994, pp. 879–891.

20 Seidel, S.Y. and T.S. Rappaport, "914 MHz path loss prediction models for indoor wireless communication in multifloored buildings," *IEEE Trans. Antennas Propagat.*, Vol. 40, No. 2, 1992, pp. 207–217.

21 Honcharenko W., H.L. Bertoni, J. Dailing, J. Qian, and H.D. Lee, "Mechanisms governing UHF propagation on single floors in modern office buildings," *IEEE Trans. Veh. Technol.*, Vol. 41, No. 4, 1992, pp. 496–504.

22 Honcharenko W., H.L. Bertoni, and J. Dailing, "Mechanisms governing propagation between different floors in buildings," *IEEE Trans. Antennas Propagat.*, Vol. 41, No. 6, 1993, pp. 787–790.

23 Dersch, U. and E. Zollinger, "Propagation mechanisms in microcell and indoor environments," *IEEE Trans. Veh. Technol.*, Vol. 43, 1994, pp. 1058–1066.

24 Clarke R.H., "A statistical theory of mobile-radio reception," *Bell Systems Technical Journal*, Vol. 47, 1968, pp. 957–1000.

25 Rappaport, T.S., et al., "Statistical channel impulse response models for factory and open plan building communication system design," *IEEE Trans. on Communications*, Vol. 39, No. 5, 1991, pp. 794–805.

26 Devasirvatham, D.M.J., "Time delay spread and signal level measurements of 850 MHz radio waves in building environments," *IEEE Trans. Antennas Propagat.*, Vol. 34, No. 2, 1986, pp. 1300–1305.

27 Rappaport, T.S. and V. Fung, "Simulation of bit error performance of FSK, BPSK, and π/4–DQPSK in flat fading indoor radio channels using measurement-based channel model," *IEEE Trans. Veh. Technol.*, Vol. 40, No. 4, 1991, pp. 731–739.

28 Kanatas, A.G., I.D. Kountouris, G.B. Kostraras, and P. Constantinou, "A UTD propagation model in urban microcellular environments," *IEEE Trans. Veh. Technol.*, Vol. 46, No. 2, 1997, pp. 185–193.

29 Katedra, M.F., J. Perez, F.S. de Adana, and O. Gutierrez, "Efficient ray-tracing techniques for three-dimensional analyses of propagation in mobile communications: application to picocell and microcell scenarios," *IEEE Antennas Propagat. Magazine*, Vol. 40, No. 2, 1998, pp. 15–28.

30 Kim, S.C., B.J. Guarino, Jr. T.M. Willis III, et al., "Radio propagation measurements and prediction using three dimensional ray tracing in urban environments at 908 MHz and 1.9 GHz," *IEEE Trans. Veh. Technol.*, Vol. 48, 1999, pp. 931–946.

31 Tsalolihin, E., N. Blaunstein, I. Bilik, L. Ali, and S. Shakya, "Measurement and modeling of the indoor communication link in angle of arrival and time of arrival domains," in *Proc. of Int. Conf. of IEEE Antennas and Propagation*, Toronto, Canada, July 19–23, 2010, pp. 338–341.

32 Yarkoni, N. and N. Blaunstein, "Prediction of propagation characteristics in indoor radio communication environments," *J. Progress in Electromagnetic Research*, PIER 59, 2006, pp. 151–174.

33 Tsalolihin, E., N. Blaunstein, I. Bilik, L. Ali, and S. Shakya, "Analysis of AOA and TOA signal distribution in various indoor environments," in *Proc. of European Conference in Antennas and Propagation* (*EuCAP*-2011), Rome, Italy, April 11–15, 2011, pp. 1746–1750.

34 Ali, L., Y. Ben Shimol, and N. Blaunstein, "Analysis of AOA-TOA signal distribution in indoor RF environments," *PIERS*-2011, Marrakech, Morocco, April 12–16, 2011, pp. 117–121.

35 Keenan, J.M. and A.J. Motley, "Radio coverage in buildings," *BT Tech. J.*, Vol. 8, No. 1, 1990, pp. 19–24.

36 International Telecommunication Union, *ITU-R Recommendation P. 1238*: "Propagation data and prediction models for the planning of indoor communication systems and local area networks in the frequency range 900 MHz to 100 GHz," Geneva, 1997.

CHAPTER SEVEN

Adaptive Antennas for Wireless Networks

As was mentioned in previous chapters, the main problem in mobile or station-ary wireless communications, satellite and aircraft (megacell), outdoor (mac-rocell and microcell), and indoor (picocell), is the additional noise factor (to the white or additive noise). This noise has two aspects to it: (1) the multiplica-tive noise caused by multipath propagation fading, delay spread (DS), and Doppler spread (see definitions in Chapter 1) and (2) the co-channel interfer-ence noise caused by interactions of information sent by different users located in the area of service and involved in the multiple access communication occurred in real time during servicing. Both of these physical phenomena degrade the grade of service (GOS), the quality of service (QOS), the capacity of the information data stream and, finally, the efficiency of wireless commu-nication networks.

Several methods have been developed during the last two to three decades to eliminate these kinds of noise factors. These methods are based on filtering [1–5], signal processing [6–12], and the so-called adaptive or smart antenna systems [13–22]. The term "smart antenna" reflects the antenna's ability to adapt to the communication channel environment in which it operates. Because both terms, "adaptive antennas" and "smart antennas," are interchangeable, from now on we will use the term adaptive antennas since it is based on analog and digital beam-forming technology [23–29]. Adaptive antennas are not only used in cellular communications, but also in many other applications such as aircraft and satellite communications, radars, and remote sensing [30–36].

The increasing demands on the operational efficiency of various wireless communication networks put a lot of technical and mathematical questions

Radio Propagation and Adaptive Antennas for Wireless Communication Networks: Terrestrial, Atmospheric, and Ionospheric, Second Edition. Nathan Blaunstein and Christos G. Christodoulou.

not only on how to eliminate noise within a channel, but also on how to increase the capacity of the information data stream inside a channel. Once has to account for limitations in the bandwidth of the existing communication networks, and increase the bit rate with small bit-error-rate (BER), the parameters that determine the QOS. According to Shannon's formula, channel capacity linearly increases with allocating new frequencies to the service or with spreading the existing bandwidth. The latter is a very complicated problem, and with a logarithmical increase of its signal-to-noise ratio (SNR), it includes additive (white) noise, multiplicative noise, and noise caused by co-channel interference in cellular networks.

From previous discussions, to predict the noise factors in environments with strong "clutter" and hard obstructive (non-line-of-sight [NLOS]) conditions is a very complicated task. During the last four decades, designers have proposed new strategies where they combined adaptive antennas with advanced signal processing to obtain effective filtering systems, which can simultaneously operate in the space, time, and frequency domains. Using such interdisciplinary mathematical and technical tools, a lot of other problems in wireless communications have been effectively solved. Here, we only point out the problems connected with multiple access communications, which deal with simultaneous service of numerous subscribers and occurs in frequency, time, code, and space domain. These multiple access communications are the frequency division multiple access (FDMA), time division multiple access (TDMA), code division multiple access (CDMA), and space division multiple access (SDMA).

The use of adaptive antennas can essentially improve the GOS, eliminate the influence of co-channel interference by increasing the carrier-to-interference ratio (C/I) and, finally, to determine with great accuracy the position of the desired subscriber, which must be serviced by a system. However, as was discussed in Chapter 1, in urban environments, a widespread antenna pattern in both azimuth and elevation domains is observed, caused by the random distribution of buildings both in the horizontal and vertical planes (see Fig. 1.15, Chapter 1).

Therefore, it is very complicated to use more directive or adaptive antennas for urban communication links. These problems are difficult to solve without knowledge, not only of the signal strength or power distribution in the space domain, that is, along the radio path between the terminal antennas, but also of the signal distribution in the separate and joint angle-of-arrival (AOA), azimuth and elevation, time-of-arrival (TOA), or delay-spread (DS) domains. All these aspects will be briefly presented in the next section. Next, we start with the architecture of adaptive antennas for different array configurations (linear, circular, and planar).

7.1. ANTENNA ARRAYS

The definition of "adaptive or smart" antenna has been used in Reference 13 to describe self-phasing antenna systems, which reradiate a radio signal in the

MIMO Communication Network

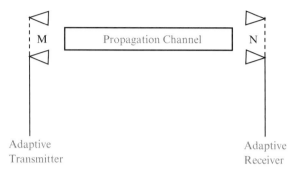

FIGURE 7.1. Scheme of MIMO channel.

direction from which it was received. Figure 7.1 shows the architecture of two arrays used in a communication system. The receiver array consists of N elements and the receiver has M elements. Such multichannel system in literature is called the multiple-input-multiple-output (MIMO) wireless channel [37].

The concept of using antenna arrays and innovative signal processing is not new to the radar and aerospace technology. Until recent years, cost-effectiveness has prevented their use in commercial systems. The advent of very fast and low-cost digital signal processors has begun to make adaptive antennas very smart and practical for cellular land- or satellite-mobile communications systems. This trend is only the beginning, and the use of smart antennas is going to accelerate in the future. Main hurdles to overcome are the costs and technological issues relating to manufacturing of a multi-antenna system.

Before we explain the principles of adaptive arrays and their operation, we proceed with a review of some of the basics of antenna arrays.

7.1.1. Antenna Array Terminology

As shown in Chapter 2, antennas in general may be classified as isotropic, omnidirectional, and directional. For antenna arrays, there are some additional terms that must be introduced, such as array factor, phased arrays, steerable beams, and so on [12–14, 38–43]. A *phased array antenna* uses an array of antennas called "elements" that combine their signals to achieve a more directive radiation pattern in some direction than others. The direction where the maximum gain would appear is controlled by adjusting the phase of the individual elements. So, in the direction where maximum gain occurs, the signals from the elements are added *in-phase* and that is the reason why an array is used to achieve more gain than a single antenna element.

An *adaptive array antenna* utilizes sophisticated signal processing algorithms to continuously distinguish between desired signals, multipath, and

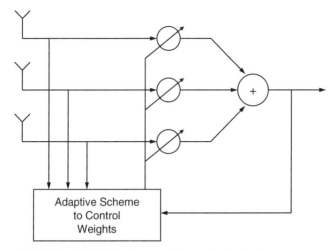

FIGURE 7.2. Block scheme of the typical adaptive antenna.

interfering signals as well as calculate their directions of arrival. The adaptive approach continuously updates its beam pattern based on changes in both the desired and interfering signal locations. The ability to smoothly track users with main lobes and interferers with nulls guarantees that the link budget is constantly maximized.

A block scheme of the typical adaptive antenna is shown in Figure 7.2. The signals received by the elements of the antenna are weighted and combined to maximize the signal-to-interference ratio [18]. A typical adaptive antenna array is shown in Figure 7.2.

The Array Factor. A plot of the array response as a function of angle is commonly referred to as the array pattern, beam pattern, or power pattern. To characterize this pattern, a new parameter called the *array factor* and denoted by $F(\varphi,\theta)$, is defined. It represents the far-field radiation pattern of an array of isotropic radiating elements in the θ and ϕ angles. The process of combining signals from different elements is known as *beamforming*. We will discuss several beamforming technologies later in this chapter.

Steering Process. For a given array, the main beam can be pointed in different directions by mechanically moving the array. This process is known as *mechanical steering*. In contrast, *electronic steering* uses the inherent delay of signals arriving at each element of the array before combining them. For narrowband signals in Figure 7.3, phase shifters are used to change the phase of signals before combining them at the output of the antenna where they arrive with their own time delay (TD) $\tau_m(\theta)$ [41].

The steering locations, which result in maximum power, yield the direction-of-arrival (DOA) estimates; that is, the steering vector contains the responses

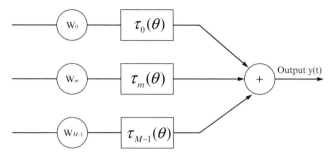

FIGURE 7.3. Summation of arriving signals with different time delays.

of all elements of the array to a narrowband source of unit power. Because the response of the array is different in different directions, a steering vector is associated with each directional source. The correlation between them depends upon the array geometry [39]. Because each component of the steering vector denotes the phase delay caused by the spatial position of the corresponding element of the array, this vector is also called the *space vector* or *array response vector*. In multipath situations in wireless communication channel, the space vector denotes the response of the array to all signals arising from the source [40, 42, 43]. The array response is steered by forming a linear combination of the element outputs, as will be shown mathematically later.

The Pattern of the Antenna Array. In an antenna array, if the distance between the elements-sensors is larger than the wavelength of radiation, several main lobes will be formed in the visible space. Conversely, if the element spacing is less than a wavelength, and all signals from elements are summed without any delay (see Fig. 7.2), then the produced array output signal will have a symmetric pattern about $\theta = 0°$ (see Fig. 7.4a). On the contrary, if in this case the output of each element is delayed in time before being summed (see Fig. 7.3), the resulting directivity pattern has its main lobe displaced at an angle ψ defined as

$$\sin \psi = \frac{\lambda \tau f}{d} = \frac{c\tau}{d}, \tag{7.1}$$

where d is the spacing between antenna array elements, $c = \lambda/T = \lambda f$ is the signal propagation velocity, and τ is the time-delay difference between neighboring-element outputs. Figure 7.4b shows the direction of the main beam and nulls for an array with $\theta = 24°, d = \lambda/2, \tau = 0.12941/f,$ and $\psi = \sin^{-1}(2\tau f) = 15°$.

7.1.2. Architecture of the Antenna Array

Next we discuss the most common geometrical configurations used in adaptive array antennas.

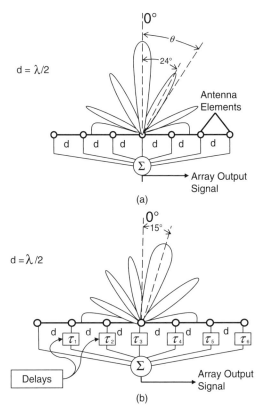

FIGURE 7.4. Antenna array: (a) all signals arrive within any delay: the main beam is oriented normally to the sensors' line; (b) each signal is delayed by its own delay that gives a shift in the main beam by 15 degrees. (Source: Reference 14. Reprinted with permission © 1967 IEEE.)

Linear Array. Let us consider a receiving antenna with a linear array of $M-1$ elements from the origin, which is uniformly spaced along the horizontal axis as shown in Figure 7.5 [23]. Let the spacing between elements be denoted by d. At each element input there is a complex signal given by $U_m = A_m e^{j\beta_m}$ (i.e., the signal input at the element m with amplitude A_m and phase $\beta_m = m\alpha$, $m = 0, 1, 2, \ldots, M-1$). Here, α is the constant phase difference between two adjacent elements.

We also assume that at the origin, the phase of the arriving ray is equal zero, and the differential distance of two rays at points $m+1$ and m is $\Delta d = md\sin\theta$. Then the array factor can be determined as [23]

$$F(\theta) = U_0 + U_1 e^{jkd\sin\theta} + U_2 e^{j2kd\sin\theta} + \ldots =$$

$$\sum_{m=0}^{M-1} U_m e^{jmkd\sin\theta} = \sum_{m=0}^{M-1} A_m e^{j(mkd\sin\theta + m\alpha)}, \tag{7.2}$$

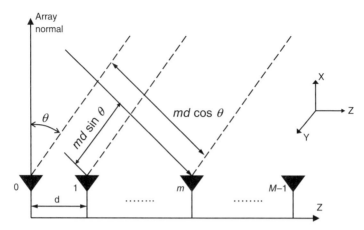

FIGURE 7.5. Linear antenna array to obtain the path difference between two neighboring elements.

or in terms of vectors and an inner product

$$F(\theta) = \mathbf{U}^T \mathbf{u}, \tag{7.3}$$

where

$$\mathbf{U} = [U_0 U_1 \dots U_{M-1}]^T \tag{7.4}$$

is the array propagation vector that contains information about the AOA of the signal. Also,

$$\mathbf{u} = \left[1 \; e^{jkd\sin\theta} \; e^{j2kd\sin\theta} \; \dots \; e^{j(M-1)kd\sin\theta}\right]^T \tag{7.5}$$

is the weigh vector with the corresponding component for each element of the array. If now $\alpha = -kd\sin\theta_0$, a maximum response of $F(\theta)$ will result at the angle θ_0; that is, the antenna beam pattern will be steered toward the wave source. Figure 7.6 shows the radiation pattern for an 8-element linear array with λ spacing between elements.

Circular Array. The same situation can be said for a circular array of equally distributed $M - 1$ elements placed on a circle of radius R, as shown in Figure 7.7.
  Here, we introduce the azimuth angle for each element m, $\varphi_m = 2m\pi/M$. The relative phase β_m at each element m with respect to the center of the array is

$$\beta_m = -kR\cos(\varphi - \varphi_m)\sin\theta. \tag{7.6}$$

Again, for a main beam directed at angles θ_0 and φ_0 in space, the phase of the complex signal $U_m = A_m e^{j\alpha_m}$ for the element m equals

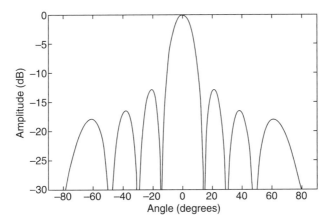

FIGURE 7.6. Field pattern of an 8-element adaptive array antenna.

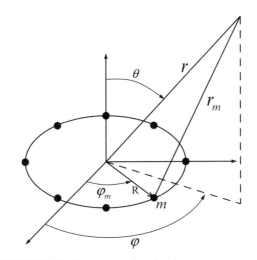

FIGURE 7.7. Geometry of a circular antenna array.

$$\alpha_m = kR\cos(\varphi_0 - \varphi_m)\sin\theta_0. \tag{7.7}$$

In this case, the array factor for the circular antenna can be presented in the following form:

$$F(\varphi, \theta) = \sum_{m=0}^{M-1} A_m e^{j[\alpha_m - kR\cos(\varphi - \varphi_m)\sin\theta]}. \tag{7.8}$$

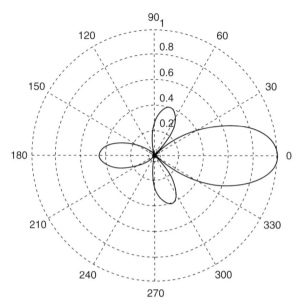

FIGURE 7.8. Radiated pattern for an adaptive circular array radius 0.8λ.

Figure 7.8 shows the radiation pattern for an adaptive circular array of radius 0.8λ. The array factor $G(\varphi,\theta)$ can be found as

$$G(\varphi, \theta) = f(\varphi, \theta) \cdot F(\varphi, \theta), \tag{7.9}$$

where $f(\varphi,\theta)$ is the element factor. This equation is usually called the *principle of pattern multiplication*, which allows us to determine the array factor of more complicated arrays as a composition of simple subarrays, and account for their mutual dependence.

Planar Array. An example of the planar array is shown in Figure 7.9 as a combination of two linear arrays, one is with $M-1$ elements and the second is with $N-1$ elements. Then, according to (7.2), the array factor for the first M-element array with the complex signal at the element m of $U_m = A_m e^{jm\alpha}$, $m = 0, 1, 2, \ldots, M-1$, is given as [23]

$$F_1(u) = \sum_{m=0}^{M-1} A_m e^{j(mkd_x \sin u + m\alpha)}, \tag{7.10}$$

where $u = \sin\theta\cos\varphi$. The array factor for the N-element array with the complex weight at the element n of $U_n = A_n e^{jn\beta}$, $n = 0, 1, 2, \ldots, N-1$, is given as [23]

$$F_2(v) = \sum_{n=0}^{N-1} A_n e^{j(mkd_y \sin v + n\beta)}, \tag{7.11}$$

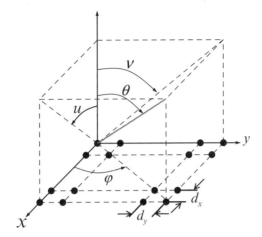

FIGURE 7.9. Geometry of a planar antenna array.

where $\nu = \sin\theta\sin\varphi$. According to the principle of pattern multiplication, the overall array factor for the rectangular array is then given by

$$F = F_1(u) \cdot F_2(v). \tag{7.12}$$

The same procedure can be used for the more complicated antenna structures such as the hexagon planar array and so on.

7.2. BEAMFORMING TECHNIQUES

The term beamforming relates to the capability of the antenna array to focus energy along a specific direction in space [18, 19, 21, 23]. Thus, in multiple access communications, a desired user must be serviced in "clutter" conditions. In this case, "clutter" means the existence of other users located in the area of service. Beamforming allows the antenna to focus energy toward a desired user only and nulls in the undesired directions (see Fig. 1.14, Chapter 1). For this reason, beamforming is often referred to as spatial filtering. Spatial filtering or beamforming was the first approach to carry out space-time processing of data sampled at antenna arrays [12].

The conventional (Bartlett) beamformer was the first to emerge during the Second World War [44]. It is a natural extension of the classical Fourier-based spectral analysis for spatial-temporal sampled data. Later, adaptive beamformers [45–48] and classical time-delay estimation techniques [48] were applied to enhance the ability to resolve signal sources that are closely spaced. From a statistical point of view, the classical techniques can be seen as spatial extensions of the spectral Wiener (or *matched*) filtering method [49]. However, the conventional beamforming approach has some fundamental limitations connected to the physical size of the aperture or the array, to the available data collection time, and to SNR. For more details, the reader is referred to References 8, 21, 50, and 51. Next, we present some aspects of *analog* and *digital* beamforming.

7.2.1. Analog Beamforming

An analog beamforming system usually consists of devices that change the phase and power of the signal emanating from its output. Figure 7.10 shows an example for creating only one beam at the output of the RF beamformer [23]. Such simple, one-beam antenna array systems, can be constructed by using microwave waveguides, microstrip structures, transmission lines, and printed microwave circuits.

Multiple-beam beamforming systems are more complex systems whose operational characteristics are based mathematically on the beamforming matrix, with the Butler matrix being the most known matrix [52]. In a beamforming matrix, an array of hybrid junctions and fixed-phase shifters are used to achieve the desired results. As an example, a Butler beamforming matrix for a four-element antenna array is shown in Figure 7.11a. This matrix uses two 45° fixed-phase shifters and four 90° phase-lag hybrid junctions with the corresponding computation links (see Fig. 7.11b).

By tracing the signal from the four ports to the array elements, one can verify that the relative phase distribution at the antenna aperture corresponds to the individual ports of the four-port Butler matrix, computed as shown in Figure 7.12 [23, 52]. An example of an array antenna pattern with elements spaced at $\lambda/2$ is shown in Figure 7.12. However, these four beams are overlapping and they are mutually orthogonal.

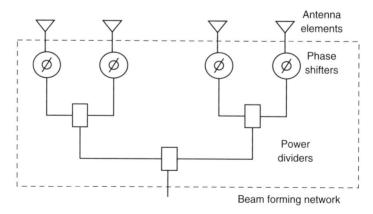

FIGURE 7.10. A simple beamformer.

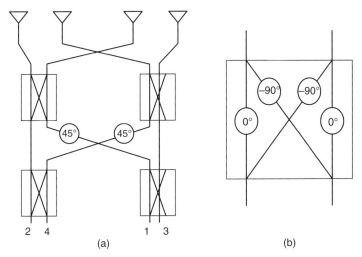

FIGURE 7.11. (a) A Butler beamforming matrix for a four-element antenna array and (b) its phasing scheme.

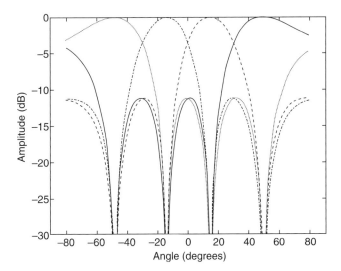

FIGURE 7.12. Computation of the antenna pattern corresponding to the beamformer in Figure 7.11. (Source: Reference 23.)

Here, we must note that the Butler matrix was developed before the fast Fourier transform (FFT) and that they are both completely equivalent. One (the Butler matrix) is used for analog beamforming, and the other (FFT) is for digital beamforming [19, 21, 23, 50–52].

Conventional beamforming is a simple beamformer, sometimes known as the delay-and-sum beamformer, with all it weights of equal amplitude.

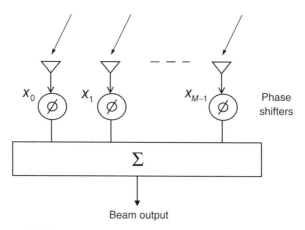

FIGURE 7.13. A beamformer with phase shifters.

As was mentioned previously, the phases of the elements are selected to steer the main beam of the array in a particular direction (φ_0, θ_0), known as the look direction. The system must be able to adapt its pattern to have lobes at $M-1$ places. This ability is known as a degree of freedom of the array. For an equally spaced linear array, this feature is similar to an $M-1$ degree polynomial of $M-1$ adjustable coefficients, with the first coefficients having the value of unity (see Equation (7.5)). The concept of a delay beamformer or phase delay is shown in Figure 7.13 [41].

Here, due to the delay of each arriving ray at an array element with respect to its neighboring element, a corresponding shift in phase occurs, with the amplitude weights remaining fixed as the beam is steered. As mentioned before, this type of array is commonly known as a *phased array*.

Null-steering beamformer is used to place nulls in the radiation pattern in specified directions. Usually, the nulls are placed in the directions of interfering signals or mobile users. In the earliest schemes [53–56], this was achieved by estimating the signal arrived from a known direction by steering a conventional beam in the direction of the desired source and then subtracting the output from each element. In this case the beam output $y(t)$ is presented by a sum of the signals x_m, $m = 0, 1, \ldots, M-1$, received from a given direction, defined by the angle θ, by each of the M elements. Each element has its own weight coefficient w_m and TD of arrival $\tau_m(\theta)$. The output of the antenna is expressed as

$$y(t, \theta) = \sum_{m=0}^{M-1} w_m x_m[t - \tau_m(\theta)]. \tag{7.13}$$

By adjusting these weights one can shape the beams. This process is very effective for canceling strong interference between subscribers.

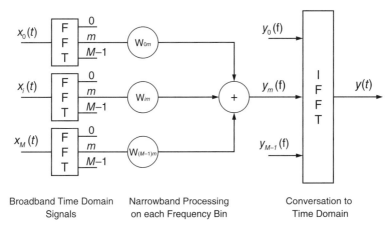

FIGURE 7.14. Summation of weights. (Source: Reference 14. Reprinted with permission © 1967 IEEE.)

Frequency-Domain Beamforming. Here, by using the direct and inverse FFT, the broadband signals from each element of the array are transformed into the frequency domain and then each frequency bin is processed by a narrowband processor structure (see details in References 57–59). The weighted signals from all elements are summed to produce an output at each bin (see Fig. 7.14).

The weights are selected by independently minimizing the mean output power at each frequency bin subject to steering-direction constraints. Thus, the weight required for each frequency bin are selected independently, and this selection may be performed in parallel, leading to a faster weight update. Various aspects of frequency-domain beamforming are reported in References 57–59 and other references.

Multiple beamforming is used to generate several beams simultaneously. These beams can be fixed in certain directions or adaptive with nulls steerable in desirable directions. This can be achieved using very complex networks of phase shifters.

In *beam-space processing* [60–62], the beamformers can distribute the signal energy to all the formed beams. One of the problems with multiple-beam beamformers is that as the number of beams is increased, the SNR of channels being carried by the individual beam decreases. This is due to additional noise introduced from the additional number of radio frequency (RF) and intermediate frequency (IF) components that must be used to increase the beamformer capacity.

7.2.2. Digital Beam Forming

Earlier ideas to use digital beamforming come from foundations in sonar [26] and radar [30] systems as a bridge between antenna technology and digital

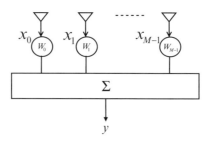

FIGURE 7.15. A simple digital beamformer.

technology. Using digital techniques, it is possible to capture RF information in the form of digital streams. Digital beamforming is based on the conversion of the incident RF signal at each antenna element into two streams of binary complex baseband signals representing an in-phase component (I) and a 90° phase-shifted or quadrature component (Q). In digital beamforming technique, the weighted signals from each element are sampled and stored, and beams are formed summing the appropriate samples [63–69]. Despite the fact that digital beamforming does not have the same direct physical meaning as analog beamforming, the same process of adaptive beamformer is used by weighting digital signals and presenting the total beam by the same array factor (7.8). Next, a simple algorithm of beamforming without any phase delays will be introduced.

Element-Space Beamforming. A simple structure that can be used for such beamforming is shown in Figure 7.15 [14, 41]. It is the same as the one sketched in Figure 7.2, but here, we introduce notations that correspond to digital processing jargon.

The output $y_n(\theta)$ at a discrete time $t = nT$ is given by a linear combination of the binary data at M sensors (also known as the *array snapshot* at the nth instant of time given by Reference 41. We define a *snapshot* as one simultaneous sampling of all array element signals:

$$y_n(\theta) = \sum_{m=0}^{M-1} w_m^* x_m(n) \tag{7.14}$$

or in inner vector form [41]

$$y_n(\theta) = \mathbf{w}^H \cdot \mathbf{x}(n), \tag{7.15}$$

where the sampling time T was omitted to simplify our discussions. Here, x_m is the signal from mth element of the array, w_m^* is the weight applied to x_m, sign "*" represents a complex conjugate, and the superscript H represents the Hermitian transpose. To relate these notations to the analog beamforming, we assume that $x_m(n) = U_m$ and $w_m^* = e^{-jmkd\sin\theta}$. Then, the total output signal $y(t)$

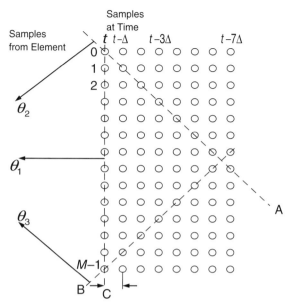

FIGURE 7.16. Obtaining the desired direction using a linear array of sensors. (Source: Reference 41. Reprinted with permission © 1997 IEEE.)

is equal to $F(\theta)$ from (7.2), that is, $y_n(\theta) = F(\theta)$. In such a manner, (7.14) describe the process, which is referred to as the *element-space beamforming*, where the binary data signals x_m are directly multiplied by a set of weights to form a beam in any desired angle.

To consider the delayed adaptive beamforming using delays, we need Equation (7.13) and represent each delay as an integer multiple of the sampling interval Δ. The process is shown in Figure 7.16 for a linear array of uniformly spaced elements, where it is desired that a beam is formed in the specific direction θ_2 [41].

The TD along θ_2 is

$$\tau_m(\theta_2) = m\Delta, \; m = 0, 1, \ldots, M-1 \tag{7.16}$$

Thus, the signal from the mth element needs to be delayed by $m\Delta$ seconds. This may be accomplished by selecting the samples for summing, as shown in Figure 7.16, by the line marked with symbol A. Similarly, a beam may be steered in a direction θ_3 by summing the samples connected by the line marked with symbol B in Figure 7.16, where the signals from the mth element need to be delayed by $(M - 1 - m)\Delta$ seconds. At the same time, the beam formed in direction θ_1, by summing the samples connected by the line marked with symbol C, does not require any delay. So, using such a process, we can only form beams in those directions that require delays equal to some integer

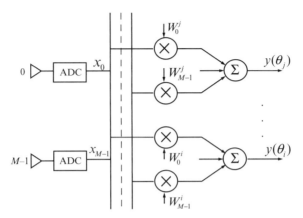

FIGURE 7.17. A simple beamformer without time delays.

multiple of the sampling interval, that is, corresponded to (7.16). The number of discrete directions where a beam can be pointed exactly increases with increased sampling. This leads to the formation of additional beams. For more information on how to form multiple beams simultaneously and synchronize the digital beamforming process, the reader is referred to work reported in References 25 and 65–70.

An example of a simple beamformer that generates an arbitrary number of simultaneous beams from M antenna elements is shown in Figure 7.17. Each beamformer creates an independent beam by applying independent weights to the array signals, that is, [28]:

$$y(\theta_i) = \sum_{m=0}^{M-1} w_m^{i*} x_m, \qquad (7.17)$$

where $y(\theta_i)$ is output of the beamformer, x_m is a sample from the mth array element, and w_m^i are the weights for forming beam at angle θ_i. By selecting appropriate weight vectors, we can implement beam steering, adaptive nulling, and beam shaping.

Space Beamforming. Instead of directly weighting the outputs from the array elements, they can be processed first by a multiple-beam beamformer to form a suite of orthogonal beams. The output of each beam is then weighted and combined to produce a desired output. This process is often referred to as space beamforming. The required multiple beamformer usually produces orthogonal beams, namely, the beamformer can be implemented by using the FFT. Thus, an M-element linear array with M overlapped orthogonal beams can be used to give [28]:

$$v(\theta_m) = \sum_{m=0}^{M-1} x_m e^{-j2\pi m\theta_m/(M-1)}, \ m = 0, 1, 2, \ldots, M-1 \qquad (7.18)$$

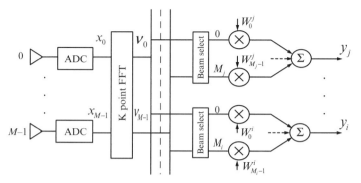

FIGURE 7.18. An FFT-based beamformer. (Source: Reference 28. Reprinted with permission © 1998 IEEE.)

where $\theta_m = \sin^{-1}[m\lambda/(M-1)d]$. Because of the fixed discrete nature of $v(\theta_m)$, the individual beam control requires the following steps:

(a) interpolation between beams in order to precise-steer the resultant beam

(b) linear combination of the output beams to synthesize a shaped beam or a low sidelobe pattern

(c) linear combination of a selected set of beams to create nulls in the direction of the interfering sources.

Thus, for space beamforming, a set of beam-space combiners to generate weighted outputs is required. In Figure 7.18 in the weighted FFT-based beamformer, the digital signal streams from the antenna elements are fed to the FFT processor, which generates M simultaneous orthogonal beams.

The role of the beam select function in Figure 7.18 is to choose a subset of these orthogonal beams that are to be weighted to form a desired signal. For example, the ith desired output may happen to be the combination of the weighted mth and $(m+2)$th beams, that is,

$$y_i = w_1^i v(\theta_m) + w_2^i v(\theta_{m+2}) = \sum_{m=0}^{M_i-1} w_m^i v(\theta_{i(m)}), \qquad (7.19)$$

where $i(m)$ is the selected beam index (i.e., $i(1) = m$ and $i(2) = m + 2$) and $(M_i - 1)$ is the number of orthogonal beams that are required to form the ith desired beam.

Two-Dimensional Beamforming. Digital beamforming technologies for mobile-satellite communications are usually based on two-dimensional planar antenna arrays [71, 72]. As was shown in Section 7.1, all algorithms, techniques, and methods for linear antenna arrays can be easily and naturally extended to two-dimensional planar arrays. Thus, for a $M \times K$ rectangular planar array, the output of the beamformer at the discrete time $t_n = n$, $y_n(\theta, \varphi)$, is given by [71, 72]

$$y_n(\theta, \varphi) = \sum_{k=0}^{K-1} \sum_{m=0}^{M-1} w_{k,m}^* x_{k,m}(n), \qquad (7.20)$$

or in the standard matrix form through inner product

$$y_n(\theta, \varphi) = \mathbf{w}^H \mathbf{x}(n) \qquad (7.21)$$

where the weight matrix is

$$\mathbf{w} = (w_{0,0}, w_{1,0}, \ldots, w_{K-1,0}, w_{0,1}, \ldots, w_{K-1,M-1})^T \qquad (7.22)$$

and the output signal matrix at each element of the planar array is

$$\mathbf{x}(n) = [x_{0,0}(n), x_{1,0}(n), \ldots, x_{K-1,0}(n), x_{0,1}(n), \ldots, x_{K-1,M-1}(n)]^T. \qquad (7.23)$$

In a similar way, the output of the beamformer at time n can be constructed for any planar array.

Adaptive Beamforming. Adaptive beamforming has been a subject of considerable interest for more than three decades, traditionally starting as other types of beamforming to be employed in sonar and radar applications. There are numerous technical papers and articles on the basic concept, special technologies, and applications of adaptive beamforming, from which more general are References 9, 14, 16, 19, 28, and 73–79.

Adaptive beamforming started with the invention of the IF *sidelobe canceller* (SLC), reported in References 13 and 15. This was the first adaptive antenna system, which was capable of nulling interference signals automatically at the antenna output. The antenna array had a typical configuration of nondelayed beamformer presented in Figure 7.4a, but with one significant difference: it contained one high-gain main-beam "dish" antenna (with weight w_0) surrounded by a linear array of several low-gain antenna sensors (with weights w_m, $m = 1, 2, \ldots, M - 1$).

Applebaum [46] developed a theoretical concept, commonly known as the Howells–Applebaum algorithm, on how to control the weights of the adaptive beamformer. The main goal of this algorithm was to maximize the SNR at the array output. For the analog SLC multibeam antenna loop, Applebaum expressed a differential adaptive processing equation given by

$$T \frac{dw_i}{dt} + w_i = G \left\{ K_i - x_i^*(t) \sum_{m=0}^{M-1} w_m x_m(t) \right\}. \qquad (7.24)$$

From this equation, one can obtain the weights in a matrix form by using

$$\mathbf{w} = \mu \mathbf{R}^{-1}\mathbf{K}, \tag{7.25}$$

where $\mathbf{R}$ is the $M \times M$ covariance matrix $\mathbf{R} = E[\mathbf{x}(t)\mathbf{x}^H(t)]$ which is formed from the expected values of the array signal correlation. In (7.24), w_i, $i = 0, 1,$ $\ldots, M-1$, is the ith weight at the ith element output, $x_i(t)$ is a signal from the ith antenna element, K_i is the component of cross-correlation matrix $\mathbf{K}$ of $x_i(t)$ with the output of the main high-gain antenna channel (with weight w_0). T is the smoothing filter time constant, and G is amplifier gain. A positive scalar μ (called the gradient step size) controls the convergence characteristic of the algorithm, that is, how fast and how close the estimated weights approach the optimal weights.

A different beamforming technique was proposed by Capon [45]. This approach leads to an adaptive beamformer with a minimum-variance distortionless response (MVDR), known also in literature as the maximum likelihood method (MLM), because it maximizes the likelihood function of the input signal vector.

Then, Reed with coworkers showed that fast adaptability is achieved by using the *sample-matrix inversion* (SMI) technique [56]. This algorithm is more convenient when fast convergence response is required in an SLC configuration [19, 56, 78]. Sometimes, it is better to use an *orthogonal lattice filter adaptive network*, often referred to as the Gram–Schmidt algorithm. According to this algorithm, each weight w_{im} represents the adaptive coefficient obtained from a one-stage Gram–Schmidt orthogonal filter and can be expressed through the nth time-sampled voltage $V_{im}(n)$ of a set of N signal data samples from the ith input as [21, 22, 78]

$$|w_{im}| = \frac{\displaystyle\sum_{n=1}^{N} V_i^*(n)V_m(n)}{\displaystyle\sum_{n=1}^{N} V_i^*(n)V_i(n)}. \tag{7.26}$$

Using this technique, the adaptive weights can be computed directly. A comparison of the convergence speed performance obtained by the Gram–Schmidt algorithm and that obtained by Howells–Applebaum algorithm made in Reference 22 has shown that the first one converses in about 30 snapshot data samples, whereas the second one achieves a signal the level of 14 dB above the receiver noise after 180 snapshots.

The SMI algorithm can be sped up by using the direct inversion of the covariance matrix $\mathbf{R}$ in (7.25). If the desired and interference signals are known a priori, then the covariance matrix could be evaluated, and the optimal solution for the weights could be computed using (7.25). As in most cases the signals are not known, they can be generated as a set of "pseudo" signals $\tilde{x}_m(t)$

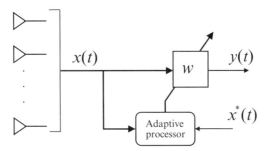

FIGURE 7.19. General adaptive beamforming.

that closely represents the real signals. Consider a general adaptive beamforming scheme in Figure 7.19 [19, 23, 78].

The choice of the weight vector **w** is based on the statistics of the signal vector **x**(t) received at the array. Basically, the objective is to optimize the beamforming response according to a prescribed criterion of the adaptive processor operation, so that the output y(t) will contain minimum contribution from noise and interference. In such a situation, the adaptive processor must continually update the weight vector to meet a new requirement imposed by the varying conditions of the signal environment. Therefore, instead of the real covariance matrix **R**, the adaptive processor "deals" with the approximate covariance matrix $\tilde{\mathbf{R}} = E[\tilde{\mathbf{x}}^*(t)\mathbf{x}(t)]$ and (7.25) becomes the Wiener–Holf equation, which is also called the optimum Wiener solution [56]:

$$\mathbf{w}_{opt} = \mu \mathbf{R}^{-1}\tilde{\mathbf{R}}. \tag{7.27}$$

In this algorithm, the weight vector is updated without a priori information, and it leads to estimates of **R** and $\tilde{\mathbf{R}}$ in a finite observation interval. These estimates are then used in (7.27) to obtain the desired weight vector. The error, due to these estimates, can be viewed as the least squares formulation of the problem. So, the weight vector derived using the SMI method can be defined as a least squares solution.

Another algorithm, the *least squares or LMS algorithm* was developed by Widrow and his colleagues [10, 14]. Because of its simplicity, the LMS algorithm is the most commonly used adaptive algorithm for continuous adaptation, and it is capable of achieving satisfactory performance under the right set of conditions. The LMS algorithm was further developed with the introduction of constraints [75, 76] that are used to ensure that the desired signals are not filtered out against the unwanted signals.

The LMS algorithm is based on the optimization method that recursively computes and updates the weight vector. It is clear that the process of successive corrections of the weight vector leads to the estimation of the mean-square error (MSE), which finally allows to obtain an optimum value for the weight vector. According to the optimization method [80], the weight vector is updated at time $n + 1$ using the following relation:

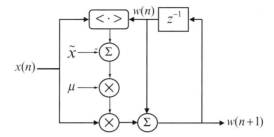

FIGURE 7.20. Schematical presentation of LMS algorithm.

$$\mathbf{w}(n+1) = \mathbf{w}(n) + \mu[\tilde{\mathbf{R}} - \mathbf{R}\mathbf{w}(n)]. \tag{7.28}$$

As in the previous algorithms, a prior knowledge of both $\mathbf{R}$ and the approximate $\tilde{\mathbf{R}}$ is not possible. Again, their instantaneous estimates are used through the error matrix

$$\mathbf{E} = \mathbf{R}\mathbf{w}_{opt} - \tilde{\mathbf{R}}. \tag{7.29}$$

Then the estimated weights can be updated as

$$\hat{\mathbf{w}}(n+1) = \hat{\mathbf{w}}(n) + \mu\mathbf{x}(n)\mathbf{E}(n)]. \tag{7.30}$$

The gain constant μ controls the convergence characteristics of the random vector sequence $\mathbf{w}(n)$. This is a continuously adaptive approach that works well when the signal environment is statistically stationary. A signal-flow scheme representing the LMS algorithm is shown in Figure 7.20 [80]. It is clear that such an algorithm is very simple and is therefore successfully used in sonar, radar, and communication applications [81–89].

However, its convergence characteristics depend strongly on the eigenvalues of the covariance matrix $\mathbf{R}$, which has a tendency to change widely with a change in the signal environment. When this occurs, convergence can be very slow. Furthermore, unlike the Applebaum's maximum SNR algorithm and the LMS algorithm, which may suffer from slow convergence, the performance of the SMI algorithm is more preferable in this situation because it is independent of the value of the eigenvalue spread. Despite the fact that Applebaum's maximum SNR algorithm and Widrow's LMS error algorithm were discovered independently and were developed using different approaches, they are basically similar. For stationary signals, both algorithms converge to the optimum Wiener solution [77].

Let us now present a simple example of an adaptive antenna processing for steering and modifying an array's beam in order to "work" only with a desired signal and to show how the complex weight coefficients are obtained to suppress the interfering signals. Consider a base station (BS) with a simple array

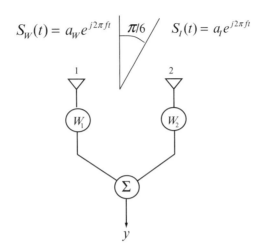

FIGURE 7.21. An adaptive array of two elements.

of two antennas separated by a distance of $d = \lambda/2$, as shown in Figure 7.21. For simplicity, let us assume that the desired signal arrives from the first mobile user at $\theta_1 = 0°$ and the interfering signal arrives from the second mobile user at $\theta_2 = \pi/6$. Both signals send their information at the same frequency f. The first desired signal, $s_w = a_w e^{j2\pi ft}$, arrives at the first and second elements with the same phase. The output y_w for the wanted signal is a linear combination of the two corresponding weight coefficients, w_1 at the first element antenna and w_2 at the second element antenna, that is,

$$y_w = a_w e^{j2\pi ft}(w_1 + w_2). \tag{7.31}$$

At the same time, the interfering signal $s_I(t) = a_I e^{j2\pi ft}$ arrives at the first and second element antenna with a phase shift of $\phi = kd\sin(\pi - 6) = 2\pi(\lambda/2)$ $(1/2)/\lambda = \pi/2$. Thus, the array output for the interfering signal at the two elements equals [23]

$$y_I = a_I e^{j2\pi ft}w_1 + a_I e^{j(2\pi ft+\pi/2)}w_2. \tag{7.32}$$

To make sure that at the output of the array there is a desired signal, we must satisfy

$$\begin{cases} \mathrm{Re}[w_1] + \mathrm{Re}[w_2] = 1 \\ \mathrm{Im}[w_1] + \mathrm{Im}[w_2] = 0 \end{cases}. \tag{7.33}$$

At the same time, to minimize the effect of the interfering signal leaving the desired signal unaffected, one must state that the array output for the interference response must be zero or as follows from (7.32)

$$\begin{cases} \mathrm{Re}[w_1] + \mathrm{Re}[jw_2] = 0 \\ \mathrm{Im}[w_1] + \mathrm{Im}[jw_2] = 0 \end{cases}. \tag{7.34}$$

Expressions (7.33) and (7.34) are then solved simultaneously to give us

$$\begin{cases} w_1 = 1/2 - j(1/2) \\ w_2 = 1/2 + j(1/2) \end{cases}. \tag{7.35}$$

With these weights, the array of two element antennas will accept the desired signal while simultaneously eliminating the interfering signal. Of course, in this simple example we assumed prior knowledge of the direction of arrivals for the interfering and desired signals, frequency, and so on. Nevertheless, this example demonstrates that a system consisting of an array of antenna elements, which is configured with complex weights, provides countless possibilities for realizing array system objectives.

7.3. ADAPTIVE ANTENNA FOR WIRELESS COMMUNICATION APPLICATIONS

The demand for mobile and fixed wireless communication continues to grow, making subscriber capacity and reliability of wireless systems a critical issue. Efficient temporal processing, such as advanced source coding, channel coding, modulation, equalization, and detection techniques, can help alleviate this problem. However, more dramatic improvements may be achieved by exploiting the spatial dimension using a smart antenna system and multiple access techniques.

Multiple access refers to the simultaneous coverage of numerous users by manipulating the transmission and reception process of signals in time, frequency, code, and space domains. As was mentioned in Chapter 3, in TDMA, each user located in the area of service obtains or transmits information in a certain period of time, called a *time slot*. In FDMA, the frequency bandwidth is divided into segments, which are then portioned among different users located in certain service area. In CDMA, each user obtains a unique random sequence of bits, that is, a unique code, generated by a generator of special random sequences (see Chapter 3).

The information waveform is spread after modulation by this code over the entire frequency bandwidth, which is allocated to all users serviced by the network. The receiver uses the same code to detect the signal with information corresponding only to the desired user by rejecting other users (having other codes) and noises (multiplicative, additive, and due to interference) that exist in the communication channel. In cellular communications, there is another access called SDMAs, which is usually used by a division of each cell in sectors using directed antennas to serve each user located in the corresponding sector (see Chapter 3). The latest form of SDMA usually employs adaptive antenna arrays based on digital beamforming technology [19, 23–30].

Here, we will consider the applications of adaptive antennas in different networks on the basis of these four types of multiple access communication [90–95]. Let us first start with applications of adaptive antenna in terrestrial communications.

7.3.1. Adaptive Antennas for Outdoor Wireless Communications

The important aim in using adaptive antenna arrays is to reject the multiplicative noise caused by multipath fading, slow or fast, to decrease the time-delay effect that occurred because of the multipath phenomena, and finally to eliminate the co-channel interference that occurs between subscribers allocated in the same frequency band (in CDMA) or that share the same time frame [74]. To overcome these impairments, an array technology, consisting of MIMO channel is usually used (see Fig. 7.1). Using M antenna elements, a significant increase of antenna gain is achieved plus a diversity gain against multipath fading, which depends on the correlation of the fading among the antenna elements. To provide a low correlation (i.e., diversity gain) between elements, there are several basic ways that can be considered: *spatial, polarization, time, frequency,* and *angle diversity* [74, 96].

For *spatial diversity*, the antenna elements are separated far enough for low fading correlation among them. The required separation depends on the obstructions surrounding the antenna such as buildings, trees, hills, and so on. There are three typical situations in the urban environment scene: when the BS antenna is higher than the overlay profile of the buildings (see Fig. 7.22a); the BS antenna is at the same level as the height of the buildings (see Fig. 7.22b); and when the BS antenna is lower than the overlay profile of the buildings (see Fig. 7.22c). Depending on the number of obstructions (scatterers) surrounding the terminal antennas, the AOA of the total signal at the receiver will spread dramatically. Thus, for the BS antenna sketched in Figure 7.22a, when only few obstructions surround the user antenna, the angular spread may be only few degrees [74], whereas for the situation in urban areas sketched in Figure 7.22b, the angular spread can exceed 10 or more degrees. In these situations, a horizontal separation of antenna elements of 10 to 20 wavelengths is required. In the third situation shown in Figure 7.22c, the angular spread can reach up to 360° and the antenna element spacing only of a quarter wavelength is sufficient [23–30].

A simpler way to obtain spatial diversity is to use two antennas separated by a distance d in space, as shown in Figure 7.23 [93]. The multipath fading is due to waves from two scatterers, A and B, separated by distance r_s, which is much smaller than the distance between antennas and the scatterers, so both antennas view the scatterer from the same direction. In such assumptions, we can present the phase difference between waves incident on the antennas in the following form: $\phi = -kd\sin\theta$.

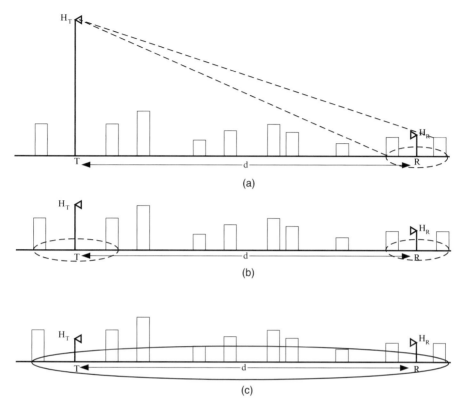

FIGURE 7.22. Three typical antenna positions with respect to rooftops level.

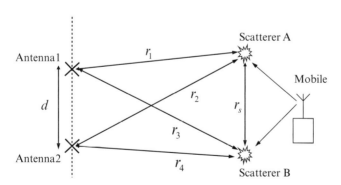

FIGURE 7.23. Spatial diversity estimation.

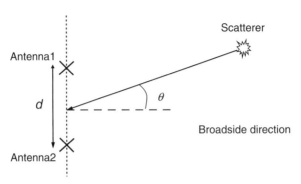

FIGURE 7.24. A single scatterer for the antenna elements separated by a distance d.

Figure 7.24 shows the path from a single scatterer at an angle θ to the broadside direction in the horizontal plane (for the horizontal antenna spacing). Assuming now that the amplitude after scattering is the same for both antennas, we get [93]

$$s_1 = a \cdot e^{j2\pi ft} \text{ and } s_2 = a \cdot e^{j(2\pi ft + \phi)}. \tag{7.36}$$

For the large number of scatterers N, these expressions can be generalized as

$$s_1 = \sum_{i=1}^{N} a_i e^{j2\pi ft} \text{ and } s_2 = \sum_{i=1}^{N} a_i e^{j(2\pi ft + \phi_i)}, \tag{7.37}$$

where a_i are the amplitudes associated with each of the scatterers. The correlation between these two signals, assuming that amplitudes from each of the scatterers are uncorrelated is given by

$$\rho_{12}(d) = \int_{0}^{2\pi} p(\theta) e^{jkd\sin\theta} d\theta, \tag{7.38}$$

where $p(\theta)$ is the PDF function of the random variable θ. This expression can be used for wide range of situations, provided a reasonable distribution for $p(\theta)$ can be found. Note that (7.38) is a Fourier transform relationship between $p(\theta)$ and $\rho_{12}(d)$. So, there is an inverse relationship between the widths of these two functions: A narrow angular distribution after multiple scattering will produce a slow decrease in the correlation with antenna spacing, which finally will limit the usefulness of space diversity. Conversely, an environment with significant scatterers, around the antenna with a wide angular distribution, will produce a decrease in the correlation with antenna element spacing. In this situation, if d goes to zero, the correlation between the antenna elements will be higher.

In many cases of *mobile-to-mobile* (*MO-MO*) communication (see Fig. 7.22c), the angular distribution of the signal after multiple scattering from obstructions can be described by a uniform PDF over $[0, 2\pi]$ with $p(\theta) = 1/2\pi$. In this case from (7.38) we get, according to Reference 93, a solution in terms of the modified Bessel function of zero order

$$\rho(d) = J_0\left(\frac{2\pi d}{\lambda}\right). \tag{7.39}$$

From Figure 7.22, for the *base-station-to-mobile (BS-MO)* communication, the angular distribution of scattering at the BS may be very different from that of the low mobile antenna. In the case of Figures 7.22a,b, with scatterers present at distance r from the BS on a ring centered around the mobile station with radius, we use instead of (7.39) the following expression [95]

$$\rho(d) = J_0\left(\frac{2\pi d}{\lambda}\frac{r_s}{r}\cos\theta\right)J_0\left(\frac{\pi d}{\lambda}\left(\frac{r_s}{r}\right)^2\sqrt{1 - \frac{3}{4}\sin^2\theta}\right), \tag{7.40}$$

where θ is the scattering angle directed from the BS to the ring of scatterers.

Comparison between (7.39) obtained for the case of MO-MO communications and (7.40) obtained for the case of BS-MO communications shows that in the second case, the spacing d between antenna elements required is much greater than that in the mobile to mobile case.

If a more compact antenna structure is used (i.e., small spacing between elements), the vertical space diversity becomes essential, as two neighboring antenna elements can be packaged together into a single vertical structure. Now, if we assume that all waves after scattering arrive at the horizontal plane, we have $p(\theta) = \delta(\theta)$, that is, the Delta-function angular spread distribution. In this case, the signals will be perfectly correlated for all separations between antenna elements. However, in more realistic conditions where waves arrive with some moderate spreading relative to the horizontal plane, the angular spread PDF can be presented according to Reference [97] as

$$p(\theta) = \begin{cases} \dfrac{\pi}{4|\theta_m|}\cos\dfrac{\pi\theta}{2\theta_m}, & |\theta| \leq |\theta_m| \leq \dfrac{\pi}{2}, \\ 0, & elsewhere \end{cases} \tag{7.41}$$

where θ_m is half of the vertical angular spread.

We notice that expression (7.40) gives more conservative correlation values with respect to (7.41), as it does not include the vertical spreading of the AOA. This is essential when we want to calculate the effect of vertical antenna element spacing, which nevertheless requires even larger spacing than the horizontal case [98].

Despite the large required spacing of antenna elements, horizontal space diversity is very commonly applied in cellular BSs to allow compensation for the low transmit power obtained from handheld portables compared with the BSs. Vertical spacing is rarely used [90–92, 95–98]. This is because of the large spacing required to obtain low cross-correlation and because the different heights of the antennas within an array can lead to significant differences in path loss for each antenna, which degrades the diversity effect. The example shown in 7.22a is more related to macrocell environments that have very narrow angular spread, whereas those in Fig. 7.22b and 7.22c are related to microcell environments, where the angular spread is larger.

For *polarization diversity*, both horizontal and vertical polarization is used. These orthogonal polarizations have low correlation, and the antenna elements can have small spacing, creating a small profile for the total antenna. This effect has been found experimentally in Reference 99, and it was shown that these two components are almost uncorrelated; thus, a pair of cross-polarized antennas can provide diversity with no spacing between them. However, polarization diversity only doubles the diversity of any antenna, and for high BS antennas, the horizontal polarization can be 6–10 dB weaker than the vertical polarization, which reduces the diversity gain [77]. Under the assumption that the vertical and horizontal components of the signal field are independently Rayleigh-distributed (i.e., we now consider the Rayleigh fading communication channel), the correlation coefficient can be presented in the following form [99]:

$$\rho = \frac{\tan^2 \alpha \cos^2 \beta - \Gamma}{\tan^2 \alpha \cos^2 \beta + \Gamma}, \tag{7.42}$$

where the cross-polar ratio Γ is defined as the ratio between the mean powers from the horizontally, E_H, and vertically, E_V, polarized signal strengths:

$$\Gamma = \frac{\langle | E_H |^2 \rangle}{\langle | E_V |^2 \rangle}. \tag{7.43}$$

Here, the fields are received by antennas inclined at an angle α with respect to the vertical axis, and the mobile is situated at an angle β with respect to the antenna bore sight.

In References 100 and 101, a new method to improve the performance of polarization diversity was proposed by using a mixed scheme where the antenna array elements were used, simultaneously, for both spatial and polarization diversity. In this case, the multiplicative correlation coefficient is approximately a product

$$\rho \approx \rho(\alpha) \cdot \rho(d, h). \tag{7.44}$$

Equation (7.44) describes the pure polarization diversity with colocated antenna elements polarized at an angle α with respect to the vertical axis, $\rho(\alpha)$, and the correlation described for copolarized antenna elements having a horizontal spacing d and vertical spacing h, $\rho(d, h)$. The correlation coefficient obtained from (7.44) can therefore be smaller than that obtained from polarization diversity or spatial diversity alone.

Time diversity can be obtained by transmitting the same signal multiple times, spaced apart in time sufficiently that the channel fading will be not correlated. From (7.39) we have the same result that can be produced by the autocorrelation function of the fading signal at a single antenna between two moments in time when the mobile antenna is in motion. This is, except with the TD, introduced by the antenna motion reinterpreted as a horizontal antenna spacing by putting the TD as $\tau = d/v$, where d is the antenna spacing, and v is the mobile velocity. However, the time diversity is rarely used in practice of wireless communications as the retransmission of information reduces the system capacity and introduces a transmission delay. Usually, this principle is applied to improve efficiency in coded modulation schemes, which apply interleaving to spread errors across fades, allowing better potential for error correction [74, 96].

In wideband channels, *frequency diversity* is used when two frequency components spaced wider than the coherent bandwidth experience uncorrelated fading. As in time diversity, the simple retransmission of information on two frequencies is not so efficient. Usually, the principle of frequency diversity is implicitly employed in some forms of equalizers [74, 96].

For *angle diversity*, adjacent narrow beams are used. The total antenna profile is small, and the adjacent beams could have received signal levels more than 10 dB weaker than the strongest beam, resulting in small diversity gain [74].

Due to low fading correlation among antenna elements, diversity gain is typically achieved in current BS by using either *selection diversity*, where selection of the antenna element (sensor) with the highest signal power is made, or *maximal radio combining*, where the procedure of beamforming (weighting and combining the received signal to maximize the SNR) is performed. This provides additional gain on the *uplink* (or *reverse link*, from mobile antenna to BS) to compensate for the higher transmit power of the BS on the *downlink* (or *direct link*, from BS to mobile antenna). Typically, only a single transmitter and receiver antenna is used on the downlink.

The *formation of multiple beams* by multiple antennas at the BS is performed to cover the whole cell size. For instance, three beams with a beamwidth of 120° each or six beams with a beamwidth of 60° each may be formed for the multiple beam pattern creation. Then, each beam is treated as a separate cell (as is done in the process of splitting cells on sectors; see References 90–94), and the frequency assignment to each user may be performed in a usual manner. Mobile users are handed to the next beam (sector) as they leave the area covered by the current beam, as is done in a normal handoff process when users cross the cell boundary [90–94].

Such multiple beams can be created by the so-called switched-beam system (SBS) [13] consisting of a beamformer, which forms the multiple nonadaptive beams; a sniffer, which determines the beam that has the best signal-to-interference-noise ratio (SINR); and a switch, which is used to select the best one or two beams for the receiver. The main advantage of SBS is the improvement in cell radio coverage on the reverse link due to the array gain and improved voice quality due to reduced interference between users who occupy the same segments in frequency or time domains. As the performed beams are narrower than the regular sector beamwidth, reduction of the interference power is obtained when the desired signal and the interfered signals are separated in angle and fall into different beams. This SINR improvement offers better information quality (digital or voice) and therefore improves capacity of data stream and spectrum efficiency of the network (see Chapter 12).

The performance of SBS depends on a number of factors, including AOA and TD spreading (see Chapters 8 and 11) caused by multipath fading, interference, and the antenna array topology. Performance gain in SBS comes from array gain, diversity gain, reduced interference, and trunking efficiency [13]. The array gain is given by M (in decibels) or by $10 \log M$, where M is a number of input antenna elements (see Fig. 7.2). The element spacing determines the degree of decorrelation of the signal across the antenna array. Therefore, for a $\lambda/2$-spacing in an M-element antenna, the diversity gain of 5 dB is preserved for the land macrocells (with radius more than 2–3 km) [90–94]. There are other system loss factors, which must be accounted, such as cusping loss, mismatch loss, beam-selection loss, and path diversity loss [13]. Cusping loss occurs because of the 2–3 dB cusp between beams. Mismatch loss refers to the use of planar beamforming in the presence of potentially nonplanar wavefronts due to local scatterers surrounding the antenna array. Beam-selection loss occurs in the presence of interference when, during signal fades, another beam containing mainly interference may be selected for processing at the receiver.

The *formation of adaptive beams* is based on adaptive beamforming, which forms independent (uncorrelated) beams at the BS. The array is used to find the location of each user, stationary or mobile, and then beams are formed to cover different groups of subscribers. As above, each beam can be considered as a co-channel cell, and thus may be able to use the same frequency, time, or code, as the case may be arranged. A typical setup involving different beams covering different moving or stationary users is shown in Fig. 7.25.

In Figure 7.26 (top), a fixed number of beams with a fixed shape cover the whole cell, whereas in Figure 7.26 (bottom), the beams are shaped to cover the traffic following any changes that may occur in traffic conditions. As the users move, the different beams cover different clusters of users, offering the benefit of transmission of energy toward the desired users. Each mobile may be covered by a separate beam, if the number of elements is large enough. Using a beam to follow the desired mobile, one can reduce the handoff problem [102, 103].

Now, let us examine the gain of the array in different urban conditions as sketched in Figure 7.22. In the first case, one of the antennas (usually, the BS antenna) is located above the rooftops (see Fig. 7.22a). In this case, only obstructions located in the vicinity of the mobile vehicle must be taken into account. Let the gain of the M-element array at the BS be G_1, and the gain of the N-element array antenna at the mobile by G_1. Then, the link gain, G_L, is given by

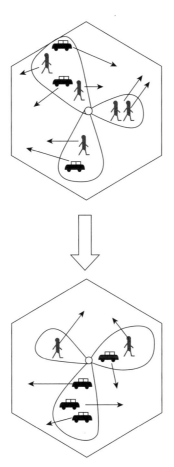

FIGURE 7.25. Typical setup of a cell of service.

$$G_L = MN \tag{7.45}$$

with a diversity gain order of *MN* [104–106]. From a range extension point of view, the situation is close to the line-of-sight (LOS) conditions.

In the second case, when the BS antenna is near the rooftop of the buildings (see Fig. 7.22b), scattering is observed around both terminal antennas. In this case, the two clusters of scatterers are widely spaced, so that they look like "point sources." This means that each antenna plus its immediate surrounding can be accounted as one effective antenna, and these two effective antennas are in the far field of each other. In this assumption, Equation (7.45) is working with a diversity order of *NM* [104–106]. In a MIMO connection (see Fig. 7.1), this is called a keyhole situation, as there is only one channel.

In the third case, both the terminal antennas are below the rooftop of the buildings (see Fig. 7.22c). In this case, both antennas are not independent and

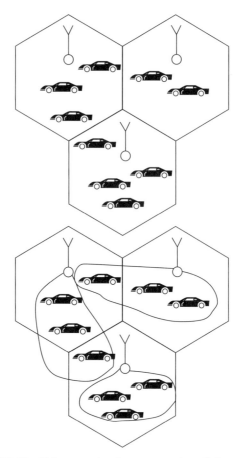

FIGURE 7.26. Using an adaptive antenna in cellular networks.

obtaining the weight parameters of the two antennas is a very complicated analytical problem. According to References 104–106, the asymptotic result for large values of M and N gives

$$G_L = \left(\sqrt{M} + \sqrt{N}\right)^2 \tag{7.46}$$

with a diversity order of MN. If N is small and M is large, the gain approaches M. Even in this worst-case scenario we can increase the accuracy of multiuser detection and service because the gain for the desired user will be close to M (for $M \gg N$), but all other users will, on average, have a gain of unity. If M is sufficiently high, this will give a good protection against multiuser interference in multiple access performance, namely, for $SDMA$.

Problems of Using Adaptive Antennas for Outdoor Communication. Unfortunately, as was mentioned in Chapter 1, by using directive antennas or adaptive antenna we cannot cancel the effects of environment, mostly in urban areas, where even from the beginning the narrow antenna beam significantly spreads in the azimuth and elevation domains after passing a couple of building rows, as it is shown in Figure 1.15 (Chapter 1). This means that the multipath component (i.e., the incoherent part of the total signal spectrum) prevails over the coherent part. Finally, in the azimuth and elevation domains, we have a wide spread of signal power distribution instead of a narrow one (see the right side of Figure 1.15).

In a typical urban case, the guiding effect along the streets is predominant, and the desired signal will arrive to the BS antenna from directions different than LOS direction shown in Figure 1.16 for user #3 (see Chapter 1). We will discuss ways to resolve this and many other problems in land communication channels by using smart antennas in Chapters 9 and 12.

7.3.2. Adaptive Antennas for Indoor Wireless Communications

There are some differences between the outdoor land-mobile and land-stationary systems and indoor systems. In indoor communications, users are usually moving on foot and use handheld portables. This results in different fading rates (negligible Doppler shift, and fading does not depend strongly on frequency, see definitions in Chapter 1). The fading rate for the indoor network is much slower than that for the outdoor network (called *flat fading*), implying that optimal combining could be implemented with the use of adaptive techniques [107–109]. The angle-spread problem in the two cases is different as well. Thus, the angular spread of signal arrival at the receiving antenna for BS antennas in indoor systems is typically isotropic 360° [110, 111]. The problem of DS is also different for the two cases [112–114].

For land outdoor communication channels, the signals generally arrive in delayed clusters due to multiple reflections and multiple diffractions from large obstructions such as hills, trees, and buildings. For indoor communications, there exist a large number of local small obstructions inside a building, which scatter and reflect rays, creating a multipath fading at the receiver. This causes the impulse response of the indoor radio channel to appear like a series of pulses [109]. The use of adaptive antennas to improve such channels is discussed in References 112 and 113. The theoretical basis for the AOA and DS signal distribution, observed experimentally in References 111–114, is given in References 112–114 by analyzing the separate one-dimensional AOA and DS distributions of the corresponding probability density functions. In Chapter 8, we will present a detailed analysis of the AOA and TOA or TD distributions, both for outdoor and indoor communication channels.

Here, some experimental data were carried out at 1.5 GHz modulated by a set of 10 ns (nanosecond) pulses with 600 ns repetition period (see Fig. 7.27 [112, 113]) in two-story office building. The transmitter was fixed in the hallway near the center of the first floor of the building with the antenna located at a height of 2 m. The receiver, with the same antenna height, was moved around

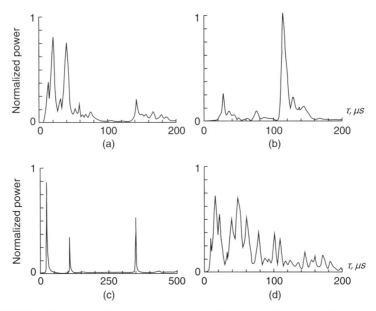

FIGURE 7.27. Normalized power versus time delay spread. See text for description. (Source: Reference 22. Reprinted with permission © 1997 IEEE.)

to collect measurements in the hallway and in several rooms on the same floor. Figure 7.27 shows one of the examples of four measured pulse responses in different locations for the receiving antenna within a building. In Figure 7.27a, two clearly separated clusters of arriving rays covering a 200-ns time span are shown. Figure 7.27b shows a 100-ns delayed echo signal that is much stronger than the first arriving rays. A strong narrow echo signal that is delayed about 325 ns, which was the largest delay of the relevant echo that was observed during this set of experiments, is represented in Figure 7.27c. Figure 7.27d corresponds to measurements done where both the transmitter and receiver were on the second floor of the building, with the receiver located in clutter conditions due to many obstructions surrounding it.

Therefore, the high density of the rays arriving at the receiver, as multiple components of the total signal caused by the multipath fading, over the entire 200-ns time axis is noticed. Results of the whole set of experiments showed that the maximum observed DS within the building is about 100–200 ns, the measured root mean square (*rms*) DS from 25 to 50 ns within the rooms, and with occasional delays of more than 300 ns within hallways. In obstructive NLOS conditions, the signal attenuation was proportional to inverse distance-power law with exponent between three and four, which is in agreement with other results obtained within buildings [107–111].

Very similar results of time DS, but using higher carrier frequency of 7 GHz, were observed in [112, 113], according to which a block scheme of the experimental setup is shown in Fig. 7.28. Note that the multipath components tend

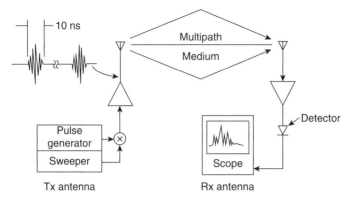

FIGURE 7.28. Experiment, described in Reference 114, for indoor communication. (Reprinted with permission ©1997 IEEE.)

to come in clusters, with the strongest cluster arriving first, and the strongest arrivals in each cluster also arriving first, as shown in Figure 7.27. As the delay time increases, the power of the clusters and the arrivals within the clusters tend to decay in amplitude until they disappear into the noise floor. Furthermore, as was observed in Reference 113, arrivals corresponding to the same cluster tend to be close in angles, whereas the clusters themselves tend to come at any angle. So, the first important result that the AOA distribution in an indoor multipath environment is not uniformly distributed was observed experimentally in Reference [113].

Most measurements carried out until very recently dealt with time domain data and did not include any data on the AOA distribution in indoor communication links. As was mentioned earlier, knowledge of the AOA associated with a multipath arrival is important because of the increasing use of multiple antenna systems with different kinds of antenna diversity and beam forming [110]. Diversity combining or adaptive array processing is used to mitigate the effects of multipath fading and multiple user servicing. Furthermore, for multiple access communications, adaptive antenna systems have the potential to allow multiple subscribers to simultaneously use the same frequency band, making efficient utilization of the system bandwidth. In order to predict the performance of such access, knowledge of both time and angle of each arrival in a multipath and multiple interference channel is required.

To satisfy the increasing demand on communication speed and ubiquity for the wireless local area networks (WLANs), wireless private branch exchanges (WPBXs), home phone-line network alliance (HomePNA), as well for Internet and data network applications, a new theoretical approach and more precise experiments were carried out, some of which are described in References 114 and 115. This research has begun recently to address the AOA signal distribution in indoor communication channels, using the statistical approach, where it was found that multipath arrivals tend to occur at various angles indoors.

For the first series of the AOA-TOA measurements, carried out at the frequency of 5.8 GHz and bandwidth of 200 MHz, the parabolic antennas with a narrow horizontal beamwidth of 4° were used at both the transmitter and the receiver. The narrow beamwidth allows energy to be directed only to or captured from a certain direction, thus allowing taking AOA measurements. At each Tx-Rx location the Tx and Rx parabolic antenna was rotated from –45° to 45° with intervals of 5° with respect to a line of reference. For LOS cases, this line of reference (or zero azimuth line) is a straight line connecting the Rx and Tx. For NLOS cases, the line of reference is the straight line connecting the Rx to the nearest door lintel of the room with the Tx.

All clockwise measurements with respect to this line of reference are considered as positive and anticlockwise measurements are considered negative. For every Tx azimuth, there are a total of 19 Rx azimuth measurements. The corresponding 19 Tx azimuth measurements and 19 Rx azimuth measurements per Tx azimuth give totally 361 measurements per Tx-Rx location.

For every Rx measured azimuth there were 19 Tx measurements of the corresponding azimuth, each with its corresponding channel impulse response. All of them then were summed up to get a single impulse response per Rx azimuth. Figure 7.29 shows the sample AOA-TOA measurements, carried out in LOS conditions, where for clarity all AOA-TOA plots are presented in relative decibel scale with respect to maximum value of instantaneous power. Also

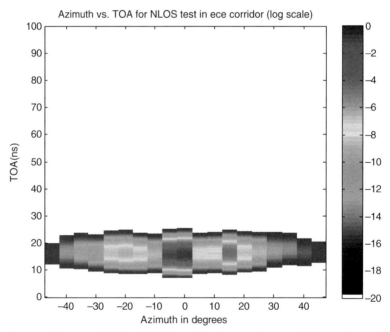

FIGURE 7.29. Measured AOA-TOA joint distribution in 3-m width corridor with 3-m separation between Tx and Rx antennas.

a threshold of –20 dB is used to remove noise and low power portions of the signal. Measured data show the cluster character of signal power distribution in the AOA domain with maximum concentrated at the LOS direction (when azimuth close to zero) and the concentration of the energy in the TOA domain around the range of 10–20 ns with maximum concentrated at the TD of 15 ns.

Further measurements were carried out in NLOS indoor conditions to analyze the AOA-TOA joint distribution for both AOA at Rx and angle of departure (AOD) from Tx. Individually, the energy of each signal arriving from every Tx for every Rx AOA were measured. The floor plan is shown in Figure 7.30a, and Figure 7.30b shows an AOA-TOA signal strength distribution.

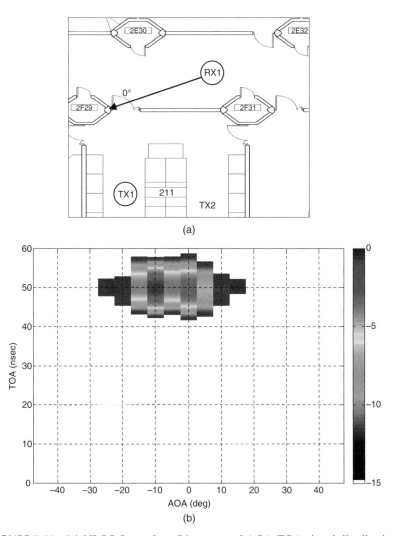

(a)

(b)

FIGURE 7.30. (a) NLOS floor plan; (b) measured AOA-TOA signal distribution.

The signal power is in dBm and a threshold is 10 dBm below the peak signal power. Figure 7.30b clearly shows that the maximum energy is coming from an AOD of -10 to $+5$ degrees. Since we use azimuth steps of $5°$, this is within acceptable region of values. The corresponding maximum in the TOA domain is much higher than in the case of LOS propagation along the corridor and is changed at the range of 50–60 ns.

7.3.3. Adaptive Antennas for Satellite-Mobile Communications

Let us now consider the use of adaptive antenna arrays in satellite communications. In such systems, the mobiles directly communicate with the satellite. In the direct communication link, multiple antenna elements may be utilized on the mobile as well as on a satellite.

Array on a Satellite. It can be assembled on board of a satellite and can provide communication in a number of ways. For instance, different frequencies may be allocated to beams covering different areas such that each area acts as a cell (called in literature *megacell* [93, 96], see also Chapter 14). This allows frequency reuse similar to land communication cellular networks [90, 91, 93, 96]. However, there is a difference between the two systems: the mobile-satellite system generates beams covering different cells rather than having different BSs for different cells as it was done in land communication systems. The antenna array system mounted on board of the satellite provides beam generation, which can be done using different possible scenarios.

A simple method is to utilize beams of fixed shape and size to cover the area of service, allowing normal handoff as the mobile pass from one cell to another or as the beams move in low-orbit satellites covering different areas (see details in Reference 116). In various European Space Agency (ESA) projects for the European geostationary satellite systems, the frequency scanning systems have been combined with phased array antennas [34, 116] assembled in a feed network to control the beam coverage area. A typical system consists of a high-gain large reflector antenna, along with an array of feed elements, placed in the focal plane of the reflector in a particular geometry to generate a limited number of fixed-shape spot beams. A particular beam is selected by choosing a combination of feed elements. The steering of beams is achieved by controlling the phases of signals prior to the feed elements. The capability of the system to generate multiple spot beams with independent power control and frequency use makes it attractive for mobile communications.

The frequency scanning system considered in References 34 and 116 uses an array of active antennas to provide a high-gain beam with the capability to steer it at any user location using frequency-dependent inter-element phase shift. The same array is used for the transmitting as well as for the receiving modes. The frequency scanning system operates as a conventional multispot system in terms of capacity and required power per channel, but requires a

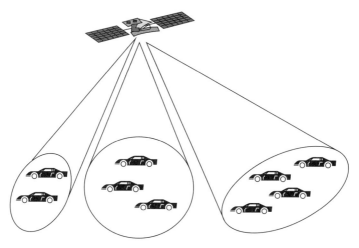

FIGURE 7.31. Coverage of different service areas using satellite antenna beams.

complex beamforming network and mobile terminal location procedure. Such a system, which uses fixed-shape beams, does not require any knowledge of the traffic conditions.

In contrast, a system generating spot beams of varying shapes and sizes (called *dynamic beams*) depends on the traffic conditions dictated by the positions of mobiles, as shown in Fig. 7.31.

To generate an arbitrarily shaped beam to be pointed at a desired location, the system architecture requires an advanced beamforming network with independent beam-steering capability [38]. These systems have been implemented in recent system developments including the GLOBALSTAR [117] and IRIDIUM [118] systems (see also Chapter 14).

Separate Beams for Each Mobile. The possibility of using separate beams for each mobile was investigated in Reference 36. A satellite acts as a relay station between mobiles and the BS, with communication between the BS and the satellite being at a different frequency than that between mobiles and the satellite. Service in such a manner allows each mobile to be tracked, and the beam is pointed toward the desired mobile, with nulls in the directions to other mobiles operating at the same frequency. Different frequencies are used to communicate with the mobiles in close vicinity to the desired one. So, this system is similar to CDMA and TDMA operational systems [90–95].

We note that the direction-finding and beamforming algorithms in this case operate in an environment different than those operating for the land BS-MO communication systems. Because of the distance involved between the satellite and the mobiles, the signals arriving from the mobiles appear more like point sources, which is not the case for the signals arriving at the BS in land communication because of the spreading of signals caused by reflections and scattering in the vicinity of mobiles.

Arrays on Mobile. When an array of antennas is mounted on a mobile to communicate directly with a satellite, the beam steering toward the satellite can be performed using simple phase shifters [119]. As the direction of the satellite with respect to the mobile varies, tracking of the satellite and adjusting of the direction of the beam is required so that the main beam points toward the satellite.

In the German TV-SAT 2 system [35], which uses multiple antennas, electronic steering was not utilized. Instead, it employs four sets of fixed beams, formed separately and switched on depending on the orientation of the mobile relative to the geostationary satellite used. The system was shown to be useful for large vehicles, such as buses and trains. At the same time, separate array elements are suggested in Reference 120 which alternately placed side by side in a planar configuration for transmit and receive mode using digital beamforming techniques for tracking two satellites in multisatellite system based on low-orbit satellites.

Test results on the characteristics and the suitability of a spherical antenna array mounted on an aircraft employing digital beamforming to communicate directly with a satellite were reported in Reference 121. An experimental development of a four-element antenna array to receive signal from the INMARSAT II F-4 geostationary satellite with a fixed position receiver show that a significant improvement in link reliability can be produced using an adaptive array at a handheld mobile compared to omni-antenna [122].

7.4. NETWORK PERFORMANCE IMPROVEMENT USING AN ANTENNA ARRAY

An antenna array is able to improve the performance of wireless communication systems by providing the capability to reduce co-channel interferences, multipath fading, and DS, and therefore, results in a better QOS, such as reduced BER and outage probability. Furthermore, its capability to form multiple beams that can be exploited to serve many users in parallel results in an increased spectral efficiency of the system. The possibility also exists to handle traffic conditions by adapting beam shapes and reducing the handoff rate, which results in increased trunking efficiency. A number of studies have been reported in the literature covering all these aspects [123–148]. Some of them are briefly mentioned in this section.

7.4.1. Reduction in Multipath Phenomena

A key issue for adaptive array antennas in wireless communication systems is their performance in multipath versus LOS environments. First of all, let us consider what happens when adaptive arrays are operating in LOS conditions. As shown in Section 7.2, in a LOS channel, an adaptive array can be used to enhance signal reception in a desired direction and to place nulls in the direction of interference.

Under these conditions, with the number of antennas much greater than a number of arriving signal multipath components (usually called *rays*), it is easy to express the array response in terms of a small number of AOAs, rather than the received signal phase at each antenna element. Techniques that exploit this fact for improved performance include the MUSIC and ESPRIT algorithms [123], which determine the DOA of the rays. Thus, such an array of M elements (called in technical literature *sensors*) can form up to $M - 1$ nulls to cancel up to $M - 1$ interferers caused by multipath components arriving at the receiver. As was shown in Reference 124, such angular domain methods can be also useful in some situations with near-LOS, such as at mobile radio BSs in flat rural environments with many (in Reference 124 it was eight) high elevated antennas.

However, in *clutter conditions* with multipath, the signals arrive from each user via multiple paths and AOAs due to multiple scattering, reflection, and diffraction phenomena. Thus, it becomes impossible to form an antenna pattern with a beam in the direction of each arriving path of the desired signal and nulls in the directions of all interfering signals, as the number of required nulls would be much greater than the number of antenna elements. Furthermore, to provide diversity gain, the elements at the BS can be spaced many wavelengths apart, which results in many grating lobes (as shown in Fig. 7.12). However, no matter how many paths each signal covers, the result is a given phase and amplitude at each antenna for each signal. Thus, there is an array response for each signal, and the performance of the array depends on the number of signals, not the number of paths, so one can do the analysis in the signal space domain rather than the angular domain. This holds true as long as the DS is small. If not, then the delayed versions of the signals must be considered as separate signals.

Hence, an adaptive array can null $M - 1$ interferers due to multipath independent of the environment; it is either LOS or clutter with multipath. It was also mentioned in Reference 37 that an important feature of adaptive arrays in multipath conditions is their ability to cancel interferers independent of the AOA. Thus, it is not sufficient if the interferer is a few meters from the desired mobile or at several kilometers from the BS. In LOS conditions, the separation of such closely spaced signals is not possible.

In a multipath environment, obstructions around antennas act as a huge reflection antenna, with the actual antennas acting as feeds, which permit the receiving array to separate the signals. If the receiving antennas are spaced far enough apart such that beams that are smaller than the angular spread can be formed, then the signals from two closely spaced antennas can usually be separated using adaptive combining techniques. They are the combination of spatial diversity of the array and channel coding [126–128], using of a combined mode of diversity combining with interference canceling [129, 130] or using the same diversity combining with adaptive equalization [131–134].

As was shown experimentally in Reference 125, the number of signals that can be separated increases with the number of receive antennas of the array. The same features were observed for the angular spread and for the density of the multipath reflections within the angular spread.

7.4.2. Reduction in Delay Spread

As mentioned in Chapter 1, delay spread (DS) is caused by multipath propagation due to multiple scattering, diffraction, and reflection from obstructions surrounding the transmitter and the receiver. In this case, the desired signal arriving from different directions gets delayed due to different travel radio paths involved. An antenna array with the capability to form beams and nulls in certain directions is able to cancel some of these delayed arrivals. There are two ways to achieve this:

(a) The *first way* occurs in the transmit mode when an antenna array focuses energy in the required direction, and finally reduces multipath components causing a reduction in the DS.

(b) The *second way* occurs in the receive mode when an antenna array provides compensation in multipath fading by (1) diversity combining, (2) adding the signals belonging to different clusters of signals after compensating for delays, and (3) by canceling delayed signals arriving from directions other than that of the main signal.

Let us briefly present some of these techniques [112, 139, 142–155].

Use of Diversity Combining. Diversity combining achieves a reduction in fading by increasing the signal level based upon the level of signal strength at different antennas [137, 139, 149], whereas in multipath cancellation methods, it is achieved by adjusting the beam pattern to accommodate nulls in the direction of late arrivals assuming to be interferences. For the latter case, a beam is pointed in the direction of the direct path or a path along which a major component of the signal arrives, causing a reduction in the energy received from other directions and thus reducing the components of multipath signal coming at the receiver. Techniques on how to identify a LOS component from a group of received multipath signals have been discussed in Reference 140.

Combining Delayed Arrivals. This technique is based on the organization of clusters of signals identically delayed within each cluster. This occurs because a radio wave originating from a source arrives at a distinct point in clusters after getting scattered and reflected from obstructions along the radio path. This occurs in the urban scene with large buildings or hills where delayed arrivals are well separated [141], as well as in indoor communications [114,

115]. We can use these clustered signals constructively by grouping them as per their delays compared with a signal available from the shortest path. Individual paths of these delayed signals may be resolved by exploiting their spatial or temporal structure (as shown in Fig. 7.27 in the time domain).

The resolution of paths using temporal structures depends upon the bandwidth of the signal compared with the coherence bandwidth of the channel, and it increases as this bandwidth increases. In a CDMA system, the paths may be resolved provided their relative delays are more than the chip period [142–145]. When these paths are well separated spatially, an antenna array may be used. This can be done, for example, by determining their directions. Spatial diversity combining similar to that used in RAKE, a receiver (see details in Reference 146), may also be employed to combine signals arriving in multipath.

The signals in each cluster may be separated by using specific information present in each signal, such as the frame identification number or the use of a known symbol in each frame utilized in a TDMA system (see details in Reference 149).

Nulling Delayed Arrivals. It was found that, using an adaptive antenna array, an essential reduction in DS is possible. Similar conclusions were obtained in Reference 148 using an experimental array of four elements mounted on a vehicle. It was shown that the array with the corresponding processing is able to null the delayed arrival in a time-division multiplexed channel. In channels where frequency-selective fading occurs, that is, where signals with different frequencies fade differently, a frequency-hopping (FH) system may also be used for eliminating degradation of signals due to fading. For information about differences between direct-sequence spread spectrum (DS-SS) systems and FH systems, the reader is referred to References 90–93 and 96. Applications of adaptive arrays in FH-communication systems are described in References 150–153, and for DS-SS-communication systems, the reader can check References 154 and 156.

7.4.3. Reduction in Angular Spread

Angular spreading occurs when a transmitted signal gets scattered or reflected in the vicinity of the source and the receiver, and a signal arrives at the receiver within range of angles, as shown in Figure 1.15 from Chapter 1. This range, called an *angular spread*, depends on the situation in the environment and on the antenna locations with respect to building rooftops. Thus, the conditions of land macrocell BS-MO communication are close to that shown in Figure 7.22a, where the BS antenna is normally high enough from the ground surface and the mobile antenna is close to the ground. In this case, the angular spread occurs only in the vicinity of the mobile. Here, a signal arrives at the BS with an angular distribution depending on the range from the two terminals and the density of obstructions surrounding the mobile antenna. Thus, the range of angle spreading becomes smaller as the distance between the mobile and the BS increases.

Experimental results indicate that the angular spreading of about $3°$ results for a distance of 1 km between terminals [156]. In Reference 157, it was shown that the AOA distribution of the signal is statistically related to the path DS. Angular spread causes space-selective fading, which means that the signal amplitude depends on the spatial location of the antenna (this effect will be also shown theoretically and experimentally in Chapter 8). Space-selective fading is characterized by the coherence distance [90–93, 96]; the larger the angular spread, the shorter the coherence distance. The latter represents the maximum spatial separation for which the channel responses at two antennas remain strongly correlated. Selection of the distribution function, however, does not appear to be critical as long as the spread is small around the main direction [156–160].

For the cases shown in Figures 7.22b and 7.22c, which usually occur in micro-cell environments for BS-MO communications (see Fig. 7.22b) or for MO-MO communications (see Fig. 7.22c), the angular spread may become wide enough (up to $45°$–$60°$, see results presented in Chapter 8) to decrease the efficiency of the adaptive antenna. Thus, a dispersion of the radio environment results in the distortion of the antenna sidelobe levels at the BS [161], as well as an increase in the correlation of fading at different antenna elements of the array [130].

The problem of fading correlations is studied in Reference 162. It was shown that by deriving the relationships between AOA, beamwidth, and correlation of fading, larger element spacing is required to reduce the correlation, especially when the AOA is parallel to the array. A correlation coefficient of fading, between the various antenna elements, greater than 0.8 can cause signals at all elements to fade away simultaneously [163]. A detailed investigation of the effect of fading correlation on the performance of the adaptive arrays to combat fading was done in Reference 130. It was shown that a correlation up to 0.5 causes little degradation of the antenna efficiency, but a higher correlation decreases its performance significantly. However, the array is able to suppress interferences, as independent fading is not required for interference suppression [130]. The aspect of AOA distribution as well as DS distribution for different land communication links will be analyzed in detail in Chapters 8 and 11.

7.4.4. Range Increase

In typical land macrocell communication links with high BS antenna and low mobile antenna (see Figs. 7.22a,b), when angular spread is small enough, a MIMO system consisting of M- and N-element antenna arrays at both ends of the link gives a link gain as described by (7.45) with a diversity gain equal to NM (see Section 7.3). It means that both the physical M-element adaptive array and a multibeam (phased array) antenna provide an MN-fold increase in antenna gain. Even if only one antenna (either BS or mobile) is in a form of an array with M elements, the increase will be in $G_L = M$ times. In the case of low elevation antennas shown in Figure 7.22c, as was mentioned above in Section 7.3, the link gain is described by (7.46) with a diversity gain of NM. If N is small and M is large, the gain will approach M.

This capability of the adaptive or multibeam antennas to increase the range of a communication link by the factor $G_L^{1/\gamma}$ is used to reduce the number of BSs required to cover a given area by a factor of $G_L^{2/\gamma}$, where γ is the propagation-loss exponent. From Chapter 5, the parameter γ is set to be somewhere between 2 and 5. The physical adaptive array antenna also provides diversity gain, and for a given array size with spatial diversity, the diversity gain increases with angular spread, and the fading correlation decreases, thus providing a greater range for the radio link. For the multibeam (phased array) antenna, however, the diversity gain is limited, as angular diversity provides only a small diversity gain. Another disadvantage of the multibeam antenna is that the antenna gain is limited by the angular spread, that is, because the antenna cannot provide additional antenna gain when the beamwidth is less than the angular spread because smaller beamwidths exclude signal energy outside the beam.

With an angular spread of $\alpha_0 = 20°$ for a 10-element adaptive antenna, the range can be increased by a factor of 2 (with respect to the single element), whereas for the same 10-beam multibeam antenna, the increase is about 1.7 [164]. This difference increases with any increase in the number of adaptive antenna elements or beams in a multibeam antenna. For example, for 30-element antenna with an angular spread of $\alpha_0 = 20°$, the range increase is 2.5 times that of a single regular antenna, whereas for the same angular spread and number of beams, the range increase of the multibeam antenna will be by a factor of 1.7. Note that these results are valid only for the uplink, where the mobile user transmits information and the BS receives it [164]. For the down-link, as the downlink frequency is different from the uplink frequency (for FDMA, GSM [combination of FDMA and TDMA, see Chapter 3], IS-95, and IS-136 systems), the same adaptive array techniques cannot be used for trans-mission by the BS and reception by the mobile antenna. Here, the multibeam antenna can be used more effectively, but to achieve diversity gain, transmit diversity must be used or the handset vehicle must have multiple antennas [165]. Although these techniques may provide less gain on the downlink than on the uplink, this may be compensated for by the higher transmit power of the BS as compared to the handset vehicle.

The results obtained in Reference 164 are valid for the uplink and for systems close to TDMA or its combinations with FDMA (i.e., GSM system). In CDMA systems, the RAKE receiver provides three-time diversity, and dif-ferent beams can be used for each output of the RAKE receiver. So, in CDMA, the multibeam antenna gives the same range increase as the adaptive array antenna. Since multibeam antennas require less complexity (with respect to weight and tracking), the multibeam antenna is preferable for CDMA systems, while an adaptive array antenna may be preferable for TDMA or GSM systems, particularly in environments with large angular spreads.

7.4.5. Reduction in Co-Channel Interference and Outage Probability

In this section, we briefly will consider the integration of an idealized adaptive antenna array into an existing cellular network and will compare it with the conventional omnidirectional BS antenna following the well-known hexagonal regular cell topology, briefly described in Chapter 3, or nonregular cell topology, briefly described in Chapter 11. For all details of cellular maps design, the reader is referred to References 90–93, 102, and 166.

Let the cluster size be given in terms of the number of cells N, which uses different frequencies compared with the *wanted* cell (i.e., the cell under service). This number or the "cluster size," N, [102, 166] , as shown in Chapter 3, is related to the co-channel reuse factor $Q = D/R_{cell}$ by:

$$N = Q^2 / 3,\tag{7.47}$$

where R_{cell} is the radius of a cell and D is the reuse distance which defines a range between cells allocated by the same frequency band (see definitions in Chapter 13). There is a limited number of N cells that are possible in a hexagonal cellular network [102, 166], that is, $N = 3, 4, 7, 9, 12, \ldots$ (see also Equation (3.2) from Chapter 3).

Co-channel interference will occur when the ratio of the received (e.g., wanted) signal envelope, S (usually called "the carrier," see Chapter 3), to the interfering signal envelope, I, is less than some protection ratio p_r (a threshold), that is [167],

$$\frac{C}{I} \leq p_r.\tag{7.48}$$

Now we consider only one co-channel cell as an i part of a cellular network, as shown in Figure 7.32. Assume that the only propagation-loss effects are proportional to distances d_{US} and d_I between the desired mobile user and each of the desired and interfering BS, respectively. Then,

$$\frac{C}{I} = \frac{d_I^\gamma}{d_{US}^\gamma} \leq p_r,\tag{7.49}$$

where γ is the loss exponent. So, for a given protection, we get [167–168]

$$\frac{d_I}{d_{US}} \leq (p_r)^{1/\gamma}.\tag{7.50}$$

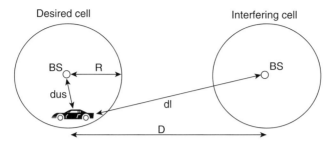

FIGURE 7.32. Co-channel interference between a desired and an interfering cell.

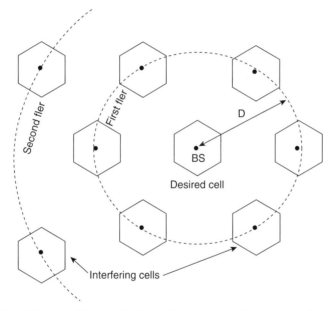

FIGURE 7.33. Scheme of how to eliminate the co-channel interference between neighboring cells operating at the same frequency band.

In the case when the desired user lies along a straight line between two BSs (the worst case for a user), the co-channel reuse ratio is

$$Q = D / R_{cell} = 1 + d_I / d_{US} = 1 + p_r^{1/\gamma}. \tag{7.51}$$

For a given protection ratio and modulation scheme, this expression defines the minimum spacing between co-channel cells in order to avoid interference. For six co-channel cell interferers (Fig. 7.33), which lie only in the first tier of co-channel cells, we have instead of (7.49) the following expression [102]:

$$\frac{C}{I} = \frac{d_{US}^{-\gamma}}{6d_I^{-\gamma}} \le p_r. \tag{7.52}$$

The co-channel reuse factor for the first tier (see definition in Figure 7.26) can be expressed as [166]

$$Q = D / R_{cell} = [6(C/I)]^{1/\gamma}. \tag{7.53}$$

To see how these parameters influence the co-channel interference occurrence, let us define some other parameters of the network. First, we assume that all users are uniformly distributed per cell with a blocking probability of service P_{bl} constant for all cells. Blocking probability is measured by the number of users/calls which cannot be served during the period of service (see also Chapter 11). The parameter A (in erlangs) defines the traffic intensity offered, where erlang is a measure of traffic intensity defining the quantity of traffic on a channel or group of channels (users) per unit time. Then, the actual traffic carried is equal to $A(1 - P_{bl})$ erlangs, and the so-called *outgoing channel usage efficiency* or *loading factor* becomes [168]

$$\eta = A(1 - P_{bl}) / n_c, \tag{7.54}$$

where n_c is the total number of channels allocated per cell.

We now assume that instead of an omnidirectional BS antenna we have an adaptive one, which generates M ideal beams with a bandwidth of $2\pi/M$, and a gain equal to that of omni antenna. Each adaptive beam will only carry the channels that are assigned to the mobiles within its coverage area. So, any mobile or group of mobiles can be tracked by using adaptive BS antennas, as shown in Figure 7.26.

As the occurrence of co-channel interference between subscribers is a statistical problem, instead of the *probability of co-channel interference*, η, we use the *outage probability*, $P(C \le I \cdot p_r)$. This probability determines the frequency of failing to obtain satisfactory reception at the mobile in the presence of interference. For identical cells, having equal probability of call blocking, there will be in average $n_c \eta$ active channels in each cell. Then, for the omnidirectional BS antenna, assuming the desired mobile is already allocated a channel, the probability of that channel being active in an interfering cell is the required outage probability, given by

$$P(C \le I \cdot p_r) = \frac{number\_of\_active\_channels}{total\_number\_of\_channels} = \frac{\eta n_c}{n_c} \equiv \eta. \tag{7.55}$$

Hence, when the desired mobile is in the region of co-channel interference for the omni terminal antennas, the outage probability is identical to the probability of co-channel interference.

For adaptive antennas with M beams per BS, we have $n_c\eta/M$ channels per beam with a uniform distribution of subscribers. Here, a desired mobile is always covered by at least one beam from the co-channel cell. Then, the outage probability is equal to the probability that one of the channels in the aligned beam is the corresponding active co-channel (i.e., the channel that has been allocated to the desired mobile). Thus, this probability equals

$$P(C \leq I \cdot p_r) = \frac{number\_channels\_per\_beam}{total\_number\_of\_channels} = \frac{\eta n_c / M}{n_c} \equiv \frac{\eta}{M}. \qquad (7.56)$$

Thus, the co-channel interference decreases with an increase in the number of antenna elements in the adaptive array or the number of beams in the multibeam antenna.

Now using a simple geometry presented in Figure 7.26, it can be shown, that for the six co-channel cells in the first tier, the outage probability at the regions of interference equals

$$P(C \leq I \cdot p_r) = \left(\frac{\eta}{M}\right)^6; \qquad (7.57)$$

that is, it decreases M^{-6} times with an increase of M. This means that there are six beams aligned onto the desired mobile at any time, and the outage probability within the region of interference is defined by the probability that the active co-channel is in each of these beams. For more details, the reader is referred to References 102 and 169–171. It is the spatial filtering capabilities of adaptive antennas that reduce co-channel interference.

In general, an adaptive array requires some information about the desired signal, such as the direction of its source, a reference signal, or a signal that is correlated with the desired signal. In situations where the precise direction of the signal is known, interference cancellation may be achieved by solving a constrained beamforming problem [38] or by using a reference signal [154, 172].

Despite the fact that the multibeam antenna is less effective than the adaptive antenna, in reducing interference in TDMA systems, for the downlink, multibeam antennas can be used at the BS in combination with adaptive arrays on the uplink. The problem is even worse in GSM or IS-136 systems, because the handsets require a continuous downlink. Therefore, the same beam pattern must be used for all users in a channel, which further reduces the effectiveness of multibeam antenna against interference.

In CDMA systems, the users' capacity or loading factor and the outage probability depend on the spreading gain and the corresponding number of equal-power co-channel interferers. Here, the multibeam antenna with M beams also reduces the number of interferers per one beam by a factor of M, increasing the user capacity M-fold. At the same time, for CDMA systems, the adaptive arrays can provide only limited additional interference suppression, because the number of interferers is generally much greater than the number of antenna elements in an array. Furthermore, as multibeam antennas are less complex than adaptive arrays and their beams need to be switched at most every few seconds versus tracking 178 Hz fading signals in adaptive arrays [74], multibeam antennas are generally preferred in CDMA systems.

We also note here that interference is typically worse on the uplink than on the downlink for two reasons. First, it is possible that the signal from an interfering mobile to be stronger than that from the desired mobile at the BS (i.e., on the uplink), whereas at the mobile, as a receiver, the signal from an interfering BS (i.e., on downlink) does not matter as the mobile chooses the BS with the strongest signal. Second, BSs are typically more uniformly spaced than the mobiles and located constantly near the center of each cell. Hence, more interference suppression on the uplink than on the downlink may be desirable.

7.4.6. Increase in Spectrum Efficiency and Decrease of BER by Using Smart Antennas

Spectrum efficiency refers to the amount of traffic a given system with certain spectrum allocation could handle [38]. An increase in the number of users of a wireless communication system, mobile or personal, causes the spectrum efficiency to increase. This measure allows us to compare the QOS of different cellular systems [167].

Following Reference 103, we denote this measure by E and determine it as the spectrum utilization through the number of channels/users per bandwidth in megahertz and per square kilometer, that is,

$$E = \frac{B_S / B_c}{B_S(N_C A_c)} = \frac{1}{B_c N_C A_c}. \tag{7.58}$$

Here, B_S is the total available bandwidth of the system, B_c is a channel spectral spacing, both in megahertz, N_C is the number of cells per cluster, and A_c is the cell area in square kilometer (km^2). We should note that in Chapter 3 we replaced a measure E by $\tilde{C}$, as it is usually denoted in the literature [91–95]. To perform a simple comparison between an omni antenna and an adaptive antenna array, we assume that an identical modulation scheme is employed in both cases. As $E \sim N_C^{-1}$, the relative spectrum efficiency can be expressed as [102, 103]

$$\frac{E_{adapt}}{E_{omni}} = \frac{N_{Comni}}{N_{Cadapt}}, \tag{7.59}$$

where N_C can be expressed as a function of the protection ratio p_r and the fading parameter Z_d (called the *fade margin*) as

$$N_C = \frac{1}{3}\left[1 + \sqrt{10^{(Z_d + p_r)/20}}\right]^2. \tag{7.60}$$

Many other studies have shown that the use of adaptive antennas can essentially increase the spectrum efficiency by increasing user capacity [173–178]. An array can also create additional channels by forming multiple beams without any extra spectrum allocation, which results in potentially extra users and thus increases the spectrum efficiency.

Using SDMA techniques, we can serve efficiently many mobile users using only two frequencies [177]. The multipath Rayleigh fading channel gives a fundamental limit on the maximum data rate and the spectral capacity of a multiple beam antenna system [178] (see also Chapter 11). It was shown in References 178 and 179 that using an antenna array in a BS, in the uplink to locate the positions of the mobile users in a cell, and then transmitting in a multiplexed manner toward different clusters of mobiles, the spectrum efficiency increases, and it depends on the number of elements in the array.

BS antenna arrays used for CDMA systems, for example, at an outage probability of 0.01, can increase the system capacity from 31, for a single antenna system, to 115 for a five-element array and to 155 for an array of seven elements [180, 181]. Also, an antenna array at the BS for TDMA system can increase the reuse factor by three, if dynamic channel assignment is utilized (see details in References 179, 182, and 183). The same BS array of L elements may lead up to an L-fold capacity improvement in an indoor-mobile communication system by allowing many users to share the same channels [184–189]. In Reference 190, using an LMS algorithm, it was found that for an omnidirectional (OM) antenna, the probability of BER is given by [190]

$$P_{eOM} = Q\left(\sqrt{\frac{3G}{U(1+8\beta)-1}}\right), \tag{7.61}$$

whereas, for an antenna array (MULT) we get

$$P_{eMULT} = Q\left(\sqrt{\frac{3GD}{U(1+8\beta)-1}}\right). \tag{7.62}$$

Here, $Q(x)$ is the standard Q-function defined in Chapter 1 of variable x, G is the processing gain of the CDMA system, $\beta = 0.05513$, and D is the directivity of the beam of the multibeam antenna system. A comparison of the BER performance of the systems using other than LMS algorithms [147, 189, 191] has shown that the maximum entropy method [189] and the spatial discrete Fourier transform method [191] provide better BER performance than the LMS algorithm. We refer the reader to these works.

Finally, we note that there are some other performance improvements that can be obtained by using an antenna array, such as an increase in transmission efficiency, reduction in handoff rate and in cross talks, improvement of dynamic channel assignment, cost-effective implementation, complexity reduction, and network implication. All these are aspects beyond the subject of this book and are well described in existing literature [160–162, 177, 179, 192–196].

7.5. CONCLUSION

Adaptive or smart antenna systems are usually categorized as either switched beam or adaptive array systems. Although both systems attempt to increase gain in the direction of the user, only the adaptive array system offers optimal gain while simultaneously identifying, tracking, and minimizing interfering signals.

The traditional switched beam method is considered as an extension of the current cellular sectorizing scheme, in which a typical sectorized cell site is composed of three 120-degree macrosectors or six 60-degree macrosectors. The switched beam approach further subdivides the macrosectors into several microsectors. The adaptive antennas take a very different approach. By adjusting to an RF environment as it changes, adaptive antenna technology can dynamically alter the signal patterns to optimize the performance of the wireless system. Usually, in the transmit mode, the adaptive array focuses energy in the required direction, which helps to reduce multipath reflections and the DS. In the receive mode, however, the array provides compensation in multipath fading by adding the signals emanating from other clusters after compensating for delays, as well as by canceling delayed signals emanating from directions other than that of the desired signal.

REFERENCES

1   Glaser, E.M., "Signal detection by adaptive filters," *IEEE Trans. Inform. Theory*, Vol. IT-7, No. 1, 1961, pp. 87–98.

2   Davisson, L.D., "A theory of adaptive filters," *IEEE Trans. Inform. Theory*, Vol. IT-12, No. 1, 1966, pp. 97–102.

3   Griffiths, L.J., "A comparison of multidimensional Weiner and maximum-likelihood filters for antenna arrays," *Proc. IEEE*, Vol. 55, 1967, pp. 2045–2057.

4   Goode, B.B., "Synthesis of a nonlinear Bayer detector for Gaussian signal and noise fields using Wiener filters," *IEEE Trans. Information Theory*, Vol. IT-13, 1967, pp. 116–118.

5   Burg, J.P., "Three-dimensional filtering with an array of seismometers," *Geophysics*, Vol. 29, 1964, pp. 693–713.

6   Haykin, S., ed., *Array Signal Processing*, Prentice-Hall, Englewood Cliffs, NJ, 1985.

7   Stuber, G.L., *Principles of Mobile Communication*, Kluwer Academic Publishers, 1996.

8   Proakis, J.G., *Digital Communication*, McGraw Hill, New York, 2001.

9   Steele, R., *Mobile Radio Communications*, IEEE Press, New York, 1992.

10 Rappaport, T.S., *Wireless Communications Principles and Practice*, Prentice Hall, New York, 1996.

11 Farina, A., *Antenna-Based Signal Processing Techniques for Radar Systems*, Artech House, Norwood, MA, 1992.

12   Krim, H. and M. Viberg, "Two decades of array signal processing research," *IEEE Signal Processing Magazine*, No. 1, 1996, pp. 67–94.

13   Paulraj, A.J. and C.B. Papadias, "Space-time processing for wireless communications," *IEEE Personal Communications*, Vol. 14, No. 5, 1997, pp. 49–83.

14   Widrow, B., P.E. Mantey, L.J. Griffits, and B.B. Goode, "Adaptive antenna systems," *Proc. IEEE*, Vol. 55, No. 12, 1967, pp. 2143–2159.

15   "Special Issue on Active and Adaptive Antennas," *IEEE Trans. Antennas Propagat.*, Vol. AP-12, March 1964; Vol. AP-24, September 1976.

16   Gabriel, W.F., "Adaptive arrays-an introduction," *Proc. IEEE*, Vol. 64, No. 2, 1976, pp. 239–272.

17   Gabriel, W.F., "Spectral analysis and adaptive array super resolution techniques," *Proc. IEEE*, Vol. 68, No. 6, 1980, pp. 654–666.

18   Monzingo, R.A. and T.W. Miller, *Introduction to Adaptive Arrays*, Wiley and Sons, New York, 1980.

19   Hudson, J.E., *Adaptive Array Principles*, Peter Peregrinus, New York, 1981.

20   "Special Issue on Adaptive Processing Antenna Systems," *IEEE Trans. Antennas Propagat.*, Vol. AP-34, March 1986.

21   Compton, R.T., Jr., *Adaptive Antennas: Concepts and Performance*, Prentice-Hall, Englewood Cliffs, NJ, 1988.

22   Gabriel, W.F., "Adaptive processing array systems," *Proc. IEEE*, Vol. 80, No. 1, 1992, pp. 152–162.

23   Litva, J. and T. Lo, *Digital Beamforming in Wireless Communications*, Artech House, 1996.

24   Pridham, R.G. and R.A. Mucci, "A novel approach to digital beamforming," *J. Acoust. Soc. Amer.*, Vol. 63, 1978, pp. 425–434.

25   Pridham, R.G., and R.A. Mucci, "Digital interpolation beamforming for low-pass and bandpass signals," *Proc. IEEE*, Vol. 67, No. 5, 1979, pp. 904–919.

26   Curtis, T.E., "Digital beam forming for sonar system," *IEE Proc. Pt. F*, Vol. 127, No. 2, 1980, pp. 257–265.

27   Mucci, R.A., "A comparison of efficient beamforming algorithms," *IEEE Trans. Acoust., Speech, Signal Processing*, Vol. ASSP-32, No. 3, 1984, pp. 548–558.

28   Van Veen, B.D. and K.M. Buckley, "Beamforming: A versatile approach to spatial filtering," *IEEE Trans. Acoustics, Speech and Signal Processing Magazine*, Vol. 5, No. 1, 1988, pp. 4–24.

29   Gocler, H.G., and H. Eyssele, "A digital FDM-demultiplexer for beamforming environment," *Space Communic.*, Vol. 10, No. 2, 1992, pp. 197–205.

30   Barton, P., "Digital beam forming for radar," *IEE Proc. Pt. F*, Vol. 127, No. 1, 1980, pp. 266–277.

31   Sakagami, S., S. Aoyama, K. Kuboi, S. Shirota, and A. Akeyama, "Vehicle position estimates by multibeam antennas in multipath environments," *IEEE Trans. Veh. Technol.*, Vol. 41, No. 1, 1992, pp. 63–68.

32   Wu, W.W., E.F. Miller, W.L. Prichard, and R.L. Pickholtz, "Mobile satellite communications," *Proc. IEEE*, Vol. 82, 1994, pp. 1431–1448.

33   Horton, C.R. and K. Abend, "Adaptive array antenna for satellite cellular and direct broadcast communications," *Proc. 3rd Int. Mobile Satellite Conf.*, Pasadena, CA, 1993, pp. 47–52.

34   Cornacchini, C., R. Crescimbeni, A. D'ippolito, et al., "A comparative analysis of frequency scanning and multispot beam satellite systems for mobile communication," *Int. J. Satellite Communic.*, Vol. 13, No. 1, 1995, pp. 84–104.

35   Schrewe, H.J., "An adaptive antenna array for mobile reception of DBS-satellites," *Proc. 44th Veh. Technol. Conf.*, Stockholm, Sweden, 1994, pp. 1494–1497.

36   Gebauer, T. and H.G. Gockler, "Channel-individual beamforming for mobile satellite communications," *IEEE J. Select. Areas Communic.*, Vol. 13, No. 3, 1995, pp. 439–448.

37   *Smart Antennas*, ed. by T.S. Rappaport, IEEE Press, Piscataway, NJ, 1998.

38   Ioannides, P. and C. Balanis, "Uniform circular arrays for smart antennas," *IEEE Antennas and Propagat. Magazine*, Vol. 47, No. 4, 2005, pp. 192–206.

39   Tsoulos, G.V. and G.E., Athanasiadou, "On the application of adaptive antennas to microcellular environments: radio channel characteristics and system performance," *IEEE Trans. Vehicular Techn.*, Vol. 51, 2002, pp. 1–16.

40   Naguib, A.F., A. Paulraj, and T. Kailath, "Capacity improvement with base-station antenna arrays in cellular CDMA," *IEEE Trans. Veh. Technol.*, Vol. 43, 1994, pp. 691–698.

41   Gorada, L.C., "Applications of antenna arrays to mobile communications. Part II: Beam-forming and direction-of-arrival considerations," *Proc. IEEE*, Vol. 85, No. 8, 1997, pp. 1195–1245.

42   Anderson, S., M. Millnert, M. Viberg, and B. Wahlberg, "An adaptive array for mobile communication systems," *IEEE Trans. Veh. Technol.*, Vol. 40, No. 2, 1991, pp. 230–236.

43   Vaugham, R.G. and N.L. Scott, "Closely spaced monopoles for mobile communications," *Radio Sci.*, Vol. 28, 1993, pp. 1259–1266.

44   Barlett, M.S., "Smoothing periodograms from time series with continuous spectra," *Nature*, Vol. 161, 1948, pp. 686–687.

45   Capon, J., "High-resolution frequency-wavenumber spectrum analysis," *Proc. IEEE*, Vol. 57, No. 8, 1969, pp. 1408–1418.

46   Applebaum, S.P., "Adaptive arrays," *IEEE Trans. Antennas Propagat.*, Vol. 24, No. 9, 1976, pp. 585–598.

47   Lacoss, R.T., "Data adaptive spectral analysis method," *Geophysics*, Vol. 36, 1971, pp. 661–675.

48   "Special Issue of Spectra from Various Techniques," in *IEEE Trans. Acoustics, Speech and Signal Processing*, Vol. ASSP-29, No. 3, 1981.

50   Wiener, N., *Extrapolation, Interpolation and Smoothing of Stationary Time Series*, MIT Press, Cambridge, MA, 1949.

50   Haykin, S., J. Litva, and T.J. Shepherd, eds., *Radar Array Processing*, Springer-Verlag, Berlin, 1993.

51   Johnson, D.H. and D.E. Dudgeon, *Array Signal Processing Concepts and Techniques*, Prentice-Hall, Englewood Cliffs, NJ, 1993.

52   Butler, J.L., "Digital matrix, and intermediate frequency scanning," in R.C. Hansen, ed., *Microwave Scanning Arrays*, Academic Press, New York, 1966.

53   Anderson, V.C., "DICANNE, a realizable adaptive process," *J. Acoust. Soc. Amer.*, Vol. 45, 1969, pp. 398–405.

54 Anderson, V.C. and P. Rudnick, "Rejection of a coherent arrival at an array," *J. Acoust. Soc. Amer.*, Vol. 45, 1969, pp. 406–410.

55 Brennan, L.E. and I.S. Reed, "Theory of adaptive radar," *IEEE Trans. Aerosp. Electron. Syst.*, Vol. AES-9, 1973, pp. 237–252.

56 Reed, I.S., J.D. Mallett, and L.E. Brennan, "Rapid convergence rate in adaptive arrays," *IEEE Trans. Aerosp. Electron. Syst.*, Vol. AES-10, 1974, pp. 853–863.

57 Weber, M.E. and R. Heisler, "A frequency-domain beamforming algorithm for wideband, coherent signal processing," *J. Acoust. Soc. Amer.*, Vol. 76, 1984, pp. 1132–1144.

58 Florian, S. and N.J. Bershad, "A weighted normalized frequency domain LMS adaptive algorithm," *IEEE Trans. Acoust., Speech, Signal Processing*, Vol. 36, 1988, pp. 1002–1007.

59 Zhu, J.X. and H. Wang, "Adaptive beamforming for correlated signal and interference: A frequency domain smoothing approach," *IEEE Trans. Acoust., Speech, Signal Processing*, Vol. 38, 1990, pp. 193–195.

60 Klemm, R., "Suppression of jammers by multiple beam signal processing," in Proceedings of the IEEE International Radar Conference, Sendai, Japan, 1975, pp. 176–180.

61 El Zooghby A., C.G. Christodoulou, and M. Georgiopoulos, "Neural network-based adaptive beamforming for one and two dimensional antenna arrays," *IEEE Trans. Antennas Propagat.*, Vol. 46, 1998, pp. 1891–1893.

62 El Zooghby A., C.G. Christodoulou, and M. Georgiopoulos, "A neural network-based smart antenna for multiple source tracking," *IEEE Trans. Antennas Propagat.*, Vol. 48, 1999, pp. 768–776.

63 Anderson, V.C., "Digital array phasing," *J. Acoust. Soc. Amer.*, Vol. 32, 1960, pp. 867–870.

64 Rudnick, P., "Digital beamforming in the frequency domain," *J. Acoust. Soc. Amer.*, Vol. 46, 1969, pp. 1089–1095.

65 Dudgeon, D.E., "Fundamentals of digital array processing," *Proc. IEEE*, Vol. 65, 1977, pp. 898–904.

66 Pridham, R.G. and R.A. Mucci, "A novel approach to digital beamforming," *J. Acoust. Soc. Amer.*, Vol. 63, 1978, pp. 425–434.

67 Mucci, R.A., "A comparison of efficient beamforming algorithms," *IEEE Trans. Acoust., Speech, Signal Processing*, Vol. ASSP-32, 1984, pp. 548–558.

68 Fan, H., E.I. El-Masry, and W.K. Jenkins, "Resolution enhancement of digital beamforming," *IEEE Trans. Acoust., Speech, Signal Processing*, Vol. ASSP-32, 1984, pp. 1041–1052.

69 Mohamed, N.J., "Two-dimensional beamforming with non-sinusoidal signals," *IEEE Trans. Electromag. Compat.*, Vol. EMC-29, 1987, pp. 303–313.

70 Maranda, B., "Efficient digital beamforming in the frequency domain," *J. Acoust. Soc. Amer.*, Vol. 86, 1989, pp. 1813–1819.

71 Chujo, W. and K. Yasukawa, "Design study of digital beam forming antenna applicable to mobile satellite communications," *IEEE Antennas and Propagation Symp. Dig., Dallas*, TX, 1990, pp. 400–403.

72 Gebauer, T. and H.G. Gockler, "Channel-individual adaptive beamforming for mobile satellite communications," *IEEE J. Select. Areas Communic.*, Vol. 13, No. 3, 1995, pp. 439–448.

73   Steyskal, H., "Digital beamforming antenna, an introduction," *Microwave J.*, No. 1, 1987, pp. 107–124.

74   Winters, J.H., "Smart antennas for wireless systems," *IEEE Personal Communic.*, Vol. 1, No. 1, 1998, pp. 23–27.

75   Griffits, L.J., "A simple adaptive algorithm for real-time processing in antenna array," *Proc. IEEE*, Vol. 57, 1969, pp. 64–78.

76   Frost III, O.L., "An algorithm for linearly constrained adaptive array processing," *Proc. IEEE*, Vol. 60, 1972, pp. 926–935.

77   Brook, L.W. and I.S., Reed, "Equivalence of the likelihood ratio processor, the maximum signal-to-noise ratio filter, and the Wiener filter," *IEEE Trans. Aerosp. Electron. Syst.*, Vol. 8, 1972, pp. 690–692.

78   Monzingo, R.A. and T.W. Miller, *Introduction in Adaptive Array*, John Wiley & Sons, New York, 1980.

79   Haykin, S., *Adaptive Filter Theory*, Prentice Hall, Englewood Cliffs, NJ, 1991.

80   Murray, W., ed., *Numerical Methods for Unconstrained Optimization*, Academic Press, New York, 1972.

81   Horowitz, L.H. and K. D. Senne, "Performance advantage of complex LMS for controlling narrow-band adaptive arrays," *IEEE Trans. Circuits Syst.*, Vol. CAS-28, 1981, pp. 562–576.

82   Iltis, R.A. and L.B. Milstein, "Approximate statistical analysis of the Widrow LMS algorithm with application to narrow-band interference rejection," *IEEE Trans. Communic.*, Vol. COM.-33, No. 1, 1985, pp. 121–130.

83   Feuer, A. and E. Weinstein, "Convergence analysis of LMS filters with uncorrelated Gaussian data," *IEEE Trans. Acoust., Speech, Signal Processing*, Vol. ASSP-33, 1985, pp. 222–229.

84   Gardner, W.A., "Comments on convergence analysis of LMS filters with uncorrelated data," *IEEE Trans. Acoust., Speech, Signal Processing*, Vol. ASSP-34, 1986, pp. 378–379.

85   Clarkson, P.M. and P.R. White, "Simplified analysis of the LMS adaptive filter using a transfer function approximation," *IEEE Trans. Acoust., Speech, Signal Processing*, Vol. ASSP-35, 1987, pp. 987–993.

86   Boland, F.B. and J.B. Foley, "Stochastic convergence of the LMS algorithm in adaptive systems," *J. Signal Processing*, Vol. 13, 1987, pp. 339–352.

87   Foley, J.B. and F.M. Boland, "A note on the convergence analysis of LMS adaptive filters with Gaussian data," *IEEE Trans. Acoust., Speech, Signal Processing*, Vol. 36, 1988, pp. 1087–1089.

88   Jaggi, S., and A.B. Martinez, "Upper and lower bounds of the misadjustment in the LMS algorithm," *IEEE Trans. Acoust., Speech, Signal Processing*, Vol. 38, 1990, pp. 164–166.

89   Solo, V., "The error variance of LMS with time-varying weights," *IEEE Trans. Signal Processing*, Vol. 40, 1992, pp. 803–813.

90   Jung, P., Z. Zvonar, and K. Kammerlander, eds., *GSM: Evolution Towards 3rd Generation*, Kluwer Academic Publishers, 1998.

91   Rappaport, T.S., *Wireless Communications: Principles and Practice*, Prentice Hall, Englewood Cliffs, NJ, 1996.

92   Prasad, R., *CDMA for Wireless Personal Communications*, Artech House, Boston-London, 1996.

93   Saunders, S.R., *Antennas and Propagation for Wireless Communication Systems*, J. Wiley & Sons, New York, 1999.

94   Blaunstein, N. and J.B. Andersen, *Multipath Phenomena in Cellular Networks*, Artech House, Boston-London, 2002.

95   Jakes, W.C., ed., *Microwave Mobile Communications*, IEEE Press, New York, 1994.

96   Barret, M. and R. Arnott, "Adaptive antennas for mobile communications," *Electron. Communic. Eng. J.*, Vol. 6, No. 2, 1994, pp. 203–214.

97   Parsons, J.D. and A.M.D. Turkmani, "Characterization of mobile radio signals," *IEE Proc. I*, Vol. 138, No. 6, 1991, pp. 549–556.

98   Turkmani, A.M.D., A.A. Arowojolu, P.A. Jefford, and C.J. Kellett, "An experimental evaluation of the performance of two-branch space and polarization diversity schemes at 1800 MHz," *IEEE Trans. Veh. Technol.*, Vol. 44, No. 3, 1995, pp. 318–326.

99   Kozono, S., T. Tsurahara, and M. Sakamoto, "Base station polarization diversity reception for mobile radio," *IEEE Trans. Veh. Technol.*, Vol. 33, 1984, pp. 301–306.

100   Vaughan, R.G., "Polarisation diversity in mobile communications," *IEEE Trans. Veh. Technol.*, Vol. 39, No. 2, 1990, pp. 177–186.

101   Eggers, P.C.F., J. Toftgard, and A.M. Oprea, "Antenna systems for base station diversity in urban small and micro cells," *IEEE J. Select. Areas Communic.*, Vol. 11, No. 7, 1993, pp. 1046–1057.

102   Swales, S.C., M.A. Beach, D.J. Edwards, and J.P. McGeehan, "The performance enhancement of multi-beam adaptive base-station antennas for cellular land mobile radio systems," *IEEE Trans. Veh. Technol.*, Vol. 39, 1990, pp. 56–67.

103   Swales, S.C., M.A. Beach, D.J. Edwards, and J.P. McGeehan, "The realization of a multi-beam adaptive base-station antenna for cellular land mobile radio systems," *Proc. of IEEE Veh. Technol. Conf.*, San Francisco, CA, 1989, pp. 341–348.

104   Andersen, J.B., "Antenna arrays in mobile communications," *IEEE Antenna Propagat. Magazine*, Vol. 42, No. 2, 2000, pp. 12–16.

105   Andersen, J.B., "Array gain and capacity for known random channels with multiple element arrays at both ends," *IEEE J. Select. Areas in Communic.*, Vol. 18, No. 11, 2000, pp. 2172–2178.

106   Andersen, J.B., "Role of antennas and propagation for the wireless systems beyond 2000," *J. Wireless Personal Communic.*, Vol. 17, 2001, pp. 303–310.

107   Winters, J.H., "Optimum combining in digital mobile radio with co-channel interference," *IEEE J. Select. Areas Communic.*, Vol. SAC-2, 1984, pp. 528–539.

108   Winters, J.H., "Optimum combining for indoor radio systems with multiple users," *IEEE Trans. Communic.*, Vol. COM-35, 1987, pp. 1222–1230.

109   Hashemi, H., "The indoor radio propagation channels," *Proc. IEEE*, Vol. 81, 1993, pp. 943–968.

110   Litva, J., A. Chaforian, and V. Kezys, "High-resolution measurements of AOA and time-delay for characterizing indoor propagation environments," in *IEEE Anten. Propagat. Soc. Int. Symp.*, 1996 Digest, IEEE, Vol. 2, 1996, pp. 1490–1493.

111 Wang, J.G., Mohan A.S., and T.A. Aubrey, "Angles-of-arrival of multipath signals in indoor environments," in *IEEE Veh. Technol. Int. Conf.*, 1996, pp. 155–159.

112 Adel, A., M. Saleh, and R.A. Valenzula, "Statistical model for indoor multipath propagation," *IEEE J. Select. Areas Communic.*, Vol. SAC-5, 1987, pp. 128–137.

113 Spenser, Q., M. Rice, B. Jeffs, and M. Jensen, "Indoor wideband time/angle of arrival multipath propagation results: a statistical model for angle of arrival in indoor multipath propagation," in *Proc. of IEEE Vehicular Technology Conf.*, Phoenix, AZ, 1997, pp. 1410–1414; pp. 1415–1419.

114 Tsalolihin, E., N. Blaunstein, I. Bilik, L. Ali, and S. Shakya, "Measurement and modeling of the indoor communication link in angle of arrival and time of arrival domains," in *Proc. of Int. Conf. of IEEE Antennas and Propagation*, Toronto, Canada, July 11–17, 2010, pp. 338–341.

115 Tsalolihin, E., I. Bilik, N. Blaunstein, and S. Shakya, "Analysis of AOA-TOA signal distribution in indoor environments," in *Proc. of 5th European Conf. on Antennas and Propagation* (EuCAP-2011), Rome, Italy, April 11–16, 2011, pp. 1746–1750.

116 Russo, P., A. D'ippolito, M. Ruggieri et al., "A frequency scanning satellite system for land mobile communications," *Int. J. Satell. Communic.*, Vol. 11, 1993, pp. 87–103.

117 Hirshfield, E., "The Globalstar system," *Appl. Microwave Wireless*, No. 4, 1995, pp. 26–41.

118 Schuss, J.J., J. Upton, B. Muers et al., "The IRIDIUM main mission antenna concepts," in *Proc. IEEE Int. Symp. Phased Array Systems Technol.*, Boston, MA, 1996, pp. 411–415.

119 Bodnar, D.G., B.K. Rainer, and Y. Rahmatsamii, "A novel array antenna for MSAT applications," *IEEE Trans. Veh. Technol.*, Vol. 38, No. 1, 1989, pp. 86–94.

120 Suzuki, R., Y. Matsumoto, R. Miura, and N. Hamamoto, "Mobile TDM/TDMA system with active array antenna," in *Proc. IEEE Global Telecommun. Conf*, Houston, TX, 1991, pp. 1569–1573.

121 Ohmori, S., Y. Hase, H. Wakana, and S. Taira, "Experiments on aeronautical satellite communications using ETS-V satellite," *IEEE Trans. Aerosp. Electron. Syst.*, Vol. 28, 1992, pp. 788–796.

122 Allnutt, R.M., T. Pratt, and A. Dissanayake, "A study in small scale antenna diversity as a means of reducing effects of satellite motion induced multipath fading for handheld satellite communication systems," in *Proc. IEEE, 9th Int. Conf. Antennas Propagat.*, Eindhoven, The Netherlands, 1995, pp. 135–139.

123 Roy, R.H., "ESPRIT, Estimation of signal parameters via rotation invariance techniques," PhD thesis, Stanford University, CA, August 1987.

124 Thomson, J.S., P.M. Grant, and B. Mulgrew, "Smart antenna array for CDMA system," *IEEE Person. Communic.*, Vol. 3, No. 5, 1996, pp. 16–25.

125 Foschini, G.J., "Layered space-time architecture for wireless communication in a fading environment when using multi-element antennas," *Bell Labs Technol. J.*, Vol. 1, No. 1, 1996, pp. 41–59.

126 Despins, C.L.B., D.D. Falconer, and S.A. Mahmoud, "Compound strategies of coding, equalization and space diversity for wideband TDMA indoor wireless channels," *IEEE Trans. Veh. Technol.*, Vol. 41, No. 3, 1992, pp. 369–379.

127 Despins, C.L.B., D.D. Falconer, and S.A. Mahmoud, "Coding and optimum baseband combining for wideband TDMA indoor wireless channels," *Canadian J. Elect. Comput. Eng.*, Vol. 16, No. 1, 1991, pp. 53–62.

128 Chang, L.F. and P.J. Porter, "Performance comparison of antenna diversity and slow frequency hopping for the TDMA portable radio channel," *IEEE Trans. Veh. Technol.*, Vol. 38, No. 2, 1989, pp. 222–229.

129 Winters, J.H., J. Salz, and R.D. Gitlin, "The impact of antenna diversity on the capacity of wireless communication systems," *IEEE Trans. Communic.*, Vol. 42, 1994, pp. 1740–1751.

130 Salz, J. and J.H. Winters, "Effect of fading correlation on adaptive arrays in digital mobile radio," *IEEE Trans. Veh. Technol.*, Vol. 43, 1994, pp. 1049–1057.

131 Balaban, P. and J. Salz, "Optimum diversity combining and equalization in digital data transmission with application to cellular mobile radio- Part I: Theoretical considerations," *IEEE Trans. Communic.*, Vol. 40, 1992, pp. 885–894.

132 Balaban, P. and J. Salz, "Optimum diversity combining and equalization in digital data transmission with application to cellular mobile radio. Part II: Numerical results," *IEEE Trans. Communic.*, Vol. 40, 1992, pp. 895–907.

133 Proakis, J.G., "Adaptive equalization for TDMA digital mobile radio," *IEEE Trans. Veh. Technol.*, Vol. 40, No. 3, 1991, pp. 333–341.

134 Ishii, N. and R. Kohno, "Spatial and temporal equalization based on an adaptive tapped-delay-line array antenna," *IEICE Trans. Communicat.*, Vol. E78-B, 1995, pp. 1162–1169.

135 Barroso, V.A.N., M.J. Rendas, and J.P. Gomes, "Impact of array processing techniques on the design of mobile communication systems," in *Proc. IEEE 7th Mediterranean Electrotechnic. Conf.*, Antalya, Turkey, 1994, pp. 1291–1294.

136 Fernandez, J., I.R. Corden, and M. Barrett, "Adaptive array algorithms for optimal combining in digital mobile communication systems," in *Proc. IEEE of Int. Conf. Antennas Propagat.*, Edinburgh, Scotland, 1993, pp. 983–986.

137 Abu-Dyya, A.A. and N.C. Beaulieu, "Outage probability of diverse cellular systems with co-channel interference in Nakagami fading," *IEEE Trans. Veh. Technol.*, Vol. 41, No. 3, 1992, pp. 343–355.

138 Goldburg, M. and R. Roy, "The impacts of SDMA on PCS system design," in *Proc. IEEE Int. Conf. Personal Communic.*, San Diego, CA, 1994, pp. 242–246.

139 Ohgane, T. Shimura, T., N. Matsuzawa, and H. Sasaoka, "An implementation of a CMA adaptive array for high speed GMSK transmission in mobile communications," *IEEE Trans. Veh. Technol.*, Vol. 42, No. 2, 1993, pp. 282–288.

140 Klukas, R.W. and M. Fattouche, "Radio signal direction finding in the urban radio environment," in *Proc. Nat. Tech. Meeting Institute Navigation*, San Francisco, CA, 1993, pp. 151–160.

141 Morrison, G., M. Fattouche, and D. Tholl, "Parametric modeling and spectral estimation of indoor radio propagation data." in *Proc. Wireless 1992*, T.R. Labs, Calgary, Alberta Canada, 1992, pp. 112–119.

142 Naguib, A.F. and A. Paulraj, "Performance of CDMA cellular networks with base-station antenna arrays," in *Proc. IEEE Int. Zurich Seminar Communic.*, Zurich, Switzerkand, 1994, pp. 87–100.

143 Turin, G.L., "Introduction to spread-spectrum antimultipath techniques and their application to urban digital radio," *Proc. IEEE*, Vol. 68, 1980, pp. 328–353.

146 Lehnert, J.S. and M.B. Pursely, "Multipath diversity reception of spread spectrum multiple access communications," *IEEE Trans. Communic.*, vol. COM-35, 1987, pp. 1189–1198.

145 Turin, G., "The effect of multipath and fading on the performance of direct sequence CDMA systems," *IEEE J. Select. Areas Communic.*, Vol. 2, 1984, pp. 597–603.

146 Price, R. and P.E. Green, "A communication technique for multipath channels," *Proc. IRE*, Vol. 46, 1958, pp. 555–570.

147 Choi, S. and T.K. Sarkar, "Adaptive antenna array utilizing the conjugate gradient method for multipath mobile communication," *Signal Process.*, Vol. 29, 1992, pp. 319–333.

148 Ohgane, T., N. Matsuzawa, T. Shimura, M. Mizuno, and H. Sasaoka, "BER performance of CMA adaptive array for high-speed GMSK mobile communication-A description of measurements in central Tokyo," *IEEE Trans. Veh. Technol.*, Vol. 42, No. 4, 1993, pp. 484–490.

149 Mizuno, M. and T. Ohgane, "Application of adaptive array antennas to radio communications," *Electron. Communic. Japan*, Vol. 77, No. 1, 1994, pp. 48–59.

150 Choi, S., T.K. Sarkar, and S.S. Lee, "Design of two-dimension Tseng window and its application to antenna array for the detection of AM signal in the presence of strong jammers in mobile communication," *Signal Processing*, Vol. 34, 1993, pp. 297–310.

151 Torrieri, D.J. and K. Bakhru, "An anticipative adaptive array for frequency-hopping communication," *IEEE Trans. Aerosp. Electron. Syst.*, Vol. 24, 1988, pp. 449–456.

152 Bakhru, K. and D.J. Torrieri, "The maximum algorithm for adaptive array and frequency-hopping communications," *IEEE Trans. Antennas Propagat.*, Vol. AP-32, 1984, pp. 919–927.

153 Torrieri, D.J. and K. Bakhru, "Frequency compensation in an adaptive antenna system for frequency-hopping communication," *IEEE Trans. Aerosp. Electron. Syst.*, Vol. AES-23, 1987, pp. 448–467.

154 Compton, R.T. Jr., "An adaptive array in a spread-spectrum communication system," *Proc. IEEE*, Vol. 66, 1978, pp. 289–298.

155 Dlugos, D.M. and R.A. Scholtz, "Acquisition of spread spectrum signals by an adaptive array," *IEEE Trans. Acoust., Speech, Signal Processing*, Vol. 37, 1989, pp. 1253–1270.

156 Trump, T. and B. Ottersten, "Estimation of nominal direction of arrival and angular spread using an array of sensors," *Signal Process.*, Vol. 50, No. 1, 1996, pp. 57–69.

157 Dasilva, J.S., B. Ni, and B.A. Fernandez, "European mobile communications on the move," *IEEE Communic. Magazine*, Vol. 34, No. 1, 1996, pp. 60–69.

158 Zettemberg, P. and Ottersten, "The spectrum efficiency of a base station antenna array system for spatially selective transmission," *IEEE Trans. Veh. Technol.*, Vol. 44, 1995, pp. 651–660.

159 Yamada, Y., K. Kagoshima, and K. Tsunekawa, "Diversity antennas for base and mobile situation in land mobile communication systems," *IEICE Trans.*, Vol. E74, 1991, pp. 3202–3209.

160 Adachi, F., M.T. Feeney, A.G. Williamson, and J.D. Parsons, "Cross correlation between the envelopes of 900 MHz signals received at a mobile radio base station site," *Proc. IEE*, Vol. 133, Pt. F, 1986, pp. 506–512.

161 Eggers, P.C.F., "Angular dispersive mobile radio environments sensed by highly directive base station antennas," in *Proc. IEEE Int. Symp. Personal Indoor Mobile Radio Communic.*, Toronto, Canada, 1995, pp. 522–526.

162 Lee, W.C.Y., "Effects of correlation between two mobile radio base station antennas," *IEEE Trans. Communic.*, Vol. COM-21, 1973, pp. 1214–1224.

163 Jakes, W.C., Jr., ed., *Microwave Mobile Communications*, Wiley and Sons, New York, 1974.

164 Winters, J.H. and M.J. Gans, "The range increase of adaptive versus phased arrays in mobile radio systems," *Proc. 28th Asilomar Conf. Signals, Systems and Comput.*, Pacific Grove, CA, 1994, pp. 109–115.

165 Seshadri, N. and J.H. Winters, "Two signaling schemes for improving the error performance of frequency-division-duplex (FDD) transmission systems using transmitted antenna diversity," *Int. J. Wireless Information Networks*, No. 1, 1994, pp. 103–109.

166 Blaunstein, N., *Radio Propagation in Cellular Networks*, Artech House, Boston-London, 1999.

167 Hammuda, H., J.P. McGeehan, and A. Bateman, "Spectral efficiency of cellular land mobile radio system," in *Proc. 38th IEEE Veh. Technol. Conf.*, Philadelphia, PA, 1988, pp. 616–622.

168 Gosling, W., "Protection ratio and economy of spectrum use in land mobile radio," *Proc. IEE*, Vol. 127, Pt. F, 1980, pp. 174–178.

169 Daikoku, K. and H. Ohdate, "Optimum channel reuse in cellular land mobile radio system," *IEEE Trans. Veh. Technol.*, Vol. VT-32, No. 2, 1983, pp. 217–224.

170 French, R.C., "The effects of fading and shadowing on reuse in mobile radio," *IEEE Trans. Veh. Technol.*, Vol. VT-28, No. 1, 1979, pp. 171–182.

171 Muammar, R. and S. Gupta, "Co-channel interference in high capacity mobile radio systems," *IEEE Trans. Communic.*, Vol. COM-30, 1982, pp. 1973–1978.

172 Windram, M.D., L. Brunt, and E.J. Wilson, "Adaptive antennas for UHF broadcast reception," *Proc. IEE*, Vol. 127, Pt. F, 1980, pp. 249–256.

173 Raith, K. and J. Uddenfeldt, "Capacity of digital cellular TDMA systems," *IEEE Trans. Veh. Technol.*, Vol. 40, 1991, No. 2, pp. 323–332.

174 Gilhousen, K.S., I.M. Jacobs, R. Padovani et al., "On the capacity of cellular CDMA system," *IEEE Trans. Veh. Technol.*, Vol. 40, No. 2, 1991, pp. 303–312.

175 Sivanand, S., "On adaptive arrays in mobile communication," in *Proc. IEEE Nat. Telesystems Conf.*, Atlanta, GA, 1993, pp. 55–58.

176 Balaban, P. and J. Salz, "Dual diversity combining and equalization in digital cellular mobile radio," *IEEE Trans. Veh. Technol.*, Vol. 40, No. 2, 1991, pp. 342–354.

177 Tangemann, M. and R. Rheinschmitt, "Comparison of upgrade techniques for mobile communication systems," in *Proc. IEEE Int. Conf. Communications*, New Orleans, LA, 1994, pp. 201–205.

178   Winters, J.H., "On the capacity of radio communication systems with diversity in Rayleigh fading environment," *IEEE J. Select. Areas Communic.*, Vol. SAC-5, 1987, pp. 871–878.

179   Lopez, A.R., "Performance predictions for cellular switched beam intelligent antenna systems," *IEEE Communic. Magazine*, Vol. 34, No. 1, 1996, pp. 152–154.

180   Suard, B., A.F. Naguib, G. Xu, and A. Paulraj, "Performance of CDMA mobile communication systems using antenna arrays," in *Proc. IEEE Int. Conf. Acoust., Speech, Signal Processing*, Minneapolis, MN, 1993, pp. 153–156.

181   Wang, Y. and J.R. Cruz, "Adaptive antenna arrays for the reverse link of CDMA cellular communication systems," *IEEE Lett.*, Vol. 30, 1994, pp. 1017–1018.

182   Winters, J.H., "Signal acquisition and tracking with adaptive arrays in the digital mobile radio system IS-54 with flat fading," *IEEE Trans. Veh. Technol.*, Vol. 42, No. 3, 1993, pp. 377–384.

183   Ohgane, T., "Characteristics of CMA adaptive array for selective fading compensation in digital land mobile radio communications," *Electron. Communic. Japan*, Vol. 74, 1991, pp. 43–53.

184   Winters, J.H., "Optimum combining in digital mobile radio with co-channel interference," *IEEE J. Select. Areas Communic.*, Vol. SAC-2, 1984, pp. 528–539.

185   Beach, M.A., P. Guemas, and A.R. Nix, "Capacity and service extension for wireless networks using adaptive antennas," *Electron. Letters*, Vol. 30, 1994, pp. 1813–1814.

186   Beach, M.A., H. Xue, R. Davies et al., "Adaptive antennas for third generation systems," in *Proc. IEE Colloquium Mobile Communications Toward Year 2000*, London, U. K., 1994, pp. 10/1–10/6.

187   Kawala, P. and U.H. Sheikh, "Adaptive multiple-beam array for wireless communications," in *Proc. IEE 8th Int. Conf. Antennas Propagation*, Edinburgh, Scotland, 1993, pp. 970–974.

188   Ganz, M.W. and R.T. Compton, Jr., "Protection of PSK communication systems with adaptive arrays," *IEEE Trans. Aerosp. Electron. Syst.*, Vol. AES-23, 1987, pp. 528–536.

189   Nagatsuka, M., N. Ishii, R. Kohno, and H. Imai, "Adaptive array antenna based on spatial spectrum estimation using maximum entropy method," *IEICE Trans. Communic.*, Vol. E77-B, 1994, pp. 624–633.

190   Liberti, J.C. and T.S. Rappaport, "Analytical results for reverse channel performance improvements in CDMA cellular communication systems employing adaptive antennas," *in Proc. Globcom'93*, Houston, TX, 1993, pp. 42–47.

191   Yim, C., R. Kohno, and H. Imai, "Adaptive array antenna based on estimation of arrival angles using DFT on spatial domain," *Electron. Communic. Japan*, Vol. 76, 1993, pp. 96–108.

192   Harbin, S.A. and B.K. Rainer, "Low-power wireless mobile communication system," *in Proc. IEEE 44th Vehicular Technology Conf.*, Vol. 1, Stockholm, Sweden, 1994, pp. 673–676.

193   Wirth, P.E., "Teletraffic implications of database architectures in mobile and personal communications," *IEEE Communic. Magazine*, Vol. 33, No. 1, 1995, pp. 54–59.

194 Jabbari, B., G. Colombo, A. Nakajima, and J. Kulkarni, "Network issues for wireless communications," *IEEE Communic. Magazine*, Vol. 33, No. 1, 1995, pp. 88–98.

195 Ivancic, W.D., M.J. Shalkhauser, and J.A. Quintana, "A network architecture for a geostationary communication satellite," *IEEE Communic. Magazine*, Vol. 32, No. 1, 1994, pp. 72–84.

196 Anderson, S., M. Millnert, M. Viberg, and B. Wahlberg, "An adaptive array for mobile communication systems," *IEEE Trans. Veh. Technol.*, Vol. 40, No. 2, 1991, pp. 230–236.

CHAPTER EIGHT

Prediction of Signal Distribution in Space, Time, and Frequency Domains in Radio Channels for Adaptive Antenna Applications

Multipath phenomena limit performance of wireless communication systems by introducing fast fading due to frequency spread in narrowband systems and by causing intersymbol interference in wideband systems due to time delay (TD) spread. Finally, a strong multiplicative noise occurs in all kinds of wireless links becoming a great problem in land communications [1–30]. As was mentioned in Chapters 1 and 5, to mitigate the effects of multiplicative noise and the noise due to interference between users in multiple access communication, the directional, sectorial, and adaptive antennas (array or multibeam) are used in one or both ends of the channel, which was defined in Chapter 7 as a multiple-input-multiple-output (MIMO) channel. Using the adaptive antenna together with the corresponding processing algorithms operating in the space, time, and frequency domains (see details in Chapter 7) allows the channel to radiate the desired energy in the desired direction or to cancel the undesirable energy from the undesirable direction (see Fig. 7.21, Chapter 7). The same method is used to minimize effect of multipath fading.

However, the operational ability of all kinds of antennas, and mostly of adaptive antennas, strongly depends on the degree of accuracy to predict propagation characteristics of the actual channel. These types of propagation

Radio Propagation and Adaptive Antennas for Wireless Communication Networks: Terrestrial, Atmospheric, and Ionospheric, Second Edition. Nathan Blaunstein and Christos G. Christodoulou.
© 2014 John Wiley & Sons, Inc. Published 2014 by John Wiley & Sons, Inc.

models, which are a priory, used to predict the angular, time, and frequency distribution of the multipath components of the total signal arriving at the receiver, are the main sources of the multipath interference [31–48]. As was mentioned in Chapter 5, realistic channel models are used for performance evaluation of different adaptive antenna solutions and for the estimation of the obtainable capacity gain. In other words, to design effectively different kinds of wireless networks with optimal user and frequency allocation and cellular planning, we need realistic spatial and temporal channel models combined with high-resolution precise experiments. The latter are required for "parameterization and validation" of such channel models [10]. Despite the importance of these aspects, a limited number of high-resolution spatial and temporal experiments are available (see Reference 3 and references therein), and only few of them use three-dimensional (3-D) measurements, accounting for not only angle-of-arrival (AOA) and time-of-arrival (TOA) but also elevation-of-arrival (EOA) distributions of the multipath components within the communication channel [5–10]. The same situation also occurs with theoretical prediction of these characteristics; only few models, simple or more complicated, exist, which can describe the mutual AOA and TOA distributions of the total signal, and at the same time, may account different situations in the corresponding environment [46–58]. Namely, in the urban scene, a predicting model must account for various built-up parameters such as height, density, and real street orientation of a building as well as position of the antenna with respect to the overlay profile of the buildings (see Fig. 7.22a–c, Chapter 7).

Furthermore, as was mentioned in Section 3.3, system gain is another important issue that has high priority in cellular network performance. In particular, since the base station (BS) output power is not limited, the *downlink* gain is not a subject of discussion; the system gain aspect in the *uplink* is a main problem. Increasing the system gain is found to be efficient by using an uplink diversity technique. At the same time, as was mentioned in Chapter 7, different types of diversity systems exist: spatial, temporal, and polarization. But the common feature related to each of them is the correlation factor between branches that defines the potential of diversity gain. Particularly, low correlation increases the utilization of the diversity gain potential.

The analysis of parameters, which influence the correlation, is based on an understanding of signal AOA and TOA distribution, which are based on the corresponding theoretical models. Figure 8.1 shows the main concept of spatial diversity technique described in Reference 48. As discussed in Chapter 7, the main idea was to combine two faded signals received by two different antennas separated in space. Here, different propagation situations influence the correlation factor in various manners. Thus, the requirement proposed by diversity designers is the ability to analyze and to simulate a potential for the diversity gain as a function of main environmental propagation parameters, such as AOA, TOA (called also the *delay spread* [DS] or *time delay* [TD]) and EOA signal distributions. As can be summarized, the detailed understanding of radio

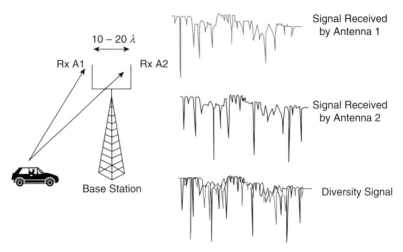

FIGURE 8.1. The explanation of spatial diversity [48].

propagation phenomena is a keystone for the development and performance assessment of different communication systems, stationary or mobile.

Below, we briefly present more realistic models on the basis of experimental and theoretical investigations of the propagation problem. We start with the simple and achieve those which are more complicated but predictable regarding corresponding experiments. Then, we present the reader with some useful recommendations on how to predict a signal power in spatial (AOA and EOA) and temporal (TOA or DS) domains based on modified stochastic approach proposed in References 55–63 and verify its accuracy by comparing it with well-known ray-tracing and ray-launching models and with the experiments carried in different European cities. Using this approach, we will present in Chapter 9 some virtual numerical experiments with adaptive antenna for the specific urban scenarios to show how the parameters of the antenna affect the signal power distribution in the space and time domains. These aspects are very actual for localization of any desired subscriber in the area of service and will be shown in the last section of this chapter.

8.1. PREDICTING MODELS FOR OUTDOOR COMMUNICATION CHANNELS

More realistic models for prediction of propagation characteristics in outdoor environments have been developed in References 42–47 based on the unified theory of diffraction (UTD) and ray-launching or ray-tracing methods using a 3-D building data. Here, the terrain can be also taken into account for portions of rays that can go around buildings.

8.1.1. VPL Model

A more general model is described in Reference 46 based on the vertical plane launch (VPL) method. The VPL technique is capable in determining the 3-D ray paths that travel between a transmitter and a receiver influenced by multiple reflections and diffractions. The VPL algorithm uses geographic vector data of buildings and the terrains that are available with a high level of accuracy. Finally, the vertical trajectory of each of the ray can be calculated analytically determining each ray path in three dimensions. To obtain this 3-D picture, both the plane of rays reflected from walls and the vertical plane of rays diffracted from rooftops were taken into account [47]. The temporal and spatial impulse power response of the channel is written in the form [46]

$$P(\tau, \varphi) = \sum_{m=1}^{M} |A_m|^2 \, \delta(\tau - \tau_m)\delta(\varphi - \varphi_m), \tag{8.1}$$

where M is the number of rays arriving at the BS, A_m is the complex amplitude of the mth ray, and τ_m and φ_m are the TD and the AOA of the mth ray, respectively. The VPL tool gives the delay, the direction, and the amplitude of each ray propagated between the BS and a mobile. Finally, this tool can also be used to define the DS and the azimuth spread (AS) (i.e., AOA spread). More specially, using VPL and the power response definition of (8.1), the DS and AS can be derived as [46]:

$$DS = \sqrt{\frac{\sum |A_m|^2 \, (\tau_m - \bar{\tau})^2}{\sum |A_m|^2}} \tag{8.2}$$

$$AS = \frac{180}{\pi} \sqrt{1 - |\bar{v}|^2}, \tag{8.3}$$

where $\bar{v}$ is the mean vector of direction of arrival and $\bar{\tau}$ is the average delay of the rays arriving from the mobile.

Using Monte-Carlo simulations, based on this site-specific 3-D ray-tracing code, the *rms* DS and the mean TD have been computed in Reference 46 and compared with experiments carried out at 900 MHz in various built-up areas of Seoul (Korea), Munich (Germany), and Rosslyn, Virginia (United States). It was shown, both numerically and experimentally, that the DS is different for different built-up areas and depends on the buildings' overlay profile and the statistical distribution of buildings over the terrain. At the same time, the simulation results showed that the cumulative distribution function (CDF) for DS is not so sensitive to the building distribution statistics compared with the CDF of AS. In most cases, the AS was very sensitive to the distribution of building heights and to antenna location with respect to the surrounding buildings. Thus, decreasing the BS antenna height from ~80 m (i.e., 5 m above the rooftops) to ~39 m (i.e., when 80% of the buildings are higher than BS) leads to an increase of the median DS from 0.13 to 0.18 μs, which is not significant,

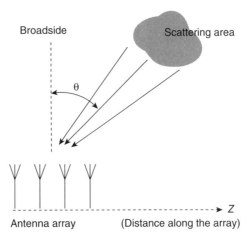

FIGURE 8.2. Scattered rays impinging on the antenna array (Source: Reference 51. Reprinted with permission @ 2002 IEEE.)

whereas the median AS is increased from 7° to 18°, that is, significantly, approximately a 2.5 factor.

8.1.2. 2-D Statistical Model of AOA Distribution

A new statistical two-dimensional (2-D) model taking in account multiple scattering from the obstructions surrounding the terminal antenna was proposed in References 49–53. This model was a way to find AOA power spectrum distribution for low-resolution antennas.

Figure 8.2 describes a particular situation, where a large number of rays are scattered from a finite area in space and arriving at the antenna. Here, the spatial derivative of the total phase along the array axis is taken as a measure of the sine of the apparent AOA. The PDF and power spectrum of the AOA were determined from the sine of the angle distribution. It is assumed [49–51] that the antenna beamwidth is much larger than an apparent direction of arrival θ, therefore the scattering area can be considered as a point source. A large number of rays emanating from finite region in space are impinging on the antenna, which has a beamwidth larger than the extent of this region. We also assume that the array depends only on distance z and that the maximum occurs when $\sin \theta = 1$. When the antenna is mechanically rotated or electronically scanned, the total signal at the receiver will be in a form of a weighted sum of all incident rays. The sum will be dependent on the direction of scan, and it will vary randomly due to the random amplitudes and phases of the incident rays. For one given position of the transmitter, there will be one apparent direction of arrival, but when the transmitter has moved to a new position (or the receiver has moved to a new position), the phases of the incident rays will be changed, and the result is a new apparent direction of arrival. If the signal plane wave is incident on the array from direction θ measured from broadside (see Fig. 8.2), then the common form of the accepted field E can be defined as [49–51]

$$E = \exp\{jkz\sin\theta\}, \tag{8.4}$$

where k is the wave number, and $\varphi = kz\sin\theta$ is the phase that varies linearly with z.

The phase gradient distribution obtained in References 49–51 allows, after simple derivations, to accept the desired θ distribution on the basis of phase gradient statistics and to obtain the power spectrum in u-domain:

$$\langle |H|^2 |u \rangle = \frac{s^2}{(u - \bar{u})^2 + s^2}, \tag{8.5}$$

where $u = \sin\theta$ and $\bar{u}$ and $\overline{u^2}$ are the first and second moments of the average power distribution measured in AOA-space; s is a measure of angular spread:

$$s^2 = \overline{u^2} - \bar{u}^2 \tag{8.6}$$

For a small array, the variation of the phase is close to linear; thus, the knowledge of the statistics of the phase gradient is the same knowledge of the distribution of $\sin\theta$, from which we can derive the statistics of θ. For small-angle results, substituting θ instead of $\sin\theta$ in (8.4) gives good results for small angles θ, when $\sin\theta$ limits to the angle θ defined in *radians* [22, 23]. Figure 8.3 shows

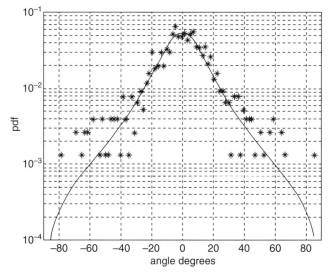

FIGURE 8.3. Distribution of angle-of-arrival obtained in References 22 and 23 compared with computations made for $s = 0.3$ in Reference 51.

the agreement with results of measurement provided in References 22 and 23 for a single-element antenna ($N = 1$) and proves the theoretically derived fact that an apparent direction of arrival is actually the instantaneous azimuth direction of signal maximum power [49–53].

Unfortunately, the presented model gives only signal power distribution in the AOA domain. Moreover, this model does not describe actual situation in the urban scene, because the distribution of power (8.4) does not take into account the actual distribution and density of obstructions around the transmitter and the receiver, as well as their position and heights with respect to the overlay profile of the obstructions.

8.1.3. 2-D Statistical Approach of AOA and TOA Distribution Prediction

A 2-D statistical approach has been proposed in References 4, 14, and 53–57, which is a measurement-based multipath approach of signal angel-of-arrival (or azimuth) (AOA) and DS distribution prediction. This model describes the probability density function (PDF) of azimuth and TD separately, and jointly as well. Moreover, according to this model, different root second central moments must be derived for different area types, so such prediction method becomes inconvenient for predicting propagation characteristics in mixing areas, making this concept just like a simple first-order approach for complicated propagation environments. This concept was proposed by References 4, 14, and 53–57, based on the statistical representation of the mobile radio channel. The signal model takes into account a multipath propagation, which can cause several replicas of the transmitted signal at the receiver. The azimuth–DS function in these cases is [4, 55]

$$h(\varphi, \tau) = \sum_{i=1}^{L} \alpha_l \delta(\varphi - \varphi_l, \tau - \tau_l), \qquad (8.7)$$

where L is the number of impinging waves from different directions φ_l and with different TD τ_l, each of them with a complex amplitude α_l. Derivation of the AS and DS in Reference 4 is based on the definition of AS and DS as a root second central moment σ_{AS} and σ_{DS} of the azimuth power spectrum $P_{AS}(\varphi)$, and the delay power spectrum $P_{DS}(\tau)$, respectively. Here $P_{AS}(\varphi)$ and $P_{DS}(\tau)$ are determined as [4, 55]:

$$P_{AS}(\varphi) = \int P(\varphi, \tau) d\tau \qquad (8.8)$$

$$P_{DS}(\tau) = \int P(\varphi, \tau) d\varphi \qquad (8.9)$$

and $P(\varphi, \tau)$ is a power azimuth–delay spectrum and according to References 4 and 55 is defined as

$$P(\varphi, \tau) \propto E\{|\alpha|^2 | \varphi, \tau\} \cdot f(\varphi, \tau), \qquad (8.10)$$

where $E\{|\alpha|^2|\varphi,\tau\}$ is the expected power of the waves conditioned on their azimuth and delay, and $f(\varphi,\tau)$ is a joint PDF. It was proved in References 4 and 55 that the AS and DS distributions are independent, that is,

$$f(\varphi, \tau) = f(\varphi) \cdot f(\tau). \tag{8.11}$$

This fact will be used below to obtain a joint PDF for modified stochastic multiparametric model. PDFs $f(\varphi)$ and $f(\tau)$ were derived empirically through the prism of the numerous experiments described in References 4 and 53–55. It was found that a Gaussian PDF matches the $f(\varphi)$, and an exponential decaying function is the best fit to the $f(\tau)$. Thus, $f(\varphi)$ and $f(\tau)$ were described as [4, 55]:

$$f(\varphi) = \frac{1}{\sqrt{2\pi}\sigma_\varphi} \exp\left\{-\frac{\varphi^2}{2\sigma_\varphi^2}\right\} \tag{8.12}$$

$$f(\tau) = \frac{1}{\sigma_\tau} \exp\left\{-\frac{\tau}{\sigma_\tau}\right\}, \tag{8.13}$$

where σ_τ and σ_φ are standard deviations of TD and azimuth distribution, respectively. Different environments were investigated in References 4 and 53–55. The PDF of the azimuth and delays were estimated from data measured for different environments. The probability distribution of the model parameters was extracted from experimental data collected during extensive measurement campaigns in the cities of Aarhus and Aalborg in Denmark and Stockholm in Sweden. The investigated environments are characterized as macrocellular urban typical (UT) and bad urban (BU) areas. The stochastic model was entirely described by the joint probability distribution of its random parameters. The random variables $(\alpha_l, \tau_l, \varphi_l)$ were assumed to be independent and identically distributed for each wave (with number L). Local realization of the delay AS function was obtained by combining L waves with random amplitudes, delays, and azimuths.

Figure 8.4 shows a histogram of the power AS obtained experimentally in the cities of Aarhus and Stockholm according to References 4 and 55. The standard deviation estimated from this experimentally obtained histogram equals $\bar{\sigma} = 6$. The solid curve in Figure 8.4 is a Gaussian function predicted theoretically according to (8.12) with the same $\sigma = 6$ like an estimated $\bar{\sigma}$. This Gaussian function provides a best fit for all results. It was also shown both experimentally and theoretically [4, 53–57] that the power AS is increased 40–50% when the BS antenna height is decreased 12 m below the rooftop level. From Figure 8.4, the incoming power is highly concentrated around 0° even though the measurement results were obtained in non-line-of-sight (NLOS) conditions. At the same time, Figure 8.5 shows a histogram of the PDS obtained experimentally in Aarhus and Stockholm for the power delay spectrum (DS) versus TD according to References 4 and 55. One can see that

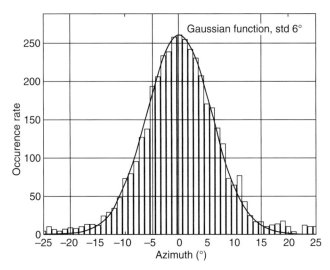

FIGURE 8.4. Histogram of the estimated azimuth with antenna located 12 m above the rooftop level obtained in References 4 and 54. (Reprinted with permission @ 2000, 2002 IEEE.)

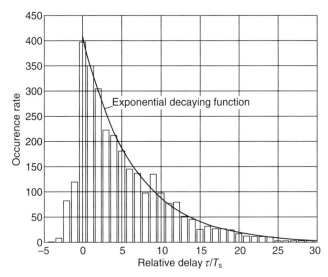

FIGURE 8.5. Relative delay τ/T_s histogram [4, 54]. (Reprinted with permission @ 2000, 2002 IEEE.)

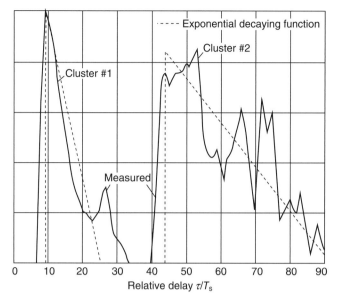

FIGURE 8.6. Relative delay τ/T_s for mixed environment [4, 54]. (Reprinted with permission @ 2000, 2002 IEEE.)

the exponential function, plotted according to (8.13), fits well to the histogram. The results presented above relate to the particular urban environment called typical urban (TU), implying uniform density of buildings. Other PDF plot, for bad urban (BU) environment, was also proposed in References 4 and 55. In the case of BU channel, the dispersion in the radio channel looks completely different. For the mixture of open area and densely built-up zones, the accepted result for TD distribution is presented in Figure 8.6 [4]. Obviously, the distribution of received signal depends significantly on the types of clusters created along radio paths from the Tx to the Rx antennas. The authors in References 4 and 55 describe this TD distribution by a two-cluster model, as was done by References 22 and 23 for indoor environments (see also Section 8.2). The suggestion is to consider the power azimuth–delay spectrum as

$$P(\varphi, \tau) = \sum_{k=1}^{2} P_k(\varphi, \tau), \tag{8.14}$$

where $P_k(\varphi, \tau)$ represents the contribution from each cluster. Situations with more than two clusters may also be described as in (8.14) by summarizing all contributions from all clusters. The disadvantage of the proposed method is the absence of the area-dependent unique parameter in the model that adapts the result according to the type of the environment.

8.1.4. Experimental-Based Model

In References 6–10, another approach, which introduces urban propagation mechanism based on high-resolution 3-D radio channel measurements, was proposed. This approach analyses the propagation using comprehensive classification of different types of wave propagation in urban scene. This classification defines three propagation classes: street-guided propagation as a first class (denoted "Class1"), direct propagation (over the rooftops) as a second class (denoted "Class 2"), and scattering from obstructions (reflection from high-rise objects) as a third class (denoted "Class 3"). To define the corresponding channel models, three different series of measurements were carried out in an urban environment. The first case corresponds to situations when the receiving BS antenna is below the rooftop; the second and the third one correspond to situations near the rooftop level and above the rooftop level, respectively. All measurements have been done in downtown of Helsinki (Finland). A typical experimental site with both terminal antennas located below the rooftop level is presented in Figure 8.7 [9, 10]. The resultant joint azimuth–DS distribution is shown in Figure 8.8 [9, 10]. Figure 8.7 and Figure 8.8 represent the situation when the transmitter (mobile) antenna is at the height of 1.5 m, that is, at the normal street level with NLOS, and the receiver (BS) antenna is *below the rooftops*. Obviously, a significant number of rays propagate within the street waveguide because of multiple reflections from walls of buildings that lie along the streets. Another arrival mechanism, but not so dominant, is the scattering from the high-rise object (such as the Theatre Tower denoted in Figure 8.7).

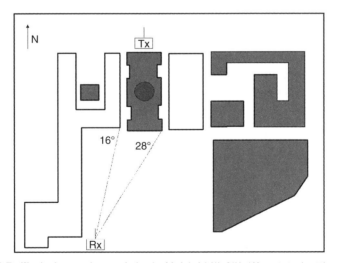

FIGURE 8.7. Typical experimental site in Helsinki [9, 10]. (Reprinted with permission © 2002 IEEE.)

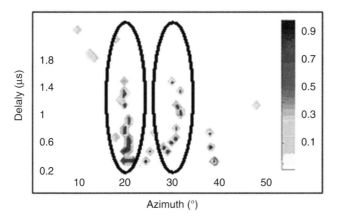

FIGURE 8.8. PDF of joint AOA-TOA distribution measured in References 9 and 10. (Reprinted with permission @ 2002 IEEE.)

From the measurement results, presented in Fig. 8.8, more than 80–90% of rays arrive at the receiver from two streets. These rays get affected by the guiding effects of streets, that is, by the multiple reflection from buildings located along the streets. Thus, propagation "Class 1" and "Class 3" are dominant in this urban area according to the urban radio channel propagation classification made in References 9 and 10. Class 1 (street-guided propagation) dominates, with over 90% of the total power for cases when the receiver (BS) antenna is below or at the same level of the rooftops.

Another experiment, when the Rx antenna is *at the rooftop level*, shown in Figure 8.9, has been carried out under the following conditions: there is no traffic on the streets, the TX antenna kit is at the street ground level (with height of 2 m), and the Rx antenna is at the rooftop level (with height of 27 m). Moreover, the 3-dB beamwidth of the Tx antenna is of 70° the azimuth and elevation directions. Finally, the Rx antenna is omnidirectional in the azimuth and with 87° in the elevation domain. A wideband (100 MHz) channel sounder has been used with the carrier frequency of 2.154 GHz, and no line-of-sight (LOS) between the Tx and Rx antenna. A pseudo-LOS distance was set at about 420 m. One wide street is placed at $\varphi = -30°$ from the main lobe of the Rx antenna (Rx is mounted at 270° from the north direction). Figure 8.10 shows the resulting image of the postprocessed measured data for the area shown in Figure 8.9. It is clear that Class 2 and Class 3 are starting to be significant (25%), but not dominant, for cases when the BS antenna is at the same level or above the average height of rooftops.

Generally speaking, the approach proposed in References 9 and 10 has demonstrated the strong dependence between the built-up profile seen from the BS and the propagation mechanisms in the corresponding urban areas. It proves that accurate prediction based on advanced simulation tool, incorporating a building geographic database, is needed to achieve network planning.

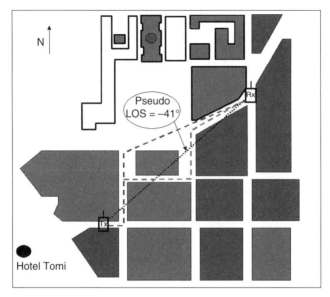

FIGURE 8.9. A microcell urban environment extracted from References 9 and 10. (Reprinted with permission @ 2002 IEEE.)

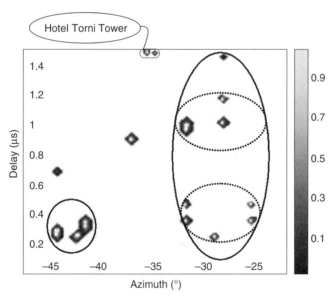

FIGURE 8.10. Measurement results in TOA-AOA plane [9, 10]. (Reprinted with permission @ 2002 IEEE.)

Unfortunately, all the above-mentioned theoretical models cannot account for all the built-up terrain features as far as overlay profile, density of buildings, and their positions and heights with respect to both terminal antennas. As shown in References 58–63, as well as in Chapter 5, the multiparametric stochastic approach can account for specific features of different urban areas and for various positions of the terminal antennas. As was shown in References 64–69, this approach can also be used as an actual predictor of AOA and DS (or TOA) signal distribution for various specific situations in the urban communication channel.

8.2. 2-D AND 3-D STOCHASTIC MULTIPARAMETRIC MODEL

The realistic statistical 2-D and 3-D model, which take into account the terrain features, such as the height profile of buildings, their density and reflection properties, were created in References 58–63. From Chapter 5, the propagation channel for UHF/X-band waves, in the built-up environment, according to these stochastic models, can be modeled by an array of randomly distributed buildings and/or natural obstacles placed on a rough terrain. The law of distribution of the obstacles is assumed to be a Poisson law and the city's relief is described by introducing the special probability functions. Using the proposed statistical multiparametric model of wave propagation in such an urban channel, the field intensity attenuation was examined, and the single and multiple scattering effects were pointed out in Chapter 6, taking into account the diffraction from the roofs of the buildings depending on the position of the receiver and transmitter antennas on the rough terrain. Using the unified stochastic approach described in Chapter 5, we employ probability theory to directly evaluate the statistical parameters of the channel and deterministic theory to obtain the correlation between the signal spectral characteristics and parameters of an urban propagation channel. We start with description of the PDF of scatterers' spatial distribution over the terrain in AOA and TOA domains. Then, we present the signal power spectra in the separate and joint AOA and TOA domains.

8.2.1. PDF Spatial Distribution in the AOA and TOA Domains

As was shown in Chapter 5, for the discrete distributed scatterers located on the rough terrain, the density of scattered points distribution can be described by Equation (5.63) and Equation (5.64). The corresponding geometry of the problem is shown in Figure 8.11. These formulas consist two separate terms, $\mu_1(r,\varphi)$ and $\mu_2(r,\varphi)$, each of which describes different situations that occur in the urban scene: the first one describes distribution of the *rare scatterers* which are distributed over the large area of the city far from the receiver located at point $B(\mathbf{r}_2)$ (see Fig. 8.11). Additional significant changes in the scatterers' distribution for the case $z_2 > h$ gives the second term, $\mu_2(r,\varphi)$, which describes

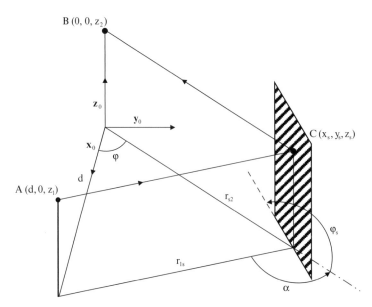

FIGURE 8.11. Geometry of single scattering from a building as a nontransparent screen.

the "illumination" of a small area of the building layer in the $\langle \rho \rangle$-region of a transmitter located at point $A(\mathbf{r}_1)$ of the transmitter (Fig. 8.11).

We now will present the total PDF, $\mu(r,\varphi) = \mu_1(r,\varphi) + \mu_2(r,\varphi)$, of scattered waves distribution over the arrival angles (AOA) and the TD (or AOA) by changing the variable r in it by the relative time τ of separately and independently scattered waves from each obstruction (i.e., single scattering from each screen, as shown in Figure 8.11). This relative time (i.e., relative to the time of the ray passing line-of-sight path between the transmitter and the receiver) determines the time of wave propagation from the transmitter to the receiver through the layer of urban building:

$$\tau = (r + \tilde{r}) / d, \tag{8.15}$$

where $\tilde{r} = (d^2 + r^2 - 2rd\cos\varphi)^{1/2}$, φ is the azimuth (or AOA) at the receiving point B, d is the LOS distance between the transmitter (point A) and the receiver (point B), as shown in Figure 8.11.

As a result of such changes, we can obtain the relative density of joint distribution of independent single scattered waves over relative TD τ and AOA φ [58, 64]:

$$\mu(\tau,\varphi) = \frac{\mu(\tau,\varphi)}{\overline{N}} = \frac{vd}{4\overline{N}}\left[\frac{(z_2 - \overline{h})}{\overline{h}}\tilde{r}\exp(-\gamma_0\tilde{r}) + \frac{hrd\gamma_0}{z_2}\tau\cdot\exp\left(-\gamma_0 r\frac{h}{z_2}\right)\right] \tag{8.16}$$

where

$$r(\tau,\varphi) = \frac{d}{2} \frac{(\tau^2 - 1)}{(\tau - \cos\varphi)}, \tilde{r}(\tau,\varphi) = \frac{d}{2} \frac{(\tau^2 - 2\tau\cos\varphi + 1)}{(\tau - \cos\varphi)}, \quad (8.17)$$

and $\bar{h}$ is the average buildings height, $\gamma_0 = 2\langle L \rangle \nu / \pi$ is the density of buildings' contours in the horizontal plane $z = 0$, ν is the building density (per square kilometer), $\langle L \rangle$ is the average length of buildings (in km), and $\bar{N}$ is the average number of independently single scattered waves:

$$\bar{N} = \int_0^\infty r\,dr \int_0^{2\pi} d\varphi \cdot \mu(r,\varphi) \quad (8.18)$$

From (8.16) it follows that the maximum of density $w(\tau,\varphi)$ is localized in a small angular interval. As was shown in Reference 64, with decrease of height of receiving antenna z_2, the maximum of the density $w(\tau,\varphi)$ spreads over the whole angle interval φ, and the probability of obtaining a larger TD τ at the receiver is decreased. We will show these features both in the AOA and TOA domains.

PDF of the Arrival Angles. From (8.16) we can easily obtain the AOA PDF distribution of waves separately and independently scattered from each non-transparent screen. Thus, the angular distribution of points which are responsible for the single scattering process at the observed point $B(\mathbf{r}_2)$ (see Fig. 8.11), can be obtained by integrating density $\mu(\tau,\varphi) = \mu_1(\tau,\varphi) + \mu_2(\tau,\varphi)$ from (8.16) by use of a new variable $\tau \in (1, \infty)$. After some straightforward derivations, following References 58 and 64, we get for the *first group* of points the AOA PDF distribution

$$\mu_1(\varphi) = \frac{\nu d^2}{2} \frac{h}{z_2} \exp\left(-\gamma_0 d \frac{h}{z_2}\right) \frac{\exp\left\{-\dfrac{\gamma_0 d(1 - \cos\varphi)(z_2 - h)}{z_2[1 + \gamma_0 d(1 - \cos\varphi)]}\right\}}{1 + \gamma_0 d(1 - \cos\varphi)} \quad (8.19a)$$

or by the approximate formula

$$\mu_1(\varphi) \approx \frac{\nu d^2}{2} \frac{(1 - \varsigma)\exp\{-\gamma_0 d(1 - \varsigma)\}}{1 + (1 + \varsigma)\gamma_0 d(1 - \cos\varphi)}, \quad (8.19b)$$

where $\varsigma = (z_2 - \bar{h})/z_2$. In this case, the normalized PDF of the event that some single scattered waves can be observed at point B, and which are characterized by angles $\varphi \in (-\pi, \pi)$, can be presented as

$$\tilde{w}(\varphi) = \frac{\mu(\varphi)}{\bar{N}} = \frac{[\mu_1(\varphi) + \mu_2(\varphi)]}{\bar{N}}. \quad (8.20)$$

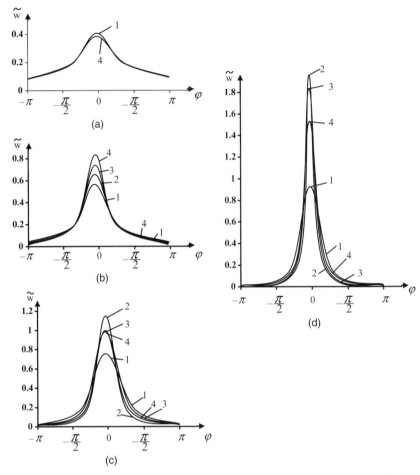

FIGURE 8.12. The normalized PDF, $\tilde{w}(\varphi)$, versus arrival angles for $\bar{h} = 20$ m, $<\rho> = 167$ m ($\gamma_0 = 6$ km^{-1}) for (a) $d = 0.5$ km, (b) $d = 1$ km, (c) $d = 2$ km, and (d) $d = 3$ km. Curve 1 is for $(z_2 - \bar{h}) = 1$ m, curve 2 for $(z_2 - \bar{h}) = 5$ m, curve 3 for $(z_2 - \bar{h}) = 20$ m, and curve 4 for $(z_2 - \bar{h}) = 100$ m.

Analysis of Equation (8.20), taking into account expressions (8.19a) and (8.19b) to determine both terms $\mu_{1,2}(\varphi)$ in it, shows that the angle distribution of the single scattered waves is concentrated near the direction $\varphi = 0$, that is, it has a unimodal sharp form, which spreads over angles φ in region $[-\pi/2, \pi/2]$ with increase of building density ν.

Thus, in Figure 8.12, the normalized PDF $\tilde{w}(\varphi)$ from (8.20) is presented versus arrival angles for single scattered waves. In numerical simulations, according to Reference 64, were taken into account the average building layer height $\bar{h} = 20$ m, the average length of buildings $<L> = 50$ m and buildings' density around the receiver (point B) and the transmitter (point A) is $\nu = 180/\text{km}$. In this case, the range of direct visibility from the observer was $\langle \rho \rangle = \gamma_0^{-1} = 167 \text{ m}$ ($\gamma_0 = 6 \text{ km}^{-1}$). In Figure 8.12a, the case of $d = 0.5 \text{ km}$ ($d/\langle \rho \rangle = 3$) is presented; in Figure 8.12b, $d = 1 \text{ km}$ ($d/\langle \rho \rangle = 6$); in Figure 8.12c, $d = 2 \text{ km}$ ($d/\langle \rho \rangle = 12$). Here, in all parts of Figure 8.12, the curves denoted by 1 are derived for $(z_2 - \bar{h}) = 1 \text{ m}$, curves 2 for $(z_2 - \bar{h}) = 5 \text{ m}$, curves 3 for $(z_2 - \bar{h}) = 20 \text{ m}$, and curves 4 for $(z_2 - \bar{h}) = 100 \text{ m}$. As follows from Figure 8.12a–c, with increasing the range between observation point B and a source placed at point A (see Fig. 8.11) from $d = 0.5 \text{ km}$ to $d = 3 \text{ km}$, a sharp "unimodal" distribution of $\tilde{w}(\varphi)$ with the maximum located along the direction to the source from the receiving point ($\varphi = 0$) is observed. The same effect is observed with increase of BS antenna height over the building layer.

Next, we investigated the normalized PDF, $\tilde{w}_2(\varphi)$, for various parameters of the height factor $\zeta = (z_2 - \bar{h})/(z_2 - z_1)$ and for inverse height factor $\tilde{\zeta} = (\bar{h} - z_1)/(z_1 - z_2)$ to analyze the role of the BS antenna height with respect to building heights (with the average height $\bar{h}$) for "illumination" of the area surrounding the receiver. For all derivations we have assumed that the height z_2 of receiving point B is higher than $(2\bar{h} - z_1) \approx 25 - 30 \text{ m}$, that is, for the case of $\zeta > 0$. Figure 8.13a illustrates the dependence of the normalized PDF $\tilde{w}_2 = \mu_2(\varphi)/\bar{N}$ over the AOAs φ for incoming reflected rays at the receiving point B, as a function of height factor ζ. As can be seen, the distribution over the incoming angles has a maximum in the direction of the source of radiation placed at point A; that is, it is unimodal. With increase of parameter ζ ($\zeta \geq 0.5$), the density of incoming angles' distribution differs from zero only for $\varepsilon|\varphi| < \varphi_1$, where $\varphi_1 = \sin^{-1}\left[(\bar{h} - z_1)/(z_2 - \bar{h})\right]$ and its maximal value quickly increases with increase of the height factor ζ.

In the inverse case, when the radiation source at point A, as a BS, is above or near the top of buildings' layer, and the receiver B is at the ground level (namely, the mobile subscriber, MS), that is, $z_2 < \bar{h} \leq z_1$, the density $\mu_2(\varphi)$ significantly differs from that described by (8.18). The corresponding equations are too complicated, and therefore, we will present only computations for the height factor $\tilde{\zeta} = (\bar{h} - z_1)/(z_1 - z_2)$. The corresponding normalized PDF of the arrival angles of incoming reflected waves, $\tilde{w}_2 = \mu_2(\varphi)/\bar{N}$, in this case is presented in Figure 8.13b.

As it follows from illustrations presented in Figure 8.13b, here, conversely, the unimodal character of $\tilde{w}_2(\varphi)$ may be obtained only for $\bar{h} = z_1$ ($\tilde{\zeta} = 0$). With decreasing transmitting antenna height z_1 at point A, the maximum of the arriving reflected rays is observed at the point B in the opposite direction from the direction of the transmitter A and a wide hole in the direction of point A is observed, the width and depth of z_1 which quickly grows with decrease of with respect to $\bar{h}$. Thus, for $\tilde{\zeta} = 0.2 - 0.4$, the maxima of the arriving rays' distribution is located near the angles $\varphi = \pi/6 - \pi/4$. With decrease of transmitter height with respect to the rooftops level (i.e., when $\tilde{\zeta} = 0.6 - 0.8$) the maximum of the reflected rays is observed from directions given by $\varphi = \pi/2 - \pi$. Hence, we may observe in this case the rays arriving in opposite directions from the

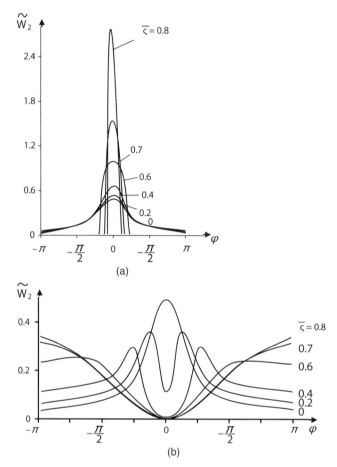

FIGURE 8.13. (a) The normalized PDF, $\tilde{w}(\varphi)$, versus arrival angles as a function of the height factor ς; (b) the normalized PDF, $\tilde{w}(\varphi)$, versus arrival angles as a function of the inverse height factor $\tilde{\varsigma}$.

direction of the source. This effect also follows from analysis of the arriving waves' correlation function distribution over the spatial angles and the energetic spectrum of field power, which will be done later.

The same features were found in numerous experiments described in References 51–52 and 53–57 separately and clearly seen also from Figure 8.3 and Figure 8.4.

PDF of the Arrival Time. Following Reference 64, let us now find the TD PDF distribution. We integrate the density $\mu(\tau, \varphi)$ over the angle variable $\varphi \in (-\pi, \pi)$. Taking into account the fact of periodicity of $\mu(\tau, \varphi)$ as a function of angles φ and its symmetry relative to $\varphi = 0$, one can describe the density of the relative TD distribution for single scattered arriving waves in the following form:

$$\mu(\tau) = 2\int_0^\pi \mu(\tau, \varphi)d\varphi = 2\left[\int_0^\pi \mu_1(\tau, \varphi)d\varphi + \int_0^\pi \mu_2(\tau, \varphi)d\varphi\right]. \quad (8.21)$$

Using a new variable $\xi = \dfrac{1 - \tau\cos\varphi}{\tau - \cos\varphi}$, $\xi \in (-1, 1)$ and the relations $\dfrac{\tau^2 - 1}{\tau - \cos\varphi} =$ $\tau - \xi$, and $\dfrac{\tau^2 - 2\tau\cos\varphi + 1}{\tau - \cos\varphi} = \tau + \xi$, we can easily derive the first integral in (8.21) and present it as

$$\mu_1(\tau) = \pi\nu\gamma_0 d^3(1 - \varsigma)\tau\sqrt{\tau^2 - 1}\exp\left\{-\gamma_0\tau\frac{(2 - \varsigma)}{2}d\right\}I_0\left(\frac{\gamma_0\varsigma d}{2}\right) \quad (8.22a)$$

and the second integral in (8.21) as

$$\mu_2(\tau) = 2\pi\nu\gamma_0 d^3\frac{\varsigma}{(1 - \varsigma)}\left[\tau\exp\{-\gamma_0\tau d\} + \frac{(\tau - 1)^2}{\sqrt{\tau^2 - 1}}\exp\left\{-\frac{\gamma_0\tau d}{2}\right\}I_0\left(\frac{\gamma_0 d}{2}\right)\right]. \quad (8.22b)$$

Here, $I_0(y)$ is the Bessel function of zero order; the parameter ς was introduced earlier as $\varsigma = (z_2 - \bar{h})/z_2$. The above expressions were obtained for the condition $\gamma_0 d \gg 1$ by taking into account the limit on the BS antenna height z_2 of $\dfrac{h}{z_2} > (\gamma_0 d)^{-1}$. Then, the total PDF of relative TD for single scattered waves can be finally presented as:

$$\mu(\tau) = 2\pi\nu\gamma_0 d^3(1 - \varsigma)\sqrt{\tau^2 - 1}\exp\left\{-\frac{\gamma_0\tau d}{2}\right\} \times$$

$$\left\{\frac{\gamma_0\tau(1 - \varsigma)}{2}\exp\left(-\frac{\gamma_0\tau d(1 - \varsigma)}{2}\right)I_0\left(\frac{\gamma_0\varsigma d}{2}\right) + \right. \quad (8.23)$$

$$\left. \frac{\varsigma}{(1 - \varsigma)}\left[\frac{\tau}{\sqrt{\tau^2 - 1}}\exp\left\{-\frac{\gamma_0\tau d}{2}\right\} + \frac{(\tau - 1)}{(\tau + 1)}I_0\left(\frac{\gamma_0 d}{2}\right)\right]\right\}$$

The PDF of the arrival of the single scattered wave distribution at the observed point B with the relative TD τ can be described by the normalized density function:

$$\tilde{w}(\tau) = \frac{\mu(\tau)}{\displaystyle\int_1^\infty \mu(\tau)d\tau}. \quad (8.24)$$

Following Reference 64, we analyze numerically Equation (8.23), which is presented in Figure 8.14a–d for the same parameters, as in Figure 8.12a–c. Results of computations show that with an increase in the range between the

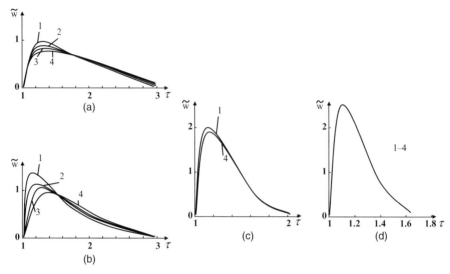

FIGURE 8.14. The normalized PDF, $\tilde{w}(\tau)$, versus arrival-time delay for the same parameters as in Figure 10.12a–d.

receiver and the transmitter from $d = 0.5$ km to $d = 3$ km (which corresponds to microcellular urban environments) and for the UHF/X-band ($\lambda = 0.1$–0.3 m), the effect of increasing the BS height z_2 (the corresponding parameter ς changes from 0.1 up to 0.8) becomes weaker and limits from nonunimodal distribution of $w(\tau)$ (Figs. 8.14a,b) to unimodal (Figs. 8.14c,d) with the maximum of the relative TD (relative to that for the LOS component) $\tau = 1.2$–1.3.

8.2.2. Signal Power Spectra in the Angle-of-Arrival Domain

Let us now examine the signal power spectrum in the urban propagation channel with randomly distributed nontransparent screens as buildings. We put the coordinate system in 2-D plane at point $B(\mathbf{r}_2)$, which now describes moving receiver, and at the moment $t > 0$ passes the point $D(\mathbf{r}_2)$ with velocity $\mathbf{v}$ in the direction determined by angle φ_0 (see Fig. 8.15). In References 58 and 64, a stochastic approach was presented by use of the Fourier transform of the total correlation function described in Chapter 6, but now converted presentation of the latter function in the Cartesian coordinates to the polar coordinates— the azimuth and the radius-vector of any point along the radio path. Following this approach, we can present the power spectrum of a signal in the azimuth domain at the receiving antenna located at point $B(\mathbf{r}_2)$ in the following form [64]:

$$W(\varphi) = \frac{\Gamma \lambda \ell_\nu \bar{h}}{16\pi^2 \left[\lambda^2 + \left(2\pi \ell_\nu \gamma_0 \bar{h} \right)^2 \right] d^3} \{ f_1(\varphi) + f_2(\varphi) \}, \qquad (8.25)$$

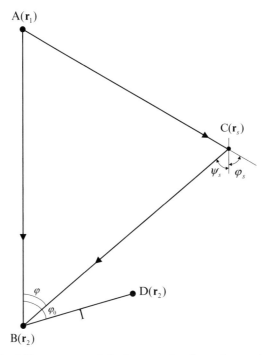

FIGURE 8.15. The 2-D geometry of single scattering from the nontransparent screen (building).

where

$$
f_1(\varphi) = \frac{2z_1^2(\gamma_0 d)^2}{(z_2 + \bar{h})} \frac{\varsigma'(1-\cos\varphi)}{\bar{h}} \frac{\exp\left[-\gamma_0 d\left(\dfrac{\bar{h}}{z_2} + \dfrac{\varsigma'}{2}\dfrac{(1+\cos\varphi)}{\left[1+\dfrac{\gamma_0 d}{2}\left(1+\dfrac{\bar{h}}{z_2}\right)(1-\cos\varphi)\right]}\right)\right]}{\left[1+\dfrac{\gamma_0 d}{2}\left(1+\dfrac{\bar{h}}{z_2}\right)(1-\cos\varphi)\right]}
$$

$$(8.26a)$$

$$
f_2(\varphi) = \frac{2\bar{h}(\gamma_0 d)}{(z_2 + \bar{h})}\left[1+\left(\frac{\bar{h}}{z_2}\right)\frac{1+(k\ell_\nu\gamma_0 d)^2}{1+(\gamma_0\varsigma'd)^2}\right]
$$

$$
\frac{\exp\left[-\gamma_0 d\left(\dfrac{\bar{h}}{z_2} + \dfrac{\varsigma'}{2}\dfrac{(1+\cos\varphi)}{\left[1+\dfrac{\gamma_0 d}{2}\left(1+\dfrac{\bar{h}}{z_2}\right)(1-\cos\varphi)\right]}\right)\right]}{\left[1+\dfrac{\gamma_0 d}{2}\left(1+\dfrac{\bar{h}}{z_2}\right)(1-\cos\varphi)\right]}
$$

$$(8.26b)$$

Here, Γ is the reflection coefficient from the building surface, ℓ_ν is height or width of building's segments (window, balcony, etc.), λ is a wavelength, z_1 is the height of the BS (point A), and z_2 is the height of the moving subscriber (point B), as shown in Figure 8.15. Here, we do not take into account the overlay profile of buildings, because this terrain feature is not very important for azimuth, TD, and frequency distribution of the signal power, that is, for 2-D stochastic model [64]. If we need to account for the overlay profile, we need instead of $\bar{h}$ in the denominator of (8.25) to put the profile function $F(z_1, z_2)$ described in Chapter 5. Here also, according to Reference 64, a new parameter is introduced, $\varsigma' = [(\lambda d/4\pi^3) + (z_2 - \bar{h})^2]^{1/2}/z_2$, which accounts for the process of diffraction from buildings, instead of the $\varsigma = z_2 - \bar{h}/z_2$, used in Reference 58, which did not take into consideration diffraction phenomenon.

The expression in (8.25) consists of two main terms f_1 and f_2 (the same as two terms μ_1 and μ_2 in joint PDF (8.16) of space-temporal distribution of scattered waves, analyzed previously). Each of them relates to a different propagation phenomenon. The term f_1 is the significant term that describes the influence of the scattering area located at the proximity of MS. The term f_2 describes the general effect of rare scatterers that are distributed uniformly in areas surrounding the BS and MS. The influence of different scatterers for the three typical cases, depending on the BS antenna height, is sketched in Figure 7.22, Chapter 7.

When both antennas are lower than the height of the buildings (see Fig. 7.22c, Chapter 7), then both components, f_2 and f_1, should be taken into account. From Equation (8.16) and Equation (8.25), as well as from (8.19a,b) and (8.26a,b), it follows that, if the BS antenna height increases up to $z_2 = \bar{h}$ (see Fig. 7.22b, Chapter 7), then $\varsigma' = \sqrt{(\lambda d/4\pi^3)}/z_2 \ll 1$ and $f_1(\varphi) \to 0$, $f_2(\varphi)_{z_2=h} > f_1(\varphi)_{z_2=h}$.

In the case of $z_2 = \bar{h}$, f_1 is close to zero, and it means that all scatterers located in the far zone from the MS, near the BS, will influence the spreading of the total signal at the BS. With an increase in the height of the BS antenna, that is, $z_2 > \bar{h}$, the influence of buildings surrounding the MS on the total signal distribution will be more significant, and f_1 becomes larger than f_2, describing the effect of scatterers located close to the MS.

When the BS antenna is above the rooftop level (Fig. 7.22a, Chapter 7), the spectrum distribution in the azimuth domain $W(\varphi)$ depends only on position and the distribution of scatterers (obstructions) close to MS. Influence of scatterers (buildings) in the proximity of MS on the signal received at the MS is increased, and it contributes more than the scatterers surrounding the BS (Fig. 7.22b, Chapter 7) and those located far from the MS.

In Reference 66, another presentation of signal spectra in the azimuth domain was shown, accounting for the relations between the average intensity of the signal, $\langle I(r) \rangle$, obtained in Chapter 5, and the signal power spectrum in AOA domain, that is,

$$\langle I(r)\rangle = \int_0^{2\pi} W(\varphi)\frac{d\varphi}{2\pi}. \tag{8.27}$$

We also will take into account that Equation (8.27) describes only the effects of the incoherent component of the power spectra of the total field caused by multiple scattering and diffraction of rays.

As discussed earlier in Chapter 5 (see also References 58–63), there is the coherent component of the total field intensity, which corresponds to an energetic spectrum with a single spectral line of $\varphi = 0$ [66]. We therefore must present the total angle energetic spectrum for the total field intensity within the built-up layer as [66]

$$W(\varphi) = \langle I_{co}\rangle\,\delta(\varphi) + \langle I_1\rangle\,\tilde{W}_1(\varphi) + \langle I_2\rangle\,\tilde{W}_2(\varphi), \tag{8.28}$$

where $\delta(\varphi)$ is the delta function which equals 1, if $\varphi = 0$ and 0, if $\varphi \neq 0$. The dependences of both coherent and incoherent components of the total field as functions of antenna heights, distance between antennas, and buildings spatial density, were described in more detail earlier in Chapter 5. Here, we only will mention that we presented the incoherent component of the total field intensity as a sum of two separate components corresponding to $\tilde{w}_1$ and $\tilde{w}_2$ from (8.24), accounting for (8.22a,b), respectively [66]:

$$\langle I_{inc}\rangle = \langle I_1\rangle + \langle I_2\rangle \tag{8.29}$$

$$\langle I_1\rangle \approx (1-\varsigma)\left[1+(1-\varsigma)^2\,\frac{1+(k\ell_v\gamma_0\overline{h})^2}{1+(\varsigma\gamma_0(\overline{h}-z_1))^2}\right]$$
$$\frac{(\gamma_0 d)^2\,\exp\{-\gamma_0 d(1-\varsigma)\}}{(2+\gamma_0 d)}\,\frac{(1-\cos\Delta\tilde{\varphi}_1)}{\sin\Delta\tilde{\varphi}_1}, \tag{8.30a}$$

and

$$\langle I_2\rangle \approx \varsigma\,\frac{(1-\cos\Delta\tilde{\varphi}_2)}{\sin\Delta\tilde{\varphi}_2}. \tag{8.30b}$$

Here we introduce the integral scales $\Delta\tilde{\varphi}_1$ and $\Delta\tilde{\varphi}_2$, which define the boundaries of the interval of angles of arrival, at which the scattered signal power falls approximately to 50%. Let us now consider the relation

$$\frac{\langle I_2\rangle}{\langle I\rangle} = \frac{\langle I_2\rangle}{\langle I_1\rangle + \langle I_2\rangle} \tag{8.31}$$

as a function of the distance between the receiver placed above the built-up layer and the transmitter placed within built-up layer. In our evaluations we

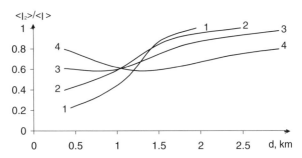

FIGURE 8.16. The normalized intensity from (10.39) for various heights of the receiving antenna; Curves 1, 2, 3, and 4 correspond to $\varsigma = 0.1$ ($z_2 = 33$ m), $\varsigma = 0.3$ ($z_2 = 43$ m), $\varsigma = 0.5$ ($z_2 = 60$ m), and $\varsigma = 0.7$ ($z_2 = 100$ m).

put $\gamma_0 = 6$ km^{-1} ($\langle \rho \rangle = 167$ m is the LOS distance [63]), the parameter of walls' roughness $l_\nu = \lambda \approx 0.1$–0.3 m, $\gamma_0 \bar{h} \approx 0.2$ ($\bar{h} = 30$ m). Results of calculations are presented in Figure 8.16 for various heights of the receiving point above the buildings' layer; curves 1, 2, 3, 4 correspond to $\varsigma = 0.1$ ($z_2 = 33$ m, $\bar{h} = 30$ m), $\varsigma = 0.3$ ($z_2 = 43$ m), $\varsigma = 0.5$ ($z_2 = 60$ m), $\varsigma = 0.7$ ($z_2 = 100$ m), respectively. From the illustrations presented in Figure 8.16, it follows that with increase of the height z_2 with respect to $\bar{h}$, at distances less than 1 km, the "illuminations" from scatterers inside the $\langle \rho \rangle$-region surrounding the transmitting antenna at the height of z_1 is predominant, but far away from the source the main contribution in the total intensity is from the other scatterers located outside the $\langle \rho \rangle$-region, that is, at distances exceeding 167 m. Moreover, the importance of these points is increased with decrease of receiving antenna height for distances more than 1–2 km.

Additional analysis of the components, $\tilde{W}_1(\varphi)$ and $\tilde{W}_2(\varphi)$, of the total angle energetic spectrum $\tilde{W}(\varphi)$ made in Reference 66, has shown that their contribution in common energetic spectrum $\tilde{W}(\varphi)$ is changed differently and non-monotonically with change of height factor ς and distance d from both points, receiving (MS) and transmitting (BS). We can show these features if we estimate the angle distribution of the signal power spectrum inside the building layer. For this purpose, we first present the normalized function $\tilde{W}(\varphi)$, which describes the average angle spectrum density of single scattered waves regarding the total average field intensity:

$$\tilde{W}(\varphi) = \frac{W(\varphi)}{\langle I \rangle}. \tag{8.32}$$

Then, we will rewrite, following References 64 and 66, Equation 8.32 rearranged in more convenient form for further computations accounting for diffraction phenomena in the 3-D case. Finally, we get

$$\tilde{W}(\varphi) = \frac{h}{z_2}\left[1+\left(\frac{h}{z_2}\right)^2 \frac{\left[1+(k\ell_\nu\gamma_0 h)^2\right]}{1+\gamma_0^2(h-z_1)^2 \dfrac{\left((\lambda d/4\pi^3)+(z_2-h)^2\right)}{z_2^2}}\right]f_1(\varphi)$$

$$+ \frac{\left((\lambda d/4\pi^3)+(z_2-h)^2\right)^{1/2}}{(h-z_1)}f_2(\varphi).$$

(8.33)

The first term consists of a function $f_1(\varphi)$, which now we can present in the following form:

$$f_1(\varphi) = \frac{(\gamma_0 d)\exp\{-\gamma_0 d(1-\tilde{\varsigma})\}}{(2-\tilde{\varsigma})} \frac{\exp\left\{-\dfrac{\gamma_0 d}{2}\dfrac{\tilde{\varsigma}(1+\cos\varphi)}{[1+\gamma_0 d(1-\tilde{\varsigma}/2)(1-\cos\varphi)]}\right\}}{[1+\gamma_0 d(1-\tilde{\varsigma}/2)(1-\cos\varphi)]}$$

(8.34)

and the second part consists of function $f_2(\varphi)$, which for the case of $\pi/2 < |\varphi| < \pi$, can be expressed as [66]

$$f_2(\varphi) = \frac{1-\cos\varphi}{2(\gamma_0 d)},$$

(8.35)

which does not depend on situations with elevations of both antennas, BS and MS, with respect to building rooftops.

In Equation (8.33), Equation (8.34), and Equation (8.35), the average height $\bar{h}$ is used for computations instead of $F(z_1,z_2)$ without any loss of generality. We examine the influence of the BS antenna height on the received signal power for the case when the moving subscriber (MS) antenna has a lower height than the buildings' rooftops ($z_2 < \bar{h}$, see Fig. 8.15) and the BS stationary antenna, as the transmitter, located above or at the same level than the rooftops ($z_1 \geq \bar{h}$). In this scenario, we introduce in Equation (8.33), Equation (8.34), and Equation (8.35) the same inverse parameter $\tilde{\varsigma} = (\bar{h}-z_1)/(z_1-z_2)$, as above. Figure 8.17a–d shows the energetic spectrum of the signal power distribution in the azimuth (or AOA) domain for various values of the parameter $\tilde{\varsigma} = (|\bar{h}-z_1|)/(z_1-z_2)$ and $z_1 \gg z_2$. Here, figure denoted by (a) corresponds to $\tilde{\varsigma} \approx 0.0(\bar{h} \approx z_1 = 20$ m, $z_2 = 2$ m), (b) corresponds to $\tilde{\varsigma} = 0.125$ ($z_1 = 18$ m), (c) is corresponds to $\tilde{\varsigma} = 0.24$ ($z_1 = 16.5$ m), (d) is corresponds to $\tilde{\varsigma} = 0.385(z_1 = 15$ m), that is, with decrease of the BS antenna height with respect to the average height of buildings. From the illustrations it is clearly seen that with decrease of the height z_1 of the transmitting BS antenna (for the low receiving MS antenna height, $z_2 = 2$ m, located at the point B, see Fig. 8.15), the main contribution for the total average angle energetic spectrum follows from the term $\tilde{W}_1(\varphi) \sim f_1(\varphi)$, which starts to determine the shape of

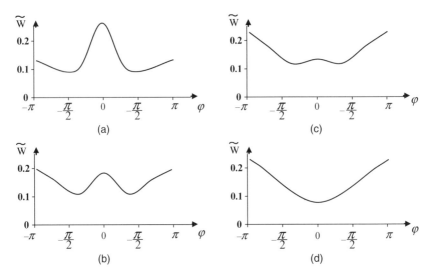

FIGURE 8.17. The normalized signal power spectrum $\tilde{W}(\varphi)$ distribution in the azimuth domain for various height factors ς.

the energetic spectrum at the ranges of the transmitter antenna height parameter $\tilde{\varsigma} > 0.1$ (at the distance $d = 1$ km). In this case, significant contributions of scattered waves arriving from the directions close to $\varphi = \pi$ are illustrated by Figures 8.17c–d. In other words, for the BS transmitting antenna (A) lower than the height of the rooftops, the waves arrive at the MS receiver (B) from directions far from direction to the source, that is, after reflections and scattering from various obstructions around both terminals.

As was also shown in Reference 66, with increase of the distance between the moving vehicle and the BS or with increase of the buildings' density, the transformation of $\tilde{W}(\varphi)$ from the sharp distribution presented in Figure 8.17a to that presented in Figure 8.17d occurs for smaller parameters $\tilde{\varsigma}$. Hence, the shape of $\tilde{W}(\varphi)$ presented in Figure 8.17a,b can be considered as a typical condition for mobile communication links with the BS antenna, as the transmitter, located at the same level as the top of the urban built-up layer and higher as the moving vehicle receiver antenna, located below the rooftops. One can also note that the dependence $\tilde{W}(\varphi)$ versus the height factor and the range between both terminals, which is shown in Figure 8.17, is close to that obtained earlier for the PDF of angular distribution for single-scattered waves $\tilde{w}(\varphi)$, which is presented in Figure 8.11, Figure 8.12, and Figure 8.13 and differs only by small value of the integral scale for $\tilde{W}(\varphi)$.

The same trend of $\tilde{W}(\varphi)$ with changes of receiving or transmitting antenna height (with increase of $\tilde{\varsigma}$ or decrease of ς) has shown from the analysis of both components of the total energetic spectrum. In fact, with a decrease of BS antenna height relative to $\bar{h}$, the probability of NLOS conditions caused by obstacles, as scatterers and reflectors, which are far from the proximity of the source (in its $\langle\rho\rangle$-region surrounding), is increased. In such a situation, these obstructions form the arriving angle dependence of the total signal spectrum: the second component $\tilde{W}_2(\varphi)$ is sharply decreased, and the contribution of the first spectrum component $\tilde{W}_1(\varphi)$ in the total power spectrum is increased. The same trend follows from results of above examinations of PDF densities $\tilde{W}_{1,2}(\varphi)$. In fact, all scatterers located in the neighborhood area around the source (within the $\langle\rho\rangle$-region around it) form the arriving angles' dependence of the second component $\tilde{W}_2(\varphi)$. Moreover, an average intensity, which corresponds to $\tilde{W}_1(\varphi)$, decreases more slowly with respect to the fast increase of $\tilde{W}_2(\varphi)$ with increase of BS antenna height.

Hence, in real situations occurring in the urban scene when the BS is located above or near the rooftops, situations mostly occur when $\tilde{W}_2(\varphi) > \tilde{W}_1(\varphi)$ and the total signal spectrum distribution can be obtained by the detailed analysis of $\tilde{W}_2(\varphi)$.

The analysis of $\tilde{W}_2(\varphi)$ distribution versus angles-of-arrival have shown the same trend as was presented in Figure 8.13b, in situations when both the BS antenna and the moving vehicle antenna are below the level of rooftops in their neighborhoods and, finally, as in Figure 8.13a for the case when the BS antenna is located higher than the rooftops).

8.2.3. Power Spectrum Distribution in the Time-of-Arrival Domain

The signal power spectrum distribution in TD domain can be obtained in the same manner [64, 66]:

$$W(\tau) = \frac{\Gamma}{8\pi^2 d^2} \frac{k\ell_\nu\gamma_0\bar{h}}{1+\left(k\ell_\nu\gamma_0\bar{h}\right)^2} \left\{ (1-\varsigma')\left[1 \right.\right.$$

$$\left.\left. + (1-\varsigma')^2 \frac{1+\left(k\ell_\nu\gamma_0\bar{h}\right)^2}{1+\left(\varsigma'\gamma_0\bar{h}\right)^2} \right] f_1(\tau) + \frac{\varsigma'}{(1-\varsigma')} f_2(\tau) \right\} \tag{8.36}$$

where:

$$f_1(\tau) = \frac{(\gamma_0 d)^2 \sqrt{\tau^2-1}}{4\tau^2} \exp\left\{ -\gamma_0\tau \frac{(2-\varsigma')}{2} d \right\} I_0\left(\frac{\gamma_0\varsigma' d}{2} \right) \tag{8.37a}$$

$$f_2(\tau) = \frac{\gamma_0 d}{2} \exp\left\{ -\frac{\gamma_0\tau d}{2} \right\}\left[\exp\left\{ -\frac{\gamma_0\tau d}{2} \right\} + \frac{\sqrt{\tau-1}}{\sqrt{\tau+1}} I_0\left(\frac{\gamma_0 d}{2} \right) \right]. \tag{8.37b}$$

In the TD domain, the same properties of the signal power spectrum, as was done for the azimuth domain, can be obtained. Thus, in the case of $z_2 \le \bar{h}$,

$$\frac{\varsigma'}{(1-\varsigma')} f_2(\tau) \gg \left[1 + \frac{1+\left(k\ell_\nu\gamma_0\bar{h}\right)^2}{1+\left(\varsigma'\gamma_0\bar{h}\right)^2} \right] f_1(\tau),$$

and the distribution of scatterers far from the BS antenna, that is, close to the MS antenna (see Fig. 7.22a, Chapter 7), does not influence on a power spectrum $W(\tau)$ in the TD domain. Then, with a height increase of the receiver antenna (when $z_2 > \bar{h}$, $\varsigma' > 1$), the main contribution follows from the first term with $f_1(\tau)$ that really describes the influence of the neighboring area around the MS. In this case, we get another condition:

$$\frac{\varsigma'}{(1-\varsigma')} f_2(\tau) << \left[1 + \frac{1 + \left(k\ell_\nu \gamma_0 \bar{h} \right)^2}{1 + \left(\varsigma' \gamma_0 \bar{h} \right)^2} \right] f_1(\tau).$$

We also note that the function $f_2(\tau)$ does not depend on parameter ς', that is, on the receiving antenna's height with respect to the rooftops. Therefore, for $\varsigma' > 0.3$, the form of the power spectrum versus TD becomes practically constant value.

Function $W(\tau)$ from (8.36) satisfies the following condition:

$$\int_1^\infty W(\tau)d\tau = <I>. \tag{8.38}$$

Analysis of the normalized power spectrum $\tilde{W}(\tau)$ of TD (with respect to the total field intensity taken from Chapter 5) is presented in Figure 8.18 versus

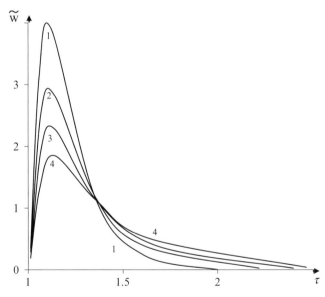

FIGURE 8.18. The normalized power spectrum $\tilde{W}(\tau)$ distribution in the time-of-arrival domain for $\tilde{\varsigma} = 0 (z_1 - \bar{h})$, $\tilde{\varsigma} = 0.05 (z_1 = 1.05 \cdot \bar{h})$, $\tilde{\varsigma} = 0.1 (z_1 = 1.11 \cdot \bar{h})$, $\tilde{\varsigma} = 0.25 (z_1 = 1.25 \cdot \bar{h})$, for $\bar{h} = 20$ m, and $\gamma_0 d = 6$.

the arrival time of the scattered total field for communication between the mobile transmitter (with the height $z_2 < \bar{h}$, where $\bar{h}$ is the average height of the built-up layer) and the stationary BS, as a receiver (with height $z_1 \geq \bar{h}$ for parameters $\tilde{\varsigma} = |z_1 - \bar{h}|/z_1 = 0$ ($z_1 = \bar{h}$, curve 4); $\tilde{\varsigma} = 0.05$ ($z_1 = 1.05 \cdot \bar{h}$, curve 3); $\tilde{\varsigma} = 0.1$ ($z_1 = 1.11 \cdot \bar{h}$, curve 2); $\tilde{\varsigma} = 0.25$ ($z_1 = 1.25 \cdot \bar{h}$, curve 1), for $\bar{h} = 20$ m and for $\gamma_0 d = d/<\rho> = 6$. As follows from the presented illustrations, with increase of receiving antenna height over the buildings' layer, the probability of direct visibility between the receiver and the neighborhood of the transmitter, which has the maximal luminance, increases. This fact corresponds to a change of role of two terms in the figure brackets in (8.36). In fact, if for $\tilde{\varsigma} = 0$ the main contribution in $W(\tau)$ follows from the first term, consisting $f_1(\tau)$. Then with increase of height z_1 (e.g., $\tilde{\varsigma} = 0.25$, $\bar{h} = 20$ m and $z_1 = 25$ m), the main contribution follows from the second term with $f_2(\tau)$, which really describes the influence of the neighboring area around the transmitter in the total scattered field coming to receiver. As can be seen from expression (8.37b), the function $f_2(\tau)$ does not depend on parameter ς, that is, on the receiving antenna's height above the rooftops.

Therefore, for $\tilde{\varsigma} > 0.2$, the form of the energetic spectrum versus TD becomes a practically constant value. At the same time, the intensity of the scattering field coming to the observer (BS) grows with increase of height factor $\tilde{\varsigma}$. Additional analysis shows that with increase of building density (parameter γ_0) or with increase of the distance between the source and observer, the transformation of spectrum $W(\tau)$ to the steady-state form takes place for smaller values of height factor $\tilde{\varsigma}$. With increase of value $\gamma_0 d$ from 6 up to 24 for $\tilde{\varsigma} > 0.2$, the form of $W(\tau)$ becomes unimodal and sharp with the maximum near the time range $\tau = (1 + 1/\gamma_0 d) = 1.2$–$1.3$. As a consequence, the energetic spectrum of times delay can be presented and derived for $\varsigma > 0.2$ ($(z_2 - \bar{h}) > 2$ m) by the following formula:

$$W(\tau) = \frac{<I> f_2(\tau)}{\int\limits_{1}^{\infty} f_2(\tau) d\tau}. \tag{8.39}$$

The integral scale of the normalized power spectrum $W(\tau)/<I>$ for the case when the maximum of $f_2(\tau)$ at the point $\tau = (1 + 1/\gamma_0 d)$ approximately equals $\gamma_0 d/3$, for the case $\tilde{\varsigma} > 0.2$ and $\gamma_0 d > 6$ can be easily estimated as $\Delta \tau_s = 3/\gamma_0 d$. Hence, the same features of signal power spectrum distribution, as we found earlier in the AOA domain, were observed in the time domain with change of BS antenna height, ranges between both terminal antennas and environmental conditions in the urban scene.

So, the proposed 3-D stochastic model, based on the general Equation (8.25) or Equation (8.33) and Equation (8.36), gives more realistic description of the propagation phenomena with respect to 2-D model obtained in Reference 58 (see results of comparison between theoretical prediction and experimental data described in References 64 and 66). Generally speaking, the 3-D model [64, 66] covers the 2-D model [58] in the case when BS antenna is located near or above the building profile height, and when the effects of diffraction from rooftops are more realistic. When the BS antenna is smaller than the building profile height, both models, the 2-D and the 3-D, can be used, but the latter gives more precise results. Moreover, this model can be used to describe street effects found in high-sensitive experiments carried out in Helsinki and described in References 9 and 10. The results of these experiments were presented earlier in Figure 8.7 and Figure 8.8. This "guide" effects were defined in References 9 and 10 as a "Class 1" of propagation phenomenon. We now will prove this statement by combining the stochastic and waveguide models, which were briefly described in Chapter 5 and more in detail in References 64–66.

8.3. 3-D STOCHASTIC MODEL ACCOUNTING FOR STRAIGHT CROSSING STREET GRID

As follows from geometry shown in Figure 8.19, and from the previous analysis presented in Sections 8.1.4 and 8.2.1, there are two distributions that are of particular interest that must be taken into account.

The *first distribution function* is $\tilde{\mu}(\tau, \varphi)$ that was obtained and analyzed in Chapter 5 and that describes the general spatial distribution of scatterers (see definitions above). We will present it in more convenient form for $\gamma_0 r \gg 1$

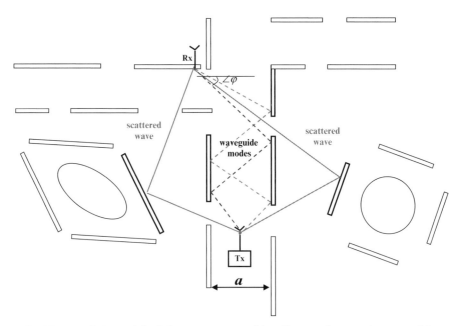

FIGURE 8.19. 2-D model of the street waveguide effects and scattering caused by a spatial random building distribution.

and in the case of $z_1 < \bar{h} < z_2$ for future numerical analysis following References 64–66:

$$\tilde{\mu}(\tau, \varphi) = \tilde{\mu}_1 + \tilde{\mu}_2 = \frac{\nu}{2}\sin^2\left(\frac{\alpha}{2}\right) \cdot \left[\frac{\gamma_0 \bar{h} d^2}{z_2} \frac{\tau(\tau^2 - 1)}{2(\tau - \cos\varphi)}\right] f_1(\tau, \varphi)$$

$$+ \frac{\nu}{2}\sin^2\left(\frac{\alpha}{2}\right) \cdot \left[\frac{(z_2 - \bar{h})d}{\bar{h}} \frac{(\tau^2 - 2\tau\cos\varphi + 1)}{2(\tau - \cos\varphi)} f_2(\tau, \varphi)\right]. \tag{8.40a}$$

For the case of $z_1 < \bar{h}$, and $z_2 < \bar{h}$ Equation (8.40a) can be reduced as [64–66]:

$$\tilde{\mu}(\tau, \varphi) = \frac{\nu\gamma_0}{2} \cdot \sin^2\left(\frac{\alpha}{2}\right) \cdot \frac{\tau d^2(\tau^2 - 1)}{2(\tau - \cos\varphi)} \cdot \exp\{-\gamma_0 d \cdot \tau\}. \tag{8.40b}$$

Here,

$$f_1(\tau, \varphi) = \exp\left[-\gamma_0 d\left(\frac{(z_2 + \bar{h}) \cdot \tau^2 - 2z_2\cos\varphi \cdot \tau + (z_2 - \bar{h})}{2z_2(\tau - \cos\varphi)}\right)\right], \tag{8.41a}$$

and

$$f_2(\tau, \varphi) = \exp\left[-\gamma_0 d\frac{(\tau^2 - 2\cos\varphi \cdot \tau + 1)}{2(\tau - \cos\varphi)}\right], \tag{8.41b}$$

where, as before, φ is the angle of multipath components of the total field in the horizontal plane (i.e., azimuth) arriving at the receiver after multiple scattering from the buildings surrounding both the transmitter and the receiver, α is the angle between lines to the receiver and to the transmitter. All other parameters are defined the same way as in Section 8.2. Here we denoted the receiver BS station antenna height by z_2, as was denoted for Equation (8.25). The above expression consists of two main terms $\tilde{\mu}_1$ and $\tilde{\mu}_2$. Each one relates to a different propagation phenomenon depending on situation with BS antenna height with respect to the overlay profile of the buildings (see Fig. 7.22, Chapter 7). In fact, because $\tilde{\mu}(\tau, \varphi)$ is a normalized signal power distribution in the AOA and TOA plane, its main terms $\tilde{\mu}_1$ and $\tilde{\mu}_2$ give the same effects as f_1 and f_2 in (8.33) and in (8.34), with a change in the BS antenna height, from z_1 to z_2 as was used for Equation (8.32).

The *second distribution function* is $\mu(r)_{wg}$ that describes a probability of the wave mode existence caused by a multislit street waveguide at distance r from the transmitter. The geometry of the problem to find the PDF, $\mu(\tau)_{wg}$, is shown in Figure 8.19. We assume that the angle φ takes discrete values with a resolution of degrees. Finally, following References 62 and 63 (see also waveguide model description in Chapter 5), and converting the PDF presentation in the space domain, $\mu(\mathbf{r})_w$, to that in the joint azimuth and TD domain, $\mu(\tau,\varphi)_{wg}$, we finally get:

$$\mu(\tau, \varphi)_{wg} = \exp\left[-2\frac{|\ln\chi|}{a'(\varphi)} \cdot \frac{d(\tau^2 - 1)}{2(\tau - \cos\varphi)}\right]. \tag{8.42}$$

Here, we use the same notations as in Chapter 5, that is, $\chi = \dfrac{<L>}{<L>+<l>}$ is a brokenness parameter, a is the street width, $<L>$ and $<l>$ are the average length of screens (nontransparent buildings lining the street) and slits (gaps between buildings and intersections of streets), r is the BS–MS distance. The definitions of $a'(\varphi)$ will be presented below. Let us examine the correlation between $\tilde{\mu}(\tau, \varphi)$ and $\tilde{\mu}(\tau, \varphi)_{wg}$. As was shown separately in References 21–23 for indoor communication links and in References 53–55 and 67–69 for outdoor communication links, these two functions are strongly independent because they describe two different physical phenomena. The first, $\mu(r,\varphi)$, relates to the random distribution of buildings, as scatterers, that are placed around both terminal antennas, and the second, $\mu(r,\varphi)_{wg}$, relates to the scatterers that lie along the streets and mainly contribute to the guiding effect (see Fig. 8.19). Then, the completed form of the joint AOA and TOA distribution of scatters (i.e., the scattered waves) can be written as

$$\tilde{\mu}_f = \tilde{\mu}(\tau, \varphi) \cdot \mu(\tau, \varphi)_{wg}. \tag{8.43}$$

Substituting the appropriate function for $\tilde{\mu}(\tau, \varphi)$ and $\tilde{\mu}(\tau, \varphi)_{wg}$, we can present (8.43) in the convenient form for future computations by accepting the density function of all single scatterers as a function of φ and τ [67–69]:

$$\tilde{\mu}_f(\tau, \varphi) = \frac{\gamma_0 \nu}{2}\sin^2\left(\frac{\alpha}{2}\right) \cdot \frac{\tau d^2(\tau^2 - 1)}{2(\tau - \cos\varphi)}\exp\{-\gamma_0 d \cdot \tau\}$$
$$\cdot \exp\left\{-2\frac{|\ln\chi|}{a'(\varphi)} \cdot \frac{d(\tau^2 - 1)}{2(\tau - \cos\varphi)}\right\}. \tag{8.44}$$

Equation (8.44) is valid for the TU situation when both the receiver and transmitter are placed below the rooftop level. Here, $a' = \sqrt{4a^4/\lambda^2 n^2 + a^2}$, where a is the street width, and n is the number of reflections from walls (number of waveguide modes). For $a > \lambda$, we have that $a' = 2a^2/\lambda n$. It was shown in References 59 and 60, for distances far from the source, only the main waveguide mode (with $n = 1$) propagates without any attenuation within the street waveguide. This result was accounted for in References 67–69 during numerical simulation of (8.44) and in comparison with experimental data obtained in built-up areas with straight crossing grid-plan streets (as Manhattan street grid).

Finally, we define the discrete spectrum of the total signal power within a broken waveguide, taking into account the geometry presented in Figure 8.19, as [67–69]:

$$W_{wg}(\tau, \varphi) = W_0 \frac{2(\tau - \cos\varphi)}{d(\tau^2 - 1)} \exp\left[-2\frac{|\ln\chi|}{a'(\varphi)} \cdot \frac{d(\tau^2 - 1)}{2(\tau - \cos\varphi)}\right], \qquad (8.45)$$

where W_0 is a signal power of the antenna in direction of direct visibility (LOS component). Now, to find the total signal power distribution, in time and azimuth domains, and accounting for guiding street effects, we need to combine Equation (8.45) and Equation (8.25) [or (8.33) and (8.35)]. That leads to:

$$W_{fin}(\varphi, \tau) = W(\varphi) \cdot W(\tau) \cdot W(\varphi, \tau)_{wg} \qquad (8.46)$$

to determine the joint 2-D distribution in the AOA and TOA (sometimes called TD) domains, respectively.

8.4. EFFECTS OF DEPOLARIZATION IN THE AZIMUTH AND ELEVATION DOMAINS

Let us now analyze of how the characteristics of the polarization ellipse influence on changes of signal energy both in the horizontal and vertical planes (i.e., in the azimuth and elevation domains, respectively) caused by the features of the built-up terrain determined in Chapter 5 by Equation (5.116), Equation (5.117), and Equation (5.118). In the elevation domain, for computation of the PDF of signal energy redistribution due to depolarization effects, $w(\theta)$, we use Equation (5.95), and that in the azimuth domain, $w(\varphi)$, we use Equation (5.115) (Chapter 5). Thus, in Figure 8.20, an ordinary PDF, $w(\theta)$, of signal power distribution in the vertical plane is presented for $\theta \in [0, 180°]$ and for $\sigma_\perp/\sigma_\parallel = 0.2$ (curve 1), 0.4 (curve 2), 0.8 (curve 3), 1 (curve 4), 2.5 (curve 5), and 5 (curve 6). It is clearly seen that $w(\theta)$ has a unimodal character for $\sigma_\perp/\sigma_\parallel < 1$ with maximum at $\theta = \pi/2$. For $\sigma_\perp/\sigma_\parallel \geq 1$, the distribution of signal energy in the vertical plane is regular and uniform, and for $\sigma_\perp \geq 5\sigma_\parallel$, $w(\theta)$ comes to have a pure bimodal character with maxima achieving proximity of vertical angles $\theta = 0$ and $\theta = \pi$.

The same tendency, from unimodal to multimodal distribution, can be seen from analysis of $w(\varphi)$ shown in Figure 8.21 for the ratio $\sigma_{x\perp}/\sigma_{y\perp}$ equals 0.2, 0.5, 0.8, 1 (curves 1, 2, 3, 4, respectively). We should note that all PDF distributions versus the angle parameters can achieve maxima only for the values of angles corresponding to $n\pi/2, n = 1, 2, 3, \ldots$. The position of these maxima and the shape of the corresponding PDF are defined by the inequalities $\sigma_\perp > \sigma_\parallel$ or $\sigma_\perp < \sigma_\parallel$, and by the ratio $\sigma_{x\perp}/\sigma_{y\perp}$. The latter ratio depends, as clearly seen from Equation (5.116), Equation (5.117), and Equation (5.118) (Chapter 5), on the features of the built-up terrain, that is, on the buildings' density and the overlay profile of buildings loscated surrounding the transmitting and receiving antennas.

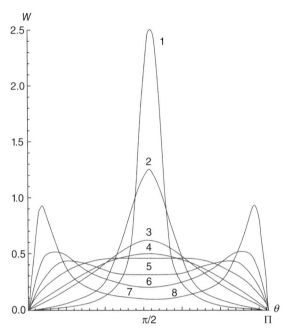

FIGURE 8.20. The ordinary PDF, $w(\theta)$, of signal power distribution in the vertical plane is presented for and for $\theta \in [0, 180°]$ and for $\sigma_\perp/\sigma_\| = 0.2$ (curve 1), 0.4 (curve 2), 0.8 (curve 3), 1 (curve 4), 2.5 (curve 5), and 5 (curve 6). 2.5 (curve 5), and 5 (curve 6).

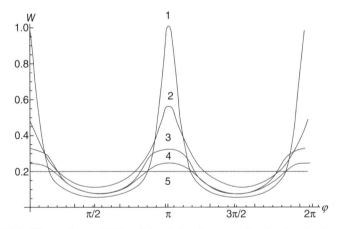

FIGURE 8.21. The ordinary PDF, $w(\varphi)$, of signal power distribution in the horizontal plane for the ratio $\sigma_{x\perp}/\sigma_{y\perp}$ equals 0.2 (curve 1), 0.5 (curve 2), 0.8 (curve 3), and 1.0 (curve 4).

8.5. PREDICTING MODELS FOR INDOOR COMMUNICATION CHANNELS

The increasing use of adaptive antennas in radio indoor communications demands an intensive study of the indoor propagation environments, including offices, buildings, warehouses, factories, hospitals, apartments, and so on. In References 21 and 22, the temporal data on indoor propagation have been collected and analyzed, from which a new statistical time-domain model for indoor propagation was created [21].

However, most measurements carried out until very recently dealt with time domain data and did not include any data on the AOA distribution in indoor communication links. As was mentioned in the previous chapter, knowledge of the AOA associated with a multipath arrival is important because of increasing use of multiple antenna systems with different kinds of antenna diversity, namely, the use of phased array beamforming [25], diversity combining or adaptive array processing (see the corresponding references in Chapter 7) to mitigate the effects of multipath fading and multiple user servicing. In a latter case, for multiple access communications, the adaptive antenna systems have the potential to allow multiple subscribers to simultaneously use the same frequency band, making efficient utilization of the system bandwidth.

In order to predict the performance of such access, knowledge of both time and angle of each arrival in a multipath and multiple interference channel is needed. Some works have begun recently to address the AOA signal distribution in indoor communication channels, and the first of them, using the statistical approach proposed in References 23 and 24, have found that the AOA distribution of the multipath components of the total signal arriving at the receiver can be accurately described by the Laplacian law [18, 22]. In Reference 25, it was found that multipath arrivals tend to occur at various angles indoors. In Reference 26, a data acquisition system was used (which is similar with that of References 18 and 22) to collect narrowband AOA data and wideband TOA data, but this was done separately without accounting for their joint distribution by measuring simultaneously signal features in the angle and time domains. In Reference 27, the authors used a rectangular antenna array to estimate both elevation and azimuth (or AOA) for major multipaths, but without accounting for the corresponding TOA of each multipath component of the total signal. Authors in Reference 28 used a rectangular adaptive array to make simultaneous measurements of angle and TOA, similar to what was done in References 18 and 22. However, as was mentioned in References 30 and 31, the experiment carried out in Reference 28 was not extensive enough to make any conclusions about the propagation channel.

To satisfy the recent increasing demand on communication speed and ubiquity for the wireless local area networks (WLANs), wireless private branch exchangers (WPBXs), home phone line network alliance (HomePNA), as well for Internet and data network applications, a new theoretical approaches and more precise experiments were carried out [32–41]. We will discuss only a few of them, which in our opinion give a realistic physical description of the channel. First of all, this is a continuation of the research, started in References 18, 22, and 23, made in Reference 36. A new statistical model for site-specific radio propagation, as a combination of geometrical optics ray-tracing description of the propagation phenomena and probabilistic description of processes occurred within a channel, may be accounted. This model was proved by high-resolution measurements carried out at 1 GHz in indoor environments, both in the azimuth and time domain.

In Reference 32 was shown that the a priori assumption concerning the independence between the propagation phenomena in the temporal and spatial domains made in References 18 and 21–23 is appropriate only under NLOS and obstructive line-of-sight (OLOS) scenarios. In LOS scenarios in indoor environments they must be investigated jointly due to dependency that exists between the spatial-temporal domains.

8.5.1. Separate AOA and TOA Signal Power Distributions

As was shown both experimentally and theoretically in References 21–23, the arrivals come in one or two large groups within a 200-ns observation window, where in each cluster an additional attenuation is observed (see Fig. 8.22 according to Reference 22). Within each cluster the arrivals are also delayed with time. In Figure 8.22, a mean envelope of a channel of three clusters is shown.

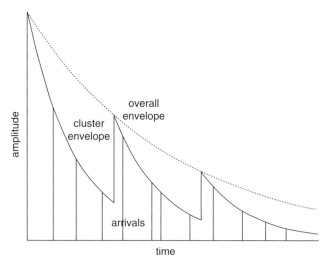

FIGURE 8.22. Cluster effect in time domain obtained in indoor communication channels according to Reference 22.

The TOA distribution corresponds to different Poisson processes [21–23]. The first process describes the arrival times of clusters defined by the cluster arrival decay time constant Γ (which varies from 30 to 80 ns) and by the cluster arrival rate A (where 1/A is changed from 16 to 300 ns) [22, 23]:

$$p(T_i \,|\, T_{i-1}) = Ae^{-A(T_i - T_{i-1})}, \tag{8.47a}$$

where T_i describes the first arrival of each cluster with number i.

The second Poisson process models the arrival times of rays within each cluster defined by the ray arrival decay time constant γ (which is varies from 20 to 80 ns) and by the ray arrival rate α (where $1/\alpha$ is changed from 5 to 7 ns):

$$p(\tau_{ij} \,|\, \tau_{(i-1)j}) = \alpha e^{-\alpha(T_{ij} - T_{(i-1)j})}, \tag{8.47b}$$

where τ_{ij} is the inter-ray arrival times, dependent on the time of the first arrival in the cluster.

At the same time, the angles-of-arrival, θ, distribution is better fitted by a zero-mean Laplacian process [21–23]:

$$p(\theta) = \frac{1}{\sqrt{2}\sigma} e^{-\left| \frac{\sqrt{2}\theta}{\sigma} \right|}. \tag{8.48}$$

Equation (8.58) is valid only for low-elevation antennas, with the standard deviation varying from $\sigma = 21.5°$ to $\sigma = 25.5°$.

However, as was shown in Reference 32, a two-dimensional (2-D) joint distribution of AOA-TOA of the signal power can be expressed as a product of marginal AOA and TOA distributions, where the AOA power spectrum is described for NLOS, OLOS, and LOS scenarios by the Laplacian law for clusters, and for arrivals in each cluster by the same Laplacian law for LOS scenario only and by uniformly distributed function over the range of $[0, 2\pi]$ for OLOS and NLOS scenarios [32], that is,

$$P_{clust}(\phi) \propto \frac{1}{\sqrt{2}\sigma_\phi} \exp\left\{ -\sqrt{2} \frac{|\phi|}{\sigma_\phi} \right\} \tag{8.49a}$$

$$\begin{vmatrix} P_{arrive}(\phi) \propto \dfrac{1}{\sqrt{2}\sigma_\varphi} \exp\left\{ -\sqrt{2} \dfrac{|\varphi|}{\sigma_\varphi} \right\}, & \text{LOS} \\[2ex] P_{arrive}(\varphi) \propto \dfrac{1}{2\pi}, & \text{OLOS and NLOS} \end{vmatrix} \tag{8.49b}$$

and TOA spectrum is described by the decayed exponential function for all three scenarios and for clusters and each arrival in clusters [32],

$$\begin{vmatrix} P_{arrive}(\text{T}) \propto \dfrac{1}{\sigma_\text{T}} \exp\left\{ -\dfrac{\text{T}}{\sigma_\text{T}} \right\} \\[2ex] P_{arrive}(\tau) \propto \dfrac{1}{\sigma_\tau} \exp\left\{ -\dfrac{\tau}{\sigma_\tau} \right\} \end{vmatrix}. \tag{8.50}$$

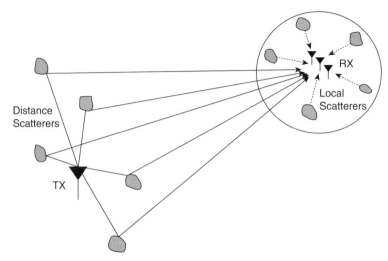

FIGURE 8.23. Geometry of obstructions around the transmitting and receiving antennas.

Here, T and τ are the TOA, ϕ and φ are the AOA, σ_T and σ_τ are the *rms* DS, and σ_ϕ and σ_φ are the angle spread for clusters and for each arrival component in clusters, respectively.

Furthermore, in Reference 32, the physical concept based on local and distance scatterers effects and the corresponding geometry based on distribution and density of scatterers surrounded both terminal antennas, Tx and Rx (see Fig. 8.23), was proposed to explain a Laplacian shape of clusters' AOA distribution at the receiver Rx. As depicted, according to Reference 32, in Figure 8.23, the Rx is surrounded by many local scatterers in its vicinity, reflections from which give rise to a wide spread of AOA in OLOS or NLOS conditions, as shown, according to Reference 32, in Figure 8.24a. Conversely, distance scatterers are located much further away from the Rx, and reflected paths arriving at the Rx are firstly from one particular direction through a much narrower angular spread. If we now assume, following Reference 32, that there are same number of scatterers at both ends of the radio path and each of the scatterers give rise to same number of paths, the higher density of paths will be observed at one particular direction (usually a direct vision path) and lower densities at other directions. This effect leads also from a Laplacian distribution with higher occupancies at the center and lower occupancies at the larger angular values, with result obtained experimentally and shown for LOS conditions in Figure 8.24b, following Reference 32.

The same Laplacian AOA distribution was obtained in Reference 41 by using the elliptical scattering model based on uniformly distributed scatterers in the elliptical regions surrounding the transmitter and the receiver antennas (see Fig. 8.23), and in References 15 and 42 for indoor and outdoor commu-

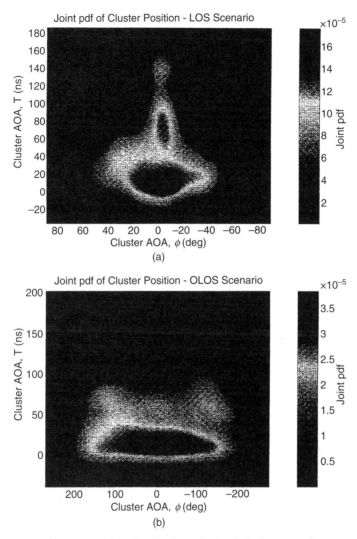

FIGURE 8.24. Joint AOA-TOA distribution obtained during experiments carried in Reference 32. (Reprinted with permission @ 2003 IEEE.)

nication channels based on geometrical LOS model in the presence of multipath phenomena. All researches, however, were based on qualitative phenomenological description of the problem without strict account for real obstruction density around antennas, the spatial distribution of obstructions, depending on antenna height with respect to building heights, and other features of the built-up terrain.

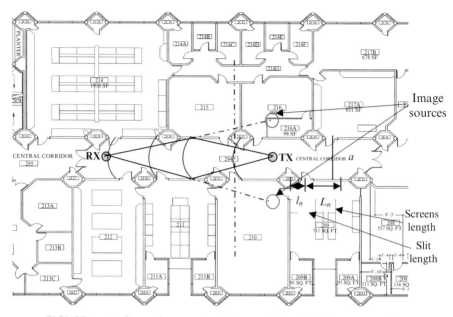

FIGURE 8.25. General geometry of the problem of indoor propagation.

8.5.2. Joint AOA-TOA Signal Power Distribution

Following results obtained in the previous sections, we, first of all, define the discrete spectrum of the total signal power along the corridor following the statistical waveguide model described by Equation (8.52) and Equation (8.55), taking into account the geometry shown schematically in Figure 8.25, where a general view on indoor environment is presented (the details of this experiment will be presented in Section 8.6). For numerical computations presented below, we convert the formulas by using the real time $t = d/c\tau$, and will use the modified formula, which describes distribution of the normalized (with respect to signal power in free space) power in the joint azimuth-time domain along the corridor in LOS conditions versus distance d between the transmitter and the receiver antenna [70]:

$$\tilde{w} = \frac{W_{wg}(t, \varphi)}{W_0} = \frac{2(t - \cos\varphi)}{d(t^2 - 1)} \exp\left[-2\frac{|\ln\chi|}{a'(\varphi)} \cdot \frac{d(t^2 - 1)}{2(t - \cos\varphi)}\right]. \quad (8.51)$$

Following Reference 70, we can rearrange a general Equation (8.46) versus t and φ by introducing the normalized power in the joint azimuth-time domain inside any room in NLOS conditions versus distance d between the transmitter and the receiver antenna:

$$\tilde{w}(t, \varphi) = \frac{1}{2}\gamma_0 \cdot t \cdot d \cdot \sin^2\left(\frac{\alpha}{2}\right) \exp\{-\gamma_0 t d\} \frac{2(t - \cos\varphi)}{(t^2 - 1)} \exp\left[-2\frac{|\ln\chi|}{a'(\varphi)} \cdot \frac{d(t^2 - 1)}{2(t - \cos\varphi)}\right].$$

$$(8.52)$$

Here, as in the above sections, a is the corridor width. φ is the angle of multipath components of the total field in the horizontal plane (i.e., the azimuth) arriving at the receiver after multiple reflections from the walls of rooms lining the corridor, χ is the brokenness parameter along the corridor, W_0 is a signal power of the transmitted antenna, d is the distance between the T-R antennas, and α is the angle between lines to the receiver and to the transmitter antennas.

Numerical experiment for general scenario of indoor environment was carried out in Reference 70 for two scenarios occurring in the indoor scene: (a) propagation along the corridor in LOS conditions, and (b) propagation inside rooms lining and crossing the corridor (see Fig. 8.25).

Propagation along the Corridor in LOS Conditions. The corresponding 3-D and 2-D plane of signal power distribution were analyzed for two variants of signal power distribution: in the joint AOA-TOA domain for different ranges between the transmitter and receiver antenna (see Fig. 8.26), and in the joint TOA-distance domain for different parameters of the azimuth φ (see Fig. 8.27).

Thus, in Figure 8.26a–e, the TOA-AOA 3-D and 2-D distributions of $w(t,\varphi)$ for $d = 1, 7, 10$, and 17 m along the corridor at real time t and the azimuth φ are presented. It is clearly seen that with increase of the distance between the Tx and Rx antennas along the corridor the effect of multiple reflections becomes stronger, and the TD of each ray n-time reflected from the walls increases with the number of reflections n.

New effects of azimuth dependence is clearly seen in Figure 8.27a–c for 3-D and 2-D TOA-distance distributions, $w(t,d)$, for various parameters $\varphi = 0°, 10°, 20°$. The characteristic extremes were found for specific azimuth and distance along the corridor, according to scheme presented in Figure 8.24.

Propagation through the Rooms in NLOS Conditions. Let us now consider a second scenario, where signal propagation takes place inside rooms lining and crossing the corridor. In this case, a numerical analysis of Equation (8.62) was done for the specific indoor area presented in Figure 8.28, where the recent experiments were carried out according to Reference 70. We do not enter now in details of the experiment; it will be done later. Now we will present only numerical results for specific radio paths shown in Figure 8.28.

In numerical computations we followed the conditions and parameters of the corresponding experiments, that is, we computed the received signal power distribution in the joint AOA-TOA domain for different angles of departure of the rotated transmitter antenna, $\varphi = 90°, 60°, 30°, 10°, 0°, -10°$. We sampled the received signal at deferent distances d from the transmitting antenna, and at each distance d we computed the received signal power distribution in 3-D AOA-distance domain. We plotted graphs of $w(t,d)$ in decibels for each $\varphi = 90°, 60°, 30°, 10°, 0°, -10°$, for $\chi = 0.9$, and $a = 2$ m, which correspond to conditions of the experiment (see the next section). In numerical experiment, the time of observation was changed from 0 to 3 seconds, and a range between

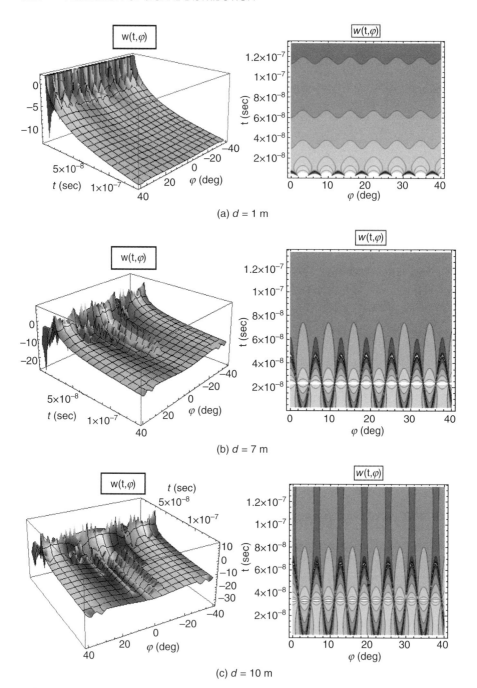

(a) $d = 1$ m

(b) $d = 7$ m

(c) $d = 10$ m

FIGURE 8.26. 3-D and 2-D distributions of $w(t,\varphi)$ along the corridor for (a) $d = 1$ m, (b) $d = 7$ m, (c) $d = 10$ m, and (d) $d = 17$ m at real time t and the azimuth φ.

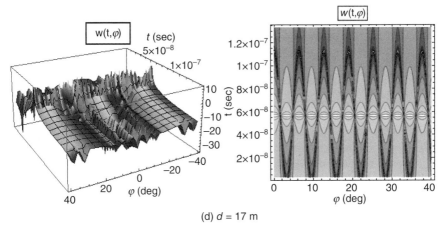

(d) $d = 17$ m

FIGURE 8.26. (*Continued*)

the transmitter (Tx) and the receiver (Rx) was changed up to 12 m for $\varphi = 90°$ (Fig. 8.29a), up to 22 m for $\varphi = 60°$ (Fig. 8.29b), up to 22 m for $\varphi = 30°$ (Fig. 8.29c), up to 33 m for $\varphi = 10°$ (Fig. 8.29d), and up to 29 m for $\varphi = 0°$ (Fig. 8.29e).

From illustrations presented in Figure 8.29a–e it is clearly seen that the power pattern distribution of radio signal passing a corridor and neighboring rooms strongly depend on the azimuth of multiray arrival. What is important to note here is that entering into the room, the energy of the signal falls down due to high absorption of room's walls, but the direction of the signal arrival (i.e., the AOA) at the receiving antenna is not essentially changed. This means that the rooms' walls attenuate the signal energy but do not change sufficiently the signal direction of arrival caused by its refraction by the material of walls. Moreover, as is also seen from presented illustrations, when radio signal penetrates the wall located at the various ranges from the transmitter, a sharp attenuation of 10–15 dB is observed with respect to propagation along the corridor in LOS conditions (as follows also comparing Figure 8.26a–e and Figure 8.29a–e).

8.6. EXPERIMENTAL VERIFICATION OF SIGNAL DISTRIBUTION IN AOA, EOA, AND TOA DOMAINS FOR OUTDOOR COMMUNICATIONS

Let us now verify a general stochastic model introduced as a combination of regular and nonregular distribution of buildings and streets in the city, using the experimental data obtained during the measurement campaign carried out in different urban and suburban environments.

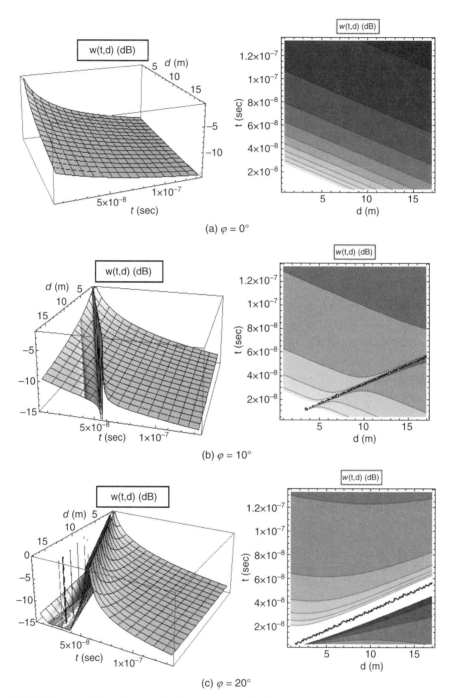

FIGURE 8.27. 3-D and 2-D distributions of $w(t,d)$ for $\varphi = 0°$ (a), $\varphi = 10°$ (b), and $\varphi = 20°$ (c) along the corridor versus the real time t and the range d.

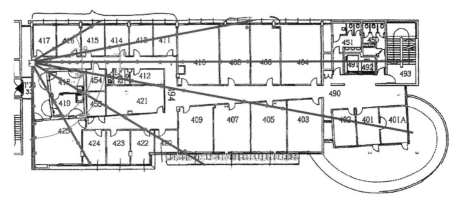

FIGURE 8.28. Topographic map of the experimental site.

8.6.1. Distance-TOA Signal Power Measurements in Tokyo City

The first cycle of experiments, concerning measurements of TD spread versus the signal power arrived at the receiver, was carried out in a typical densely built-up Japanese residential suburban area, all details of which can be found in Reference 65.

Figure 8.27 shows a map of the experimental area. Multipath data were obtained in most of the roads wide enough to allow the measurement vehicle to pass through. The total travel length for TD profile measurements was about 12.5 km, and the data were recorded at 1-m interval. The part of the route where the receiver input level has exceeded −91 dBm is depicted by the bold line in Figure 8.30, which corresponds to a propagation loss of 110 dB. As it follows from Figure 8.30, the bold line exactly presents the LOS area. Figure 8.31 shows a scatter plot of the receiver input power versus DS (mean over each 10 m). The line in Figure 8.31 presents the regression line, as a best fit of the cumulative experimental data obtained from all routs. The empirical expression of this line, obtained from experiments versus the received power P (in dBm), is as follows [65]:

$$\sigma_\tau[\text{ns}] = -56.0 - 1.49 \cdot P\,[\text{dBm}] \tag{8.53a}$$

or through the propagation loss L (in dB) is

$$\sigma_\tau[\text{ns}] = -83.6 + 1.49 \cdot L\,[\text{dB}]. \tag{8.53b}$$

Using now Equation (8.45) and Equation (8.46), we can present the corresponding signal power distribution in the TOA (or TD) domain in a general

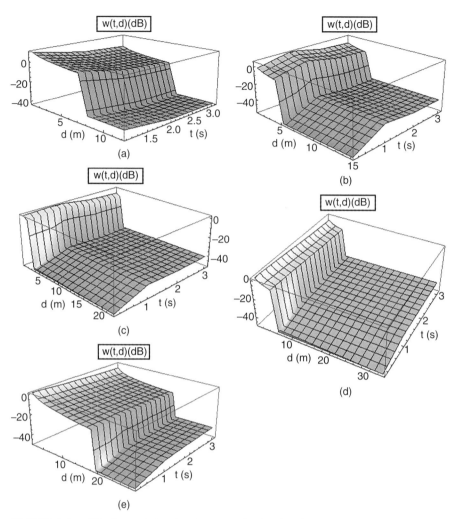

FIGURE 8.29. 3D-plane of the signal power distribution in the AOA-distance domain for $\chi = 0.9$ and (a) $\varphi = 90°$; (b) $\varphi = 60°$; (c) $\varphi = 30°$; (d) $\varphi = 10°$; and (e) $\varphi = 0°$.

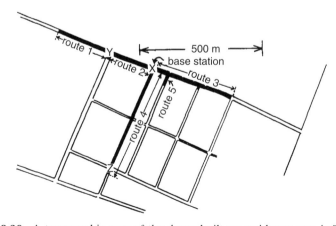

FIGURE 8.30. A topographic map of the dense built-up residence area in Tokyo city.

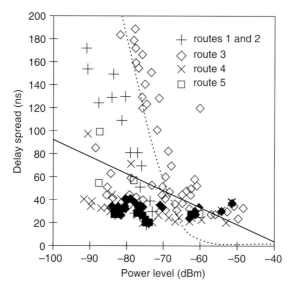

FIGURE 8.31. Scatter print of the measured data obtained in Reference 65 along the different routes (denoted by different signs), the empirically simulated data (continuous line), and computed for route 3 according to a general stochastic model (dotted line).

form accounting for the combined stochastic multiparametric and street-waveguide models:

$$W_{total}(\tau) = W(\tau) \cdot W_{wg}(\tau). \tag{8.54}$$

Thus, in Figure 8.31, as an example, a dotted line is presented according to numerical computations of Equation (8.52) along route 3. It can be seen that the theoretical predicting results shown by the dotted line coincides well with the measured data carried out along route 3 (the same good coincidence was obtained for intersection between route 1 and 2).

The second cycle of measurements was carried out in the dense urban area of Tokyo city near railway station. A map of the experimental area near the Tokyo Station is shown in Figure 8.32, where the bold lines indicate the text routes. Again, the straight line in Figure 8.33 presents a best fit of the cumulative experimental data obtained from all routes. The empirical expression of this line, obtained from experiments, versus the received power P (in dBm), is as follows [65]:

$$\sigma_\tau[\text{ns}] = -165.8 - 5.793 \cdot P[\text{dBm}] \tag{8.55a}$$

or through the propagation loss L (in dB) is

$$\sigma_\tau[\text{ns}] = -285.1 + 5.793 \cdot L[\text{dB}]. \tag{8.55b}$$

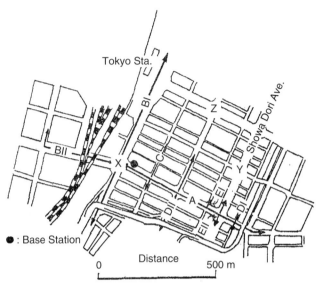

FIGURE 8.32. A topographic map of the dense built-up area near the Tokyo Railway Station.

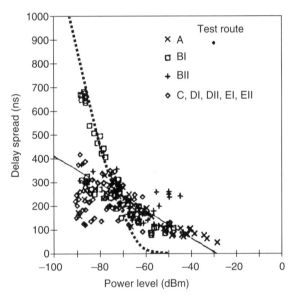

FIGURE 8.33. Scatter print of the measured data obtained in Reference 65 along the different routes (denoted by different signs), the empirically simulated data (continuous line), and computed for route BI according to a general stochastic model (dotted line).

The corresponding regression straight line is shown in Figure 8.33 passing through the scatter plot of experimental data obtained along different routes. In Figure 8.33, we present again numerical computations carried out in Reference 65 in the form of straight dotted line which corresponds to route BI. We find from Figure 8.33 that the theoretical results shown by dotted line coincides well with measurements carried out along route BI (the same good agreement was found separately along routes BII and A) [65]. Above comparison with experimental data allows us to suggest that the proposed general stochastic model might be a good predictor for the behavior of signal power distribution in the TD domain in the LOS, quasi-LOS, and NLOS conditions.

8.6.2. Joint Azimuth, Elevation, and Time Delay Measurements in Helsinki

The first cycle of experiments were carried out in downtown Helsinki [9, 10], a wide band single-channel sounder with RF-switch has been used to perform spatial channel measure. Built-up environment around the experimental area is shown in Figure 8.7 for the cases when the BS antenna is *below than building rooftops*. The receiving antenna was installed on the rooftop level (left low corner on the map). The transmitting antenna is placed in such a way that there is no LOS between the receiver and the transmitter (Tx11 position on Fig. 8.7). Parameters of built-up terrain have been calculated on the basis of detailed maps shown in Figure 8.7.

AOA-TD Measurement Results Affected by Antenna Pattern. The results of simulation based on a new model described by general Equation (8.54) are shown on Figure 8.34. They repeat the results of measurements of area shown

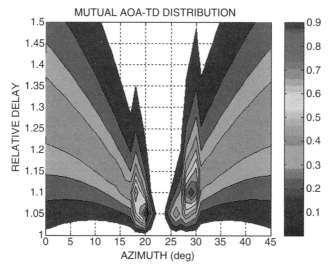

FIGURE 8.34. Simulation results of the multiparametric model that takes into consideration the guiding street effects: the antenna is below the rooftop level.

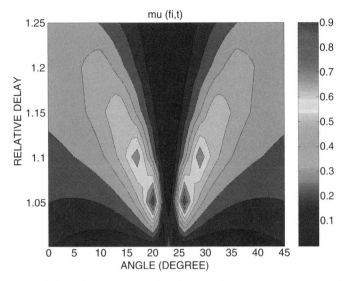

FIGURE 8.35. Simulation results of the multiparametric model only.

in Figure 8.7, which are presented in Figure 8.8. The guiding effect of two streets shown in Figure 8.8 is clearly seen from both measured data and results of computations. A model strictly predicts the maximums and the distribution of signal AOA and TOA (or TD) distributions. At the same time, results obtained according to a single stochastic model, described by (8.50) (see Fig. 8.35), are not in the precise agreement with the measured data seen in Figure 8.8.

The same results were obtained by computation of joint signal power spectrum distribution in the AOA-TD plane, defined by (8.56), where a combination of a single stochastic and a single waveguide model was taken into account. The results of numerical simulation for the experimental site depicted in Figure 8.7 are shown in Figure 8.36a,b for 2-D case and 3-D case, respectively. The illustrations of numerical computations, presented by these figures, also showed that the influence of scatterers surrounding LOS direction is more significant (see Fig. 8.7) on the received field distribution than the direct LOS arrivals (called *pseudo-LOS* components).

Nevertheless, Figure 8.34 and 8.36a,b, show that despite the fact that the dominant number of rays (more than 80%) arrive with the guiding effect of streets, there are number of arrivals (about 15–20%), which are coming from different directions with symmetry to the pseudo-LOS azimuth. In addition, the maximal number of arrivals has minimal TD and is coming closest to the pseudo-LOS direction street canyons. This result is clearly understood by use of the analysis of Equation (8.54) and Equation (8.56) for the cases of Tx and Rx antennas that are below the rooftop level.

Let us now describe the second cycle of experiments in area of downtown Helsinki that is depicted in Figure 8.9, where a BS antenna is located *at the*

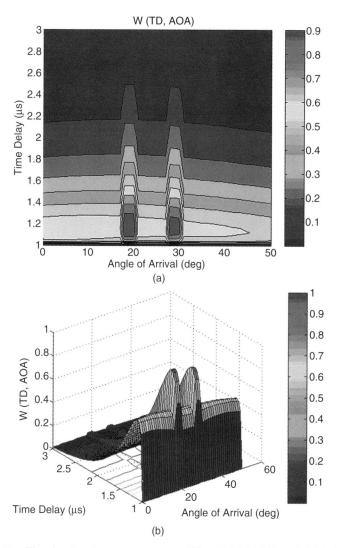

FIGURE 8.36. The simulated power spectrum, $W(\tau,\varphi)$, (a) in 2-D and (b) in 3-D plans.

same level as the rooftops. Figure 8.37 shows the resulting image of simulations of joint PDF of AOA-TOA distribution according to Equation (8.54) for the case of experiment presented in Figure 8.9. Figure 8.38a and Figure 8.20b show results of computations of joint TOA-AOA signal power spectrum distribution according to Equation (8.56) for 2-D plane and 3-D plane, respectively. Once again, a good agreement between measured data presented by Figure 8.10 and simulation results is observed. It was also obtained on the basis of results of simulation; the number of scatterers, influencing on the number of arrivals from Tx direction, exponentially reduces with increase of azimuth and TD. This

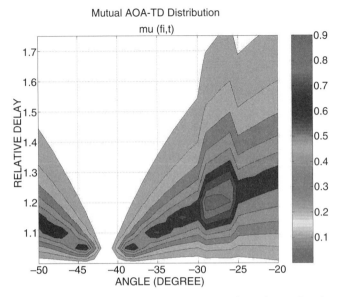

FIGURE 8.37. Simulation results: The antenna is at the rooftop level.

result was obtained experimentally in References 1, 4, 8, 9–11, and 20 without any satisfactory theoretical explanation of the influence of urban street orientation on the results of field distribution at the receiver.

At the same time, we must note here that the proposed modified stochastic model described by (8.54) and (8.56), presented only in the azimuth plane, cannot show properly two groups of arrivals as in Figure 8.10, characterized by common azimuth range from $-25°$ to $-30°$ and different TD (see comparison theory with experiment shown in Figure 8.39). Each of the two groups describes different ray travel distance, when all arrivals in the group have the same route. The proposed combined stochastic model takes in account consolidation of arrivals across the waveguides, assuming straight-crossing-street-grid plan. The urban scene, presented in Figure 8.9, does not support our basic assumption completely, and that is the reason for the differences mentioned earlier. Using only azimuth plane, the proposed model cannot describe sharp fluctuations at elevation domain. Owing to this reason, we cannot see matching for the arrivals from the high tower (Hotel Torni), which is measured in References 9 and 10 and presented in Figure 8.39. The reason is due to our basic assumption that the building height distribution is homogenous between h_1 and h_2, which are the minimum and maximum building heights in the area of investigation. Therefore, we start now to investigate signal distribution in the elevation plane, following results obtained in References 67–69.

Joint AOA-EOA Measurement Results Affected by Antenna Pattern. When we consider a smart antenna, it is important to understand that we deal with a

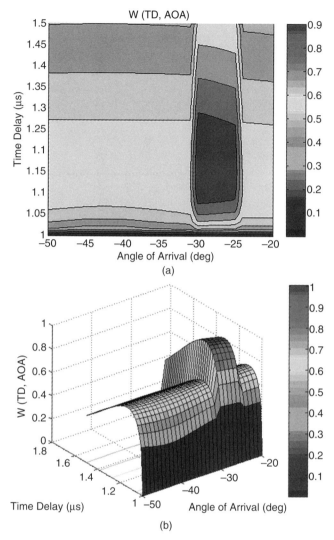

FIGURE 8.38. The simulated power spectrum, $W(\tau,\varphi)$, (a) in 2-D and (b) in 3-D planes.

directional, and not with an omnidirectional or isotropic antenna radiation pattern. Scanning the literature, we have found that the well-known von Mises PDF satisfies our requirements to describe the behavior of the directional antenna. Von Mises introduced this PDF in 1918 to study the deviations of measured atomic weights [63]. Recently, this distribution was introduced for statistical modeling and analysis of angular variables [71, 72]. We analyzed all antenna directivity and tilt effects by assuming a priori omnidirectional antenna pattern in the AOA or azimuth domain and a variable antenna pattern, according to von Mises distribution, in the elevation-of-arrival (EOA) domain.

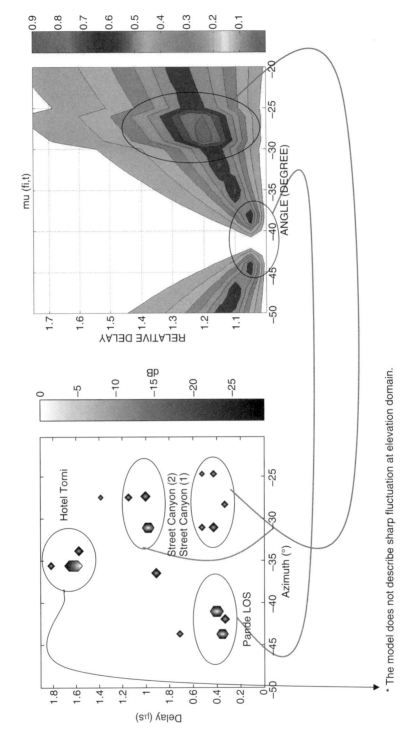

FIGURE 8.39. Comparison with experiments in References 9 and 10.

* The model does not describe sharp fluctuation at elevation domain.

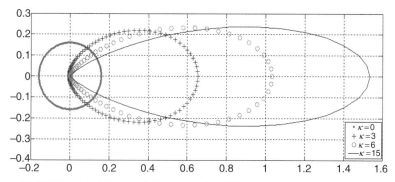

FIGURE 8.40. The von Mises PDF for different antenna directivities.

Let the variable θ represent the elevation angle and β represent antenna tilt ($\beta < 0$ corresponds to direction of tilt downward, $\beta > 0$ is for direction of tilt upward). The von Mises distribution is defined as [67–69]:

$$p(\theta) = \frac{1}{2\pi I_0(\kappa)} \exp[\kappa \cos(\theta - \beta)], \tag{8.56}$$

where $I_0(\kappa)$ is the zero-order modified Bessel function, and κ can be regarded as an antenna directivity parameter. Particularly, the antenna directivity κ is one of the basic antenna parameters, and it was defined in Chapter 2. This general idea of antenna pattern introduced by von Mises PDF, is presented in Figure 8.40. The solid thick line relates to the omnidirectional isotropic pattern in the vertical domain with $\kappa = 0$ value; the (+), (o), and the thin solid lines relate to $\kappa = 3$, $\kappa = 6$, and $\kappa = 15$ values, respectively.

Various antenna patterns influence, in different manners, the AOA and TOA signal distribution at the receiver. To reflect this influence, let us examine the PDF, $p(\theta)$, from (8.56) and the PDF, $\mu_{fin}(\tau, \varphi)$, from (8.44). These two distributions are independent, as was shown in References 67–69; therefore, their product can be written as:

$$\mu_{total}(\tau, \varphi, \theta) = \mu_{fin}(\tau, \varphi) \cdot p(\theta). \tag{8.57}$$

The obtained Equation (8.57) is a general description of the adaptive antenna pattern in the vertical and horizontal planes, and fully presents the AOA, EOA, and TOA joint distribution of signal affected by the array of obstructions surrounding the BS-MS terminal antennas.

The corresponding signal intensity distribution in the EOA domain can be easily obtained, taking into account the same derivation algorithm as was done for AOA and TOA domains, accounting for their mutual independency. Following the same procedure of derivation mentioned in References 68 and 69, we get

$$W(\theta) = \frac{W_0}{2\pi I_0(\kappa)} \exp[\kappa\cos(\theta - \beta)], \tag{8.58}$$

where, as above, W_0 is the signal power of the isotropic antenna in LOS conditions. Equation (8.58), combined with Equation (8.25), or with Equation (8.36), describes the joint signal power spectrum in EOA-AOA and EOA-TD plane, respectively.

The results of joint AOA-EOA distribution measured in the urban scene, depicted in Figure 8.7, are presented in Figure 8.41. It was measured that the arrivals were coming homogeneously over elevation angles of $-4°$ to $+2°$ with concentration of rays near the street canyons caused by guiding effect. Figure 8.42 shows results of simulation of the general PDF (8.57), accounting for

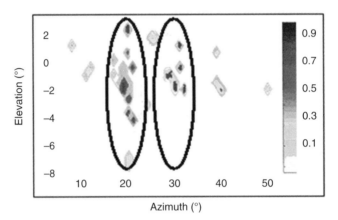

FIGURE 8.41. Measurement results: The antenna is below the rooftop level (EOA-AOA)

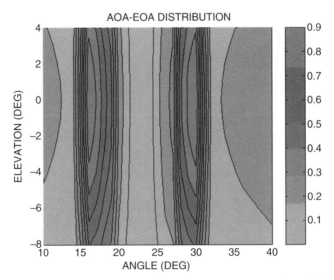

FIGURE 8.42. Simulation results of the joint PDF of AOA-EOA distribution.

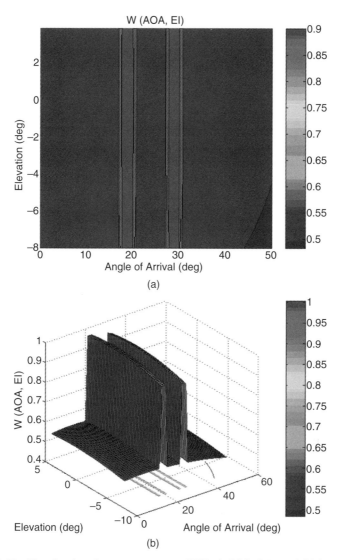

FIGURE 8.43. The simulated power spectrum, $W(\theta, \varphi)$, (a) in 2-D and (b) in 3-D planes.

(8.44) and (8.56), and Figure 8.43a,b shows results of simulation of the total signal power spectrum, as combination of (8.46) and (8.58), in the EOA-AOA plane for the urban experiment scenario depicted in Figure 8.7.

As above for AOA-TOA joint distribution, here we also obtained an agreement between measurement and simulation results in the AOA-EOA plane. Simulation results as well as experimental data show the guiding effect from two parallel streets presented in Figure 8.7 and Figure 8.8. Furthermore, they show that the maximum number of arrivals is concentrated around zero elevation angles. This result explains the over-rooftop-propagation effect observed during experiments (see Fig. 8.41 and References 9 and 10). It means that the buildings, which are placed in close proximity to the receiver (Rx), define the elevation AOAs at the Rx antenna. In addition, a significant number of arrivals spread around elevation angle of zero, when there is the guiding effect at the azimuth domain. This result explains the multiple reflections of rays passing through the street canyon. So, we can fully explain the guiding effect of streets and concentration of rays near the elevation angles of zero by using a modified stochastic approach which combines the two models, the multiparametric and waveguide, by using corresponding combinations of Equation (8.32), Equation (8.44), Equation (8.54), Equation (8.56), and Equation (8.58), depending on what plane, azimuth, elevation, or TD we need to investigate.

Better results of measurements of joint EOA-AOA distributions were obtained in the experimental site, where antenna was located at the rooftops level (see Fig. 8.9). Thus, results of measurements of joint EOA-AOA signal power distribution are shown in Figure 8.44. The corresponding simulations of joint PDF of EOA-AOA distribution and those for normalized signal power (to signal power along LOS direction) are presented in Figure 8.45, and in Figure 8.46a (2-D case) and b (3-D case), respectively. These results explain the over-rooftop propagation (in our examples Rx antenna is at the rooftop level) experimentally observed in References 9 and 10. It means that the height of the buildings, which are placed in the closest receiver area, define the elevation angles of the arrival at the Rx antenna. In addition, a significant number of arrivals spread around the elevation angle of zero are observed, when there is the guiding effect in the azimuth domain. The guiding effect may

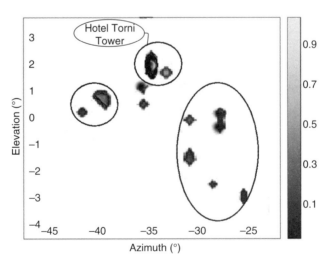

FIGURE 8.44. Measured data in the AOA-EOA plane [9,10]. (Reprinted with permission @ 2002 IEEE.)

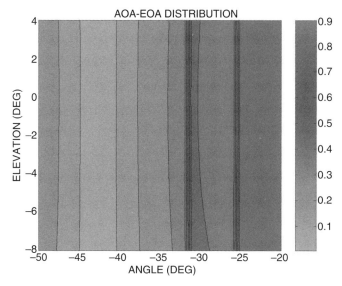

FIGURE 8.45. Joint PDF distribution in AOA-EOA plane.

be explained by the multiple reflections of rays passing through the street canyon, as observed in References 9 and 10. It is clearly seen from Figure 8.46a,b that the main signal energy arrives at the receiver after diffraction caused by the right-side buildings (with respect to Tx), localized at the azimuth range of $20°$–$30°$ (for whole range of elevation angles of $0° \pm 10°$). These results are in a good agreement with measurements [9, 10], according to which the effects of diffraction from the building blocks located in the right side (from Tx) are predominant. The distance between these buildings is about 350–400 m from Rx2, which corresponds to TD spread of 1.2–1.3 μs. These buildings are located $-25°$–$30°$ from location of the transmitter in the azimuth plane (see Fig. 8.9).

8.6.3. TOA-Distance and AOA-Distance Measurements in Aalborg University

Let us introduce the AOA measurements made at Aalborg University, Denmark, an urban scene which is shown in Figure 8.47 according to Reference 52. The area is composed of several buildings connected by glass-covered bridges and walkways. The buildings are two-story high, about 8 m. The Tx antenna is omnidirectional in the azimuth plane, and it is located 4 m above the ground. The Rx antenna has 3-dB beamwidth in both polarizations, and it is located on the ground level. There is no LOS between the Tx and Rx antennas. The Tx–Rx distance is about 100 m. The measurements were made using 80-MHz wideband channel sounder at a carrier frequency of 1.845 GHz. The

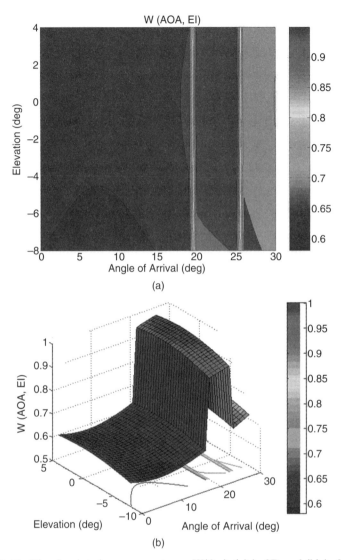

FIGURE 8.46. The simulated power spectrum, $W(\theta, \varphi)$, (a) in 2D and (b) in 3D planes.

following parameters, extracted from Figure 8.47, describe the urban scene. The parameters $\gamma_0 = 3$, $\nu = 50$, $\lambda = 0.16$, $\chi = 0.5$, and $d = 0.1$ define the input parameters for numerical simulation of the stochastic model described by Equation (8.50), Equation (8.51), Equation (8.52), Equation (8.53), Equation (8.54), Equation (8.55), and Equation -(8.56).

Azimuth-Time Delay Signal Power Distribution. Let us now analyze and compare the theoretical approach proposed earlier according to Equation

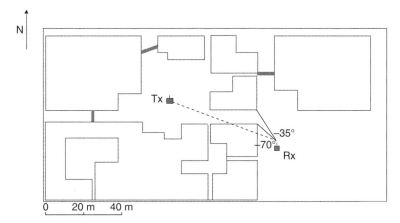

FIGURE 8.47. Aalborg University as a microcell environment. (Source: Reference 52. Reprinted with permission @ 2001 IEEE.)

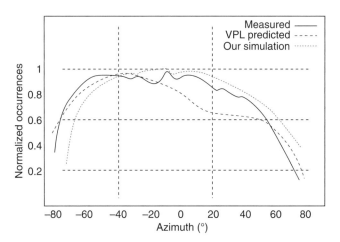

FIGURE 8.48. VPL model and measured data compared with the stochastic model prediction.

(8.50), Equation (8.51), Equation (8.52), Equation (8.53), Equation (8.54), Equation (8.55), and Equation (8.56) with the results of the VPL ray-tracing algorithm [46, 47], which where compared with AOA measurements made at Aalborg University, Denmark [52]. We also compare our results with those from Reference 52. The results of simulation presented in Figure 8.48 compared against measurement results [52] and against the VPL model prediction [46, 47]. There is a good fit between the measured VPL predicted and the proposed model-simulated results. However, the stochastic model shows better agreement achieved in the azimuth range of 0° to 40°, where VPL prediction results differ from the measured data. Thus, the difference between measurements and VPL model prediction can achieve in this range of up to 30% accuracy, whereas the stochastic model predicts with accuracy 95% of the experimental data. Comparing the VPL model, the stochastic model simulation, and the measured data results, we have found that the stochastic model gives better prediction over the entire range of the considered points (from −80° to +80°). The mean error, with respect to the measured data, equals 0.16 for our model simulation and 0.3 for the results based on VPL technique; that is, our model in these ranges is twice as precise.

Next, we present some important simulation results that define the joint distribution of AOA and TOA (or TD) at the receiver. We analyze the distribution using the results presented separately for AOA in References 9 and 10, and using the built-up terrain data presented in Reference 52. In Figure 8.49, from the concentration of rays near the direction of the two streets, which are placed between -40° and 0°, according to map shown in Figure 8.47, we can see the existence of a strong street guide effect. This effect plays a significant role for the case when both the Tx and Rx antennas are below the rooftop level (as was mentioned in References 9 and 10, more than 80% of rays arriving at the receiver antenna are generated by the guiding effect of straight streets). It is interesting also to note that there are ray arrivals on the azimuth of −70° (pseudo- LOS direction), and these arrivals have the normalized energy (compared with free space) of about 0.9. Apparently, these pseudo-LOS arrivals present over rooftop diffracted components with low TD and narrow angular

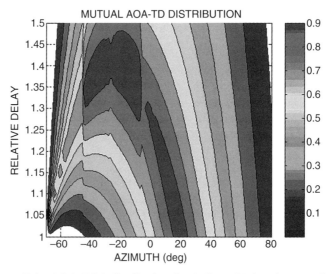

FIGURE 8.49. Joint AOA-TOA distribution for Aalborg University experimental site.

dispersion, but they do not contribute significantly to the Rx received power, as shown in Figure 8.47.

This fact states again that all specific features of the built-up terrain and situations with antennas are important and can change significantly the spatial and temporal characteristics of the signal arriving at the antenna. These effects are very important for predicting effective operational characteristics of smart wireless systems (using adaptive antennas).

Azimuth-Distance Signal Dependence. Let us now analyze the behavior of expressions (8.40) to (8.46), obtained by the proposed stochastic approach, using the conclusions of experiments from References 4, 9, 10, 46, 52, 56, and 57. It was shown in References 56 and 57 that the AS depends on the distance between the Tx and Rx antenna in a specific manner. We show here a similar dependence described by (8.44). In References 46, 52, 56, and 57, the angular spread is reduced when the distance is increased for certain situations.

We analyzed two cases to confirm the empirical results obtained in References 56 and 57. In the first case, which is shown in Figure 8.50, there is no street guide effect because the street plan grid does not have a rectangular form. In the second case shown in Figure 8.51, there is a straight-cross-street grid. The pseudo-LOS azimuth between Tx–Rx antennas is about 23°. Different lines depict different distances between them. A curve denoted by "*" is for the distance 0.2 km. The "o" line is for 0.6-km, and the "+" line is for the 2-km distance between the antennas. Here, only the distance parameter is varied in Equation (8.54), Equation (8.55), and Equation (8.56) to investigate its influence. Other parameters have typical values for urban scene:

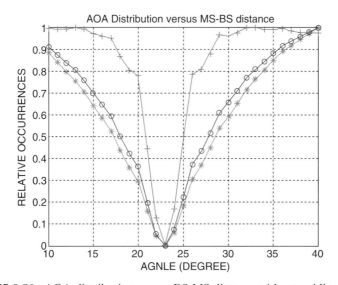

FIGURE 8.50. AOA distribution versus BS-MS distance without guiding effect.

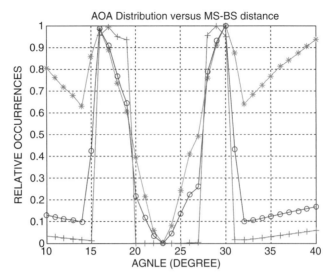

FIGURE 8.51. AOA distribution versus BS-MS distance accounting waveguide effect.

$\gamma_0 = 4 \ km^{-1}(\nu = 120$ buildings per km^2 and $\bar{\rho} = 250 \ m$), $\lambda = 0.13 \ m$ a $\chi = 0.5$.

Figure 8.50 and Figure 8.51 show that the angular spread is decreased with an increase in separation distance between the antennas. Figure 8.51 shows an interesting phenomenon: As the distance between the MS and BS antennas increases, the street-guide effect becomes more dominant. It is intuitively obvious, when the MS is far enough from the BS, that the probability of arrival of multiple reflected rays across randomly distributed scatterers is low. The same results were observed in Helsinki [9, 10] for the urban scene with a grid-street plan.

8.7. EXPERIMENTAL VERIFICATION OF SIGNAL DISTRIBUTION IN JOINT AOA-TOA AND TOA-DISTANCE DOMAINS IN INDOOR COMMUNICATIONS

Several series of measurements have been carried recently and described in References 73–76 in the frame of joint collaboration between the Ben-Gurion (Israel) and Massachusetts (USA) Universities, and the corresponding comparison with theoretical approach proposed in References 64 and 67 was done. Results of theoretical and experimental work described in References 73–76 were compared with those obtained in References 77–89.

The series of experiments were carried recently at the University of Massachusetts (USA) in the frame of joint experimental and theoretical investigations between two groups of researchers from the USA and Israel. All

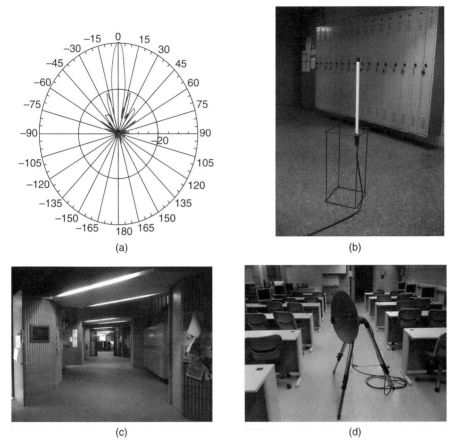

FIGURE 8.52. (a) Parabolic antenna beam pattern in polar coordinates: azimuth (in degrees) versus loss (in decibels), (b) monopole vertical-directed Tx antenna, (c) typical straight corridor LOS conditions between Tx and Rx antennas, (d) parabolic Rx antenna in a typical indoor room in NLOS conditions with respect to the Tx antenna.

measurements were conducted in the frequency domain following the procedure in References 21–23. The special vector network analyzer (VNA) was used to measure the 801 discrete values in the frequency domain with a center frequency of 5.8 GHz and a bandwidth of 200 MHz. The VNA was calibrated before each measurement set with all connected components. In Figure 8.52a,b,d, the pattern of the vertical-oriented directional monopole antennas with a gain of 10 dBi and beamwidth of 8° in elevation (Figure 8.52a,b), and highly directive parabolic antennas with a gain of 27 dBi and horizontal and vertical beamwidth of 4° (Figure 8.52d) are shown. The typical indoor LOS and NLOS environments are presented in Figure 8.52c and d, respectively. This means that the measurements were collected in three sites: in the first two sites, measurements were collected in straight corridors of 1.3 m and 3 m width characterized by the LOS propagation conditions (see Fig. 8.52c). In the third site, measurements were collected in the complex NLOS scenario with propagation between rooms and the corridor (see Fig. 8.52d).

We will not present here all experiments accumulating the reader's attention on the joint AOA-TOA and TOA-distance domains, as was done in previous sections for outdoor communication links. At the same time we should notice that so-called elevation domain signal distribution in indoor communication links is not so actual because of low-elevated antennas and low heights of the corridors and rooms.

The accuracy of the proposed statistical indoor radio propagation model (see Equation (8.51) and Equation (8.52)) is evaluated using the collected measurements in the following scenarios: (1) the straight-corridor LOS scenario where the Tx and the Rx stations are equipped with directive parabolic antennas (see Fig. 8.58c), (2) the NLOS scenario where the Tx is located inside the room and the Rx is located at various positions along the corridor (see Fig. 8.52d). The accuracy of the proposed statistical model was also evaluated via specular reflection analysis of the DS characteristics measured in two corridors with 2-m and 3-m width, respectively.

8.7.1. Verification of the Predicting Model in the Along-Corridor Scenario

The measurements in the AOA-TOA domain were collected using parabolic antennas at both the transmitter and the receiver. The measurements of the TOA were obtained similarly to the DS measurements. The TOA measurements were obtained using the directionality of the parabolic antennas that provides measurement accuracy of $4°$, determined by the antenna horizontal beamwidth. The direct path between the Rx and the Tx along the corridor was used as a reference line in scenarios with the LOS propagation conditions, and the line between the Rx and the nearest opening to the space where the Tx is located was used as a reference line in scenarios with the NLOS propagation conditions. For every tested Rx–Tx distance, the orientation of the parabolic antennas was changed in steps of $5°$ within the range from $-45°$ to $45°$ with respect to the reference line. The measurements were collected at 19 azimuth orientations of the Tx and the Rx antennas, providing $19 \times 19 = 361$ total measurements for every Rx–Tx scenario. The channel impulse response at every Rx orientation was obtained as a sum of impulse responses for 19 tested Tx orientations. Figure 8.53 shows the representative example of the collected experimental data in the joint AOA-TOA domain obtained via signal power density plot as a function of the Rx azimuth.

Specular Reflections along the Corridor. First, the data that was collected in the straight corridor scenario is used to evaluate the ability of the proposed statistical model to predict the specular reflection. Two types of corridors with width of $a = 1.5$ m and $a = 3$ m, and different waveguide brokenness properties were considered. The measurements in the AOA domain that were induced

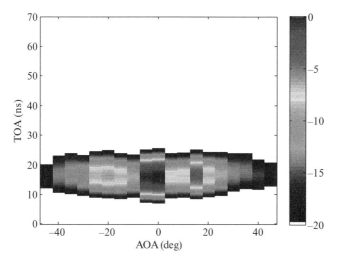

FIGURE 8.53. Example of the collected measurements in the AOA-TOA domain in the scenario of corridor with 3-m width in LOS conditions. The color bar represents normalized signal intensity in decibels.

TABLE 8.1. Specular Reflection Angles Obtained Using the Statistical Model for the Corridor of $a = 1.5$-m width

Distance (m)/Modes	$n = 1$	$n = 2$	$n = 3$
3	26.6°	45°	56.3°
5	16.7°	31°	42°
9.4	9.1°	17.7°	25.6°

TABLE 8.2. Specular reflection angles obtained using the statistical model for the corridor of $a = 3$-m width.

Distance (m)/Modes	$n = 1$	$n = 2$	$n = 3$
7	23.2°	40.6°	52.1°
11	15.3°	28.6°	39.3°

by the specular reflection from the corridor walls were collected. Considering distances from 3.5 to 9.4 m between the Tx and the Rx, and considering propagation of only three significant modes $n = 1, 2, 3$, the specular reflection angles were obtained using stochastic indoor waveguide model proposed in the previous sections. Table 8.1 and Table 8.2 show the angles of specular reflection in the two considered corridors, respectively.

Figure 8.54a,b and Figure 8.55a–d show the collected measurements of the specular reflection that correspond to results in Table 8.1 and Table 8.2,

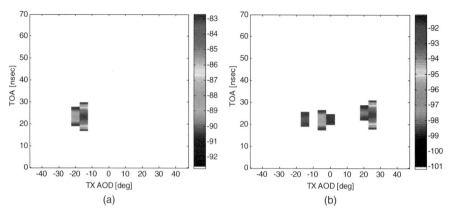

FIGURE 8.54. Specular reflection analysis using collected measurements in AOA-TOA domain in a scenario with straight corridor of $a = 1.5$ m width with Rx-Tx distance of $d = 5$ m and two significant propagation modes: (a) $n = 1$, AOA $= 15°$, and (b) $n = 2$, AOA $= 30°$. The color bar represents normalized signal intensity in [dB].

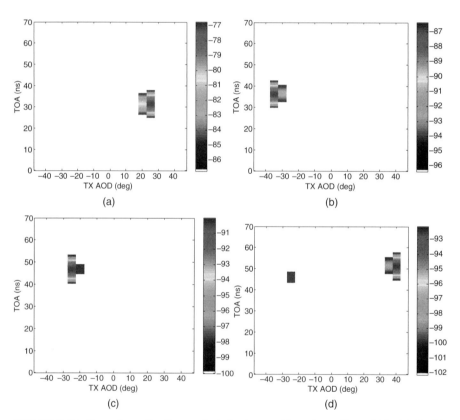

FIGURE 8.55. Specular reflection analysis in the corridor of $a = 3$ m width: (a) Rx-Tx distance of 7 m, $n = 1$, AOA $= 20°$; (b) Rx-Tx distance of 7 m, $n = 2$, AOA $= -40°$; (c) Rx-Tx distance of 11 m, $n = 2$, AOA $= -30°$; (d) Rx-Tx distance of 11 m, $n = 3$, AOA $= -40°$. The color bar represents normalized signal intensity in dBm.

respectively. Subplots (a) and (b) in Figure 8.54 show the collected measurements for two significant modes, $n = 1$ and $n = 2$, respectively, that correspond to results in Table 8.2. Figure 8.54 shows that, in the scenario where the roughness of the corridor walls is low compared with the wavelength of the transmitted signal (see Fig. 8.54c), the considered scenario can be modeled as propagation through a corridor with completely flat walls, which is characterized by specular reflection. Thus, as was mentioned in Chapter 5, the low roughness of surfaces is defined as a critical height of surface protuberances, $h_c = \lambda/8\cos\theta_i$, where λ is a wavelength and θ_i is angle of incidence. For the wavelength of about 5 cm in the indoor environment considered [73], the critical height h_c varies in a range between 1 cm for an AOA of 40° and 4 cm for an AOA of 10°, whereas the physical roughness of the walls lining the corridor was between 1 and 2 cm.

Subplot (a) in Figure 8.55 shows the collected measurements for mode $n = 1$, subplots (b) and (c) show the collected measurements for mode $n = 2$ and subplot (d) shows the collected measurements for mode $n = 3$. As it follows from the illustrations presented in Figure 8.54 and Figure 8.55, the larger AS of AOA measurements occurs in wider corridor scenarios.

Now we will analyze the same two scenarios with the corridor widths of $a = 1.5$ m and $a = 3$ m numerically based on the corresponding stochastic corridor-waveguide model described by Equation (8.51) and compare them with the measured data. Thus, Figure 8.56 shows the distribution of the received signal intensity in the AOA-TOA domain in scenario with corridor width of $a = 1.5$ m. It shows a strong dependence of AS on the Rx–Tx distance within a sector between −45° and 45°. This phenomenon can be explained by the relation between the Rx–Tx distance and the corridor width. Note that the waveguide modes with the lower number n that experience smaller numbers of reflections have higher intensity and lower AOA-TOA (they are concentrated around direct path in azimuth and in delay that corresponds to this direct path).

This result is predicted by (8.51) where the relation between the field intensity at each AOA and the Rx–Tx distance was established. This analysis clearly demonstrates that the collected measurements contain distinct first and second wave-mode reflections. The calculated discrete spectrum of the average specular field intensity is presented by filled circles. It can be seen that Figure 8.56 illustrates an accurate fit between the result predicted by the proposed statistical model and the collected measurements. Note that the calculated spectrum in the AOA-TOA domain is discrete due to the specular reflection consideration. The mismatch between the measured and the theoretically predicted data was evaluated via the mean square error (MSE) criterion and equals 2.1 dB with standard error deviation of 5 dB which are lower with respect to data obtained by other models [21–23].

Next, Figure 8.57a–c shows the distribution of the received signal intensity in the AOA-TOA domain in the scenario with corridor a width of $a = 3$ m. In this scenario, the MSE of the calculated distribution of the received signal

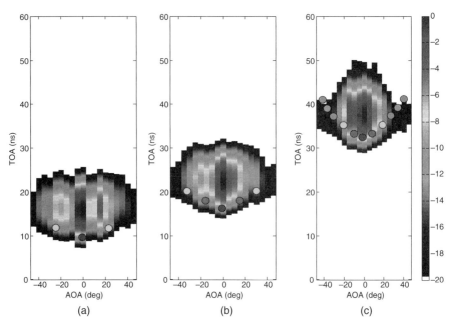

FIGURE 8.56. Collected and calculated intensity distribution in the AOA-TOA domain in the corridor of $a = 1.5$-m width with (a) Rx-Tx distance of 3 m, (b) Rx-Tx distance of 5 m, and (c) Rx-Tx distance of 9.5 m. The color bar represents normalized signal intensity in decibels.

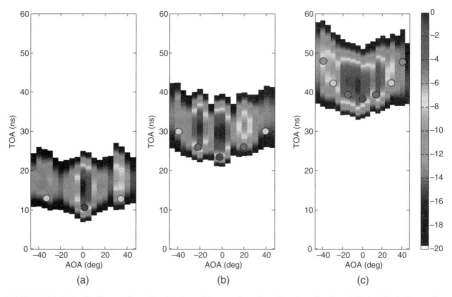

FIGURE 8.57. Collected and calculated intensity distribution in the AOA-TOA domain in the corridor of $a = 3$-m width with (a) Rx-Tx distance of 4 m, (b) Rx-Tx distance of 7 m, and (c) Rx-Tx distance of 11 m. The color bar represents normalized signal intensity in decibels.

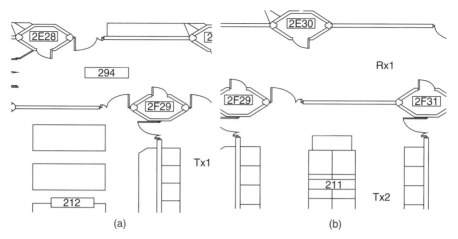

FIGURE 8.58. Floor plans in scenarios A (plot (a)) and B (plot (b)), respectively, for NLOS conditions between the transmitting (Tx) and receiving (Rx) antennas.

energy in AOA-TOA domain (values in the filled circles in Fig. 8.57) with respect to the collected measurements equals 2.3 dB with standard error deviation of 4.5 dB. Comparing data presented in Figure 8.56 and Figure 8.57, we can notice that at low mode numbers, the received energy is more concentrated around the zero-azimuth direction in the AOA domain in the scenario with a narrower corridor, and that propagation in the wider corridor induces the larger azimuth and TD spreads (see Fig. 8.57).

Verification of the Predicting Model in NLOS-Condition Scenarios. The ability of the proposed statistical model to represent the indoor characteristics in the NLOS propagation conditions is discussed next. The NLOS propagation conditions were created in the indoor scenario that consists of the corridor with adjacent rooms, as shown in Figure 8.58. Accounting for conditions mentioned earlier, we again consider small roughness of the concrete walls in the considered scenario. Subplots (a) and (b) in Figure 8.58 show the floor plans of two considered scenarios (denoted as A and B), where two adjacent rooms were selected for the experiment. In scenario A, the Tx was placed in one room and Rx in the corridor toward the east direction. In scenario B, the Tx was placed in the second room and the Rx in the corridor toward the west direction.

In Figure 8.59, the measured data for scenario A and B are presented by plots (a) and (b), and the data computed according to Equation (8.52) are by plots (c) and (d), respectively.

From the plots (a) and (b), shown in Figure 8.59, it is clearly seen that the majority of the transmitted energy propagates through the corridor waveguide and then from one room to another via openings, such as doors. Comparing plots presented by Figure 8.59a,c with plot (a) from Figure 8.58, and then plots presented by Figure 8.59b,d with plot (b) from Figure 8.58, we can notice that

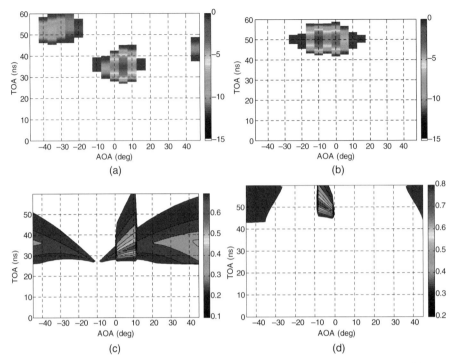

FIGURE 8.59. Plots (a) and (b) present collected measurement data, and plots (c) and (d) present calculated data in scenarios A and B, respectively. The color bar represents normalized signal intensity in decibels.

the AOA of the received energy corresponds to the location of the openings in the floor plan. Comparing plots (a) with (c), and (b) with (d) from Figure 8.65, we can notice a good fit between the collected measurements and the measurements predicted by the proposed statistical model.

TOA-Distance Signal Distribution along the Corridor. The influence of the Rx–Tx distance and the corridor width on the DS is analyzed in scenarios with LOS propagation conditions within straight corridors with different widths. The height of the considered corridor was $h_c = 4$ m, and the tested Rx–Tx distances were much shorter than the critical distance, L_c, where the reflections from the ceiling and the floor can affect pure waveguide propagation. This distance was estimated in the present experiments as $L_c \approx 28$ m [73, 74]. The influence of the front and the back walls in the corridor was eliminated by careful selection of the Rx and the Tx locations.

Figure 8.60a,b shows the measured and modeled DS in two considered scenarios with indoor corridors of different widths. In these scenarios, a phase-difference between the direct and the reflected waves which is mainly determined by the corridor width and the wall brokenness parameter is small and

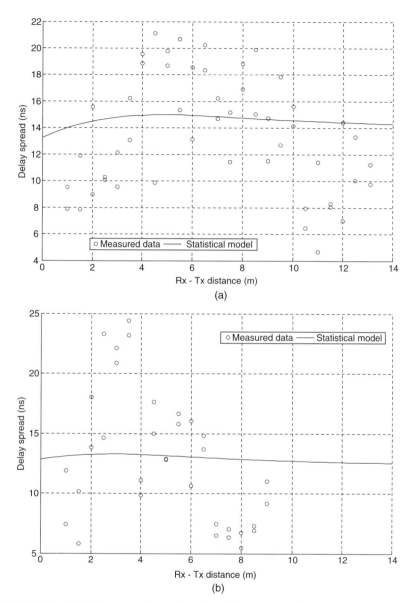

FIGURE 8.60. (a) Delay spread as a function of the Rx-Tx distance in the LOS propagation conditions along the corridor of $a = 3$ m width and with waveguide brokenness parameter $\chi = 0.9$. (b) The delay spread as a function of the Rx-Tx distance in the LOS propagation conditions along the corridor of $a = 1.5$ m width and with brokenness parameter $\chi = 0.9$.

approximately constant. Therefore, the DS is also small and approximately constant. The proposed statistical model represents a slight increase in DS for small Rx–Tx ranges. This behavior can be explained by the corridor geometry in scenarios where the Rx–Tx ranges are shorter than the corridor width a. In these scenarios, every small change in the Rx–Tx distance makes a significant change in the TOA of the direct propagation mode and the corresponding loss when compared with the reflected modes. In scenarios where the Rx–Tx distance is much larger than the corridor width, the above-described phenomenon is absent or become weak enough due to the similarity between TOAs of the direct and the reflected waves.

Finally, we can state that comparative analysis of measured and computed data presented in Figure 8.60a,b shows an accurate fit between the measured and predicted DS versus the distance between Tx and Rx along the corridors. Note that high variation in the measured DS can be explained by (1) limited resolution of measurement equipment in the time domain and (2) the fast fading phenomenon that induces high variations in the received power when the Rx motion is in steps that are comparable to the wavelength. In addition, we note that the measured and theoretically predicted DS fits to the results reported in References 86–89.

8.8. SIGNAL POWER SPECTRA DISTRIBUTION IN DOPPLER SPREAD DOMAIN

Now we will talk about the spectral characteristics of the received (or transmitted) signal in order to describe the Doppler spectrum for moving the subscriber (MS) antennas in various built-up areas.

8.8.1. Spatial Signal Distribution

Let us consider the spectral properties of the signal strength spatial variations. We can present the spectral function $\tilde{W}(q, \varphi_0)$ for two typical situations in the urban scene regarding the elevations of the two terminal antennas with respect to the average height of the buildings.

For the *first case*, when the transmitter/receiver is placed *at the rooftops level or below it*, that is, $z_1 \leq \bar{h}$ we get [58, 90]

$$\tilde{W}(X, \varphi_0) = \frac{2 \cdot sh\chi \cdot (ch\chi - X\cos\varphi_0)}{k[1-X^2]^{1/2}[(ch\chi - X\cos\varphi_0)^2 - (1-X^2)\sin^2\varphi_0]}, \quad (8.59)$$

where $X = p/k$, $X \in (-1, 1)$; p is the spatial wave vector, $k = 2\pi/\lambda$; $sh\chi$ and $ch\chi$ denote hyperbolic sine and cosine, respectively; and the parameter χ accounts for the density of buildings and the range between the antennas:

$$\chi = \ln\left[\left(1 + \frac{1}{\gamma_0 d}\right) - \left(\left(1 + \frac{1}{\gamma_0 d}\right)^2 - 1\right)^{1/2}\right]. \quad (8.60)$$

The dependence of $\tilde{W}(X, \varphi_0) / \tilde{W}(0, \varphi_0)$ as a function of the normalized parameter X for $\varphi_0 = \pi/2, \pi/3, \pi/4, \pi/\pi$ (curves 1, 2, 3, 4, respectively) for $\gamma_0 d = 10$ (built-up area with high density of buildings around the receiver and transmitter, or the terminals are placed far from each other), is presented in Figure 8.61a. The nonsymmetrical spectrum $\tilde{W}(X, \varphi_0)$ is observed for the orientation angle φ_0 deviations from $\pi/2$ (φ_0 is the angle between the scatterer and radio path between terminal antennas, see Fig. 8.15). This phenomenon can be

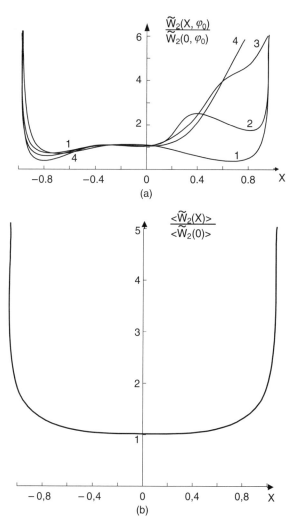

(a)

(b)

FIGURE 8.61. The dependence $\tilde{W}(X, \varphi_0) / \tilde{W}(0, \varphi_0)$ as a function of normalized parameter X (a) for $\varphi_0 = \dfrac{\pi}{2}, \dfrac{\pi}{3}, \dfrac{\pi}{4}, \dfrac{\pi}{6}$ (curves 1, 2, 3, 4) for $\gamma_0 d = 10$. (b) The same dependence of $<\tilde{W}(X)> / <\tilde{W}(0)>$, but for φ_0 is regularly distributed from 0 to 2π.

understood if we again return to the results of the analysis presented previously, where the PDF and the signal power AOA distributions are unified and symmetrical relative to the wave path, which is directed strictly to the transmitter. If we consider that the position of a segment of the scatterer relative to radio path is not fixed, and that it can be oriented with equal probability anywhere in space, that is, the angle φ_0 is the angle regularly distributed within the angle interval $[0, 2\pi]$, we immediately obtain the same case which is described in References 58 and 90. According to expressions obtained there, we get

$$<\tilde{W}(X)>=\frac{2}{X[1-X^2]^{1/2}}. \tag{8.61}$$

From this formula it follows that the spectrum of spatial frequencies for the case of $z_1 \leq \bar{h}$ does not depend on the range d between the BS and MS antennas and on the building contours density γ_0 at the plane $z = 0$. This result also follows from the dependence of $<\tilde{W}(X)>/<\tilde{W}(0)>$ shown in Figure 8.61b.

In the *second case*, when the transmitter/receiver is above the level of rooftops, that is, $z_1 > h$, an increase of the illumination area ($\bar{\rho}$-area) surrounding the receiver/transmitter is observed. In this case, the signal spectrum of normalized spatial frequencies ($X = q/k \in \{-1, 1\}$)can be determined as:

$$<\tilde{W}(X)>=\frac{2(1-X\cos\varphi_0)}{X[1-X^2]^{1/2}}. \tag{8.62}$$

In Figure 8.62, the dependence $\tilde{W}(X, \varphi_0)/\tilde{W}(0, \varphi_0)$ as a function of normalized parameter X for $\varphi_0 = \pi/2, \pi/3, \pi/4, \pi/6$ (curves 1, 2, 3, 4, respectively) for $\gamma_0 d = 10$ is shown. Only the slight deformation of the classical U-shape (shown in Figure 8.61b) is observed in the case of the high-elevated BS antenna.

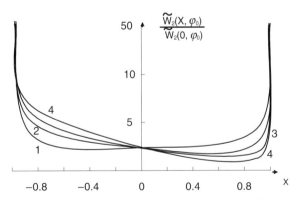

FIGURE 8.62. The same dependence $\tilde{W}(X, \varphi_0)/\tilde{W}(0, \varphi_0)$, as in Figure 8.61a, but calculated for the limited interval $q \in [-k, k]$. BS antenna is higher than the rooftops.

8.8.2. Signal Power Distribution in Doppler Shift Domain

In built-up areas, the spatial distribution of signal strength fully determines the properties of temporal signal distribution obtained at the receiver. As indicated in References 58 and 90, the energy spectrum of the signal temporal deviations, which are created by the interference of waves arriving at the receiver, relates to the spectrum of spatial variations of signal strength through the following relationship:

$$\tilde{W}_\tau(\tau, \varphi_0) = \frac{1}{v}\tilde{W}(\tfrac{\omega}{v}, \varphi_0). \tag{8.63}$$

Here, v is the velocity of the transmitter (or receiver) traveling from point B to point D (segment $|BD| = \ell$, see Fig. 8.15) during the time $\tau = \ell/v$, where the velocity v is related to the maximal Doppler frequency f_{dm} via the expression $v = f_{dm}c/f$; f is the radiated frequency; c is the speed of the light; φ_0 is the angle between the direction of movement of the mobile antenna and the direction from point B to point A (see Fig. 8.15). Using this relation between the signal spectra in the space and time domains, we can examine their distribution in various built-up areas with randomly distributed buildings.

Let us consider two typical situations in the urban scene. For the *first case* the stationary transmitter/receiver antenna is *at the roofs level or below it*, that is, $z_1 \le \bar{h}$. In this case one can obtain the following expression for the spectral function of signal temporal fluctuations [90]:

$$\tilde{W}_\tau(\omega, \varphi_0) = \frac{2 \cdot sh\chi \cdot \omega_d}{\sqrt{\omega_d^2 - \omega^2}} \frac{(\omega_d \cdot ch\chi - \omega \cdot \cos\varphi_0)}{\left[(\omega_d \cdot ch\chi - \omega \cdot \cos\varphi_0)^2 - (\omega_d^2 - \omega^2)\sin^2\varphi_0\right]}, \tag{8.64}$$

where $\omega_d = kv = \dfrac{v}{c}\omega = 2\pi f_{dm}$.

The frequency dependence of power spectrum, described by expression (8.64) is the same as the one presented in Figure 8.61a, that is, very complicated and depends on several of the built-up terrain factors, such as the parameter χ, the direction of moving vehicles (on φ_0) and their speed v (i.e., on $\omega_d \propto v/\lambda$).

For the *second case*, when the antenna height is *higher than the rooftops* of buildings, that is, $z_1 > \bar{h}$, and $\gamma_0 d \ge 10$, we have [90]

$$\tilde{W}_\tau(\omega, \varphi_0) \approx \frac{2(\omega_d - \omega \cdot \cos\varphi_0)}{\omega_d \sqrt{\omega_d^2 - \omega^2}}. \tag{8.65}$$

The power spectrum, in the Doppler shift domain, in this case can be presented by the function

$$\tilde{W}_\tau(\omega, \varphi_0) \approx \left[\left(\omega_0 \frac{v}{c}\right)^2 - \omega^2\right]^{-1/2}, \tag{8.66}$$

which is close to the classical Clark's "U-shaped" Doppler spectrum distribution described in References 18 and 20. Generally speaking, the effect of asymmetry, which follows from (8.59) or (8.64), depends on the angle φ_0 and the influence of the built-up terrain, accounted by the parameter χ.

Based on Clark's theoretical framework, Jakes [76] obtained the Doppler power spectral function, close to (8.66), for the mobile multipath channel with fading. Thus, the normalized Doppler power spectrum following Reference 76 can be presented in our notations as

$$\tilde{W}(f) \approx \frac{1}{\pi f_{d_m} \left[1 - \left(\dfrac{f}{f_{d_m}} \right)^2 \right]^{1/2}} . \tag{8.68}$$

Parallel, Gans [77] and then Haas [78], using directional antenna description, had obtained nonuniform normalized Doppler spectral density, which is close to that obtained in Reference 76 and Reference 20, which has limits f_1 and f_2 in the full Doppler bandwidth $f_d \in [-f_{d_m}, f_{d_m}]$, that is,

$$\tilde{W}(f) \approx \frac{1}{\left[\cos^{-1} \left(\dfrac{f_1}{f_{d_m}} \right) - \cos^{-1} \left(\dfrac{f_2}{f_{d_m}} \right) \right] f_{d_m} \left[1 - \left(\dfrac{f}{f_{d_m}} \right)^2 \right]^{1/2}} , \tag{8.69}$$

where $-f_{d_m} < f_2 \le f \le f_1 < f_{d_m}$.

The same "band-limited" power spectral density (PSD) distribution has obtained Aulin [20] in the so-called quasi 3-D model (sometimes called in the literature the 2.5-D model), where the Doppler spectrum does not cover the full Doppler bandwidth $f_d \in [-f_{d_m}, f_{d_m}]$, as it is clearly seen from Figure 8.63, where the 2-D "U-shaped" Clark's model according to Reference 18 presented together with the 2.5D "band-limited" Aulin's model according to Reference 20. The later approximate model is usually used in the more realistic cases for mobile communication links whose antenna heights are less than the heights of the local surrounding obstructions. In this case, the majority of arriving multipath rays travel in the vertical plane in a nearly horizontal direction with the elevation angles θ_i having a mean value close to $0°$.

As it is clearly seen from Figure 8.63, the 2-D Doppler PSD obtained by Clark limits to infinite at $|f| = f_{d_m}$, whereas the quasi 3-D Doppler PSD obtained by Aulin is actually constant for $f_{d_m} \cos \theta_m \le |f| \le f_{d_m}$, where θ_m is the maximum elevation angle that does not exceed $30°$ [20]. This PSD was derived in Reference 20 and equals:

$$\tilde{W} = \frac{1}{4 f_{d_m} \sin \theta_m} = const. \tag{8.70a}$$

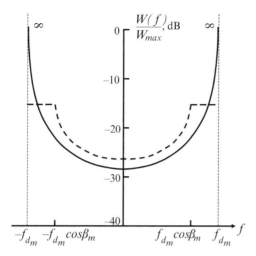

FIGURE 8.63. 2-D "U-shaped" Clark's Doppler PSD compared with quasi 3-D "sband-limited" Aulin's Doppler PSD in the Doppler shift domain.

In the range of $-f_{d_m}\cos\theta_m \leq f \leq f_{d_m}\cos\theta_m$, the quasi 3-D (or 2.5-D) PSD equals [20]:

$$\tilde{W} = \frac{1}{f_{d_m}} \left[\frac{\pi}{2} - \sin^{-1} \left(\frac{2\cos^2\theta_m - 1 - (f/f_{d_m})^2}{1-(f/f_{d_m})^2} \right) \right], \qquad (8.70b)$$

which clearly seen from Figure 8.63.

We notice that the same asymmetry in PSD distribution in the Doppler shift domain, shown in Figures 8.61 and 8.62, was obtained also in References 71, 79–85, and 91, by using the stochastic 2-D- and 3-D-multiray models described by Gaussian and Rician statistics that covers both the Aulin and the Clark theoretical frameworks [18, 20]. We will compare our stochastic approach, where PSD is described by Equation (8.64) and Equation (8.65), with other statistical analytical models later, by introducing also in the further discussions the results obtained in the next paragraph.

8.8.3. Modulated Signal Distribution in the Doppler Domain

All models, mentioned earlier in this chapter, deal with propagation of mono-chromatic radio signals through the urban communication links, but not with modulated radio signals, frequency, or phase. The models presented in Chapter 6 and in the previous sections were concentrated mostly on monochromatic radio signal propagation in various terrestrial communication links, mixed residential, suburban and urban. There, the monochromatic radiation of radio signals and the energetic characteristics of the multiray fields, caused by multiple scattering, reflection, and diffraction, were investigated in the space, angle of arrival (AOA and EOA), time delay (TD or AOA), and Doppler shift domains [18–20,71–72, 76–85, 91]. Continuation of this explanation for signal power fluctuations in the Doppler frequency domain was analyzed in the previous section also for monochromatic signals, accounting for two scenarios occurring in mobile communication links: BS antenna was elevated below and above the rooftop level according to References 90 and 92.

As follows from the presented analysis, the spectra of signal power differ from the classical U-shaped spectra obtained in the 2-D statistical model of Clarke [18], and then in the 2.5-D and 3-D models of Aulin [20], and close to those obtained experimentally [19, 72] and theoretically [71, 91]. At the same time, the stochastic model evaluated in References 73 and 90 and presented in the previous section gave the same results as in References 19, 71, and 91 for some particular scenarios occurring in the urban scene.

In the present section, we analyze propagation of frequency (FM) (or phase PM) modulated signals, for which we should obtain information on how each spectral component of the signal is changed along the radio path between the BS and any MS, as well as on distortion of the received modulated signal defined by K-factor as a ratio of the coherent and incoherent components of the signal passing the outdoor channel with fading. We examine these aspects in the framework of the correlation analysis, according to which the distortion of the signal spectrum can be estimated by deriving a correlation function of the spectral components of the FM signal, its intensity, as a response of the channel with fading, the scale of frequency correlation between spectral components of the modulated signal, and, finally, the K-parameter deviations in the frequency domain. All these characteristics are analyzed based on the stochastic multiparametric model described in detail in Chapter 5 and in the previous sections.

Signal Characteristics in the Frequency Domain. The same model of the terrain, as was described in Chapter 5, as well as the geometry of single scattering presented in Figure 8.15, are used to describe propagation of the modulated baseband signal (e.g., without the radiated frequency f_0, see definition in Chapter 1), the bandwidth of which $\Delta\omega = |\omega - \omega_0|$ is limited by

$$|\omega - \omega_0| \leq \Delta\Omega << \omega_0. \tag{8.71}$$

After scattering from the nontransparent screen (e.g., building) at an arbitrary point $C(\mathbf{r}_s)$, as shown in Figure 8.15, the radiated signal arrives at point $B(\mathbf{r}_2)$ of the moving vehicle (usually the MS), which during a period of $\Delta\Delta t = \ell/|\mathbf{v}|$

passes from point $B(\mathbf{r}_2)$ to any arbitrary point $D(\mathbf{r}_2)$, defined by the angle φ_0. Here $\nu \equiv |\mathbf{v}|$ is the value of the MS speed, and ℓ is the range $|BD|$ passed by MS during the time period of Δt.

Unlike the situation described in References 58–69 for propagation of the monochromatic signal, we now present here a strength of the independent single-scattered wave by introducing in it the frequency ω as a new variable, that is, describing now the modulated signal

$$U(\mathbf{r}_2, \omega) = 2i \frac{\omega}{c} \int_{S_2} \{Z(\mathbf{r}_2, \mathbf{r}_s, \mathbf{r}_1) \cdot \Gamma(\varphi_s, \mathbf{r}_s) \cdot \sin\varphi_s \cdot \tilde{G}(\mathbf{r}_s, \mathbf{r}_1) \cdot \tilde{G}(\mathbf{r}_2, \mathbf{r}_s)\} dS. \quad (8.72)$$

Equation (8.72) was written for close radio traces with NLOS conditions by omitting the LOS component directly arriving from point A to point B (see Fig. 8.15) for estimating the pure effect of multipath component (i.e., the effects of fading). In (8.72) $Z(\mathbf{r}_2, \mathbf{r}_s, \mathbf{r}_1)$ is the shadow function defined and evaluated in Chapter 6, where all other functions and parameters are also described.

To obtain now frequency-selective properties of the outdoor radio channel, we consider the signal radiated by the source located at point $A(\mathbf{r}_1)$ through the Fourier integral

$$s(t) = \frac{1}{2\pi} \int_{\Omega} S(\omega) \exp\{-i\omega t\} d\omega. \quad (8.73)$$

If so, we can now construct the received signal at point $B(\mathbf{r}_2)$ as follows:

$$s(\mathbf{r}_2, t) = \frac{1}{2\pi} \int_{\Omega} S(\omega) \cdot U(\mathbf{r}_2, \omega) \exp\{-i\omega t\} d\omega, \quad (8.74)$$

where $U(\mathbf{r}_2, \omega)$ is described by (8.72).

Signal Intensity in the Frequency Domain. The intensity of the modulated signal $\langle I(\mathbf{r}_2, \omega_1, \omega_2) \rangle = \langle U(\mathbf{r}_2, \omega_1) \cdot U(\mathbf{r}_2, \omega_2) \rangle$ can be presented using (8.72) and modified results obtained in Reference 93, in the following manner:

$$\langle I(\mathbf{r}_2, \omega_1, \omega_2) \rangle$$
$$= \frac{1}{(2\pi)^2} \int_{\Omega} d\omega_1 \int_{\Omega} d\omega_2 \cdot S(\omega_1) \cdot S^*(\omega_2) \cdot K_U(\omega_1, \omega_2) \cdot \exp\{-i(\omega_1 - \omega_2)\}, \quad (8.75)$$

with the correlation function of two different frequencies ω_1 and ω_2

$$K_U(\omega_1, \omega_2) = \langle U(\mathbf{r}_2, \omega_1) \cdot U^*(\mathbf{r}_2, \omega_2) \rangle. \quad (8.76)$$

Next, we analyze the correlation function $K_U(\omega_1,\omega_2)$ at the same point $\mathbf{r}_2$, but for two different frequencies, ω_1 and ω_2. We should note that this function is limited by the region of its definition:

$$D = \{\omega_1 \in (\omega_0 - \Delta\Omega, \omega_0 + \Delta\Omega) \cup \omega_2 \in (\omega_0 - \Delta\Omega, \omega_0 + \Delta\Omega)\}$$

since beyond this region the integrand in (8.83) limits to zero due to a limited width of the spectral function $S(\omega)$ of the radiated signal. For the convenience of future derivations, let us denote $\omega_2 = \omega$, $\omega_2 = \omega_1 + \Delta\omega$, and $K_U(\omega,\omega + \Delta\omega) \equiv H(\omega,\Delta\omega)$ where $\omega \in \Omega$ and $|\Delta\omega| \leq 2\Delta\Omega$.

Correlation Function and Scale of Correlation of the Modulated Signal. To analyze now the correlation function, we use the same approach described in previous sections by introducing the polar coordinates r and φ at the origin of point B, and by using again the function $W(\tau)$ which describes a distribution of signal energy of the radiated field at the receiver in the TOA domain, defined for $\gamma_0 d \gg 1$ in previous section by Equation (8.36):

$$H(\omega, \Delta\omega) = \int_1^\infty W(\tau) \cdot \exp\{i\Delta k \cdot d \cdot \tau\} d\tau. \tag{8.77}$$

Substituting now (8.36) into (8.77), we, after straightforward computations made in Reference 93. will obtain a simple presentation of the correlation function of two spectral components of the total signal in the frequency domain:

$$H(\omega, \Delta\omega) = \langle I(\mathbf{r}, \omega)\rangle \left(1 - i\frac{2\Delta\omega}{c\gamma_0}\right)^{-3/2}. \tag{8.78}$$

In the above expression, only the term $\langle I(\mathbf{r},\omega)\rangle$ depends on frequency ω and does not depend on $\Delta\omega$. If so, we can introduce the coefficient of the frequency modulation of the total signal, $R\langle\Delta\omega\rangle = H(\omega,\Delta\omega)/\langle I(\mathbf{r},\omega)\rangle$ and estimate the effect of urban terrain features on spectral characteristics of the signal in the frequency domain. Thus, using (8.78), we can easily find that this coefficient equals

$$R(\Delta\omega) = \left(1 - i\frac{2\Delta\omega}{c\gamma_0}\right)^{-3/2}. \tag{8.79}$$

Introducing now, as in previous sections, the average distance of LOS, $\langle\rho\rangle = \gamma_0^{-1}[km]$, we can conclude that the modulation coefficient $R(\Delta\omega)$ depends only on one parameter, $\langle\tau\rangle = 2\langle\rho\rangle/c$, that is, on the time during which the radio signal passes the urban channel over the range of $d = 2\langle\rho\rangle$.

At the same time, the integral scale of the frequency correlation of two spectral components can be determined as

$$\Delta\omega_k = \int\limits_0^\infty |R(\Delta\omega)|^2 \, d(\Delta\omega). \tag{8.80}$$

For the coefficient described by expression (8.79), this scale can be easily estimated as

$$\Delta\omega_k = \frac{\gamma_0 c}{2} = \frac{c}{2\langle\rho\rangle} = \langle\tau\rangle^{-1}. \tag{8.81}$$

It is clearly seen that the frequency-correlated scale depends only on the 1-D density of building contours, γ_0, which are located surrounding the mobile vehicle antenna.

Experimental Verification of the Theoretical Model. Let us estimate the integral scale from (8.81) for future practical applications. Thus, for most cities with $\nu = 50$–150 buildings per square kilometer (km^{-2}) with the average length (or width) of buildings $\langle L \rangle = 50$–150 m (according to numerous experiments carried out in different urban environments [46–69]), we find that γ_0 changes in a range from 5 to 12 km^{-1}, and $\langle\rho\rangle = \gamma_0^{-1} \approx 100 - 1000$ m. Then, by using (8.81), we get that Δf_k varies from 25 to 100 kHz. These results are in a satisfactory agreement with measured data obtained in earlier experiments carried out in Birmingham (England) [91] in the frequency domain, where $\Delta f_k \approx 20$–40 kHz was found by analyzing the spectrum of multiplicative noise occurring in the urban time- and frequency-dispersive channel with fading.

The same good agreement with above estimations was found in experiments carried out in Tokyo city described in detail in Reference 93. In both experimental sites, according to the topographic maps of the built-up terrain, the buildings' density was estimated varied from $\gamma_0 = 6$ km^{-1} to 8 km^{-1}, that is, $\langle\rho\rangle = \gamma_0^{-1} \approx 125$ m $\div 175$ m. At the same time, the TD spread was estimated in Reference 6 as $\langle\sigma_\tau\rangle_I \approx 0.1 \div 0.2$ μs (for the first experimental site of Tokyo, see Fig. 8.31) and $\langle\sigma_\tau\rangle_{II} \approx 0.1 \div 0.5$ μs (for the second experimental site of Tokyo, see Fig. 8.33). Accounting for the relations between the coherence bandwidth and the TD spread, $\Delta f_c \approx 0.02/\sigma_\tau$ [48, 61], we get that $\Delta f_{cI} \approx 40$ kHz and $\Delta f_{cII} \approx 80$ kHz, respectively. From these estimations, it follows that if the coherent bandwidth of the specific urban multipath channel Δf_c is lower than Δf_k estimated above theoretically, strong frequency selective fast fading occurs in such a channel, and we should wait at the receiver a strong distortion of the modulated baseband signal with information data. And, conversely, we expect flat fast fading with small distortion of the information signal, when the coherent bandwidth Δf_c exceeds Δf_k.

The Rician K-factor in the Frequency Domain. For analyzing how the fading parameter of fading, called K-factor, is changed in the frequency domain, we will rearrange the previous formulas to find signal power distribution in the

Doppler shift domain. As above, we will consider, following Reference 93, two usually occurring scenarios with BS antenna elevation: above and below the rooftops.

When the BS antenna is *above the rooftop level*, the spectrum of the normalized signal power distribution $\tilde{W}(\omega, \varphi_0) = W(\omega, \varphi_0) / \langle I(\mathbf{r}, \omega) \rangle$ is limited by the interval $\omega_d \in [-\omega_{d\max}, \omega_{d\max}]$, and within this interval can be determined for $\gamma_0 d \gg 1$ as [93]

$$\tilde{W}(q) = \frac{2(1 - q\cos\varphi_0)}{\omega_{d\max}[1 - q^2]^{1/2}}, q = \frac{\omega_0}{\omega_{D\max}}, \tag{8.82}$$

where $\omega_{d\max} = 2\pi \cdot f_{d\max} = 2\pi \cdot v/\lambda = 2\pi \cdot f_0 \cdot v/c$ is the maximum Doppler shift, depending on the velocity v and on the radiated frequency $f_0 = \omega_0/2\pi \equiv c/\lambda$. The root in the denominator in (8.82) does not depend on the features of the terrain and the range from the BS antenna. We should mention that (8.82) fully corresponds to results obtained above for the case of the uniform and probability-equal Clark's distribution [18] of signals in the azimuth domain, that is, for $PDF(\varphi) = 1/2\pi$. The spectrum described by (8.82) is only slightly asymmetric with respect to the classical U-shaped [18] (see also Fig. 8.61b) due to the dependence on AOA, because for the high-elevated BS antenna (with respect to rooftops), the terrain features do not affect the total power spectrum shape and form.

In the case of consideration, following Reference 93, we obtain the following expression of the Rician parameter in the frequency domain following its definition [48, 61]:

$$K(q) = \frac{1}{4[1 - q\cos\varphi_0]}, \tag{8.83}$$

which depends only on MS antenna velocity (through the maximum Doppler shift or the parameter q) and the azimuth of the arriving signal φ_0 regarding to radio path.

In the second limited case where the BS antenna *is below or at the same height as the building rooftop*, the normalized power spectrum in the frequency domain $\tilde{W}$, can be presented by the following expression [93]:

$$\tilde{W}(q, \varphi_0) = \frac{2 \cdot sh\chi \cdot (ch\chi - q\cos\varphi_0)}{\omega_{d\max}[1 - q^2]^{1/2}[(ch\chi - q\cos\varphi_0)^2 - (1 - q^2)\sin^2\varphi_0]}; \tag{8.84}$$

that is, it has the same form as the signal power spectrum described by (8.74). Then, in the case of the low-elevated BS antenna, we can, as above, obtain the Ricean K-parameter, according to its definition [48, 61]

$$K(q) = \frac{[(ch\chi - q\cos\varphi_0)^2 - (1 - q^2)\sin^2\varphi_0]}{4 \cdot ch\chi \cdot (ch\chi - q\cos\varphi_0)}. \tag{8.85}$$

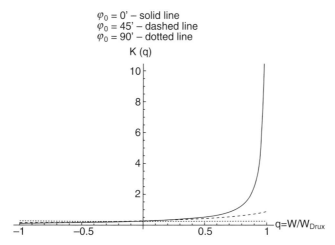

FIGURE 8.64. K-parameter as a function of q for $0 \leq \varphi_0 \leq \pi/2$: $\varphi_0 = 0$ (solid line), $\varphi_0 = \pi/4$ (dashed line), $\varphi_0 = \pi/2$ (dotted line).

In the following section we analyze numerically the Ricean K-parameter for two scenarios of the BS antenna elevation compared with building roofs: above and below the rooftop.

High-Elevated Antenna. The results of computations are shown in Figure 8.64, which were made for the first scenario of the high-elevated BS antenna, described by expression (8.82), and different orientations of MS antenna with respect to the MS antenna path (denoted by continuous, dashed and dotted curves, respectively).

It is seen that for a high-elevated BS antenna with respect to building roof-tops and for the case when the MS antenna moves toward the radiopath ($\varphi_0 = 0$) the shape of the function $K(q)$ follows the well-known classical U-shaped distribution of the signal power spectra $W(q)$ obtained in References 18 and 20 and limited at $q = 1$ to infinity. With increase of AOA ($0 < \varphi_0 < \pi/2$) with respect to the direction of radio path between the BS and MS antennas, the limit of the shape of the function $K(q)$ for $q = 1$ becomes smaller and at $\varphi_0 = \pi/2$ the function $K(q)$ is a constant equal to $K = 1/4$. With increase of the angle φ_0 in the range of $0 < \varphi_0 < \pi/2$, the effect of LOS component decreases as compared with NLOS component of the total signal, and the increase of multipath fading phenomenon should be expected: the parameter K decreases dramatically from the case of $\varphi_0 = 0$ to the case of $\varphi_0 = \pi/2$.

Low-Elevated Antenna. Now we will analyze the case when the BS antenna is below or at the same level with rooftops. To understand effects of the terrain features, as well as of speed and orientation of MS antenna with respect of radio path (angle φ_0), we computed the function $K(q)$ described by (8.85) for

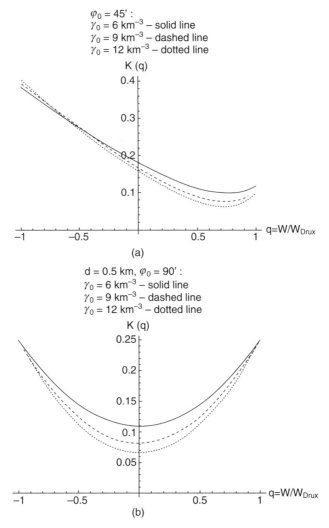

FIGURE 8.65. (a) K- parameter as a function of q for $\varphi_0 = \pi/4$ and for $\gamma_0 = 6,9,12$ km^{-1}; the distance between antennas is $d = 0.5$ km. (b) K- parameter as a function of q for $\varphi_0 = \pi/2$ and for $\gamma_0 = 6,9,12$ km^{-1}; the distance between antennas is $d = 0.5$ km.

angles $\varphi_0 = \pi/4, \pi/2$ and for different scenarios of the density of the buildings, $\gamma_0 = 6, 9, 12$ km^{-1}, which corresponds to small, moderate, and big cities, respectively, according to definitions made in References 47 and 61. Thus, in Figure 8.65a,b and Figure 8.66a,b, each separate graph is given for the specific scenario occurring with the build-up area accounting for different buildings' densities for distances between the BS and MS antennas of $d = 0.5$ km (see Fig. 8.65a,b) and of $d = 2$ km (see Fig. 8.66a,b), which are actual for the practi-

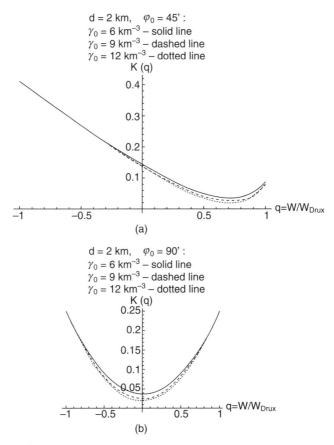

FIGURE 8.66. (a) K- parameter as a function of q for $\varphi_0 = \pi/4$ and for $\gamma_0 = 6,9,12$ km^{-1}; the distance between antennas is $d = 2$ km. (b) K- parameter as a function of q for $\varphi_0 = \pi/2$ and for $\gamma_0 = 6,9,12$ km^{-1}; the distance between antennas is $d = 2$ km.

cal configuration of WiFi and WiMAX systems, respectively, for NLOS conditions. Here the curves corresponded to the different buildings' densities: $\gamma_0 = 6$ km^{-1}, $\gamma_0 = 9$ km^{-1}, and $\gamma_0 = 12$ km^{-1} are denoted by continuous, dashed and dotted curves, respectively.

As it is seen from the illustrations shown in Figure 8.65a,b, at the close distances from the BS antenna, due to high density of buildings surrounding two terminal antennas, the fading parameter K is always less than unity (or sometimes close to it) for different orientation of the MS antenna with respect to the radio path ($0 \leq \varphi \leq \pi/2$) and for all values of normalized frequency $|q| < 1$ (or velocities). Furthermore, with increase of building density from $\gamma_0 = 6$ km^{-1} to $\gamma_0 = 12$ km^{-1}, this effect becomes more evident. This result allows us to conclude that with increase of density of buildings surrounding both terminal antennas, the multipath (NLOS) component (e.g., the incoherent part) of the signal power spectrum always exceeds (or is sometimes the same as) the LOS component (e.g., the coherent part) of the signal power spectrum.

This effect becomes stronger with increase of the range between the terminal antennas to 2 km (see Fig. 8.66a,b). Moreover, far from the BS antenna, where effects of multipath (NLOS) become predominant (with respect to LOS), the parameter K of fading becomes always less than unity, as follows from Figure 8.57a,b, and the effects of building density become more and more actual. The knowledge of the magnitude of the K-factor depending on velocity of the moving terminal antenna enables designers of urban communication links to estimate the *fade margin* in link budget design and the radius of cells in cellular maps performance (see Chapter 11).

The above-mentioned results allow us to state that the building density features and the distance between the BS and MS antennas, in the case when one of them or both are below the rooftops, can significantly change effects of fading occurring in the urban by distortion of the classical U-shape of the signal power spectra and Rician K-factor. A new main result discussed earlier (compared with those obtained in Reference 90) is the significant deviations of the K-factor of fading in frequency domain observed with changes of the densities and the distance between both low-elevated terminal antennas, BS and MS. This is an important result for the designers of advanced wireless networks operated in built-up environments, because the K-parameter of fading fully describes contributions of the multipath (NLOS) component in the modulated signal passing the urban channel with fast fading caused by movements of the MS antenna in different directions (defined by φ_0) with respect to the original radio path AB between the transmitter and the receiver (see Fig. 8.15). Strong fading effects can be estimated also via the correlation scale described by (8.81), comparing it with the bandwidth of coherency Δf_c of the outdoor multipath channel with fading defined in Chapter 1.

REFERENCES

1 Paulraj, A. and C. Papadias, "Space-time processing for wireless communications," *IEEE Signal Processing Magazine*, Vol. 14, No. 1, 1997, pp. 49–83.

2 Ertel, R.B., P. Cardieri, K.W. Sowerby, T.S. Rappaport, and J.H. Reed, "Overview of spatial channel models for antenna array communications systems," *IEEE Personal Communic.*, Vol. 5, No. 1, 1998, pp. 10–22.

3 Martin, U., J. Fuhl, I. Gaspard et al., "Model scenarios for direction-selective adaptive antennas in cellular communication systems-scanning the literature," *Wireless Personal Communic.*, Vol. 11, No. 1, 1999, pp. 109–129.

4 Pedersen, K., P. Mogensen, and B. Fleury, "A stochastic model of the temporal and azimuthal dispersion seen at the base station in outdoor propagation environments," *IEEE Trans. Veh. Technol.*, Vol. 49, No. 2, 2000, pp. 437–447.

5   Fuhl, J., J.P. Rossi, and E. Bonek, "High resolution 3-D direction-of-arrival determination for urban mobile radio," *IEEE Trans. Antennas Propagat.*, Vol. 44, No. 4, 1997, pp. 672–682.

6   Kuchar, A., J.P. Rossi, and E. Bonek, "Directional macro-cell channel characterization from urban measurements," *IEEE Trans. Antennas Propagat.*, Vol. 48, No. 1, 2000, pp. 137–146.

7   Kalliola, K., H. Laitenen, L. Vaskelainen, and P. Vainikainen, "Real-time 3D spatial-temporal dual-polarized measurement of wideband radio channel at mobile station," *IEEE Trans. Instrum. Measurements*, Vol. 49, No. 3, 2000, pp. 439–448.

8   Kalliola, K. and P. Vainikainen, "Characterization system for radio channel of adaptive array antennas," *Proc. Int. Symp. Personal Indoor Mobile Radio Conf. (PIRMC'97)*, Helsinki, Finland, 1997, pp. 95–99.

9   Laurila, J., K. Kalliola, M. Toeltsch et al., "Wide-band 3-D characterization of mobile radio channels in urban environment," *IEEE Trans. Antennas Propagat.*, Vol. 50, No. 2, 2002, pp. 233–243.

10   Roeltsch, M., J. Laurila, K. Kalliola et al., "Statistical characterization of urban spatial radio channels," *IEEE J. Select. Areas Communic.*, Vol. 20, No. 3, 2002, pp. 539–549.

11   Libetri, J.C. and T.S. Rappaport, *Smart Antennas for Wireless Communications*, Prentice-Hall, Englewood Cliffs, NJ, 1999.

12   Martin, U., "Spatio-temporal radio channel characteristics in urban macrocells," *Proc. IEE Radar, Sonar Navigation*, Vol. 145, No. 1, 1998, pp. 42–49.

13   Pedersen, K., P.E. Mogensen, and B. Fleury, "Dual-polarized model of outdoor propagation environments for adaptive antennas," in *Proc. IEEE Veh. Technol. Conf.* (VTC'99), Houston, TX, 1999, pp. 990–995.

14   Pedersen, K., P.E. Mogensen, and B. Fleury, "Power azimuth spectrum in outdoor environments," *IEE Electron. Letters*, Vol. 33., 1997, pp. 1583–1584.

15   Liberti, J.C. and T.S. Rappaport, "A geometrical based model for line-of-sight multipath radio channels," in *Proc. of Veh. Technol. Conf.*, Atlanta, GA, 1996, pp. 844–848.

16   Fuhl, J., A.F. Molisch, and E. Bonek, "Unified channel model for mobile radio systems with smart antennas," *IEE Proc. Radar, Sonar Navigation*, Vol. 145, No. 1, 1998, pp. 32–41.

17   Petrus, P., J.H. Reed, and T.S. Rappaport, "Geometrically based statistical channel model for macrocellular mobile environments," in *Proc. of IEEE Global Telecommunic.*, London, UK, 1996, pp. 1197–1201.

18   Clarke, R.H., "A statistical theory of mobile-radio reception," *Bell Syst. Technol. J.*, Vol. 47, No. 6, 1968, pp. 957–1000.

19   Petrus, P., J.H. Reed, and T.S. Rappaport, "Effects of directional antennas at the base station on the Doppler spectrum," *IEEE Communic. Letters*, Vol. 1, No. 2, 1997, pp. 40–42.

20   Aulin, T., "A modified model for the fading signal at a mobile radio channel," *IEEE Trans. Veh. Technol.*, Vol. 28, No. 3, 1979, pp. 182–203.

21   Saleh, A. A.M., and R.A. Valenzula, "A statistical model for indoor multipath propagation," *IEEE J. Select. Areas Communic.*, Vol. 5, No. 2, 1987, pp. 128–137.

22 Spencer, Q., M. Rice, B. Jeffs, and M. Jensen, "Indoor wideband time/angle of arrival multipath propagation results," in *Proc. of IEEE Vehicular Technology Conf.*, 1997, pp. 1410–1414.

23 Spencer, Q., M. Rice, B. Jeffs, and M. Jensen, "A statistical model for angle of arrival in indoor multipath propagation," in *Proc. of IEEE Vehicular Technology Conf.*, 1997, pp. 1415–1419.

24 Turin, G.L., F.D. Clapp, T.L. Johnston, S.B. Fine, and D. Lavry, "A statistical model of urban multipath propagation," *IEEE Trans. Veh. Technol.*, Vol. 21, No. 1, 1972, pp. 1–9.

25 Lo, T. and J. Litva, "Angles of arrival of indoor multipath," *Electron. Letters*, Vol. 28, No. 18, 1992, pp. 1687–1689.

26 Guerin, S., "Indoor waveband and narrowband propagation measurements around 60.5 GHz in an empty and furnished room," in *Proc. of IEEE Veh. Technol. Conf.*, IEEE, 1996, pp. 160–164.

27 Wang, J.-G., A.S. Mohan, and T.A. Aubrey, "Angles-of-arrival of multipath signals in indoor environments," in *Proc. of IEEE Veh. Technol. Conf.*, IEEE, 1996, pp. 155–159.

28 Litva, J., A. Ghaforian, and V. Kezys, "High-resolution measurements of AOA and time-delay for characterizing indoor propagation environments," in *IEEE Antennas and Propagation Society International Symposium 1996 Digest*, IEEE. Vol. 2, 1996, pp. 1490–1493.

29 Devasirvatham, D., C. Banerjee, M. Krain, and T.S. Rappaport, "Multi-frequency radiowave propagation measurements in the portable radio environment," in *Proc. IEEE ICC'90*, 1990, pp. 1334–1340.

30 Molkdar, D., "Review on radio propagation into and within buildings," *Proc. IEE – H*, Vol. 138, 1991, pp. 61–73.

31 Keenan, J. and A. Motley, "Radio coverage in buildings," *British Telecom Technol. J.*, Vol. 8, 1990, pp. 19–24.

32 Chong, C.-C., C.-M. Tan, D.I. Laurenson et al., "A new statistical wideband spatio-temporal model for 5-GHz band WLAN systems," *IEEE J. Select. Areas Communic.*, Vol. 2, No. 2, 2003, pp. 139–150.

33 Hassan-Ali, M. and K. Pahlavan, "A new statistical model for site-specific indoor radio propagation prediction based on geometric optic and geometric probability," *IEEE Trans. Wireless Communic.*, Vol. 1, No. 1, 2002, pp. 112–124.

34 Fortune, S., D. Gay, B. Kernighan et al., "WISE design of indoor wireless systems: Practical computation and optimization," *IEEE Comput. Sci. Eng.*, Vol. 2, No. 1, 1995, pp. 58–69.

35 Hassan-Ali, M. and K. Pahlavan, "Site-specific wideband and narrowband modeling of indoor radio channel using ray-tracing," in *PMIRC'98*, Boston, MA, 1998.

36 Valenzula, R., O. Landron, and D. Jacob, "Estimating local mean signal strength of indoor multipath propagation," *IEEE Trans. Veh. Tech.* Vol. 46, 1997, pp. 203–212.

37 McKown, J. and R. Hamilton, "Ray tracing as a design tool for radio networks," *IEEE Network Magazine*, No. 1, 1991, pp. 27–30.

38 Fortune, S., "Algorithms for the prediction of indoor radio propagation," http://cm.bell-labs.com/cm/cs/who/sjf/pubs.html, 1998.

39   Dietert, J.E. and B. Rembold, "Stochastic channel model for outdoor applications based on raytrace simulations," in *Proc. of Millennium Conf. Antennas and Propagat. (AP2000)*, Davos, Switzerland, 2000.

40   Chong C.-C., D.I. Laurenson, and S. McLaughlin, "Statistical characterization of the 5.2 GHz wideband directional indoor propagation channels with clustering and correlation properties," in *Proc. IEEE Vehicular Technol. Conf. (VTC 2002-Fall)*, Vancouver, BC, Canada, Vol. 1., 2002, pp. 629–633.

41   Ertel, R.B. and J.H. Reed, "Angle and time of arrival statistics for circular and elliptical scattering models," *IEEE J. Select. Areas Communic.*, Vol. 17, 1999, pp. 1829–1840.

42   Constantinou, C.C. and L.C. Ong, "Urban radio propagation: A 3-D path-integral wave analysis," *IEEE Trans. Antennas Propagat.*, Vol. 46, No. 2, 1998, pp. 266–270.

43   Kanatas, A.G., I.D. Kountouris, G.B. Kostraras, and C.C. Constantinou, "A UTD propagation model in urban microcellular environments," *IEEE Trans. Veh. Technol.*, Vol. 46, No. 1, 1997, pp. 185–193.

44   Erceg, V., S.J. Fortune, G. Ling et al., "Comparisons of a computer-based propagation prediction tool with experiment data collected in urban microcellular environments," *IEEE J. Select. Areas Communic.*, Vol. 15, 1997, pp. 677–684.

45   Kim, S.C., B.J. Guarino, Jr., T.M. Willis III et al., "Radio propagation measurements and prediction using three dimensional ray tracing in urban environments at 908 MHz and 1.9 GHz," *IEEE Trans. Veh. Technol.*, Vol. 48, 1999, pp. 931–946.

46   Liang, G. and H.L. Bertoni, "A new approach to 3-D ray tracing for propagation prediction in cities," *IEEE Trans Antennas Propagat.*, Vol. 46, 1998, pp. 853–863.

47   Bertoni, H., *Radio Propagation for Modern Wireless Systems*, Prentice Hall, New Jersey, 2000.

48   Saunders, S.R., *Antennas and Propagation for Wireless Communication Systems*, J. Wiley & Sons, New York, 1999.

49   Andersen, J.B., S.L. Lauritzen, and C. Thommesen, "Distribution of phase derivatives in mobile communications," *Proc. IEE H*, Vol. 137, 1990, pp. 197–201.

50   Andersen, J.B., and I.Z. Kovacs, "Power distributions revisited," *COST-273*, Guildford, January 2002.

51   Andersen, J.B. and K.I. Pedersen, "Angle-of-arrival statistics for low resolution antennas," *IEEE Trans. Antennas Propagat.*, Vol. 50, No. 3, pp. 391–395.

52   Kloch, C., G. Liang, J.B. Andersen, G.F. Pedersen, and H.L. Bertoni, "Comparison of measured and predicted time dispersion and direction of arrival for multipath in a small cell environment," *IEEE Trans. Antennas Propagat*, Vol. 49, No. 9, 2001, pp. 867–876.

53   Pedersen, K.I., P.E. Mogensen, and B. Fleury, "Spatial channel characteristics in outdoor environments and their impact on BS antenna system performance," *Proc. Int. Conf. on Vehicular Technologies, VTC'98*, Ottawa, Canada, pp. 719–723, May 1998.

54   Pedersen, K.I., P.E. Mogensen, B. Fleury et al., "Analysis of time, azimuth and Doppler dispersion in outdoor radio channels," *Proc. ACTS Mobile Communication Summit '97*, Aalborg, Denmark, 1997, pp. 308–313.

55   Algans, A., K.I. Pedersen, and P.E. Mogensen, "Experimental analysis of the joint statistical properties of azimuth spread, delay spread, and shadow fading," *IEEE J. Select. Areas Communic.*, Vol. 20, No. 3, 2002, pp. 523–531.

56   Pedersen, K.I., P. Mogensen, and B.H. Fleury, "A stochastic model of the temporal and azimuthal dispersion seen at the base station in outdoor propagation environments," *IEEE Trans. Antennas Propagat.*, Vol. 49, 2000, pp. 437–447.

57   Pedersen, K.I., P. Mogensen, and B.H. Fleury, "Experimental analysis of the joint statistical properties of azimuth spread, delay spread, and shadow fading," *IEEE J. Select. Areas Communic.*, Vol. 20, No. 3, 2002, pp. 523–531.

58   Ponomarev, G.A., A.N. Kulikov, and E.D. Telpukhovsky, *Propogation of Ultra-Short Waves in Urban Environments*, Tomsk, Rasko, USSR, 1991.

59   Blaunstein, N., "Prediction of cellular characteristics for various urban environments," *IEEE Antennas and Propogat. Magazine*, Vol. 41, No. 6, 2000, pp. 135–145.

60   Blaunstein, N., D. Katz, D. Censor et al., "Prediction of loss characteristics in built-up areas with various buildings' overlay profiles," *IEEE Anten. Propagat. Magazine*, Vol. 43, No. 6, 2001, pp. 181–191.

61   Blaunstein, N., "Wireless communication systems," in *Handbook of Engineering Electromagnetics*, ed. R. Bansal, Marcel Dekker, New York, 2004.

62   Yarkoni, N., N. Blaunstein, and D. Katz, "Link budget and radio coverage design for various multipath urban communication links," *Radio Science*, Vol. 42, 2007, pp. 412–427.

63   Katz, D., N. Blaunstein, M. Hayakawa, and Y.S. Kishiki, "Radio maps design in Tokyo City based on stochastic multi-parametric and deterministic ray tracing approaches," *J. Antennas and Propag. Magazine*, Vol. 51, No. 5, October, 2009, pp. 200–208.

64   Blaunstein, N., "Distribution of angle–of–arrival and delay from array of building placed on rough terrain for various elevation of base station antenna," *Journal of Communic. and Networks*, Vol. 2, No. 4, 2000, pp. 305–316.

65   Hayakawa, M., D. Katz, and N. Blaunstein, "Signal power distribution in time delay in Tokyo City experimental sites," *Radio Sci.*, Vol. 43, RS3006, 2008, pp. 1–9.

66   Blaunstein, N., N. Yarkony, and D. Katz, "Spatial and temporal distribution of the VHF/UHF radio waves in built-up land communication links," *IEEE Trans. Antennas and Propagat.*, Vol. 54, No. 8, 2006, pp. 2345–2356.

67   Blaunstein, N. and E. Tsalolihin, "Signal distribution in the azimuth, elevation and time delay domains in urban radio communication links," *IEEE Antennas and Propagation Magazine*, Vol. 46, No. 5, 2004, pp. 101–109.

68   Blaunstein, N., M. Toeltsch, C. Christodoulou, et al., "Azimuth, elevation and time delay distribution in urban wireless communication channels," *J. Antennas and Propag. Magazine*, Vol. 48, No. 3, 2002, pp. 425–434.

69   Blaunstein, N., M. Toulch, J. Laurila, et al., "Signal power distribution in the azimuth, elevation and time delay domains in urban environments for various elevations of base station antenna," *IEEE Trans. Antennas and Propagation*, Vol. 54, No. 10, 2006, pp. 2902–2916.

70   Tsalolihin, E., N. Blaunstein, I. Bilik, L. Ali, and S. Shakya, "Measurement and modeling of the indoor communication link in angle of arrival and time of arrival

domains," in *Proc. of Int. Conf. of IEEE Antennas and Propagation*, Toronto, Canada, July 19–23, 2010, pp. 338–341.

71 Abdi, A., J. Barger, and M. Kaveh, "A parametric model for distribution of the angle of arrival and the associated correlated function and power spectrum at the MS," *IEEE Trans. Veh. Technol.*, Vol. 51, No. 3, 2002, pp. 425–434.

72 Jenison, R. and K. Fissell, "A comparison of the von Mises and Gaussian basis function for approximating spherical acoustic scatter," *IEEE Trans. Neural Networks*, Vol. 6, No. 5, 1995, pp. 1284–1287.

73 Tsalolihin, E., N. Blaunstein, I. Bilik, L. Ali, and S. Shakya, "Measurement and modeling of the indoor communication link in angle of arrival and time of arrival domains," in *Proc. of Int. Conf. of IEEE Antennas and Propagation*, Toronto, Canada, July 19–23, 2010, pp. 338–341.

74 Tsalolihin E., N. Blaunstein, I. Bilik, and S. Shakya, "Analysis of AOA-TOA signal distribution in various indoor environments," *European Conference on Antennas and Propagation (EuCAP-2011)*, Rome, Italy, April 11–15, 2011, pp. 1746–1750.

75 Ali, L., Y. Ben Shimol, and N. Blaunstein, "Analysis of AOA-TOA signal distribution in indoor RF environments." *PIERS-2011*, April 12–16, 2011, Marrakesh, Morocco, p. 740.

76 Ben-Shimol, Y. and N. Blaunstein, "Localization and positioning of any subscriber in indoor environments on the basis of analysis of joint AOA-TOA signal distribution," *Proc. of Int. Conference on Antennas and Propagation*, September 12–17, Torino, Italy, pp. 1420–1423.

77 Gans, M.J., "A power-spectral theory of propagation in the mobile-radio environment," *IEEE Trans. on Vehicular Technology*, Vol. 21, No. 1, 1972, pp. 27–38.

78 Haas, E., "Aeronautical channel modeling," *IEEE Trans. on Vehicular Technology*, Vol. 51, No. 2, 2002, pp. 254–264.

79 *Digital Land Mobile Radio Communications*, COST207, Final Report, 1989.

80 Erceg, V., K.V.S. Hari, and M.S. Smith, *Channel Models for Fixed Wireless Applications*, IEEE 802.16a-03/01, 2003.

81 Wyne, S., T. Santos, A.P. Singh, F., Tufvesson, and A.F. Molisch, "Characterization of a time-variant wireless propagation channel for outdoor short-range sensor networks," *IET Communications*, Vol. 4, No. 3, 2010, pp. 253–264.

82 Vatalaro, F. and A. Forcella, "Doppler spectrum in mobile-to-mobile communications in the presence of three-dimensional multiple scattering," *IEEE Trans. on Vehicular Technology*, Vol. 46, No. 1, 1997, pp. 213–219.

83 Kasparis, C., P. King, and B.G. Evans, "Doppler spectrum of the multipath fading channel in mobile satellite systems with directional terminal antennas," *IET Communications*, Vol. 1, No. 6, 2007, pp. 1089–1094.

84 Qu, S. and T. Yeap, "Three-dimensional scattering model for fading channels in land mobile environment," *IEEE Trans. on Vehicular Technology*, Vol. 48, No. 3, 1999, pp. 765–781.

85 Qu, S., "An analysis of probability distribution of Doppler shift in three-dimensional mobile radio environments." *IEEE Trans. on Vehicular Technology*, Vol. 58, No. 4, 2009, pp. 1634–1639.ss

86 Tholl, D., M. Fattouche, and R.J.C. Bultitude, "A comparison of two radio propagation channel impulse response determination techniques," *IEEE Trans. on Antennas and Propagation*, Vol. 41, No. 4, 1993, pp. 515–517.

87 Award, M., K Wong, and Z. Li, "An integrated overview of the open literature's empirical data on the indoor radiowave channel's delay properties," *IEEE Trans. on Antennas and Propagat.*, Vol. 56, No. 5, 2008, pp. 1451–1468.

88 Alsindi, N., B. Alavi, and K. Pahlavan, "Measurement and modeling of ultrawideband TOA-based ranging in indoor multipath environments," *IEEE Trans. on Vehic. Technology*, Vol. 58, No. 3, 2009, pp. 1046–1058.

89 Curry, M., B. Koala, and Y. Kuga, "Indoor angle of arrival using wide-band frequency diversity with experimental results and EM propagation modeling," *Proc. on IEEE-APS Conference on Anten. and Propagat*, 2000, pp. 65–68.

90 Blaunstein, N. and Y. Ben-Shimol, "Spectral properties of signal fading and Doppler spectra distribution in urban communication mobile links," *Wireless Communic. and Mobile Computing*, Vol. 6, No. 1, 2006, pp. 113–126.

91 Tepedelenlioglu, C., A. Abdi, M. Kaveh, and K. Pahlavan, "Estimation of Doppler spread and signal strength in mobile communications with applications to handoff and adaptive transmission," *J. Wireless Communications and Mobile Computing*, Vol. 1, No. 2, 2001, pp. 221–242.

92 Blaunstein, N. and Y. Ben-Shimol, "Frequency dependence of path loss characteristics and link budget design for various terrestrial communication links," *IEEE Trans. on Antennas and Propagat.*, Vol. 52, No. 10, 2004, pp. 2719–2729.

93 Blaunstein, N., Katz, D., M. Hayakawa, "Spectral properties of modulated signal in the Doppler domain in urban multipath radio channels with fading," *IEEE Trans. Antennas and Propagat.*, Vol. 58, No. 8, 2010, pp. 2795–2800.

CHAPTER NINE

Prediction of Operational Characteristics of Adaptive Antennas

In adaptive antenna performance, as mentioned in Chapters 7 and 8, it is very important to predict the position of a desired subscriber within a certain area. To obtain information about the direction of the subscriber's signal as well as his range, designers of wireless networks need to carry out some special experiments [1–14]. The direction of the subscriber is determined by the distribution of the recording signal in the azimuth, elevation, and time delay domain, as shown in Figure 9.1. Thus, the range to any subscriber located within the cell of service, as well as the error of its estimation, can be totally determined from the joint time delay, azimuth, and elevation (for subscriber located at any height as shown in Figure 9.2) distribution of the arriving signal. Furthermore, when the adaptive antenna "searches" for the mobile subscriber (MS) or mobile vehicle (MV), it is very important to obtain information about its velocity, based on the signal distribution in the Doppler shift domain. In Chapter 8, we showed that this can be done theoretically using the proposed unified stochastic multiparametric approach combined (or not, depending on the built-up terrain scenario) with the street-waveguide model [1–3, 15–23] as shown in Figure 9.3.

In the following sections, we first verify theoretical results described in Chapter 8 by using more precisely arranged experiments, carried out in different urban sites that focus our attention on adaptive antenna operational

Radio Propagation and Adaptive Antennas for Wireless Communication Networks: Terrestrial, Atmospheric, and Ionospheric, Second Edition. Nathan Blaunstein and Christos G. Christodoulou.
© 2014 John Wiley & Sons, Inc. Published 2014 by John Wiley & Sons, Inc.

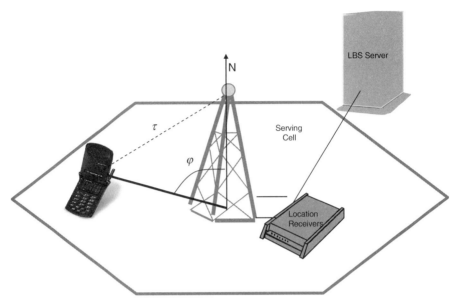

FIGURE 9.1. Schematical presentation of the subscriber localization by knowledge of the azimuth (φ) and time delay (τ) determination.

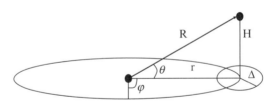

FIGURE 9.2. Main geometrical parameters for the subscriber positioning definition.

characteristics in the space, angle, time, and frequency domains. Then, we present the two statistical approaches. The first model is based on information about the maximum radio signal strength indicator (RSSI), minimum time delay, and the direct azimuthal relationship to the desired subscriber, obtained through the analysis of the topographic map of the area of service and the combined stochastic model.

The second model is based on the error minimization of the position of any subscriber located within various urban environments. These models are verified experimentally and compared with our stochastic model.

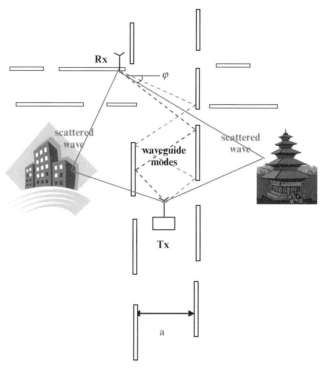

FIGURE 9.3. 2-D model of the urban scenario consisting combination of the straight crossing streets and buildings and other obstructions randomly distributed on the terrain.

9.1. EXPERIMENTAL VERIFICATION OF SIGNAL DISTRIBUTION IN AZIMUTH, ELEVATION, TIME DELAY, AND DOPPLER SHIFT DOMAIN

9.1.1. Signal Azimuth Distribution

Here, we consider the case of a specific urban environment consisting of straight crossing streets with buildings randomly lining each street according to Poisson law (see Chapter 5). Additionally assume the existence of buildings and other obstructions randomly distributed on the rough terrain, as is schematically presented in Figure 9.3. To verify this general stochastic approach, we will present here results of measurements carried out in the microcell urban environment of the small town of Kefar-Yona; conditions of this experiment will be briefly presented in Chapter 10 (see also References 1 and 2). The measurements were carried not only in the space domain, but also in the azimuth domain. The test environment was that of a typical small urban region of three-to five-story brick buildings with approximately uniform heights $h = 8 - 10$ m and with a right-angle crossing-street plan with buildings randomly surrounded according to Poisson's law (as is presented in Figure 9.4).

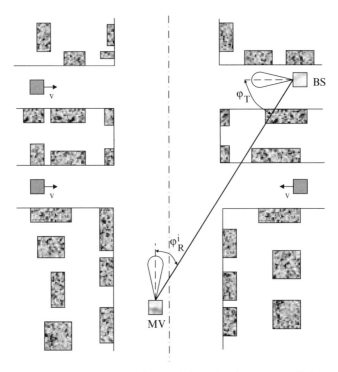

FIGURE 9.4. Geometry of the experimental site of Kefar Yona and the experiment.

In such an urban scenario, we will verify the predicted theoretical results described in Chapter 8 via series of 1 experiments that will be mentioned in Chapter 10. Thus, using narrow-beam directed antennas, the pulse signals were sent by using directive rotating antennas arranged at the transmitter and the receiver, at radio paths up to 1.5–2 km. The height of the base station (BS) transmitter antenna was changed from $z_2 = 6$ m (as in the first series of experiment) to $z_2 = 12$ m, and therefore the MV receiver antenna became $z_1 = 2$ m. During its movement, the MV passed streets with different orientations with respect to the BS antenna, with various densities and overlay building profiles. The receiving antenna with mechanical rotations on the vehicle had registered N multipath components of the wideband signal, each of which corresponded to one of the possible angular positions on the axis of the MV directive antenna. Because the antenna loop width was initially smaller than the integral scale width $\tilde{\theta}$ of the energetic spectrum $\tilde{W}(\varphi)$ described in Chapter 8, we can discuss the registration of realizations, having delta-function kind of shape on the angle energetic spectrum $\tilde{W}(\varphi)$. The schematic geometry of the experiment is shown in Figure 9.4, where φ_T is the angle between the transmitter antenna axis and the direction to the receiver, φ_R^i is the current angle of observation of the vehicle receiving antenna, $i = 1°, 2°, \ldots, 360°$. According to the conditions

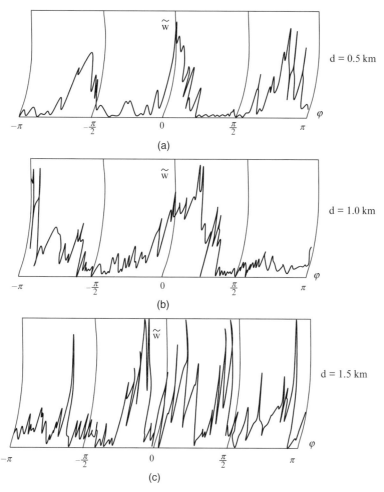

FIGURE 9.5. The results of measurements of the normalized signal power spectrum, $\tilde{w} = W(\varphi)/W(\varphi_T)$ for various ranges between the terminal antennas: $d = 0.5$ km (a), $d = 1.0$ km (b), $d = 1.5$ km (c).

of the experiments, $\varphi_R^i = \Delta\varphi(i-1)$, $\Delta\varphi = 1°$. The same methodology of measurements in the azimuth domain was described in Reference 3.

In Figure 9.5, results of measurements of the normalized signal power spectrum, $W(\varphi)/W(\varphi_T)$, where $W(\varphi_T) = W$max, are presented for the case when the height of the BS antenna was lower than the level of the average rooftop $\bar{h}$ (i.e., $z_2 = 6$ m $< \bar{h} = 8$ m), for the following distances from BS: (a) $d = 0.5$ km, (b) $d = 1.0$ km, and (c) $d = 1.5$ km. All data are presented for the case when the BS antenna axis is directed to the MV receiver antenna, that is, $\varphi_T \equiv 0°$. As follows from these measurements, when the two terminal antennas are lower than the rooftop level, the maximal level of the signal power at small ranges from the source is concentrated near the direction to the source. Simultaneously, some multipath signal components with power of the same order are registered from angles close to back scattering and reflection (see Fig. 9.5a,

$$\varphi_R^i \in \left(-\pi, -\frac{\pi}{2}\right) \cup \left(\frac{\pi}{2}, \pi\right).$$

This effect is increased with an increase of range between the two terminal antennas. At far zones from BS, multiray effects are observed, with quasi-uniform distribution of signal spectrum within the range of $[0, 2\pi]$ angles to the BS antenna, and with minimum energy from the direction to the MV (see Fig. 9.5b,c). Quite a different picture is observed for the case of the BS antenna higher than the overlay building profile, for example, $\bar{h} = 8\,\text{m} < z_2 = 12\,\text{m}$. In this case, as shown in Figure 9.6a–c, measured for the same conditions and ranges between both antennas, despite the fact that far from the BS, due to an

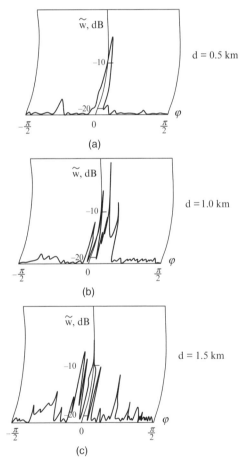

FIGURE 9.6. The same as in Fig. 9.5a–c, but for the base station antenna higher than building rooftops.

increase in the number of obstructions, the multipath components of the total signal are also present with approximately the same power. The maximum of the power spectrum is concentrated near the direction to the BS. All these features are predicted using the proposed muiltiparametric stochastic approach described in Chapter 8 (see a detailed description of this approach in References 15–23).

In general, with an increase in the BS antenna height and for low elevated MV antenna, the area, where all obstructions are located around the MV, becomes smaller, that is fewer obstructions are involved in the multiple scattering and reflection process. It means that for a high BS antenna the multiray phenomena become weaker. These results are seen in Figure 9.5a–c and in Figure 9.6a–c. To obtain enough statistical data for a theoretical prediction and better verification, averaged results of measurements of the pulse signal power spectrum distribution were obtained in References 16–19 by using approximately 10 local positions of the MV.

The normalized signal power azimuth spectrum $\tilde{W}(\varphi) = W(\varphi)/W_{max}$ (in decibel) is obtained and is presented by a continuous curve in Figure 9.7 for $z_2 = 5$ m (top figure) and for $z_2 = 12$ m (bottom figure), respectively. Here, the results of the calculations according to the proposed combined 3-D multiparametric model, which takes into account effects of diffraction, street orientations,

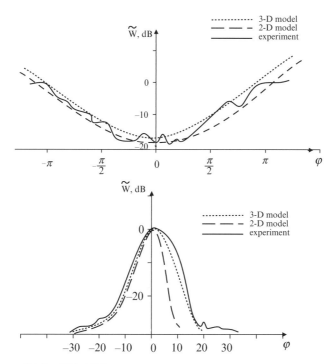

FIGURE 9.7. $\tilde{w}(\varphi) = W(\varphi)/W_{max}$ versus φ for $z_2 = 5$ m (top panel) and $z_2 = 12$ m (bottom panel).

and the overlay building profile (see Chapters 5 and 8), are presented in both figures by a dotted curve. The same calculations according to the regular 2-D stochastic model obtained in Reference 3 without accounting for all these propagation phenomena are presented by dashed curves.

All parameters of the calculations are mentioned above and also follow from the topographical map and the geometry of the experiments of the tested area described in References 1, 2, 21, and 22. As follows from the presented illustrations, the proposed model can, with great accuracy, predict the asymmetric shape of the signal power azimuth spread (PAS) distribution obtained experimentally, due to diffraction and nonuniform height distribution of the buildings. These effects cannot be explained by using the statistical model developed in References 3–7.

9.1.2. Signal Distribution in Time Delay Domain

The same dependence on the antenna height, predicted theoretically in Chapter 8, was obtained during experimental investigations of the signal power spectrum in the time delay domain. The method of our experimental investigation of the time delay spectrum distribution is based on recording the wideband channel response at the same 10 points of the MV schematically presented in Figure 9.4 by recording of the unit pulse responses of the channel. The transmitter pulse duration was 60 ns at the level of –3 dB from the maximum of the signal amplitude. The same geometry of the experiment and the same parameters of both antennas (MS and BS) and of the built-up terrain were used [1, 2, 16]. The results of measurements using both low terminal antennas of 6 and 2 m (see above) are shown in Figure 9.8 for the range between both terminals: (a) $d = 0.5$ km, (b) $d = 1.0$ km, and (c) $d = 1.5$ km.

As follows from the given illustration in the time domain, multipath pulse propagation, the number of incoming pulses exceeds 10–15 at all ranges from the BS, because of the effects of multiple scattering and reflection from the obstructions surrounding the moving vehicle. The impulse (wideband) signal power distribution over the time delay is quasi-homogeneous and unified. A different picture is observed for the high BS antenna ($\sim$12 m) with respect to building roofs. From the results of the normalized power spectrum distribution measurements in the time delay domain (presented in Figure 9.9a,b), it follows that the shape of the spectrum has an obvious maximum that corresponds to the direct pulse arriving from the BS at the moving vehicle receiving antenna. Only a few arriving pulses as multipath components due to multiple reflection and scattering are distributed close to the initial pulse, and their power is less than that of the main pulse. Here, three different positions of antennas, denoted by 1, 2, 3 near each curve, are presented. We also present by dashed curves in Figure 9.9a,b the theoretical results for various situations (denoted by 1 in Fig. 9.9a and 2, 3 in Fig. 9.9b, respectively). Again, as for the spectrum distribution in the azimuth domain, the theoretical predictions are in agreement with the experimental data.

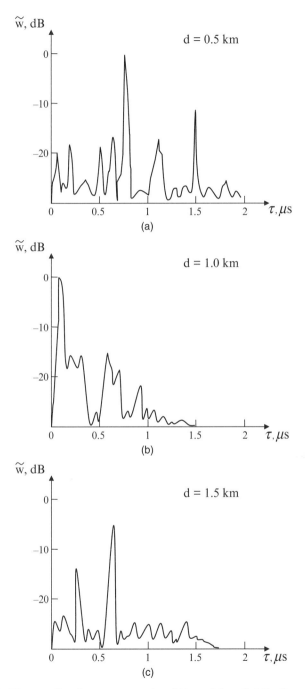

FIGURE 9.8. The results of measurements of time delay distribution for the range between both terminals: (a) $d = 0.5$ km, (b) $d = 1.0$ km, and (c) $d = 1.5$ km.

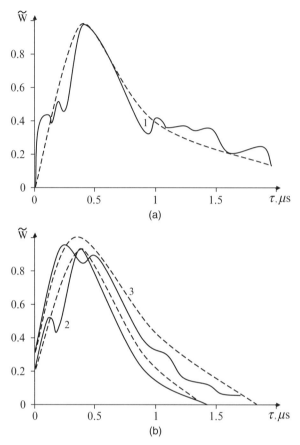

FIGURE 9.9. Comparison between measurements (continuous curves) and theoretical prediction (dashed curves) for time delay distributions for three position of vehicle antenna denoted by 1, 2 and 3, which corresponds the antenna height of 2, 5, and 12 m, respectively.

9.1.3. Signal Distribution in Doppler Shift Domain

To verify of the theoretical signal power spectrum distribution in the Doppler shift domain, a special experiment was carried at $f = 920$ MHz using the same experimental site described in References 1, 2, and 16 (see Fig. 9.4). This tested site contains mostly three five-floor buildings, which are uniformly distributed around both terminal antennas along the crossing streets and are nonuniformly distributed around the antennas. The range between the mobile antenna and the fixed BS antenna varied from 200 m to 2 km. The transmitter antenna was assembled at the top of the mobile vehicle at the height of 2 m; the receiver antenna was assembled on a crane at the height of 30 m, and then at the height of 3 m above the roof on the concrete building. The mobile vehicle speed varied from 10 to 40 km/h.

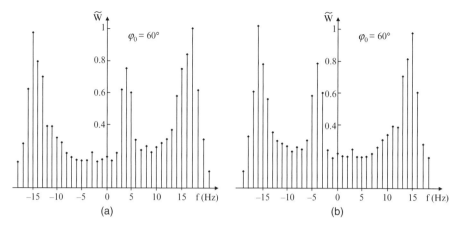

FIGURE 9.10. Power spectrum distribution in the Doppler shift domain obtained experimentally for low-elevated BS and vehicle antennas.

In Figure 9.10a,b, one of the measured samples of the signal power spectrum distribution, relative to that for $\varphi_0 = 0°$, is presented for the mobile trajectory oriented to the radio path between the terminal antennas upon the angle $\varphi_0 = 60°$ and for mobile speed of 32 km/h, from which we get that $f_d = \dfrac{v}{\lambda}\cos\varphi_0 \approx 15Hz$ in the case when both antennas are below the rooftop ($z_1 = 2$ m $< z_2 = 3$ m $< \bar{h}_b = 8 - 10$ m). Figure 9.10a,b correspond to the opposite directions of the same mobile vehicle along the experimental path. Both the spectra illustrated in Figure 9.10a,b are the mirror transformation of each other. This is because of the opposite direction of the movement of the vehicle antenna. The same picture follows from an experiment when the receiver antenna was above the rooftops ($z_2 = 30$ m) for opposite directions of movements of the vehicle antenna. It is shown in Figure 9.11a,b.

In the latter case, the mobile vehicle path was oriented along the angle $\varphi_0 = 65°$ to the radio path between terminal antennas. Again, asymmetry of the spectrum shape corresponds to the direction of the moving vehicle. Moreover, because of the existence of the line-of-sight (LOS) component of the total signal (BS antenna is higher than building roofs), the sharp component is clearly seen in the signal spectrum at the frequency, which corresponds to source movement in free space $f_{d0} = v/\lambda\cos\varphi_0 \approx 7$ Hz. In both cases presented in Figure 9.10 and Figure 9.11, the power spectrum is limited by the maximum Doppler frequency, which is around $\pm(15-17)$Hz.

A comparison between the experimentally obtained signal spectrum distribution in the Doppler shift domain and the Doppler power spectrum distribution obtained from theoretical analysis is presented in Chapter 8, which allow us to conclude that the data obtained experimentally fully confirm the theoretically predicted results. This makes the model extremely useful for fixed wireless access applications, where the subscriber antenna deployment can be made on rooftops. At the same time, taking into account effects of shadowing

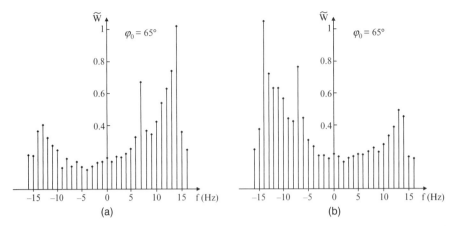

FIGURE 9.11. Power spectrum distribution in the Doppler shift domain obtained experimentally for a high-elevated BS antenna.

and multipath fading, the algorithm of how to evaluate these effects for specific situations in an urban environment is described in Chapter 10. We can improve the accuracy of the proposed stochastic approach even though some of the subscribers will be located in the shadow zones behind buildings. It means that, for a uniform angle-of-arrival (AOA) distribution of the signal, an additional slow- or long-term fading in the link budget, due to shadowing, can be estimated according to methods described in Reference 18 (see also Chapter 10).

We obtained from the proposed stochastic model a general description of the signal power distribution in the AOA (called "Power Azimuth Spread" [PAS]) and time delay (called "Power Delay Spread" [PDS]) domains for various BS antenna elevations with the continuous transition to both limiting cases of multipath phenomena. Either the diffraction or the multiple scattering component of the total field tends to be predominant, and the PAS and PDS distributions are closer to the classical Laplacian theory or to the Gaussian theory, respectively [16]. This fact was obtained experimentally and theoretically in References 3–12 and briefly described in Chapter 8. The proposed stochastic model predicts with great accuracy the nonuniform and nonregular distribution of the narrowband (CW) and wideband (pulse) signal power spectra in the azimuth, elevation, time delay, and Doppler shift domains for the case where BS and moving vehicle antennas are lower than the rooftops [15–21, 24]. By increasing the BS antenna height compared with the rooftops of the building, the power spectra shape in the angle, time, and frequency domains tend to approach the regular, uniform distribution.

Using the unified combined statistical multiparametric and cross-street waveguide model, we obtain clear relations between the parameters of the built-up terrain, the conditions of propagation, and the multipath signal characteristics, investigated in Chapters 1, 5, and 8. These characteristics of radio channel cannot be obtained from the simple empirical, semi-empirical and the so-called ray-tracing models proposed by References 4–12, as well as from those that are ever based on the statistical description of propagation effects [3, 13, 14, 24, 25].

9.2. PREDICTION OF ADAPTIVE ANTENNA CHARACTERISTICS BASED ON UNIFIED STOCHASTIC APPROACH

In Section 8.2 (see Chapter 8), we presented results of numerical simulations of the proposed unified stochastic approach [1–3, 15–21] through comparisons with real experiments carried out in downtown Helsinki [13, 14, 24, 25]. An agreement between the theoretical prediction and measured data for different built-up areas allows us to create a virtual numerical experiment with the specific antennas assembled at the BS. In our virtual numerical experiment, we will use, for example, the experimental site Rx1, shown in Figure 8.7, in which we will change the BS antenna directivity κ, azimuth φ, and tilt β, and will analyze how the changes to these parameters change the signal power distribution in such an urban area.

9.2.1. Tilt-Dependence of the Base Station Antenna

In this case we will take into account four different tilts of the Rx antenna $\beta = -20°, -10°, 10°, 20°$. Here, again, condition of $\beta > 0$ corresponds to the tilt up and $\beta < 0$ is to the tilt down from the horizon. Figure 9.12a–d shows these variations in 2-D plane and 3-D plane, respectively. From changes of the antenna tilt from a negative to a positive direction in the vertical (elevation) plane, it is clearly seen that when the tilt is directed down from horizon ($\beta < 0$), most of the energy arriving at the receiving (Rx) antenna (see Fig. 9.12a,b) is located in the direction of the receiver (i.e., around the pseudo-LOS direction). At the same time, when the tilt is directed up from the horizon ($\beta > 0$), most of the energy arrives from the areas located far from the Rx antenna (see Fig. 9.12c,d). This is caused by multipath components of the total signal because of the propagation along the two streets.

9.2.2. Azimuth-Dependence of the Base Station Antenna Maximum Loop

Now we will turn the array of the antenna by the maximum azimuth direction angle of $\varphi_0 = 5°$ and $50°$ from the north direction. The simulated results are shown in Figure 9.13a,b by the 2-D and 3-D radio maps for $\varphi_0 = 5°$ and $\varphi_0 = 50°$, respectively.

Figure 9.13a shows the case when the Rx antenna is turned left from the original direction and most of the energy is received from the azimuth direction of $\varphi_0 = 5°$; that is, more energy comes from street #1 and less energy is arriving from street #2 (see Fig. 9.14). This result differs from that in the real situation shown by Figure 8.34 and Figure 8.36 (see Chapter 8). In the situation described by Figure 9.13b, the energy arriving from the azimuth direction

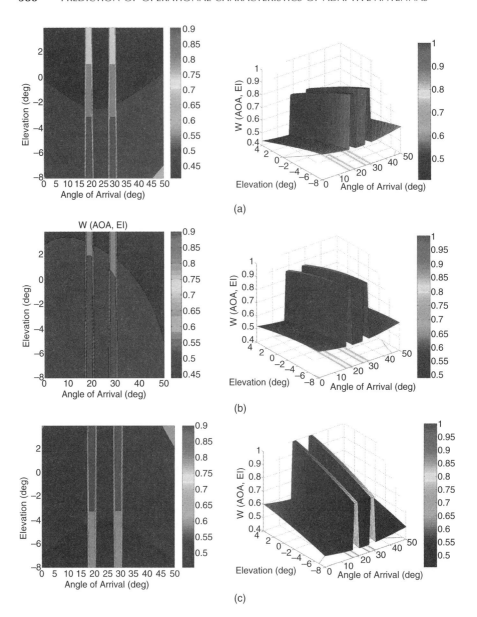

FIGURE 9.12. (a) The relative power, $\tilde{w}(\theta,\varphi)$, $\beta = -20°$: 2-D and 3-D plane. (b) the relative power, $\tilde{w}(\theta,\varphi)$, $\beta = -10°$: 2-D and 3-D plane. (c) The relative power, $\tilde{w}(\theta,\varphi)$, $\beta = 10°$: 2-D and 3-D plane. (d) The relative power, $\tilde{w}(\theta,\varphi)$, $\beta = 10°$: 2-D and 3-D plane.

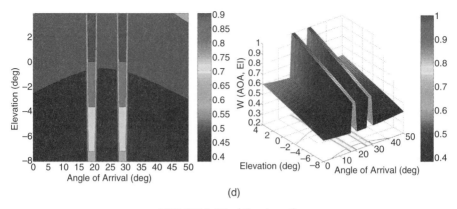

(d)

FIGURE 9.12. (*Continued*)

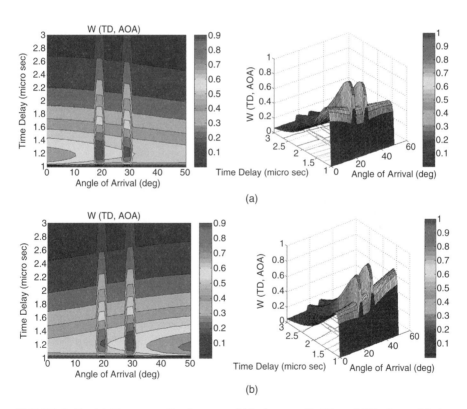

(a)

(b)

FIGURE 9.13. (a) The normalized power, $\tilde{w}(\theta, \varphi)$, $\varphi_0 = 5°$: 2D and 3D plane. (b) the relative power, $\tilde{w}(\theta, \varphi)$, $\varphi_0 = 50°$: 2-D and 3-D plane.

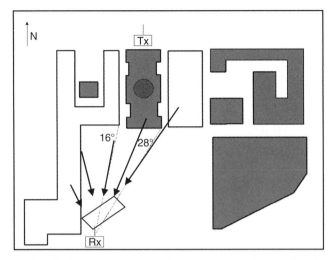

FIGURE 9.14. Detailed map for the situation where the azimuth direction is $\varphi_0 = 5°$.

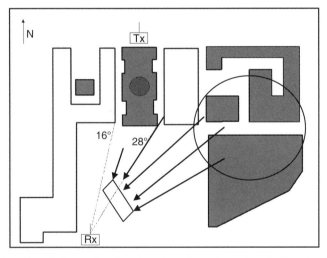

FIGURE 9.15. Detailed map for the situation where the azimuth direction is $\varphi_0 = 50°$.

$\varphi_0 = 50°$ is about 0.8–0.9 (compared with W_{max}); that is, it is of the same order with the ray energy arriving from street #1. In this situation, as shown by Figure 9.15, a significant part of the total energy comes from directions far from street #1 and street #2. The "side" effect, when the antenna is oriented as shown in Figure 9.14, gives the strong influence on signal energy azimuth redistribution as in the cases described by Figure 9.13b and Figure 9.15.

9.2.3. Directivity-Dependent Effects of the Base Station Antenna

Next, we analyze the normalized signal power spectrum $w(\theta, \varphi)$ for different values of the directivity of the antenna $\kappa = 10, 20, 30$. Figure 9.16a–c shows these cases, respectively. From this virtual numerical experiment, we can understand how the increment of directivity of the Rx antenna affects the decrease of the arriving power of the multipath components in the total signal power received by the BS antenna. It is clearly seen that with an increase in the directivity of the antenna κ in the vertical (elevation) plane, most of the energy arrives at the BS antenna from the direction closest to the zero degree elevation angle, working as a spatial filter to eliminate multipath components arriving from other direction in the elevation-of-arrival (EOA) plane.

9.3. POSITIONING OF THE DESIRED MOBILE SUBSCRIBER IN URBAN ENVIRONMENTS

As was mentioned earlier, there are several advanced approaches to find the position of any subscriber located in the area of service, which we propose to the reader to use for the purposes of increasing the efficiency of the wireless network on the basis of employment of adaptive (smart) multibeam antennas. We will consider them separately. There are three approaches that have been introduced recently for localization of any BS or any mobile subscriber (MS) antenna in built-up areas for various scenarios of BS antenna elevations with respect to buildings' height profile. First of all, we start with a practical approach based on the adaptive multibeam antenna applications.

9.3.1. Positioning of the Subscriber Based on Adaptive Antenna Application

As was discussed in Chapters 1 and 8, the terrain features, such building profiles and their density affect essentially the signal distribution in the AOA, the EOA, and time-of-arrival (TOA) [or time delay] domains. Basically, the corrupted information is sent via each channel for each desired user and introduces strong multiplicative noise into the channel. In such a situation, each subscriber obtains its own specific signal-to-noise ratio (SNR). We explain this again schematically in Figure 9.17, where each subscriber located under different angles with respect to the transmitting antenna is influenced by different buildings, profiles, and their densities, characterized by a number of buildings lining each azimuth direction, their density, the maximum and minimum height of building in each direction, and so forth.

Therefore, it is important to obtain *a priori* distribution of the signal for each subscriber in joint AOA-TOA domains (actual mostly for mobile communication), or in joint AOA, TOA, and EOA domains (mainly for the personal communication, when the subscriber is located inside the building). This methodology is based on the knowledge of the azimuthal direction to the

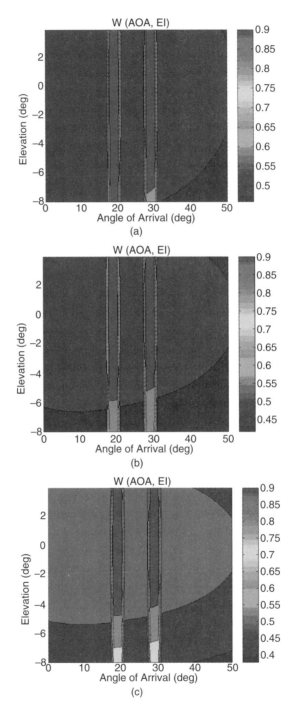

FIGURE 9.16. (a) The normalized power, $\tilde{w}(\theta, \varphi)$, $\kappa = 10$. (b) The same as in panel a, but for $\kappa = 20$. (c) The same as in panel a, but for $\kappa = 30$.

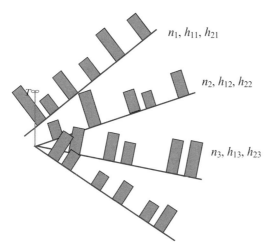

n_1, h_{11}, h_{21}

n_2, h_{12}, h_{22}

n_3, h_{13}, h_{23}

FIGURE 9.17. Schematical presentation of the specific overlay building profile for different directions of the beam of the adaptive antenna.

desired subscriber, where the maximum of RSSI (or SNR), as well as the minimum time delay is measured at the receiver. So, maximization of the signal power for the desired direction with the minimization of the time delay (with respect to other multipath components arrived at the receiver) allow us to find the real direction and the distance of the desired user located in the area of service. The obtained parameters fully depend on information about features of built-up terrain, which can be obtained a priori from the corresponding topographic map of the area of service.

We present below a special experiment carried out recently in a medium-size town with mainly three- to four-story buildings, where two antennas with low elevation (1.5 m each) and the subscriber antenna at a height of 1 m where used (see Fig. 9.18). Each sectoral antenna, transmitting and receiving, having four beams for each sector with velocity of scanning (from one beam to another) of 50–100 ns and operating at the frequency of 5.8 GHz were used in such an experiment. The tested area with dimensions 250–250 meters, was filled by several buildings. As it is shown in Figure 9.18 (left side), the first antenna found the subscriber with an accuracy of 1–2 m. For the second antenna, shown in Figure 9.19, where clutter conditions were more complicated, the accuracy of the positioning was 6–10 m. We can state that the proposed approach, based on utilization of two multibeam antennas and on signal intensity distribution in joint distance AOA-TOA domain, allows us to obtain the position of any subscriber with accuracy of several meters.

9.3.2. Positioning of Any MS by One BS Antenna Using Statistical Approach

This approach is based on the stochastic multiparametric model, particularly, on the use of the multiray components of the total signal distribution in the

(a)

(b)

(c)

FIGURE 9.18. View on (a) the transmitting antenna, (b) the receiving antenna, and (c) the moving subscriber transmitter/receiver antenna.

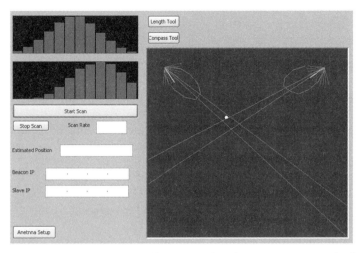

FIGURE 9.19. The left side picture: histogram of each antenna pattern obtained over the azimuthally scanning angle with a step of 1.5°; the right side picture: positioning of the subscriber (denoted by the bold point) with two scanning antennas.

joint AOA-TOA domains. This distribution is presented in Chapter 8 by Equation (8.50) and Equation (8.51) for low- and high-elevated terminal antennas, respectively.

Based on the results of the stochastic model, in Reference 26, the maximum-likelihood (ML) and minimum distance measure (MDM) approaches were proposed for any MS localization from single BS. Moreover, in Reference 27 was shown that the ML approach is optimal when the stochastic model accurately represents the propagation conditions for the specific scenarios occurring in the urban scene, and it is sensitive to the modeling errors.

Combining this approach with the MDM method, we can overcome the major drawbacks of the ML method [27]. Then, following References 26 and 27 (see also Chapter 8), it was supposed that the resulting distribution of AOA-TOA measurements that are received at the BS via multipath replicas, can be parameterized by hypothesized MS locations, $\mathbf{p}_j = [x_j, y_j]$, in plane (x,y), and, finally, can be expressed for the above-the-rooftop level in LOS propagation conditions as

$$f_a(t, \phi; \mathbf{p}_j) = \frac{\nu d(\mathbf{p}_j)}{4} \left[\frac{(h_R - \bar{h})}{\bar{h}} \tilde{r}(\mathbf{p}_j) e^{-\gamma_0 \tilde{r}(\mathbf{p}_j)} + \frac{\bar{h} r(\mathbf{p}_j) d(\mathbf{p}_j) \gamma_0}{h_R} \tau(\mathbf{p}_j) e^{-\gamma_0 r(\mathbf{p}_j) \frac{\bar{h}}{h_R}} \right].$$

(9.1)

Note that when the BS and the MS antennas are below the rooftop level in non-line-of-sight (NLOS) propagation conditions, Equation (9.1) can be rewritten as follows (see Chapter 8):

$$f_b(t, \phi; \mathbf{p}_j) = 0.5\gamma_0\nu\sin^2\left(\frac{\alpha}{2}\right)t(\mathbf{p}_j)c\exp[-\gamma_0 t(\mathbf{p}_j)c]$$
$$\exp\left[-2\frac{|\ln\chi|}{a'(\phi(\mathbf{p}_j))}\cdot\frac{t^2(\mathbf{p}_j)c^2 - d^2}{2(t(\mathbf{p}_j)\cdot c - d(\cos\phi(\mathbf{p}_j)))}\right]. \qquad (9.2)$$

All parameters presented in Equation (9.1) and Equation (9.2) are fully described in Chapters 5 and 8.

The proposed MS localization approach, which is motivated by a stochastic multiparametric model of the RF propagation conditions in dense urban environment, is the following [26]:

- We should find the maximum of the likelihood (ML) function described by (9.1) or Equation (9.2), that is, to find minimum error in distance to the desired mobile subscriber (MS).
- In order to find the MS location, the urban environment is discretized into J hypothesized MS locations [27], and (10.1) and/or (10.2) are used to obtain the statistical models of AOA-TOA measurements, $f(t, \phi; \mathbf{p}_j)$ $\forall j = 1, \dots, J$. Note that the family of these distributions depends on the source locations $\{\mathbf{p}_j, \forall j = 1, \dots, J\}$, and the proposed localization approach interprets it as a location likelihood function, parameterized by the source location.

In References 26 and 27, the problem of MS localization was formulated in the framework of target classification. The proposed classification process consists of two stages, where during the first, training stage, the statistical models of the hypothesized MS locations are obtained. Figure 9.20 schematically shows the training stage. Note that due to the statistical nature of the adopted propagation model, this training process does not involve any data collection. During the second, localization stage, the location likelihood function in the AOA-TOA measurements space, $\{f(\mathbf{x}; \mathbf{p}_j), j = 1, \dots, J\}$, was obtained for all potential MS locations. This likelihood function might be multimodal, and therefore, a set

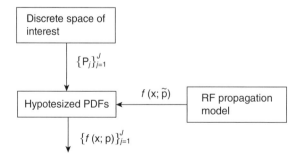

FIGURE 9.20. Schematic representation of the training process.

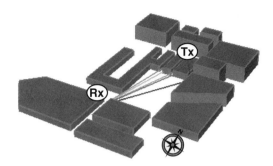

FIGURE 9.21. Scene I of the tested area in Helsinki, Finland. 3-D layout rearranged from Figure 8.7, accounting for buildings' profile.

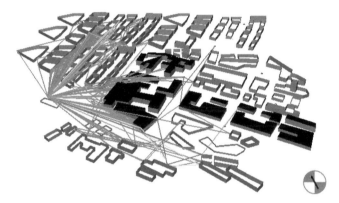

FIGURE 9.22. Scene II of the tested area in Helsinki, Finland. 3-D layout rearranged from Figure 8.9, accounting for buildings' profile.

of local maxima, and the set of corresponding source locations, can be obtained according to the proposed algorithm presented in Figure 9.20.

To verify such an approach, the experimental areas in Helsinki were investigated by rearranging the 2-D pictures of the Site I and the Site II, presented in Figure 8.7 and Figure 8.9 (see Chapter 8), into 3-D building array pattern, as shown in Figure 9.21 and Figure 9.22, respectively, by accounting for the overlay profile of the buildings obtained from the topographic map of both experimental sites.

Below, the performance of the proposed MS localization approach is evaluated using the simulated and the collected measurements in two urban scenes. Figure 9.21 shows urban Scene I with an approximate area size of 400 m × 400 m. This scene was adopted from the measurement collection campaign in Helsinki, Finland [13, 14, 24, 25]. Figure 9.22 shows the urban Scene II with an approximate area size of 900 m × 900 m. This scene contains randomly

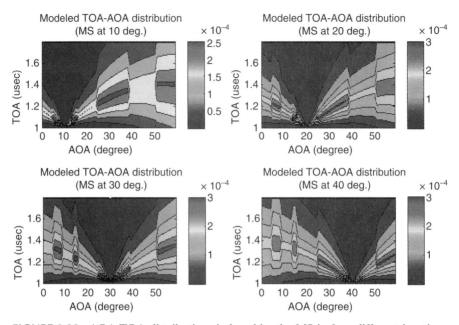

FIGURE 9.23. AOA-TOA distributions induced by the MS in four different locations with fixed distance from the BS, and the azimuth angles of 10°, 20°, 30°, 40°.

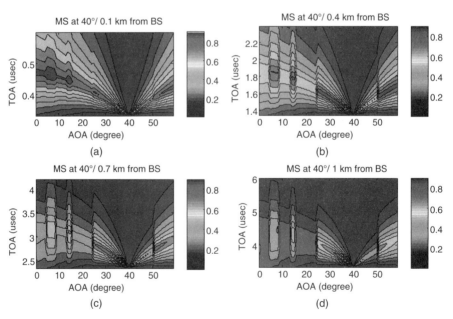

FIGURE 9.24. Received signal distributions in the TOA-AOA domain induced by the MS located at the azimuth of 40° and distances of [0.1, 0.4, 0.7, 1] km from the BS.

oriented buildings, arrays of buildings, and straight crossing streets. Considering the partition size of $0.01 \mu sec \times 1°$ in polar coordinates, the areas in Scene I and Scene II in the AOA-TOA space $\Omega = (-45° < \phi < -25°, 0.1 < d < 0.9$ [km]) were discretized into $J = 9 \times 21 = 189$ hypothesized MS locations. Space discretization determined the minimal achievable resolution of 100 m in time (range) domain, and of 1° in azimuth domain. Every one of the hypothesized MS location, $\mathbf{p}_j = [x_j, y_j]$, was found by using the statistical model described now by (9.1) and (9.2).

Note that the number of partitions in the area of interest determines the maximal achievable localization accuracy. In this work, the number of the partitions J was selected to provide the sufficient range of localization error values for reasonable performance evaluation [26, 27].

An omnidirectional antenna with 4 dBi gain for the MS, and a directional antenna with the horizontal beamwidth of 120° and 17 dBi gain for the BS were simulated. A dynamic range of 150 dB was considered for an appropriate balance between the transmitting power and the receiver sensitivity. The MS antenna height of 1.5 m above the ground level was simulated in all scenarios. Both scenes were simulated with the wavelength of $\lambda = 0.14$ m, and the average building length of $\bar{L} = \pi / 80$ km. The density of $\nu = 150$ and $\nu = 200$ buildings in square kilometer was simulated in Scene I and in Scene II, respectively. The building height was considered to be uniformly distributed with the mean height of $\bar{h} = 15$ m. The discontinuity parameter of $\chi = 0.8$ and $\chi = 0.75$ was simulated in Scene I and in Scene II, respectively.

Using the estimated above parameters of the built-up terrain, the relation between the MS geographical location in the urban Scene I in Figure 9.23 and the corresponding received signal distributions in the TOA-AOA domain simulations were investigated in Reference 26. Thus, considering polar coordinates with the BS in its origin, Figure 9.23 shows the received signal distribution, induced by the MS transmission from the locations with the fixed distance of 0.4 km from the BS and different azimuth angles of [10°, 20°, 30°, 40°]. Next, Figure 9.24 shows the received signal distribution, induced by the MS transmission from the locations with the fixed azimuth of 40° and different distances from the BS of [0.1, 0.4, 0.7, 1]km. Notice that in order to investigate the received signal distributions induced by the MS transmission from different ranges (different TOAs), the TOA-axis in subplots (a)–(d) in Figure 9.24 were selected to be different.

Finally, Figure 9.25 shows the received signal distributions, induced by (a) random scattering, (b) one waveguide, (c) two waveguides, and (d) local scattering in the MS proximity in the above-the-rooftop propagation scenario. All these scenarios are shown schematically in Figure 9.26. It can be emphasized that Figure 9.23, Figure 9.24, and Figure 9.25 support our conjecture that the MS transmission from different geographical locations induces significantly different received signal distributions in the TOA-AOA domain that could be exploited for the MS localization.

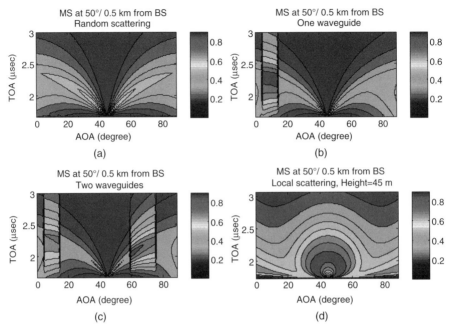

FIGURE 9.25. Received signal distributions in the TOA-AOA domain induced by (a) the random scattering, (b) one waveguide, (c) two waveguides, and (d) local scattering in the above the rooftop propagation.

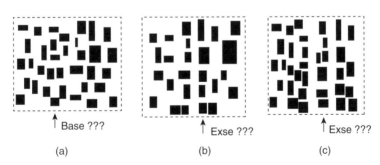

FIGURE 9.26. Three scenarios of buildings' layout: (a) randomly distributed buildings, (b) existence of one-street waveguide, and (c) existence of two-street waveguides.

Next, we present some practical aspects obtained from this statistical approach. Figure 9.27 shows the AOA-TOA likelihood function. One can notice the multimodal nature of the 2-D likelihood function. This multimodality can be explained by the fact that when the MS and BS antennas are below the rooftop level, the MS illuminates both local and distant scatterers (via the multi-slit waveguides), whose reflections are received at the BS. As a result,

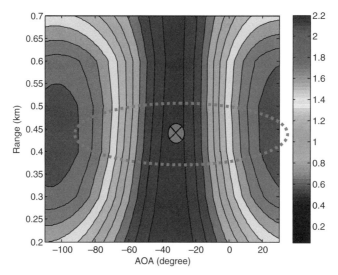

FIGURE 9.27. AOA-TOA likelihood functions, when BS is below the rooftop level. The true MS position marked by a cross is at AOA=−41° (the estimated AOA is 37°), and at the distance (true and estimated) of 420 m from the BS.

multiple subsets of scatterers are illuminated, inducing multimodal behavior of the likelihood function in the AOA-TOA domain. Note that this multimodality is a natural behavior of the maximum likelihood (ML) decision rule (i.e., maximum probability to localize a subscriber) in case of NLOS propagation conditions due to the fact that this ML result shows the distribution of scatterers that have maximum likelihood to be illuminated by MS and those scatterers are distributed in space domain, surrounding the MS.

Moreover, according to References 26 and 27, the accuracy of MS location in Scene I scenario was estimated as $\varphi = -37°$ and $d = 0.4$ km. Comparing this result with the true MS location at $\varphi = -41°$ and $d = 0.4$ km, the localization performance of the proposed statistical approach are evaluated using the RMS localization error [26, 27]:

$$RMSE(x_{MS}, y_{MS}) = \sqrt{(x_{MS} - \hat{x}_{MS})^2 + (y_{MS} - \hat{y}_{MS})^2} \, [m]. \qquad (9.3)$$

For the considered scenario shown in Figure 9.21, the localization RMSE of 32 m was achieved. Note that here the simulated model is derived for a straight-crossing street configuration, and different street structures are expected to generate different spreading phenomena.

Next, a scenario corresponding to that shown in Figure 9.22, where the BS antenna was placed on the rooftop level, was analyzed in Reference 26. Figure 9.28 shows the likelihood function of the AOA-TOA measurements. The localization mean square error in the distance domain, RMSE = 48 m, was achieved

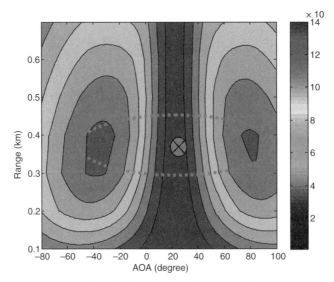

FIGURE 9.28. ML AOA-TOA likelihood functions, when BS is on the rooftop level. The true MS position marked by a cross is at AOA=−23°, and at the distance of 310 m from the BS. The *RMSE* = 48 m.

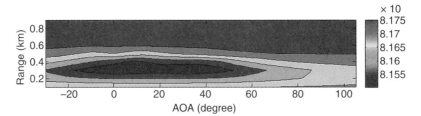

FIGURE 9.29. The AOA-TOA likelihood functions, when BS is above the rooftop level. The true MS position is around AOA = 22°, distance of ∼300 m (true and estimated) from the BS.

in this scenario; that is, the accuracy in this scenario is approximately less than twice than that in the case of the first scenario. An additional scenario shown in Figure 9.24 for Scene II, where the BS antenna is placed above the rooftops, was simulated in Reference 26, and is presented in Figure 9.29. The true MS position was considered to be at AOA = 22° and at distance of 300 m from the BS.

9.3.3. Positioning of Any MS Based on Shannon's Capacity Bound

Let us briefly describe this approach and present some important results on how to minimize the localization error of any MS located within a specific area of service by using one to three BSs operated within the array of elements of the adaptive multibeam antenna.

The framework of this statistical approach is called *information theoretic bound* (ITB) and was evaluated and proposed by Buck for the acoustic source localization problem using passive sonar [28, 29]. We briefly will follow Reference 30 because in this work the proposed ITB concept was adapted to the MS positioning in the urban environment using one to three BS antennas with various configurations of their location.

The ITB concept interprets the MS positioning as a pure communication system, in which the MS location is simply a "message" that is encoded by the propagation channel response, that is, by the propagation features of the outdoor channel through which this message was sent. The BS, as the receiver, uses the observed codeword to estimate the MS location through the Shannon capacity bound related to the entropy and the probability of error that were found through the data processing theorem combined with source-channel coding [30]. In other words, the codeword for the desired *message* (i.e., desired MS) represents the receiving by BS signal in the joint AOA-TOA domain induced by the urban propagation conditions (i.e., the channel response). This aspect is fully described in Chapters 5 and 8.

The procedure described in Reference 30 allows for minimizing the error of MS localization and predicting the location of the MS as a desired codeword "embedded" in the information data flow arriving at the receiver as "noise" (caused by multiple rays coming from many obstructions located in area of service surrounding the desired MS).

In such interpretations of information theory (IT) via the "physical layer" interpretation, that is, the radio propagation in urban environments for various elevations and configurations of BS antennas, the theoretical approach proposed in Reference 30 differs from the well-known statistical approaches of the source localization developed separately by other researchers [29, 31–37, 64].

Discretization of the Urban Environment Area. According to the algorithm proposed in Reference 30, the MS location in the 2-D coordinate domain $S = \{x, y\}$ is described by a random vector $\mathbf{W} = [x, y]^T$ that is characterized by the corresponding PDF, $f_w(\mathbf{w})$, and by the specific realizations of this random vector, $\mathbf{w}_j, \forall j = 1, 2, \ldots, J$, which are randomly distributed at the rough built-up terrain. Formulating the MS localization problem using IT concepts, proposed by Reference 28, the authors in Reference 30 have divided uniformly the area of interest S of the size $D \times R$ into separate partitions, $v_n, \forall n = 1, 2, \ldots, N$, having hexagonal shape. In such a procedure, a partition $V(\mathbf{W})$ was assigned to every possible MS location $\mathbf{W}$ and was modeled as a discrete random variable uniformly distributed in area S with its PDF, $f_v = (V)$, and having the following outputs: $V = v_1, V = v_2, \ldots, V = v_N$, where N is number of partitions. In such definitions and geometrical consideration, the radius of the hexagon-shaped partition will be $\Delta d = \sqrt{2(D \times R) / 3\sqrt{3}N}$.

Stochastic Approach of the Urban Propagation Conditions. According to the above procedure, Equation (9.1) and Equation (9.2) can be now parameterized by every MS location, defined by its possible realizations, $\mathbf{w}_j$, for the corresponding partition v_n, defined by PDF $f_v(V)$, that is,

- for the BS antenna elevated below the rooftop level:

$$f(t, \phi, \mathbf{w}_j) = \frac{0.25\gamma_0 v t(\mathbf{w}_j)c\beta(\phi, \mathbf{w}_j)\left[(t(\mathbf{w}_j)c)^2 - d^2(\mathbf{w}_j)\right]\exp[-\gamma_0 t(\mathbf{w}_j)c]}{t(\mathbf{w}_j)c - d(\mathbf{w}_j)\cos\phi(\mathbf{w}_j)}$$
$$\times \exp\left\{-\frac{\left[(t(\mathbf{w}_j)c)^2 - d^2(\mathbf{w}_j)\right]|\ln\chi|}{[t(\mathbf{w}_j)c - d(\mathbf{w}_j)\cos\phi(\mathbf{w}_j)]a'}\right\} \tag{9.4}$$

- for the BS antenna elevated above the rooftop level:

$$f(t, \phi, \mathbf{w}_j) = \frac{0.25\gamma_0 v t(\mathbf{w}_j)c\beta(\phi, \mathbf{w}_j)\left[(t(\mathbf{w}_j)c)^2 - d^2(\mathbf{w}_j)\right]\left(\dfrac{\bar{h}}{h_R}\right)}{t(\mathbf{w}_j)c - d(\mathbf{w}_j)\cos\phi(\mathbf{w}_j)}$$
$$\times \exp\left\{-\gamma_0 \frac{\left[\tilde{r}(t, \phi, \mathbf{w}_j) + \dfrac{\bar{h}}{h_R}(t(\mathbf{w}_j)c)^2 - d^2(\mathbf{w}_j)\right]}{2[t(\mathbf{w}_j)c - d(\mathbf{w}_j)\cos\phi(\mathbf{w}_j)]}\right\} \tag{9.5}$$
$$+ \frac{v\beta(\phi, \mathbf{w}_j)(h_R - \bar{h})\tilde{r}(t, \phi, \mathbf{w}_j)}{2\bar{h}}\exp\{-\gamma_0 \tilde{r}(t, \phi, \mathbf{w}_j)\}$$

Here, $t(\mathbf{w}_j) = \tau(\mathbf{w}_j)d(\mathbf{w}_j)/c$ is TOA (in seconds), $\phi(\mathbf{w}_j)$ is AOA (in degrees), $d(\mathbf{w}_j)$ is the straight line distance between the MS location, $\mathbf{w}_j$, and the BS location (in kilometers), $\chi = \bar{L}/(\bar{L}+\bar{l})$, $a' = ((4a^2/\lambda^2 n) + a^2)^{1/2}$ is the effective width of the street waveguide, where a is the real street width, and n determines number of reflections or the wave modes inside the waveguide (usually not exceeding $n = 2$, see Chapter 5):

$$\beta(\phi, \mathbf{w}_j) = \sin^2\left[0.5\sin^{-1}\left(\frac{d(\mathbf{w}_j)\sin(\phi(\mathbf{w}_j))}{\tilde{r}(\mathbf{w}_j)}\right)\right], \tag{9.6a}$$

$$\tilde{r}(t, \phi, \mathbf{w}_j) = \frac{c^2 t^2(\mathbf{w}_j) - 2ct(\mathbf{w}_j)d(\mathbf{w}_j)\cos\phi(\mathbf{w}_j) + d^2(\mathbf{w}_j)}{2[ct(\mathbf{w}_j) - d(\mathbf{w}_j)\cos\phi(\mathbf{w}_j)]}. \tag{9.6b}$$

The first term in the product of (9.4) describes scattered waves from buildings randomly distributed surrounding the MS and the BS, and the exponent describes guiding effects caused by a grid of crossing-street waveguides. As was shown in Chapter 5, if $\chi \to 1$, a strong guiding effect dominates the urban propagation conditions. Conversely, if $\chi \to 0$, the random scattering from buildings is predominant in urban propagation conditions. The same type of scattering from buildings, randomly distributed around the MS and the BS is described the first term in (9.5), whereas the second term in (9.5) describes the effects of local scattering in the MS proximity. These propagation features were analyzed in detail in Chapter 5.

The ITB Framework. Using these two signal distributions for two scenarios of the BS antenna elevation, one can derive the bound on the partition size that provides necessary conditions to achieve a minimum localization error defined by the IT probability, P_e^L, as a probability that one or more errors are made out of the L decisions on the MS transmitter partitions, that is,

$$P_e^L = P\{\widehat{\mathbf{V}}^L \neq \mathbf{V}^L\}, \tag{9.7}$$

where $\mathbf{V}^L = [V_1, \ldots V_L]^T$ is the vector of a sequence of partitions. As was mentioned earlier, the vector $\mathbf{W}^L = [\mathbf{w}_1, \mathbf{w}_2, \ldots, \mathbf{w}_L]^T$, as a sequence of independent randomly distributed MS locations (i.e., messages) in area of interest S, maps into sequence of partitions $\mathbf{V}^L = [V_1, \ldots V_L]^T$. If so, the sequence of L transmitted codeword, $\mathbf{X}^L = [\mathbf{x}(\mathbf{w}_1), \mathbf{x}(\mathbf{w}_2), \ldots, \mathbf{x}(\mathbf{w}_L)]^T$, results at the BS in a sequence of received codeword $\mathbf{Y}^L = [\mathbf{y}(\mathbf{w}_1), \mathbf{y}(\mathbf{w}_2), \ldots, \mathbf{y}(\mathbf{w}_L)]^T$ (with "noise" caused by multipath).

At the final step of the algorithm, the number of bits that is required to describe the area of interest is estimated through the entropy $H(V)$ of the partition random variable V [28, 29]:

$$H(V) = -\sum_{n=1}^{N} f_v(v_n) \log_2 f_v(v_n). \tag{9.8}$$

Combining the results of source-channel coding and the data processing theorem, one can also find the bound on partition entropy, H, that provides the minimum probability of the error, P_e^L,

$$H(V) = I(V, \widehat{V}) \leq I(\mathbf{X}, \mathbf{Y}). \tag{9.9}$$

Then, using (9.9) and the expression of entropy via the hexagon area [30],

$$H(V) = \log_2 \frac{2R \times D}{3\sqrt{3}(\Delta d)^2}, \tag{9.10}$$

we can present the bound of partition size that provides minimum of P_e^L as

$$(\Delta d) \geq \sqrt{\frac{2}{3\sqrt{3}}} (R \times D)^{1/2} 2^{-0.5I(X,Y)}. \tag{9.11}$$

Finally, the authors in Reference 30 present the partition entropy (9.9) through the propagation characteristics of urban environment the specific covariance

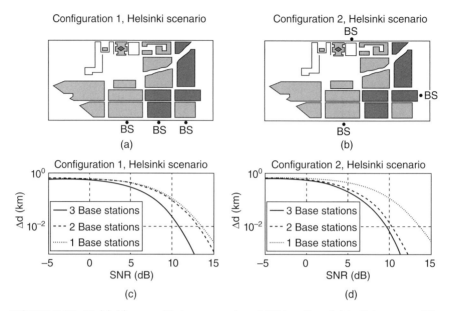

FIGURE 9.30. Helsinki area with two scenarios of BS location: (a) in lining one of the outer streets; (b) triangulate scenario. (c) Accuracy of positioning of any MS by use of one, two, and three BS antennas in panel a. (d) Accuracy of positioning of any MS by use of one, two, and three BS antennas in panel b.

matrices that include Equation (9.4) or Equation (9.5) (depending on the scenario of the BS antenna elevations) and obtain the achievable lower bound on the partition size that provides minimum probability of the error, P_e^L.

IT-Bound Framework Verification in Urban Scene. Some numerical experiments for several scenarios presented in Reference 30 are shown in Figure 9.21 and and Figure 9.22. We will show an example where the IT localization bound was analyzed for a network of spatially distributed BSs. The sensitivity of the IT localization bound to the number of BSs in the network and their configuration in the network were investigated in this numerical experiment. In Figure 9.30, the upper subplots, (a) and (b), show two simulated network configurations where three BSs were located: (a) in a line along the outer street of the area of interest and (b) at three different sides of outside the area of interest. The bottom subplots, (c) and (d), show the localization bounds for the simulated networks configurations with one, two, and three BSs in the corresponding network.

It is clear from the presented subplots that a minimum error in MS localization (i.e., minimum localization bound) can be achieved with an increase in the number of BSs in the network. Thus, the accuracy of MS localization by using one or two antennas for scenario (c) is several times weaker with respect to that when three BSs are employed. The same tendency is observed in scenario (b) using one BS and presented by subplot (d) in Figure 9.30. However, in this scenario of a more diverse network configuration, even two-BS network is enough to achieve a good accuracy in MS localization. Thus, the comparison between the two configurations of the network presented in subplots (c) and (d) demonstrates that significant improvement in the MS localization can be associated with the spatial diversity in the network BS locations. In terms of the IT bound framework, this means that the geographical diversity corresponds to different encoding of the transmitted message and allows to obtain the minimum probability of the error in the partition size estimation where the desired MS is located.

SUMMARY

Let us briefly summarize the advantages of the proposed methods of localization of any mobile subscriber (MS) by using only one BS with sectorized or multibeam antennas via existing and recently performed localization techniques. Thus, some alternative approaches for localization of any MS located in area of service in built-up environments were proposed [38–53]. The motivation behind these investigations is due to the increasing demand for multiuser servicing with steady growth of the number of mobile subscribers [38], by novel capabilities of 4-G WiMAX networks [43] and long-term evolution (LTE) technology [44], which finally, will satisfy the mandatory requirements of the Federal Communication Committee (FCC) [Phase Enhanced 911] for quick and precise location identification of wireless 911-seviced subscribers [39], as well as for location-based services [54, 55]. Thus, the recently performed deterministic techniques were dealt with the special function that described urban environment in the BS proximity and allowed to achieve a good MS location performance [39]. However, the proposed method requires exact knowledge of the urban environment surrounding the BS and does not account for the BS antenna height. As was shown in Chapters 5 and 10, as well as in References 1–7 and 15–20, the latter aspect strongly affects the accuracy of MS localization due to the different "response" of the built-up channel (e.g., propagation environment) on scenarios with below-the-rooftop and above-the-rooftop BS antennas.

At the same time, in References 40–42, by using simplified 3-G multiple-input-multiple-output (MIMO) model, the guiding effects along the straight-crossing streets was ignored, which, as was shown in References 1, 2, 16, and 17, can be one of the major propagation phenomena in the below-the-rooftop propagation conditions.

Existing conventional geometric-based models that use triangulation technique are based on the estimation of location-dependent parameters for multiple spatially distributed BS antennas [45–52]. Achieving high MS localization accuracy, the triangulation method requires the availability of the LOS propagation conditions between all three BS antennas and the desired MS antenna. As was shown by numerous experiments carried out in built-up environments [1–7, 15–20, 45, 53–58], the geometric techniques cannot be usually used in a common practice in NLOS-limited urban and suburban scenarios.

The above proposed localization approach consists of two stages: training and classification. During the *training stage*, the area of service was described into J virtual locations of desired MSs, each of which was represented by the statistical model of the urban propagation conditions from this virtual location to the BS. This was done without any data collection of the built-up terrain-surrounded MS and BS. In the proposed stochastic approach, all parameters are obtained analytically from the area map, and therefore, the proposed training stage does not require any data collections in the urban area of interest. This is a main advantage of the proposed technique with respect to other statistical models described in References 3–7 and 38–42, which obtain the statistical parameters from the site-specific or scenario-specific measurements and involve intensive data collection campaigns [59–63]. Lastly, in this stage, the height of BS antennas is taken into account with respect to building profile heights, describing all conditions: below, at the same level, or above any rooftop propagation conditions. During the *classification stage*, the decision on the MS location was made by finding the statistical approach that fits with the received measurements best, based on the ML and MDM methods for the desired MS localization. The major benefits of the proposed approaches are that they can be implemented using a single BS or multiple BSs. Moreover, it does not require the network-based localization methods [45–52] or the methods of identification and mitigation of the measurements in NLOS propagation conditions, as were proposed in References 56–59.

Other statistical approaches presented in Reference 30 propose an information theoretical framework, developed by Buck [28, 29], for the performance of the localization of any subscriber, moving or stationary, by using one to several BSs located in an urban environment. This approach interprets the desired MS location with accuracy from several tens of meters (for one BS antenna) to 10 ms (for three simultaneously operating BSs). Because the approach is based on the proposed stochastic model of radio propagation in multipath urban environments and uses in estimation of lower bound the binary error metrics, it gives better prediction of any MS position in an urban environment with those statistical frameworks that are based on the conventional techniques of global bounds estimation, such as MSE metric of Zakai-Ziv [31] and Weinstein–Weiss [32], and so on (see also References 33–37).

On the other hand, from Figure 9.27, Figure 9.28, Figure 9.29, and Figure 9.30, one can see that the two theoretical approaches based on the stochastic multiparametric model of radio propagation in multipath urban environments fully agree with the practical approach that uses adaptive multibeam antennas. Furthermore, the latter approach gives for each MS the localization accuracy not exceeding several meters with respect to several tens of meters predicted by the two theoretical approaches.

REFERENCES

1   Blaunstein, N. and M. Levin, "VHF/UHF wave attenuation in a city with regularly spaced buildings," *Radio Sci.*, Vol. 31, No. 2, 1996, pp. 313–323.

2   Blaunstein, N. and M. Levin, "Propagation loss prediction in the urban environment with rectangular grid-plan streets," *Radio Sci.*, Vol. 32, No. 2, 1997, pp. 453–467.

3   Ponomarev, G.A., A.N. Kulikov, and E.D. Telpukhovsky, *Propagation of Ultra-Short Waves in Urban Environments*, Rasko, Tomsk, Russia, 1991.

4   Pedersen, K., P.E. Mogensen, and B. Fleury, "Power azimuth spectrum in outdoor environments," *IEE Electron. Letters*, Vol. 33., 1997, pp. 1583–1584.

5   Pedersen, K., P.E. Mogensen, and B. Fleury, "Dual-polarized model of outdoor propagation environments for adaptive antennas," in *Proc. IEEE Veh. Technol. Conf.* (VTC'99), Houston, TX, 1999, pp. 990–995.

6   Pedersen, K.I., P. Mogensen, and B.H. Fleury, "Experimental analysis of the joint statistical properties of azimuth spread, delay spread, and shadow fading," *IEEE J. Select. Areas Communic.*, Vol. 20, No. 3, 2002, pp. 523–531.

7   Algans, A, K.I. Pedersen, and P. Mogensen "Experimental analysis of the joint statistical properties of azimuth spread, delay spread, and shadow fading," *IEEE J. Select. Areas Communic.*, Vol. 20, No. 3, 2002, pp. 523–531.

8   Ertel, R.B., P. Cardieri, K.W. Sowerby, T.S. Rappaport, and J.H. Reed, "Overview of spatial channel models for antenna array communications systems," *IEEE Personal Communic.*, Vol. 5, No. 1, 1998, pp. 10–22.

9   Fuhl, J., J.P. Rossi, and E. Bonek, "High resolution 3-D direction-of-arrival determination for urban mobile radio," *IEEE Trans. Antennas Propagat.*, Vol. 44, No. 4, 1997, pp. 672–682.

10 Martin, U., J. Fuhl, I. Gaspard et al., "Model scenarios for direction-selective adaptive antennas in cellular communication systems-scanning the literature," *Wireless Personal Communic.*, Vol. 11, No. 1, 1999, pp. 109–129.

11 Paulraj, A. and C. Papadias, "Space-time processing for wireless communications," *IEEE Signal Processing Magazine*, Vol. 14, No. 1, 1997, pp. 49–83.

12 Kuchar, A., J.P. Rossi, and E. Bonek, "Directional macro-cell channel characterization from urban measurements," *IEEE Trans. Antennas Propagat.*, Vol. 48, No. 1, 2000, pp. 137–146.

13 Kalliola, K., H. Laitenen, L. Vaskelainen, and P. Vainikainen, "Real-time 3-D spatial-temporal dual-polarized measurement of wideband radio channel at mobile station," *IEEE Trans. Instrum. Measurements*, Vol. 49, No. 3, 2000, pp. 439–448.

14 Kalliola, K. and P. Vainikainen, "Characterization system for radio channel of adaptive array antennas," *Proc. Int. Symp. Personal Indoor Mobile Radio Conf.* (*PIRMC'97*), Helsinki, Finland, 1997, pp. 95–99.

15 Blaunstein, N., "Distribution of angle–of–arrival and delay from array of building placed on rough terrain for various elevation of base station antenna," *Journal of Communic. and Networks*, Vol. 2, No. 4, 2000, pp. 305–316.

16 Blaunstein, N., "Average field attenuation in the nonregular impedance street waveguide," *IEEE Trans. on Antennas and Propagat.*, Vol. 46, No. 6, 1998, pp. 1782–1789.

17   Blaunstein, N. and E. Tsalolihin, "Signal distribution in the azimuth, elevation and time delay domains in urban radio communication links," *IEEE Antennas and Propagat. Magazine*, Vol. 46, No. 5, pp. 101–109, 2004.

18   Blaunstein, N., M. Toeltsch, Ch. Christodoulou, et al., "Azimuth, elevation and time delay distribution in urban wireless communication channels," *Antennas and Propagat. Magazine*, Vol. 48, No. 2, pp. 112–126, 2006.

19   Blaunstein, N., M. Toulch, J. Laurila, et al., "Signal power distribution in the azimuth, elevation and time delay domains in urban environments for various elevations of base station antenna," *IEEE Trans. Antennas and Propagation*, Vol. 54, No. 10, 2006, pp. 2902–2916.

20   Blaunstein, N., "Prediction of cellular characteristics for various urban environments," *IEEE Antennas and Propagat. Magazine*, Vol. 41, No. 6, 1999, pp. 135–145.

21   Blaunstein, N. and Y. Ben-Shimol, "Spectral properties of signal fading and Doppler spectra distribution in urban communication mobile links," *Wireless Communic. and Mobile Computing*, Vol. 6, No. 1, 2006, pp. 113–126.

22   Blaunstein, N., Katz, D., and M. Hayakawa, "Spectral properties of modulated signal in the Doppler domain in urban multipath radio channels with fading," *IEEE Trans. Antennas and Propagat.*, Vol. 58, No. 8, 2010, pp. 2795–2800.

23   Blaunstein, N., D. Katz, and M. Hayakawa, "Spectral properties of modulated signal in the Doppler domain in urban radio channel with fading," *European Conference on Antennas and Propagation 2010*, April 12–16, 2010, Barcelona, pp. 234–237.

24   Toeltsch, M., J. Laurila, K. Kalliola et al., "Statistical characterization of urban spatial radio channels," *IEEE J. Select. Areas Communic.*, Vol. 20, No. 3, 2002, pp. 539–549.

25   Laurila, J., K. Kalliola, M. Toeltsch et al., "Wide-band 3-D characterization of mobile radio channels in urban environment," *IEEE Trans. Antennas Propagat.*, Vol. 50, No. 2, 2002, pp. 233–243.

26   Tsalolihin, E., I. Bilik, and N. Blaunstein, "A single-base-station localization approach using a statistical model of the NLOS propagation conditions in urban terrain," *IEEE Trans. on Vehicular Technology*, Vol. 60, No. 3, 2011, pp. 1124–1137.

27   Bilik, I., K. Adhikari, and J.R. Buck, "Information theoretic bounds on mobile source localization in a dense urban environment," in *Proc. IEEE SAM Workshop*, 2010, pp. 109–112.

28   Buck, J.R., "Information theoretic bounds on source localization performance," *Proc. IEEE SAM*, 2002.

29   Meng, T. and J. R. Buck, "Rate distortion bounds on passive sonar performance," *IEEE Trans. Signal Processing*, Vol. 58, No. 1, 2010, pp. 326–336.

30   Bilik, I., K. Adhikari, and J. R. Buck, "Shannon capacity bound on mobile station localization accuracy in urban environments," *J. Select. Areas of Communic.*, Vol. 59, No. 12, 2011, pp. 6206–6216

31   Zakai, M. and J. Ziv, "Improved lower bounds on signal parameter estimation," *IEEE Trans. on Information Theory*, Vol. 21, No. 1, 1975, pp. 90–93.

32   Weinstein, E. and A. Weiss, "Fundamental limitations in passive time-delay estimation—Part II: Wide-band systems," *IEEE Trans. Acoustics, Speech, and Signal Processing,* Vol. 32, No. 5, 1984, pp. 1064–1078.

33 Kaemarungsi, K. and P. Krishnamurthy, "Modeling of indoor positioning systems based on location fingerprint," *Proc. INFOCOM*, Vol. 2, 2004, pp. 1012–1022.

34 Miao, H., K. Yu, and M. Juntti, "Positioning for NLOS propagation: algorithm derivation and Cramer-Rao bounds," *IEEE Trans. on Veh. Technol.*, Vol. 56, no. 5, 2007, pp. 2568–2580.

35 Kikuchi, S., A Sano, and H. Tsuji, "Blind mobile positioning in urban environment based on ray-tracing analysis," *EURASIP Journal of Applied Signal Processing*, Vol. 26, 2006, pp. 1–12.

36 Nezafat, M., M. Kaveh, H. Tsuji, and T. Fukagawa, "Subspace matching localization: a practical approach to mobile user localization in microcellular environments," *Proc. VTC*, 2004, pp. 5145–5149.

37 Zhao, L., G. Yao, and J. Mark, "Mobile positioning based on relaying capability of mobile stations in hybrid wireless networks," *Ins. Elect. Eng. Communic. Proc.*, Vol 153, No. 5, 2006, pp. 762–770.

38 Poretta, M., P. Nepa, G. Manara, and F. Giannetti, "Location, location, location," *IEEE Vehicular Technol. Magazine*, Vol. 1, No. 1, 2008, pp. 20–29.

39 Poretta, M., G. Nepa, G. Manara, et al., "A novel single base station location technique for microcellular wireless networks: description and validation by a deterministic propagation model," *IEEE Trans. Veh. Tech.*, Vol. 53, No. 5, 2004, pp. 1502–1514.

40 Pedersen, K., P. Mogensen, and B. Fleury, "A stochastic model of the temporal and azimuthal dispersion seen at the base station in outdoor propagation environments", *IEEE Trans. Veh. Technol.*, Vol. 49, No. 2, 2000, pp. 437–447.

41 Foschini, G. J., "Layered space-time architecture for wireless communication in a fading environment when using multi-element antennas", *Bell Labs Technol. J.*, Vol. 1, No. 1, 1996, pp. 41–59.

42 Allnutt, R. M., T. Pratt, and A. Dissanayake, "A study in small scale antenna diversity as a means of reducing effects of satellite motion induced multipath fading for handheld satellite communication systems", in *Proc. IEEE, 9th Int. Conf. Antennas Propagat.*, Eindhoven, The Netherlands, 1995, pp. 135–139.

43 Ergen, M., *Mobile Broadband—Including WiMAX and LTE*, Springer, New York, 2009.

44 Dahlman, E., S. Parkvall, J. Sköld, P. Beming, *3G Evolution—HSPA and LTE for Mobile Broadband*, Acad. Press, 2008.

45 Sun, G., J. Chen, W. Guo, and K. Liu, "Signal processing techniques in network-aided positioning," *IEEE Signal Processing Magazine*, Vol. 22, No. 4, 2005, pp. 12–23.

46 M. Vossick, L. Wiebking, P. Gulden, J. Weighardt, C. Hoffmann, and P. Heide, "Wireless local positioning," *IEEE Microwave Magazine*, Vol. 3, 2003, pp. 77–86.

47 Zhao, Y., "Standardization of mobile phone positioning for 3G systems," *IEEE Com. Magazine*, Vol. 40, 2002, pp. 108–116.

48 Ahonen, S., and P. Eskelinen, "Mobile terminal location for UMTS," *IEEE Aerospace Electron. Syst. Mag.*, Vol. 18, No. 2, 2003, pp. 23–27.

49 Zhao, L., G. Yao, and J. Mark, "Mobile positioning based on relaying capability of mobile stations in hybrid wireless networks," *Proc. Inst. Elect. Eng.—Commun.*, Vol. 153, No. 5, 2006, pp. 762–770.

50 Sayed, A., A. Tarighat, and N. Khajehnouri, "Network-based wireless location," *IEEE Signal Processing Magazine*, Vol. 22, No. 4, 2005, pp. 24–40.

51 Gustafsson, F. and F. Gunnarsson, "Mobile positioning using wireless networks," *IEEE Sig. Proc. Magazine*, Vol. 22, No. 5, 2005, pp. 41–53.

52 Kleine-Ostmann, T. and E. Bell, "A data fusion architecture for enhanced position estimation in wireless networks," *IEEE Commun. Lett.*, Vol. 5, No. 8, 2002, pp. 343–345.

53 Chan, Y., W. Tsui, H. So, and P. Ching, "Time-of-arrival based localization under NLOS conditions," *IEEE Trans. Veh. Technol.*, Vol. 55, No. 1, 2006, pp. 17–24.

54 McGuire, M., K. Plataniotis, and A. Venetsanopoulos, "Location of mobile terminals using time measurements and survey points," *IEEE Trans. Veh. Tech.*, Vol. 52, No. 4, 2003, pp. 999–1011.

55 Al-Jazzar, S. and J. Caffery, "ML and Bayesian TOA location estimators for NLOS environments," *Proc. IEEE Veh. Tech. Conf.,* 2002, pp. 1178–1181.

56 Ma, C., R. Klukas, and G. Lachapelle, "A nonline-of-sight error-mitigation method for TOA measurements," *IEEE Trans. Veh. Tech*, Vol. 56, No. 2, 2007, pp. 641–651.

57 Miao, H., K. Yu, and M. Juntti, "Positioning for NLOS propagation: Algorithm derivation and Cramer-Rao Bounds," *IEEE Trans. on Vech. Technol.* Vol. 56, No. 5, 2007, pp. 2568–2580.

58 Zhu, X., M. Shi, J. Zhang, X. Tao, and P. Zhang, "A scattering model based non-line-of-sight error mitigation algorithm via distributed multi-antenna," *Proc. IEEE PIMRC*, 2007, pp. 18–22.

59 Klukas, C., G. Lachapelle, "A nonline-of-sight error-mitigation method for TOA measurements," *IEEE Trans. on Vech. Technol.* Vol. 56, No. 2, 2007, pp. 641–651.

60 Kikuchi, S., A. Sano, and H. Tsuji, "Blind mobile positioning in urban environment based on ray-tracing analysis," *EURASIP Journal of Applied Signal Processing*, Vol. 2006, 2006, pp. 1–12.

61 Battiti, R., M. Brunato, and A. Villani, "Statistical learning theory for location fingerprinting in wireless LANs," *Tech. Rep. DIT-02-0086, Dept. Inform. Telecomun., Universita di Trento*, 2002.

62 Hightower, J. and G. Borriello, "Location systems for ubiquitous computing," *Computer*, Vol. 34, 2001, pp. 57–66.

63 Guvenc, I. and C. Chong, "A survey of TOA based wireless localization and NLOS mitigation techniques," *IEEE Communic. Surveys Tuts.*, Vol. 11, No. 3, 2009, pp. 107–124.

64 Tsalolihin, E., I. Bilik , and N. Blaunstein, "Mobile user location in dense urban environment using unified statistical model," *European Conference on Antennas and Propagation 2010* (EuCAP 2010), April 12–16, 2010, Barcelona, Spain, pp. 157–158.

CHAPTER TEN

Multipath Fading Phenomena in Terrestrial Wireless Communication Links

For all cellular radio communication networks, land, atmospheric, and iono-spheric, the process of transmitting information must be accompanied by the knowledge of parameters of both the transmitting and receiving terminal antennas at the ends of the radio link. It is also important to have some knowledge of the statistical parameters of the channel and their correlations in the space, time, and frequency domains, in order to evaluate the performance of the radio system considered. As was mentioned in Chapter 1, the important statistical characteristics that must be predicted is the path loss, slow, and fast fading, which allow radio network designers to predict strict link budgets, and to obtain the full radio coverage of areas of service, that is, to create radio maps of service areas.

In order to avoid measuring channel statistics for all operating environments and networks designs, we proposed in Chapters 5 and 8 a unified stochastic approach for multipath radio channel description based on real physical phenomena, such as multiple reflections, diffraction, and scattering from various nontransparent obstructions (trees, hills, houses, and buildings) located in the terrain. All of them produced the relevant effects when compared to measurements reported in the literature for specific rural, forested, mixed residential, suburban, and urban environments.

Radio Propagation and Adaptive Antennas for Wireless Communication Networks: Terrestrial, Atmospheric, and Ionospheric, Second Edition. Nathan Blaunstein and Christos G. Christodoulou.
© 2014 John Wiley & Sons, Inc. Published 2014 by John Wiley & Sons, Inc.

In Section 10.1, we compare the results of theoretical predictions of the loss characteristics in the space domain, with experiments carried out in various land environments. Here, we also focus our attention on estimating the accuracy of the statistical description of the built-up terrain on radio signal spatial attenuation and its frequency dependence. Furthermore, we analyze the advantages and limitations of the proposed stochastic approach concerning mostly urban environments with very complicated built-up terrain configurations. Section 10.2 presents a theoretical and experimental analysis of slow and fast fading based on the unified algorithm estimating these phenomena in various communication links, taking into consideration both the classical methods and the unified stochastic multiparametric model. Then, the typical scenarios occurring in the built-up environments are discussed by converting the resulting equations of the stochastic model, obtained in Chapter 5, into simple "straight-line" mathematical descriptions. In Section 10.3, the role of Rician K-factor in multipath phenomena differentiation is briefly described both in space, angle of arrival (i.e., azimuth), and time of arrival (i.e., time delay) domains. Section 10.4 deals with radio coverage and radio map constructions accounting for terrain features and overlay profile effects of the buildings, that is, the 3-D configuration of built-up terrains. These terrain features have not been utilized until now in the physical analysis of the problem of radio propagation in various land communication links.

10.1. PREDICTION OF LOSS CHARACTERISTICS FOR TERRESTRIAL RADIO LINKS

As was mentioned in Chapter 5, the proposed 3-D multiparametric model is the stochastic approach that combines a statistical description of the terrain and overlay profile of the buildings, with a description of the signal strength and average intensity based on the theoretical description of propagation phenomena in homogeneous and inhomogeneous media (see Chapters 4 and 5). Here, specific features of land communication channels based on multiple reflections, diffraction, and scattering are described, caused by various obstructions as well as accounting for their vertical and horizontal reflection characteristics (see Chapter 5).

It should be pointed out that by using the multiparametric stochastic approach, we cannot give exact, point-to-point description of signal strength surrounding each individual obstruction, which can be easily obtained by the use of ray-tracing approaches [1–4]. In the stochastic approach, we strive to express some average properties of obstructions that take into account scattering (or *diffuse reflections*) from the rough wall surfaces and diffraction mechanisms from building rooftops and corners along the propagation paths. With all the advantages of the proposed physical–statistical method to describe the terrain relief and overlay profile of the buildings, as well as the average signal intensity or path loss, it is clear that it cannot explain some extreme cases of receiver antennas positioned within deep shadow regions surrounded by tall buildings. For such cases, the strict "knife-edge" models, mentioned in Chapter 4 (see also References 2 and 5–13), are better predictors of the local signal strength. As will be shown in Section 10.2, in some experimental sites where antennas were located well within the shadow zone caused by nearby buildings, the median error and standard deviations between theoretical predictions and measured data are significant, (around 15–20 dB). Therefore, the validity and accuracy of the multiparametric stochastic approach will be established via experimental data.

10.1.1. Statistical Distribution of Buildings in Urban Environments

In Chapter 6, by introducing the stochastic approach, we have assumed that the number of rays N, which arrive at the receiver during time t after multiple diffuse reflections by randomly distributed buildings lining the streets or the terrain, can be described by the Poisson distribution described in Chapter 5 (see also References 14–20). An array of buildings (nontransparent scatterers) was presented there as an "ordinary and simple flow of positive virtual pulses," where each pulse represents the real building location at the terrain. Such mathematical description of an array of buildings randomly distributed around both terminal antennas allowed us, in Chapter 5, to determine the probability of direct visibility between antennas and other statistical functions. Finally, we used them to derive the average field intensity as a superposition of coherent and incoherent components of the total signal average intensity. Poisson's law strictly describes the real situation of building distribution in the urban environment, with randomly distributed buildings as nontransparent screens. To verify this law for real situations in the urban scene, a statistical analysis of topographic maps of built-up areas that correspond to the experimental sites mentioned in References 14–32 was carried out.

If the randomly distributed buildings create a flow of randomly distributed positive virtual pulses placed at the smooth plane, then the distribution of such pulses of amount m, at the limit segment d along the radio path must be described according to Poisson's law by the following probability function:

$$P_d(m) = \frac{(\mu)^m}{m!} \exp\{-\mu\}, \tag{10.1}$$

where the mean number of positive virtual pulses (μ) within the segment d can be presented in the following form [14, 19, 20]:

$$\mu = \int_0^l \gamma_0(\mathbf{r}_{1\perp} + \mathbf{q}_{\perp}l)dl, l \in (0,d]. \tag{10.2}$$

As defined in Chapter 5, $\mathbf{r}_{1\perp}$ is the projection of the vector to the transmitter at the ground plane $z = 0$ (see Fig. 5.15, Chapter 5); $\mathbf{q}_{1\perp} = (\mathbf{r}_{2\perp} - \mathbf{r}_{1\perp}) / d$ is a unit vector at the ground plane $z = 0$; $d = |\mathbf{r}_{2\perp} - \mathbf{r}_{1\perp}|$, where $\mathbf{r}_{1\perp}$ and $\mathbf{r}_{2\perp}$ are the projections of the vector to the transmitter and to the receiver at the ground plane $z = 0$, respectively; parameter γ_0 determines the density of the building contours at the plane $z = 0$ and was defined by (5.13) or (5.53) in Chapter 5.

We have used only similar areas with the quasi-homogeneous building density of 80–100 buildings per square kilometer, the average building length of 70–100 m, and the amount of building floors varied from 5 to 15. Finally, all open areas consisting of parks and gardens were eliminated from consideration. Using topographic maps of various experimental sites, a set of virtual radio paths for each position of the base station antenna (or the radio port) were constructed around it from $0°$ up to $360°$, with the step of $5°$. Thus, for radio paths with ranges $d \in (0, 200\ \text{m})$, a number of crossings (outcomes) with buildings along each radio path for all experimental sites were 920, for radio paths with ranges $d \in (200\ \text{m}, 500\ \text{m})$ this amount was 550, and for radio paths with ranges $d \in (500\ \text{m}, 1000\ \text{m})$ it was 360. We summarized this statistics in Figure 10.1a,b,c by the corresponding continuous curves for each range: 200 m, 500 m, and 800 m, respectively. Then using this statistics and the parameters of each topographic map of the concrete experimental site, the mean value μ was estimated and used for calculation of the actual Poisson's distribution (10.1). Results of these calculations are presented in Figure 10.1a, b, c by dashed curves.

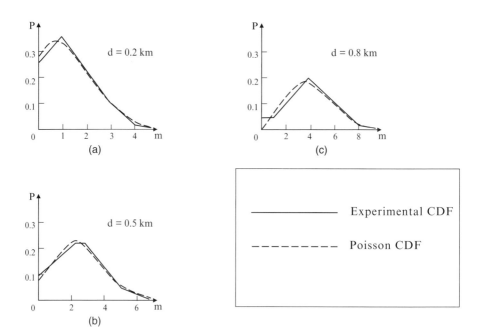

FIGURE 10.1 (a–c) Buildings' distribution above terrain for each tested radio path: 200 m, 500 m, and 800 m, respectively. Continuous curves describe statistical data; dashed curves describe actual Poisson distribution according to (10.1).

Here, as in Chapter 5, we consider an ordinary and simple flow of scatterers, that is, each single ray scattered as independent without crossings with other rays or other scatterers. Notice that only rays with the level of –10 dB below the maximum were taken into account in our statistical analysis. Moreover, for four similar experimental sites of Beer-Sheva, Tel-Aviv, Ramat-Gan, and Jerusalem, the topographic maps were described in more detail compared to other experimental sites. A statistical analysis of the number of waves N arriving at the receiver for each concrete radio path, which corresponds to the number of scatterers, was carried out. Figure 10.2a,b show the distribution of rays after multiple scattering. Poisson's law can explain the statistical distribution for about 70% of rays for radio links that vary from 300 m to 500 m (Fig. 10.2a), and for about 90% of rays for radio links that vary from 800 m to 1.5 km (Fig. 10.2b).

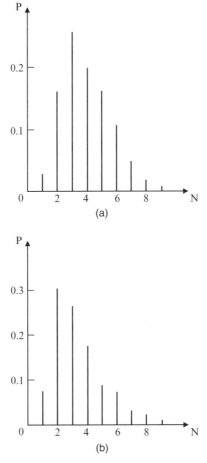

FIGURE 10.2. Probability laws of rays' number for links: (a) from 300 to 500 m and (b) from 800 to 1.5 km.

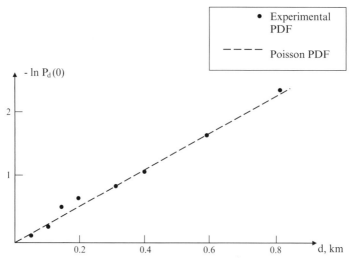

FIGURE 10.3. Resul t s of the statistical analysis of direct visibility versus the range d between base station and each virtual subscriber; points corresponds to the statistical experiment, dashed curve is obtained according using Equation (10.3).

Simultaneously, the same statistical analysis from the experimental sites mentioned earlier was carried out to verify situations of direct visibility between antennas along each radio path. The results of this statistical analysis are depicted by points in Figure 10.3 in a logarithmic scale versus the range between the real base station and virtual subscriber placed at the range d from this station.

Using Equation (5.52) from Chapter 5, which describes the probability of direct visibility, and rewriting it in the following form,

$$P_d(0) = \exp(-\gamma_0 d), \tag{10.3}$$

where for each experimental site and for the concrete range d, the parameter γ_0 was estimated, we finally calculated this probability function presented in Figure 10.3 by the dashed curve. From the results of statistical analysis of the topographic map of various built-up areas and those calculated using Equations (10.1) and (10.3), a significant deviation from Poisson's law is observed, specifically for radio links with ranges between the antennas that are less than 50–100 m. Beyond these ranges (important for micro and macrocell planning), these deviations are negligible.

10.1.2. Influence of Terrain Features on Loss Characteristics

The accuracy of the theoretical prediction depends on the range of variance of each parameter that describes the terrain features. Thus, in rural forested, mixed residential, and urban areas, the variations of the tree and building densities, the average height, and reflection properties significantly affect the signal path loss. Therefore, it is very important to estimate all these parameters accurately.

Let $\bar{\Theta}$ be a vector of the set of parameters we are using to describe the terrain features

$$\bar{\Theta} = \{\Gamma, \gamma_0, l_h, l_v, \delta_1, \delta_2\} \tag{10.4}$$

and $\Delta\bar{\Theta}$ be the vector of errors in each of those parameters. Here, $\delta_1 = (\bar{h} - z_1)$ and $\delta_2 = (z_2 - \bar{h})$. All parameters of the terrain are defined in Chapters 5 and 8. Now, let Θ_i denote one of these parameter, and $\Delta\Theta_i$ be an error in that parameter. It is clear that an error in the path loss (ΔL_i) introduced by the ith parameter, where other five parameters are constant, can be written in the following form:

$$\Delta L_i = \frac{\partial L}{\partial \Theta_i} \Delta\Theta_i. \tag{10.5}$$

Following expressions for the total path loss immediately have come from (10.5) that

$$\frac{\partial L}{\partial \Theta_i} = -\frac{10}{\ln 10} 10^{\frac{L}{10}} \frac{\partial(<I>_{co} + <I>_{inc})}{\partial \Theta_i}. \tag{10.6}$$

Let us now analyze the variations of the total path loss according to (10.6) for the mixed residential areas with vegetation, according to (5.29) and (5.30), and for the urban areas, according to (5.76) and (5.78). In this case we get

$$\frac{\partial <I>}{\partial \Theta_1} \equiv \frac{\partial <I>}{\partial \Gamma} = \frac{1}{\Gamma} <I> \tag{10.7}$$

$$\frac{\partial <I>}{\partial \Theta_2} \equiv \frac{\partial <I>}{\partial \gamma_0} = -\gamma_0 <I_{co}> -2\gamma_0 <I_{inc}> [K_1 + K_2] \tag{10.8}$$

$$\frac{\partial <I>}{\partial \Theta_3} \equiv \frac{\partial <I>}{\partial l_h} \equiv \frac{\partial <I_{inc}>}{\partial l_h} = <I_{inc}> K_3 \tag{10.9}$$

$$\frac{\partial <I>}{\partial \Theta_4} \equiv \frac{\partial <I>}{\partial l_v} \equiv \frac{\partial <I_{inc}>}{\partial l_v} = <I_{inc}> K_4 \tag{10.10}$$

$$\frac{\partial <I>}{\partial \Theta_5} \equiv \frac{\partial <I>}{\partial \delta_1} = <I_{inc}> K_5 + <I_{co}> K_6 \tag{10.11}$$

$$\frac{\partial <I>}{\partial \Theta_6} \equiv \frac{\partial <I>}{\partial \delta_2} \equiv \frac{\partial <I_{inc}>}{\partial \delta_2} = <I_{inc}> K_7. \tag{10.12}$$

Here,

$$K_1 = \frac{(2\pi\ell_h)^2}{\lambda^2 + \left[2\pi\ell_h\bar{L}\gamma_0\right]^2}, \quad K_2 = \frac{\left[2\pi\ell_v(\bar{h} - z_1)\right]^2}{\lambda^2 + \left[2\pi\ell_v\gamma_0(\bar{h} - z_1)\right]^2}$$

$$K_3 = \frac{1}{l_h} - \frac{8\pi^2\gamma_0^2}{\lambda^2 + \left[2\pi\ell_h\bar{L}\gamma_0\right]^2}, \quad K_4 = \frac{1}{l_v} - \frac{8\pi^2\gamma_0^2(\bar{h} - z_1)^2}{\lambda^2 + \left[2\pi\ell_v\gamma_0(\bar{h} - z_1)\right]^2}$$

$$K_5 = -\frac{4\pi^2\gamma_0^2 l_v^2}{\lambda^2 + \left[2\pi\ell_h\gamma_0(\bar{h} - z_1)\right]^2}, \quad K_6 = -\frac{\gamma_0 d}{z_2 - z_1}$$

$$K_7 = \frac{(z_2 - \bar{h})}{\left[(\lambda d / 4\pi^3)^2 + (z_2 - \bar{h})^2\right]}.$$

(10.13)

Without a loss of generality, we replaced the profile function $F(z_1, z_2)$ in the earlier equations, with its simple approximation $(z_2 - \bar{h})$, valid for a quasi-homogeneous distribution of building heights (when the parameter of building profile $n = 1$) (see Chapter 5). As seen by computations of the earlier equations, which are presented in Table 10.1, deviations of the terrain and antenna location parameters over the range of $\pm(30-50\%)$ lead to deviations in the average path loss in the wide range, mostly due to variations of the parameter of building contour density γ_0. These variations of path loss can reach $\pm(7.0-15.0)$ dB. At the same time, mistakes in the terminal antenna elevation can vary the path loss only by 3–5 dB, and become irrelevant if the parameters of the building walls or the parameters of the terrain are changed in $\pm(30-50\%)$; on 1–2 decibels only. Hence, the analysis presented earlier shows that by having full information about the tested area and conditions of the experiment, one can precisely predict the loss characteristics in the area of service.

In the following section, we will estimate the influence of built-up relief and terrain features on the frequency dependence of the total signal intensity within various kinds of land communication channels.

10.1.3. Frequency Dependence of Signal Power in Various Built-Up Areas

As was shown in References 33–39, the total field attenuation increases in built-up areas with an increase in the radiated frequency. Other researchers

TABLE 10.1. The Accuracy of the Initial Parameters of the Multiparametric Model

Parameter	Value	Variations	Path Loss Variations
γ_0	0.1–1 km^{-1}	0.05–0.5 km^{-1}	$\pm(7.0-15.0)$ dB
Γ	0.6–0.9	0.1–0.3	$\pm(0.5-1.5)$ dB
l_h, l_v	0.5–0.2 m	0.1–0.5 m	$\pm(1.0-3.0)$ dB
δ_1, δ_2	3–7 m	1–2 m	$\pm(3.0-5.0)$ dB

have obtained the same conclusions later [10–14]. Such dependence given in energy units can be presented in the following form [10–14, 33–39]:

$$\langle I_{total} \rangle \sim f^{-p}. \tag{10.14}$$

The loss factor, p, varies from 0.2 to 2.5, with an increase in frequency from 100 MHz to 3 GHz at the radio ranges that do not exceed 8–10 km. Beyond this range, a weak frequency dependence of signal strength loss is observed experimentally [33–36].

Now, let us present the frequency dependence of the signal intensity for various kinds of terrains, mixed residential, and urban, taking into account the detailed information about the spatial distribution of buildings and natural obstructions (hills, trees, vegetation etc.).

Results of signal frequency dependence (10.14) obtained from numerous experiments can be explained obviously using results of the theoretical prediction of the total field frequency dependence described by (5.49) for the grid-street scene, by (5.29) and (5.30) for mixed residential, and (5.76) to (5.79) for suburban and urban areas.

In fact, for urban and suburban areas, we consider in (5.76) that $\lambda^2 < \left(2\pi l_v \gamma_0 (\bar{h} - z_1) \right)^2$, then

$$\langle I_{inc1} \rangle \sim \lambda^{3/2} (\sim f^{-3/2}), \text{ if } (\lambda d / 4\pi^3) > \left(z_2 - \bar{h} \right)^2 \tag{10.15a}$$

and

$$\langle I_{inc1} \rangle \sim \lambda^1 (\sim f^{-1}), \text{ if } (\lambda d / 4\pi^2) > \left(z_2 - \bar{h} \right)^2. \tag{10.15b}$$

In the case, when we deal with (5.78)

$$\langle I_{inc2} \rangle \sim \lambda^4 (\sim f^{-4}), \text{ if } (\lambda d / 4\pi^3) > \left(z_2 - \bar{h} \right)^2 \tag{10.16a}$$

and

$$\langle I_{inc2} \rangle \sim \lambda^3 (\sim f^{-3}), \text{ if } (\lambda d / 4\pi^2) < \left(z_2 - \bar{h} \right)^2. \tag{10.16b}$$

Depending on the relationship between $\langle I_{inc1} \rangle$ and $\langle I_{inc2} \rangle$, the total field intensity is changed with frequency as

$$\langle I_{total} \rangle \sim f^{-1} - f^{-4}. \tag{10.17}$$

If now, in (5.76) $\lambda^2 > \left(2\pi l_v \gamma_0 (\bar{h} - z_1) \right)^2$, then,

$$\langle I_{inc1} \rangle \sim \lambda^{-1/2} (\sim f^{1/2}), \text{ if } (\lambda d / 4\pi^3) > \left(z_2 - \bar{h} \right)^2 \tag{10.18a}$$

and

$$\langle I_{inc1} \rangle \sim \lambda^{-1}(\sim f^1), \text{ if } (\lambda d / 4\pi^3) < (z_2 - \bar{h})^2. \tag{10.18b}$$

In the case of (5.78), then,

$$\langle I_{inc2} \rangle \sim \lambda^0(\sim f^0), \text{ if } (\lambda d / 4\pi^3) > (z_2 - \bar{h})^2 \tag{10.19a}$$

and

$$\langle I_{inc2} \rangle \sim \lambda^{-1}(\sim f^1), \text{ if } (\lambda d / 4\pi^3) < (z_2 - \bar{h})^2. \tag{10.19b}$$

Again, comparing $\langle I_{inc1} \rangle$ and $\langle I_{inc2} \rangle$, we get for the total field intensity the following frequency dependence:

$$\langle I_{total} \rangle \sim f^0 - f^1. \tag{10.20}$$

From (10.17) and (10.20), we can see that the average signal intensity frequency dependence is not constant and depends on the situation in the urban scene.

For mixed residential areas mostly close to rural environments, if we consider now in (5.29) that $\lambda^2 < [2\pi\ell_h\bar{L}\gamma_0]^2$, that is also $\lambda^2 < (2\pi l_v\gamma_0(\bar{h} - z_1))^2$, then,

$$\langle I_{inc1} \rangle \sim \lambda^{5/2}(\sim f^{-5/2}), \text{ if } (\lambda d / 4\pi^3) > (z_2 - \bar{h})^2 \tag{10.21a}$$

and

$$\langle I_{inc1} \rangle \sim \lambda^2(\sim f^{-2}), \text{ if } (\lambda d / 4\pi^3) > (z_2 - \bar{h})^2. \tag{10.21b}$$

Here, comparing (10.21a) and (10.21b), we get

$$< I_{total} > \sim f^{-2} - f^{-5/2}. \tag{10.22}$$

If now in (5.29), $\lambda^2 > (2\pi l_v\gamma_0(\bar{h} - z_1))^2$, that is also $\lambda^2 < [2\pi\ell_h\bar{L}\gamma_0]^2$, then

$$\langle I_{inc1} \rangle \sim \lambda^{-3/2}(\sim f^{3/2}), \text{ if } (\lambda d / 4\pi^3) > (z_2 - \bar{h})^2 \tag{10.23a}$$

and

$$\langle I_{inc1} \rangle \sim \lambda^{-2}(\sim f^2), \text{ if } (\lambda d / 4\pi^2) < (z_2 - \bar{h})^2. \tag{10.23b}$$

Again, comparing (10.23a) and (10.23b), we finally get

$$\langle I_{total} \rangle \sim f^{3/2} - f^2. \tag{10.24}$$

From (10.22) to (10.24), the signal average intensity is also not constant and depends on the terrain parameters in the mixed residential area.

From the earlier equations, it is clear that the frequency dependence of the total average signal intensity is changed from $\lambda^{-0.5} - \lambda^{-1.5}$ ($f^{0.5} - f^{1.5}$) for the low part of the high frequency (HF) band, and to $\lambda^{0.5} - \lambda^{1.5}$ ($f^{-0.5} - f^{-1.5}$) for the higher part of the very high frequency (VHF)/ultra high frequency (UHF) band with continuous transaction through $\sim\lambda^{0}$ (no dependence) around $f = 70$–90 MHz. These dependences can vary widely depending on the propagation scenarios that occurred in the built-up environment, the building density, the average height of the building layer with respect to terminal antennas, the building average length, the distance between antennas, and so on.

Figure 10.4 shows an example where the field loss relative to that in free space is presented, versus the radiated frequency for the antenna heights of $z_1 = 2$ m and $z_2 = 90$ m, and for the ranges between antennas of $d = 0.5, 1.0, 2.0, 5.0, 10.0$ km. The continuous curves are a result of calculations according to Equation (5.76), Equation (5.77), and Equation (5.78), taking into account the various situations in the urban areas, described in References 14–20 and 23–33.

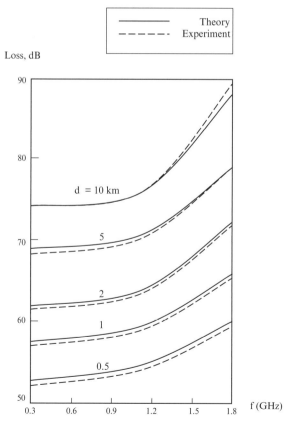

FIGURE 10.4. Relative signal loss versus radiated frequency for $z_1 = 2$ m, $z_2 = 90$ m, and for $d = 0.5, 1.0, 2.0, 5.0$, and 10.0 km.

The dashed curves are the result of measurements of the median signal power obtained for each corresponding experiment described in References 33–39. The results of theoretical prediction are close to those obtained by the corresponding measurements. Moreover, the frequency dependency of the signal loss, obtained analytically for various land environments, covers all results of this dependence obtained experimentally that allows designers of the land wireless networks to predict various situations in the radio communication links for each subscriber located within the area of service.

10.1.4. Experimental Verification of the Stochastic Approach

Here, we focus our attention on estimating the accuracy of the proposed stochastic approach by verifying the results of the theoretical prediction against numerous measurements carried in different land communication links, mixed residential, urban, and suburban areas.

Pass Loss in Forested Areas. Let us, first of all, discuss the accuracy of the proposed stochastic model that was introduced before for the description of path loss in the rural forested environments, which describes the large-scale slow signal decay in forested areas. For the considered UHF/L-band, the wavelength is in the range of 0.1 to 0.5 m, that is, it is larger than the physical dimensions of the scatterers in the canopy layer, but smaller than the tree trunk diameters. One can expect different scattering mechanisms in these two forest layers, and therefore, the vegetation density parameter should also be considered separately. The accuracy of the proposed path loss empirical model for the different behavior of the two major vegetation layers (tree trunks and canopy) was analyzed in Reference 21. As for the stochastic model, it is valid only if the distances between trees and the tree trunk diameters are larger than the wavelength, that is, for the UHF and higher frequency band, which corresponds to the assumptions presented earlier. It accounts for the absorption effects by introducing the reflection parameter Γ, which is less than unity for non-perfectly conductive (e.g., dielectric) trees surfaces and which also satisfies the real forest conditions.

We now compare the results obtained from the proposed stochastic model with experimental data, as well as with the empirical model, both described in details in Reference 22. In this comparison, we made the following assumptions. First, the transmitter (TX) and receiver (RX) antenna losses were not considered in the data post-processing. Second, the TX antenna gain patterns account for a maximal 2 dB variation and were not included in this model. Third, the RX antenna patterns were not documented and were considered to be the ideal ones for a dipole $\lambda/2$-antenna, with no significant gain variation at the practical orientation. All parameters of both antennas are presented in Table 10.2 [22].

Using the parameters of the forested terrain [22], one can approximate the average short range forest path loss to be within ±6 dB from measures obtained

TABLE 10.2. UHF Measurement Set-up and the TX and RX Antennas

Carrier Frequencies	961 MHz and 1900 MHz (Vertical Polarization)
Bandwidth	500 KHz
TX antennas:	At 3-, 6-, and 12-m heights
–3 dB Beamwidth	961 MHz: 50° H-plane/40° E-plane
	1900 MHz: 60° H-plane/20° E-plane
Gain	Approx. 12.1 dBi gain
Transmit power	961 MHz: −1.5 dBm
	1900 MHz: −5.7 dBm
RX antennas	@ 2.5-m height
	on ground plane
Spatial sampling	Approx. 1.3 complex IR/m
TX-RX distance range	100–1500 m

TABLE 10.3. Vegetation (in Leaves) Components along the Measurement Route in the Forested Area

Symbol	Vegetation Type	Trunk Diameter (m) D_i	Tree Density ($1/15 \times 15$ m) η	Height Canopy/Trunk (d) H_c / H_t ($H_c + H_t = H_f$)
D1	Deciduous 1	0.60	6–8	8/10
D1	Deciduous 2	0.35	10–12	5/10
L_v	Low trees	0.20–0.30	1–5	5–10/-

from a relatively flat terrain. At high frequencies, the predominant power loss is due to a $\rho^{-4.3}$ range dependency and the absorption/scattering attenuation in the different forest structures present in the region between the TX and RX antennas.

According to the fact that the empirical model is very close to measurements (with accuracy ±6 dB [22]), it is appropriate to estimate the accuracy of the proposed stochastic approach through the prism of measured data. We present in Figure 10.5a,b,c the relative path loss of the received signal calculated according to Equation (5.19), Equation (5.20), Equation (5.21), and Equation (5.22) within forested areas, the parameters of which are described in Table 10.3, and presented in the figures by a family of continuous curves for a density of the trees, γ_0, variations from $\gamma_0 = 10^{-3}$ to $\gamma_0 = 10^{-2}$ m^{-1} for the corresponding average trunk diameter presented in Table 10.3 and for an

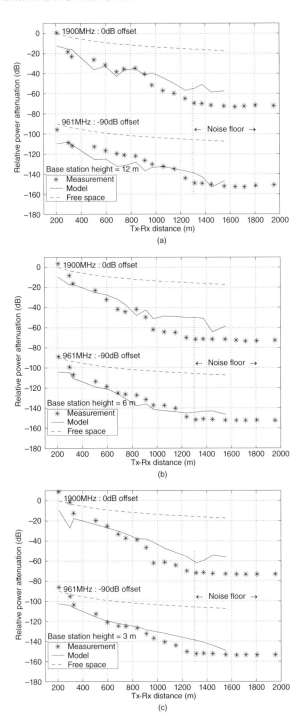

FIGURE 10.5. (a) Path loss versus distance between Tx and Rx antennas for base station height of 12 m and for radiated frequencies of $f = 1900$ MHz (top continuous curve) and $f = 961$ MHz (bottom continuous curve) for various densities of trees in areas investigated according to (5.19)–(5.22). Asterisks "*" plotted represent experimental data obtained according to experiments described in References 21 and 22; dashed curves represent free space attenuation for the each frequency separately. (b) The same as in panel a, but for base station height of 6 m. (c) The same as in panel a, but for base station height of 3 m.

average value of γ_0. All the experimental data are plotted by the asterisks "*", which correspond to those presented in References 21 and 22. Figure 10.5a,b,c present the results of calculations according to 3-D stochastic model obtained for two radiated frequencies, $f = 961$ MHz (bottom continuous curve) and $f = 1900$ MHz (top continuous curve), and for TX antenna heights of 12, 6 and 3 m, respectively. The horizontal axis in all figures indicates the distance from the transmitter antenna on the route as measured by the ground antenna in the experiment. For comparison, the free space attenuation is also presented by the dashed curves for the both radiated frequencies.

As follows from these figures, the proposed stochastic model, based on the statistical description of different forested areas with their specific distribution and density of trees, is also very close (with an accuracy that does not exceed 6–8 dB) to the measurements, that is, the same accuracy obtained from the empirical model [21]. Moreover, the accuracy of theoretical prediction is increased with increase of TX antenna height.

Path Loss in Mixed Residential Areas. The measurements were taken in three locations in Poland: Lipniki, Koscierzyna, and Tarczyn [23, 26]. These are typical rural neighborhoods, with one or two story buildings surrounded by vegetation. The frequency used was 3.5 GHz. The measurements were made with a transmitter antenna, at h_T, of 40 m and 80 m, above the average tree and building heights, whereas the height of the receiver antenna, h_R, was between 3 and 15 m. Between 5 and 8 points were measured in each of the locations (see References 23 and 26).

The average parameters of the obstructions were estimated in the following manner. For the reflection coefficient, Γ, the average values of measure for the brick walls and wooden surfaces of trees were used, that is of 0.3–0.4. As for the correlation scales, ℓ_h and ℓ_v for trees, they are of the order of tens of centimeters, whereas for one–two floor houses they are of the order of 1–1.5 m. In the case of uniformly distributed obstructions, we took ℓ_h and ℓ_v to be between 0.5 and 1 m in our calculations. The same procedure was used to obtain the minimum, average, and maximum obstruction contours density, γ_0. Thus, using the topographical map of the mixed area, we divided it into regions of 1 km^2 area each, and in each region, we estimated the density of the obstructions. In regions with pure vegetation, the parameter γ_0 varies from

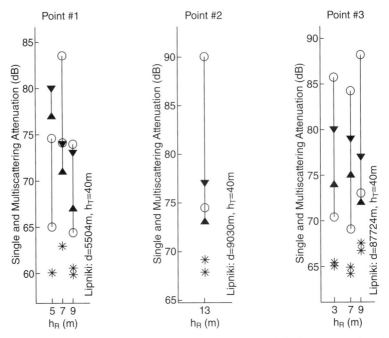

FIGURE 10.6. Average field attenuation versus receiver height for a transmitter height of 40 m. Wide thick segments represents experimental data. Circles and asterisks "*" represent calculations according to (5.19)–(5.20) and (5.29)–(5.30), respectively.

$\gamma_0 = 0.01$ km^{-1} to $\gamma_0 = 0.1$ km^{-1}. In regions where buildings are predominant, it ranges between $\gamma_0 = 1$ km^{-1} to $\gamma_0 = 3$ km^{-1} (see References 23 and 26). Finally we obtained that the average parameter of obstruction density over the terrain varies between $\gamma_0 = 0.1$ km^{-1} to $\gamma_0 = 1$ km^{-1}. These values will be used later in numerical computations of the path loss based on Equations (5.29) and (5.30) for single-knife diffraction (5.12), and Equation (5.19), Equation (5.20), Equation (5.21), and Equation (5.22) for multiple scattering without diffraction.

In Figure 10.6, Figure 10.7, and Figure 10.8, the path loss (in dB) is presented as a function of the receiver antenna height calculated according to equations mentioned earlier: for the case of single scattering with diffraction (denoted by circles), and for the case of multiple scattering without diffraction (denoted by asterisks "*"). The experimental data are presented here by a thick line that connects the minimum and maximum measured values. We also connected the circles and asterisks by thin lines spanning the range between possible results from the minimum value (the bottom circle calculated for $\gamma_0 = 0.1$ km^{-1}) and the maximum value (the top circle calculated for $\gamma_0 = 1$ km^{-1}).

In these figures, the distance d between the transmitting and receiving antennas, as well as the base station antenna height (h_T) is given in meters.

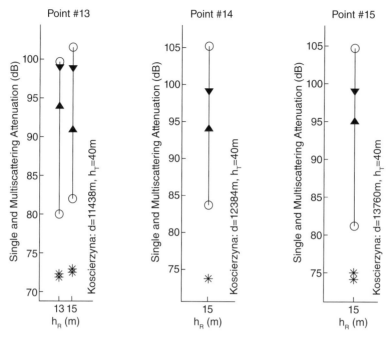

FIGURE 10.7. The same as in Fig. 10.6, but for the other experimental site.

Thus, for points 1–3, the measured data are close to those obtained numerically. On the one hand, the deviation between the calculated and measured values does not exceed 2–5 dB (see Fig. 10.7a, the Lipniki experimental site). On the other hand, results of calculations made using a model that does not take diffraction into account give a much larger deviation from the experimental data, about 8–10 dB (for points 1–3).

The same behavior can be observed at the other sites, as presented in Figure 10.9b (for the Koscierzyna experimental site) and Figure 10.8a,b (for the Tarczyn experimental site), respectively.

From the earlier illustrations, one can see that the difference between the theoretical model without diffraction effects and the experimental data can exceed 25–30 dB. At the same time, the model that includes diffraction from the rooftops and corners of the houses can predict the signal intensity attenuation, with an accuracy equivalent to that of measurements. The deviation between them does not exceed 2–5 dB, an effect which depends on the density of the houses surrounding both terminal antennas, as well as on the height of the antennas.

Path Loss in Urban and Suburban Areas. In each experiment carried out in Israel, Sweden, Portugal, and Japan, the stand-alone radio port unit (RPU)

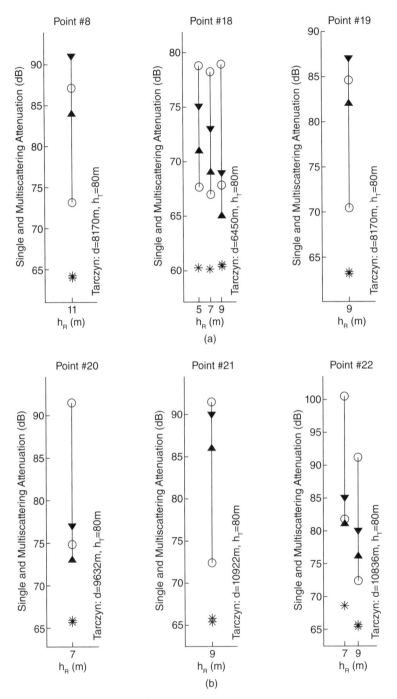

FIGURE 10.8. (a) The same as in Fig. 11.6, but for another experimental site and for a transmitter antenna height of 80 m. (b) The same as in Fig. 10.6, but for the other experimental site and for the a transmitter antenna height of 80 m.

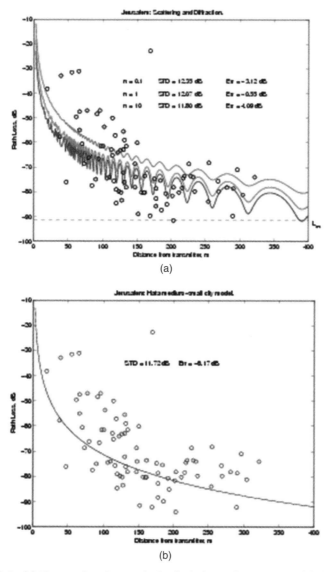

FIGURE 10.9. (a) The results of numerical calculations of average total field according to (5.76) and (5.78)–(5.81) (continuous curves) for $n = 0.1$, 1, and 10; the results of measurements are denoted by circles. (b) The same as in panel a, for comparison with Hata model (continuous curve).

plays the role of the transmitter. The fixed access unit (FAU) was used as the receiver, which during the experiment was moved from point to point. According to the FAU specification, its measurement accuracy in all experiments does not exceed 3–4 dB. The investigated built-up terrain was different: in Stockholm and Tokyo, it was characterized as quasi-smooth, whereas in Jerusalem and Lisbon, it was characterized as hilly.

Experiments in Jerusalem. We will start by examining the experiments in Jerusalem according to References 17 and 18. The notion of the medium urban area is relevant to Jerusalem's propagation conditions. Two or three samples there were taken at each experimental point along the vehicle route and the average values based on these measurements have been found. To determine received signal strength indication (RSSI) values from the earlier expressions of average intensity of the field, we have to multiply the sum of (5.76) to (5.78) by the term $\sim \lambda^2$ according to References 19 and 20, which finally gives us (5.79). The major problem is the nonflat terrain profile of Jerusalem (i.e., the existence of substantial height differences between relatively close points in the area). In this situation, the accuracy of the theoretical prediction is reduced. In addition, the complex terrain can affect the distance of direct visibility, and this influence, which is the diffraction phenomena, has to be taken into consideration. To overcome these difficulties, we took into account the ground height to determine the actual FAU height as a function of its location. We added the ground height to the building height as well, and then determined the average building height.

From the topographical map, we found that we can approximate the different built-up layer profiles obtained for each experimental site by the polynomial functions (5.57), with parameter n lying between 0.1 and 10. From the topographical map of Jerusalem's experimental site, we obtained the following parameters of the built-up terrain: the building density $\nu = 103.9$ km^{-2}, the average building length $\bar{L} = 18$ m, and the average building height (not including the local ground height) $h = 8.3$ m. All the local ground heights were determined by using the global positioning system (GPS). The measurements were made at 930 MHz bandwidth using the transmitting antenna with a height $z_2 = 42$ m [17, 18]. The results of numerical calculations are presented in Figure 10.9a by oscillating curves for $n = 1$ (uniform built-up terrain), $n = 0.1$, and $n = 10$ (non-uniform built-up terrain with a predominant number of small and tall buildings, respectively).

In Figure 10.9a, the measurement results denoted by circles are compared with the multiple scattering and diffraction model according to Equation (5.76), Equation (5.77), Equation (5.78), and Equation (5.79). The indicated number pair is the standard deviation value (STD) and the following prediction error (Err) between two points sets (theory and measurements). We can define them as follows [17, 18]:

$$Err_{model} = R_i - r_i, [dB] \tag{10.25a}$$

$$\langle Err_{\text{model}} \rangle = \frac{1}{N} \sum_{i=1}^{N} (R_i - r_i), [\text{dB}] \qquad (10.25\text{b})$$

$$STD_{\text{model}} = \sqrt{\langle (Err_{\text{model}} - \langle Err_{\text{model}} \rangle)^2 \rangle}, [\text{dB}] \qquad (10.25\text{c})$$

Here, N is the set dimension and R_i and r_i are the theoretically obtained and measured path loss values, respectively.

From Figure 10.9a, the poorer convergence between the theoretical prediction and experimental data can be improved by at least 3–7 dB by taking into account the real built-up layer relief. In fact, as was shown earlier, depending on the parameter n, from (5.58) and both antenna evaluations, this can be improved up to 25–35 dB. Thus, for the conditions of the experimental site in Jerusalem and of both antennas (see References 17 and 18), we have the following: for $n = 0.1$, the additional excess loss is of –5–7 dB; for $n = 1$, it is of –10–12 dB; and for $n = 10$, it is of –15–17 dB. These results cover all experimental measurements. Taking into account these factors in the proposed parametric model, one can give the same and more accurate predictions of loss characteristics within a channel than the Hata's small–medium empirical model (see Chapter 5). In fact, the comparison between the experimental results obtained in Jerusalem (denoted in Fig. 10.9b by circles) and the calculations according to Hata's small–medium model (denoted by the curve) is presented in Figure 10.9b. In this case, the value of the standard deviation is at the same level as that obtained by using the parametric model, but the absolute value of mean error is higher, with respect to the parametric model.

We have physically clear relations between the parameters of environment, terrain profile, both antennas elevation, and the loss characteristics of the signal within the urban communication channel. Moreover, the multiparametric model is relatively simple and it does not need to be calibrated every time. What was missing in these experiments was that we had no real profile of the built-up terrain in Jerusalem [17, 18]. On the contrary, in other experimental city sites, the actual terrain relief and the building layer profile (the concrete parameter n) have been used to compare the experimental data with our theoretical prediction.

Experiments in Stockholm. As was mentioned in References 40 and 41, the built-up area of the experimental site can be characterized as high, dense, or obstructed. According to the topographical map of the experimental site [40, 41], the height of buildings distribution versus the number of buildings along the radio path from base station (BS) to each vehicle's position was estimated and presented in Figure 10.10a. From this, the complementary cumulative distribution function (CCDF) (see definition in Chapter 5) was obtained along each route of the moving vehicle from the base station to the path-end denoted as zip1 to zip10 (see Fig. 10.10b). According to the experimental data and topographical map of the site described in References 40 and 41, the building

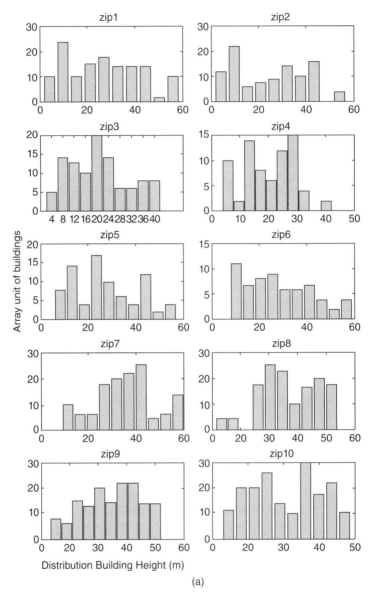

(a)

FIGURE 10.10. (a) Distribution of buildings along each radio path. (b) The corresponding PDF of buildings' profile distribution.

contour density parameter, γ_0, was estimated by using (5.53) as about 8 km^{-1}–12 km^{-1} with the mean value of 10 km^{-1}. The built-up profile parameter n was estimated from (5.58). Each individual vehicle path is presented in Figure 10.10b and is changed from 0.65 (for the vehicle end position denoted as zip8) to 1.45 (for point zip6). All experiments were carried out at the carrier frequency of 1.8 GHz.

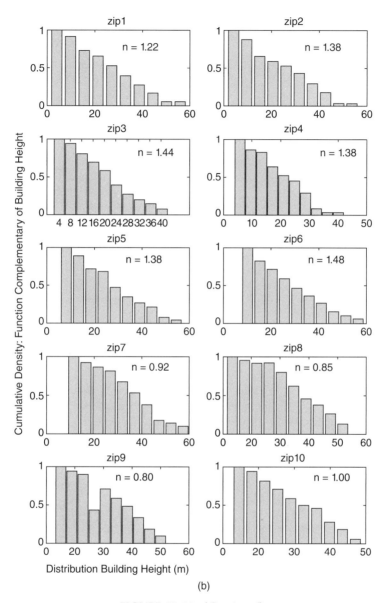

FIGURE 10.10. (*Continued*)

The numerical simulation results for the total field intensity attenuation, according to the proposed multiparametric model, are presented as a 3-D picture in Figure 10.11. The comparison between theoretical predictions, using Equation (5.76), Equation (5.77), Equation (5.78), Equation (5.79), Equation (5.80), and Equation (5.81), and the corresponding experimental data, shows that the median error and the standard deviation do not exceed 6 dB

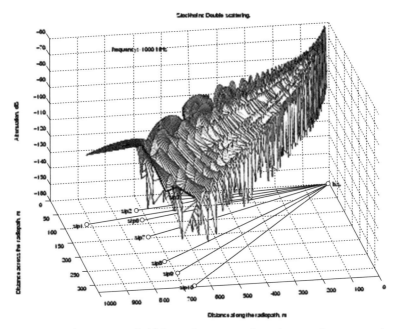

FIGURE 10.11. The average field intensity attenuation along and across each radio path.

(Err = 5.2 dB, STD = 5.6 dB). We must note that close results were also obtained in other cities with various terrain profiles (see References 17–20).

Experiments in Tokyo. In the measurements and for numerical simulation for the area of the Tokyo University of Electro-Communications, we took into account the overlay profile of real buildings for the BS antenna located at the roof of the highest building and for each location of the subscriber antenna along five roads (denoted from Rx1 to Rx5 in Fig. 10.12).

The experimental condition for computations using ray-tracing approach is summarized in Table 10.4, including the number of reflection and diffraction in the following computations. For each road, we have calculated the density of buildings v/km^2, the average length of buildings (or width depending on the orientation to antenna direction) $\bar{L}$, and then the buildings' contour density above the urban terrain $\gamma_0 = 2\bar{L}v/\pi = 8 \sim 12\,[\mathrm{km}^{-1}]$ or $8 \cdot 10^{-3} \sim 1.2 \cdot 10^{-2}\,[m^{-1}]$.

The corresponding distribution of γ_0 in the experimental site is shown in Figure 10.13. Here, γ_0 is denoted along the vertical axis and the 2-D range of experimental site is depicted in the horizontal plane. The parameter ℓ_v (see definitions in Chapter 6) was changed according to the analysis of the topographic map and the corresponding buildings' features (windows, balconies etc.), from 2 to 3 m. The parameter Γ was estimated to vary from 0.5 to 0.85.

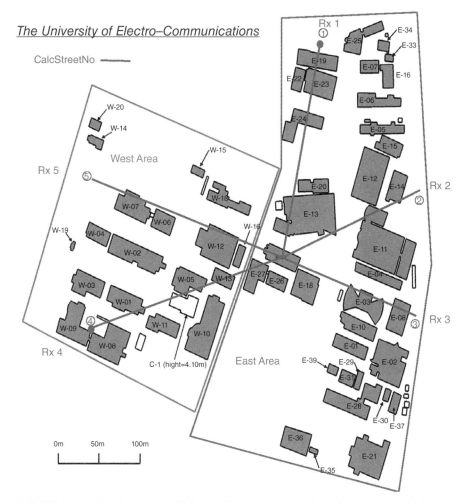

FIGURE 10.12. Testing area of Tokyo university; five roads of the moving vehicle are denoted as Rx1 to Rx5.

TABLE 10.4. Experimental Data for Ray tracings Simulation

Calculation method	Imaging Method
Frequency	5GHz($\lambda = 0.06$ m)
BS and MS polarization	Vertical
Maximum of scattering number	Reflection 2
	Diffraction 2 (line Rx1 and Rx2)
	Diffraction 1 (line Rx3 and Rx4)
Height of BS and MS	BS: 50.0 m
	MS: 1.5 m

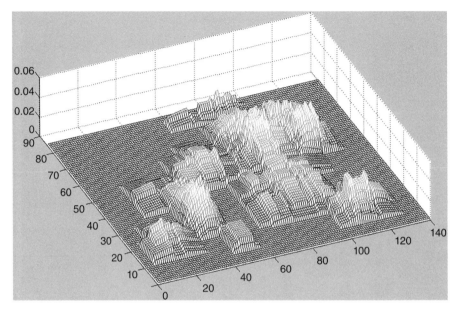

FIGURE 10.13. Distribution of building contours density in the testing area of Tokyo university.

As for the overlay profile of the buildings, it changes depending on the view on BS antenna from each mobile subscriber (MS) antenna placed for the corresponding radio path. Thus, from the MS placed at the path Rx1 (the upper right line from the BS antenna depicted in Figure 10.12), the buildings' profile will have a view shown in Fig. 10.14a. Figure 10.14b illustrates the same profile seen from MS antenna moving along the path Rx2 (depicted in Fig. 10.14 by the straight line along the right direction of the switched clock). The same profiles are shown in Figure 10.14c,d for MS antenna moving along the path Rx3 to path Rx4 (the corresponding paths following right switched clock). Using now the corresponding profiles shown in Figure 10.14a,b,c,d, the parameter n of the overlay profile of buildings and their average height for four locations of subscriber moving antennas was found. According to estimations, the parameter of buildings profile was changed from $n = 0.67$ to $n = 0.35$.

The corresponding radio coverage for the four radio paths depicted in Figure 10.12, the profiles of which are presented in Figure 10.14a,b,c,d, are shown in Figure 10.15a,b,c,d, respectively.

Because the ray-tracing approach gave the best fit to measured data, let us compare the results obtained by the stochastic model and those using the ray-tracing approach. As was previously mentioned, we show the ray-tracing results only for the four specific radio paths, Rx1 to Rx4, as shown in Figure 10.12. In the computations on the basis of two approaches, we took the frequency of 5 GHz ($\lambda = 0.06$ m) and have assumed that all of the buildings in Figure 10.12

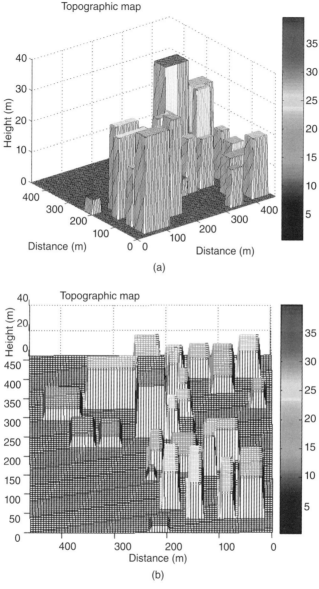

FIGURE 10.14. (a)–(d) View on the overlay profile of buildings from the position of the user located at points Rx1 to Rx4, respectively.

are composed of concrete, so that we take the walls' electric properties such as $\varepsilon_r = 6.76$, $\sigma = 2.3 \cdot 10^{-3}(S/m)$ and those for the ground surface are assumed to be $\varepsilon_r = 3.0$, $\sigma = 1.0 \cdot 10^{-4}(S/m)$. These electric properties of the concrete buildings are consistent with the assumption made for the absolute value of the reflection coefficient estimated as $\Gamma = 0.85$.

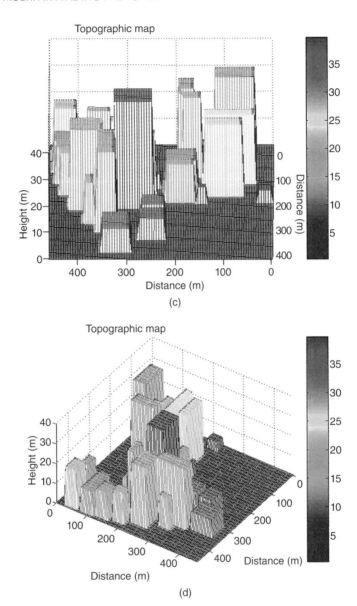

FIGURE 10.14. (*Continued*)

In Figure 10.16a,b,c,d, the ray-tracing results are presented for every 10 m of radio path; the same was done for the stochastic model. Here we compare stochastic and ray-tracing approaches where the latter is based on double reflection (R2) and double diffraction (D2), which correspond to results presented for other cities (see References 19, 20, 29, and 30). At the same time, we have used the R2 and

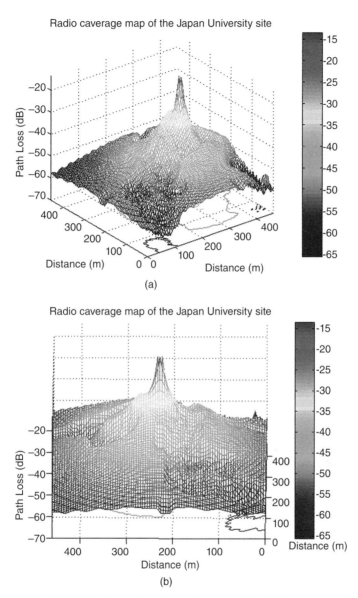

FIGURE 10.15. (a)–(d) Radio pattern from the array of buildings as seen from point Rx1 to point Rx4, respectively.

one-time diffraction (D1) for the paths of Rx3 and Rx4. A comparison between the results obtained from the stochastic model and those from the ray-tracing approach is shown in Figure 10.16a,b,c,d for the paths Rx1–Rx4, respectively. Generally speaking, the agreement between the stochastic and ray-tracing approaches is good, and the stochastic result is considered to reflect the general

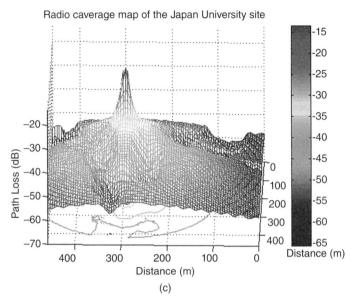

(c)

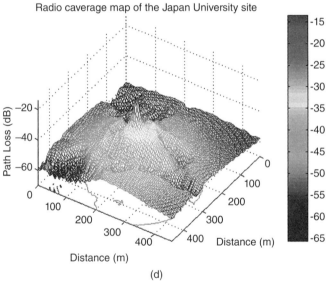

(d)

FIGURE 10.15. (*Continued*)

trend obtained by the ray-tracing estimation. At shorter distances ($d < 50$ m), we notice a rather weaker agreement between the stochastic and ray-tracing results, as suggested by a discrepancy between both models, whereas at a distance larger than 50–70 m, they are closer to each other. This can be explained by the fact that far from the BS antenna, randomization of the field strength,

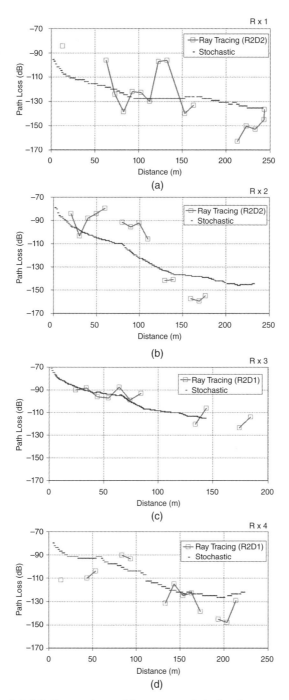

FIGURE 10.16. (a)–(d) Comparison of the proposed stochastic approach (continuous curves) with the ray tracing approach (discrete segments) for points Rx1 to Rx4, respectively.

caused by the rough built-up terrain, becomes dominant and the multiray effect, described by the ray-tracing model, fully coincides with that obtained using the stochastic multi-parametric model.

Experiments in Lisbon. To understand and assess the accuracy of the present stochastic approach, and to find its limits for the prediction of loss characteristics, the specific experiments were carried out in Lisbon (Portugal) [24, 25]. As mentioned in References 24 and 25, the Lisbon terrain profile is hilly with various densities of buildings and vegetation surrounding the base station antenna, as a transmitter, which was of the height of 93 m assembled at the roof of the tower, as a taller building in the tested area. The radiated frequency was of the 3.5 GHz band.

As for a mobile vehicle antenna, a view of which is shown in Figure 10.17, it was equipped with the crane. Using the crane, it was possible to take measurements in different heights of the vehicle antenna. Each points measurement was taken starting with the crane elevated to a position in which there is line of sight with the transmitter, down in 1 m increments to a height were the signal level reached the noise floor.

FIGURE 10.17. A measurement moving vehicle, as the receiver, equipped with a crane; the antenna is visible on the basket.

Two typical examples of these profiles in various experimental sites (identified by numbers at the top of the figure) are presented in Figure 10.18a,b. In the experiments, the receiver antenna was mounted on a crane and changed its height from a line-of-sight (LOS) position down to a position behind the building in non-line-of-sight (NLOS) conditions, when the signal amplitude was reduced to the noise level. The clearance between terminals, the transmitter, and receiver is shown in each figure by introducing the first Fresnel zone, as an ellipse, clearly showing the extent of obstruction within such an elliptical cross-section. By reducing the receiver antenna height, we continuously move from conditions of full clearance (LOS) to NLOS conditions, that is, from fully illuminated paths to shadow zones due to the obstructing building.

Figure 10.19a,b,c show a comparison of the measured and theoretically predicted path loss in some experimental points in Lisbon. Each figure shows the following:

- The measured path loss, denoted by a thick line bar connecting the minimum and maximum loss measured at that point.
- The results of the model described in Chapter 5 using (5.80)–(5.81) with (5.76), (5.78) and (5.79), and with the values of the parameters taken from a topographic map of Lisbon city. These results are depicted by asterisks, connected by a thin line indicating the results for high, medium, and low values of the obstacle density ν in km^{-2}.
- The results of various "knife-edge" models, described in Reference 17, indicated by circles. For the sites point #44 and #51, we used a single diffraction model according to Lee's approximate equations (see Chapter 5). For points #4, #25, and #41, we used a double diffraction model based on the same equations, and finally for points #5, #9, #16, we used the triple diffraction model based on the empirical approach described in Reference 17 based on the same Lee's "knife-edge" model.

From all the figures presented, it is clear that the performance of the present stochastic model (with accuracy of 1–3 dB) is satisfactory as long as the receiver antenna height does not decrease to the extent that it is put in the local shadow zone in the vicinity of high obstructing buildings. In such cases, the knife-edge diffraction models are better predictors for the signal attenuation in such regions.

10.1.5. Advantages and Limitations of 3-D Stochastic Multiparametric Approach

First, let us consider the advantages of the proposed stochastic model. As was mentioned in References 24–32, the proposed 3-D multiparametric model is the stochastic approach that combines statistical description of the terrain and the built-up overlay with a description of the signal intensity according to (5.76)–(5.81), taking into account the effects of various obstructions according to their vertical and horizontal geometrical parameters distribution.

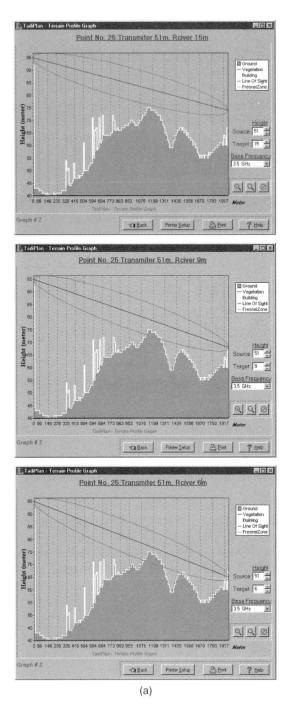

(a)

FIGURE 10.18. (a) The typical examples of the profiles of various experimental sites identified as #25. (b) The same as in panel a, but for experimental site #16.

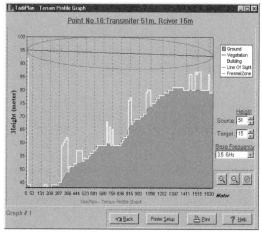

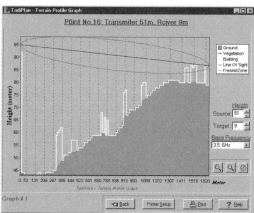

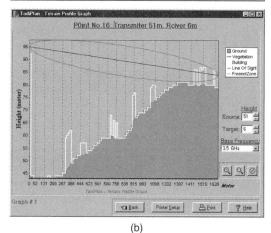

(b)

FIGURE 10.18. (*Continued*)

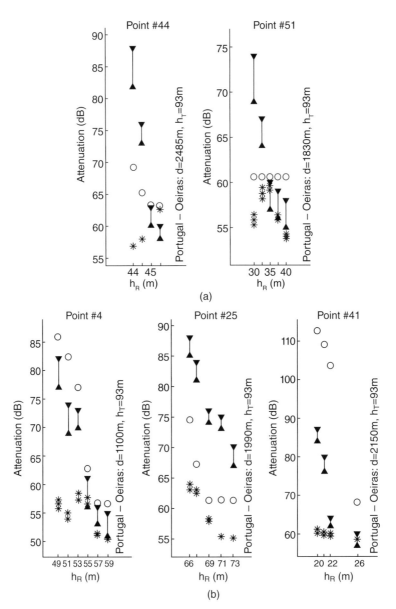

FIGURE 10.19. (a) Signal intensity decay versus the height of the vehicle antenna; the experimental data, the theoretical prediction according to (5.76)–(5.79) and the 2-D "knife-edge" model are denoted by segments, asterisks "*" and circles, respectively. (b) The same as in panel a, except the circles indicate results of calculations by using the double diffraction "knife-edge" model. (c) The same as in panel a, but circles indicate results of calculations by use of three-time diffraction "knife-edge" model.

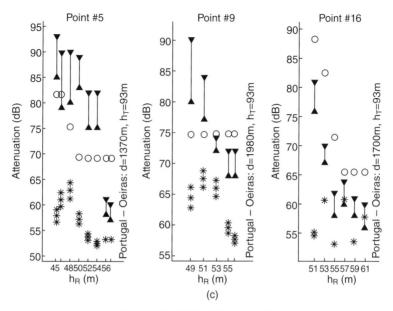

FIGURE 10.19. (*Continued*)

Figure 10.20a is a 2-D cross-sectional view of a 3-D urban relief profile, with buildings indicated by black blocks. The positions of the antennas are shown as well. The purpose of Figure 10.20 is to explain the physical motivation for the definition of $P_h(z)$. We are interested in the probability for a full clearance between antennas, connected with strong signal intensity. The less the electromagnetic radiation is obstructed, that is, fewer high buildings are encountered, the higher this probability will be. Note that we do not describe the "local" deterministic expected signal strength for the individual receiving antenna position within the urban communication channel. We strive to express some average properties that will take into account secular reflection from building tops and diffraction mechanism, along the propagation route. Evidently, this stochastic approach must be proved via experimental data.

In Figure 10.20a, we put the "1" value in areas of location of the buildings (black pixels) and "0" value elsewhere. The averaging on horizontal "slices" will produce a grey scale appearance (see Fig. 10.20b). Below h_{min} we obtain the darkest grey, and above h_{max} we have only white. Finally, the correspondence of this description to $P_h(z)$ is shown in Figure 10.20c.

The present averaging method can be carried out over an area of urban scene, as it is done on the corresponding 2-D cross-section shown in Figure 10.20a. Although a 2-D model ignores "side looking" effects, such as horizontal specular reflections and diffraction mechanisms, a 3-D approach accounts these effects very well. Furthermore, the shape of CCDF $(P_h(z))$ distribution inside the overlay of the buildings, depicted in Fig. 10.20c at the right side of

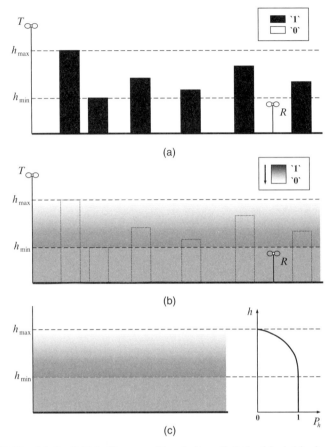

FIGURE 10.20. (a) the 2-D built-up profile of city relief; the black blocks correspond to sign "1," the gaps between them correspond to sign "0," (b) the 3-D case taking into account "side looking" effects, and (c) the 3-D distribution of inside the building layer.

the picture depends on the relief parameter n. Thus, the height profile function $P_h(z)$ describes a wide range of city building profiles: from one-level height close to $h1$ (for $n \gg 1$) or to $h2$ (for $n \ll 1$) to various levels of buildings heights hi, including a quasi-homogeneous distribution of hi with equal probability from $h1$ to $h2$, when $n = 1$, for which the average building height equals $\bar{h} = (h_1 + h_2)/2$, where $h_1 = h_{min}$ and $h_2 = h_{max}$.

Hence, the proposed multiparametric stochastic model is a 3-D model that uses data for terrain and buildings' overlay geometries to calculate the pertinent CCDF. As was mentioned in Chapter 5, this CCDF determines the probability of an event for each observer located in the built-up layer. This is a general approach that takes into account the terrain input parameters, such as the height distribution of the buildings, their density, and spatial distribution over the ground surface.

At the same time, the principal limit of the proposed stochastic model was found during its evaluation. Thus, it was found that the *critical boundary*, beyond which one can use only 2-D deterministic multiple diffraction or empirical "knife-edge" models, depends mainly on the angle-of-arrival distribution depicted in Figure 10.21.

Using our knowledge about signal power distribution as a function of the angle-of-arrival, described in Chapter 8 following Reference 28, we can estimate the range of critical angles between the ray arriving from the base station and the top of specific buildings. In Figure 10.21, the choice of the geometry is shown, for which the angle $\theta = \tan^{-1}[d / (h_B - h_R)]$ is defined. In the case where the random distribution of the angles of diffraction occurs from roofs of the randomly distributed buildings, one can determine, following to the geometry presented in this figure, the virtual angle $\theta' = \tan^{-1}[d / (h_B - h_R - h'_R)]$. Beyond this boundary of *critical angles*, we cannot use the proposed statistical approach and must account the additional shadowing effects.

10.1.6. Experimental Verification of the Waveguide Crossing-Street Model

We now will consider the case of specific urban environment with rectangular grid-plan streets, that is, the scenario occurs in the urban scene with straight

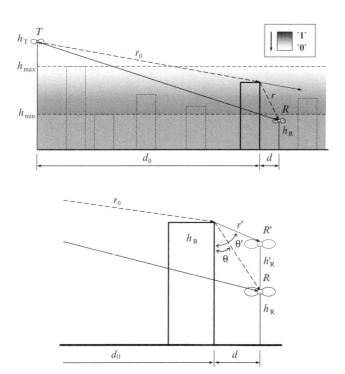

FIGURE 10.21. Geometry of the shadowing from building's roof.

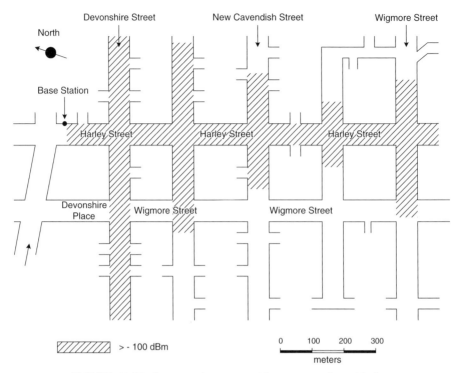

FIGURE 10.22. Layout of a street with a rectangular grid plan.

crossing streets and with building randomly lining each street according to Poisson law discussed in details in Chapter 5, as is schematically presented in Figure 10.22. As was expected in References 42–51, in such scenario, both radio port antennas are located at the street level heights *below* the rooftops. Thus, tests described in References 49 and 50, carried out in New York City, indicate that the subscriber at the street level, moving radially from the base station, or on the streets parallel to these, may receive a signal 10–20 dB higher than that received when moving on the perpendicular streets. Thus, we observe waveguide channeling effect of field energy due to the buildings lining the street, the heights of which are higher than the height of moving subscriber antennas, described mathematically in Chapter 5.

However, as was shown in experiments carried in References 48–50, the channeling effect is more significant in the microcellular areas (up to 1–2 km away from the transmitting antenna), becoming negligible at distances above 5–10 km, that is, in the macrocell areas. As pointed out in References 42–48, it was very difficult to obtain a strict theoretical framework or treatment of this channeling phenomenon. A simple theoretical approach, proposed by References 42 and 45, to represent the relative field strength by the density of arrows along the various streets, indicates only a way in which field strength may vary in an urban area because of street orientation, but not the real field strength distribution between streets and their intersections. According to this approach, the total path loss at the crossing-street level is a simple arithmetic summation of path loss at the radial street where the transmitter is located, and of path loss from the intersection to the side street.

Experiments carried out in the crossing-street area of Central London at 900 MHz and 1.7 GHz [43, 46] have shown a complicated 2-D shape of microcell radio coverage, similar to a Christmas tree with the base station, as a transmitter, located near the foot of the tree (see Fig. 10.22). This complicated redistribution of field energy among the rectangular crossing streets cannot be understood using a simple geometric optic model, even taking into account diffraction from the building corners, as was done in References 12 and 13. Later, we will show the reader via the corresponding experimental tests that the waveguide rectangular-crossing-street model can be successfully used for the prediction of spatial signal intensity coverage in built-up areas having a grid-plan pattern of crossing–straight streets for microcells with effective radius that does not exceed 2–3 km.

To verify this crossing-waveguide statistical approach, we will present here results of measurements carried out in microcell urban environment of the small town of Kefar-Yona; conditions of this experiment are fully described in References 17 and 18. The measurements were carried not only in the space, but also in the azimuth domain, about this subject we will talk later. The tested environment was a typical small urban region of three- to five-story brick buildings, with approximately uniform heights $h = 8$ m – 10 m and with a right-angle crossing-street plan with buildings randomly surrounded with Poisson law (as schematically presented in Figure 10.23).

The BS sectoral antenna was arranged first at the distance of 4–5 m from the corner of a building lining a street, as depicted in Figure 10.23. Its height in this series of experiment was arranged at the height of $z_2 = 6$ m and was operated in the frequency band $f_c = 902 - 928$. The mobile directive radio-port antenna moved along the straight crossing streets in the middle of the road (each position of mobile vehicle [MV] is depicted schematically in Figure 10.23 by numbers II, III, IV, . . .). The MV antenna was lower than rooftop level ($z_1 = 2 - 3$ m) and changed its distance from the stationary BS antenna in the range of 10 m–500 m. In Figure 10.23, the main radial street where the base station was located, is denoted by A_1, other radial and straight-crossing streets are denoted by A_i and B_j, $i, j = 1, 2, 3, . . .$, respectively. Field intensity measurements in decibels (dB) relative to intensity in free space at the range $r = 100$ m were obtained to estimate the total signal intensity attenuation along the crossing streets. For this purpose, the actual dielectric properties of the brick walls of buildings ($|R_n| = 0.73 - 0.8$) and the real distribution of buildings along the street level ($\chi = 0.5 - 0.6$) for the radial and crossing street widths (from 10 to 20 m), were taken into account.

The measured relative intensity of the received signal in dB is presented as a set of points near each curve in Figure 10.24 for the different positions of

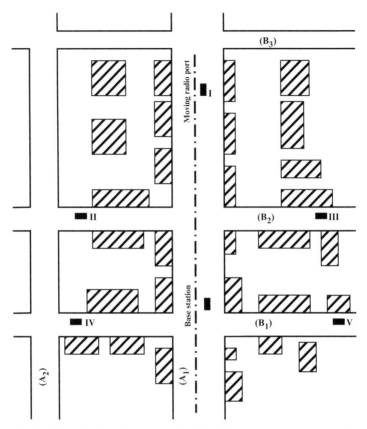

FIGURE 10.23. The simplified scheme of a Kfar Yona area as a rectangular-crossing street grid. The main radial street is noted by A_1; other streets are noted by A_i and $B_i, i = 1, 2, \ldots$; the positions of moving radio port are noted by Roman numbers I, II, II, ...

the MV antenna: at the main and the first radial streets A_1 and A_2, and at the first (B_1), second (B_2), and third (B_3) crossing streets, respectively (see Fig. 10.24). The main continuous curve in Figure 10.24 represents calculations of relative signal intensity inside the main street (A_1), according to street waveguide model described in Chapter 5, the second continuous curve (A_2) and the dashed curves $(B_i, i = 1,2,3)$ represent the crossing-waveguide model described in References 17 and 18 and briefly mentioned in Chapter 5, on the basis of geometrical portraiture illustrated by Figure 5.8b or in Figure 10.23 for the parameters of the street taken from the topographic map and from the experimental measurements: $a_i \approx 20$ m (radial street widths), $b_i \approx 10$ m, $i = 1, 2, 3$ (side street widths); $|R_n| = 0.75$ for the radial street, $|R_n| = 0.7$ for the side streets; $\chi_1 \approx 0.5$ for the radial streets and $\chi_2 \approx 0.6$ for the side streets, the average buildings' height is $\bar{h}_b \approx 9 - 10$ m. As seen from the comparison between theoretical and experimental results, presented in Figure 10.24 by

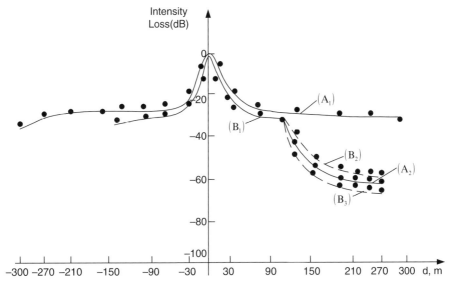

FIGURE 10.24. A signal intensity loss (in decibel) along the crossing-street grid measured in Kefar Yona area using the notations depicted in Fig. 11.25; measured data are depicted by points and computed data are depicted by continuous and dashed curves, respectively.

curves and dots, respectively, the waveguide-street model give satisfactory description (with the accuracy of 1–2 dB) of radiowave propagation along the radial waveguides in LOS conditions between the BS and MV antennas, that is, in directions of direct visibility between the terminal antennas. Even in intersections and in crossing perpendicular streets denoted by B_i, the crossing-street model gives results comparable to experimentally obtained data with a sufficiently good accuracy of 4–5 dB. This also satisfies the real experimental conditions and accuracy of measurements with error of 2–3 dB.

Furthermore, using results obtained both theoretically and experimentally, and depicted in Figure 10.24, we can show the accuracy of the predicting algorithm proposed in Chapter 5 on how to estimate the total field attenuation both along the radial and side streets, on the basis of the theoretical framework described propagation phenomena in the urban areas having straight crossing-street grid pattern. Thus, using Equation (5.51) and the corresponding equations for the path loss:

(a) along the radial street

$$L_r = 10\log\frac{J}{J_0} = 10\log\left\{\left(\frac{a\log 2}{y|\log\chi\,|\,R_n\,\|}\right)\right\} \qquad (10.26)$$

(b) along the side street after intersection

$$L_s = 10\log\frac{J}{J_0} = 10\log\left[\left(\frac{a\log 2}{y_0\,|\log\chi_1\,|\,R_n\,\|}\right)\cdot\left(\frac{b\log 2}{|z|\,|\log\chi_2\,|\,R_n\,\|}\right)\right]. \quad (10.27)$$

Taking into account the parameters estimated earlier, we get: for a distance of 90 m along the main radial street (that corresponds to the first intersection, see Fig. 10.23), and then for $|z| = 10$ m from the first intersection along the side street (denoted if Fig. 10.23 by B_1), according to (10.27) we get $L_s \approx -8.5 - 19.5 = -28$ dB, whereas the experiment gives -30–31 dB (see Fig. 10.24).

From comparison between the theoretical prediction based on the pure street waveguide and crossing-street waveguide models and the experimentally obtained data, we notice that, with accuracy of 3–4 dB, we can predict the radio signal intensity attenuation for each subscribed located in the areas of service, having knowledge of the signal power distribution in the joint path loss–distance domain.

10.2. LINK BUDGET DESIGN FOR VARIOUS LAND ENVIRONMENTS

The main goal of this analysis is to predict, according to the existing stochastic approach and the corresponding statistical distributions (Gaussian, Rayleigh, and Rician), a set of relevant parameters of signal power budget design in RF wireless networks, for different kinds of terrestrial environments: rural forested, mixed residential, and urban.

10.2.1. Existing Methods of Link Budget Design

First, we determine parameters of a communication link budget according to well-known concepts [42–46], and then we derive all parameters for the same link budget following the multiparametric stochastic approach and also on the classical statistical presentation of fading phenomena. According to References 42–46, the link power budget, that is, the *total path loss* inside the communication link consists three main terms that satisfy three independent statistical processes (see Chapter 1).

There are three independent characteristics of the signal power decay: the *median path loss* or the mean signal power decay along the radio path ($\bar{L}$), the *slow fading* or the characteristic of shadowing (L_{SF}), and the characteristic of *fast fading* (L_{FF}), which yield

$$L_{Link} = \bar{L} + L_{SF} + L_{FF}. \quad (10.28)$$

As was shown in References 42–46, these characteristics in general vary as a function of propagation range between terminal antennas, operating frequency, spatial distribution of natural and man-made obstructions surrounding these antennas, vehicle speed, and antenna height with respect to obstructions, and so forth.

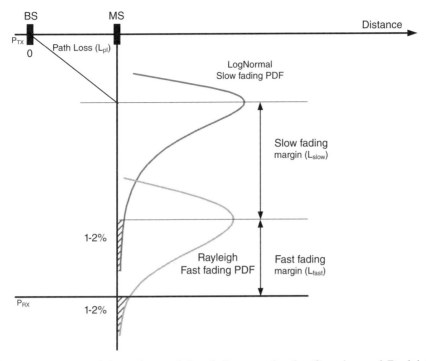

FIGURE 10.25. Path loss, slow and fast fading margins for Gaussian and Rayleigh PDFs. (Source: Reference 43. Reprinted with permission © 1995 IEEE.)

First, we describe a well-known approximate concept of link budget estimation that is based on numerous experimental data obtained for various terrestrial communication links [43]. According to this concept, the expected median signal power at the MS must be derived. This is also used for determining the radio coverage of a specific BS and the interference tolerance for the purpose of cellular map construction. Usually, when designing the network power budget and the coverage area pattern, the slow and fast fading phenomena are taken into account, as schematically demonstrated in Figure 10.25 by introducing a shadow fading margin $L_{SF} = 2\sigma_{Sh}$ and a fast fading margin (L_{FF}) as functions of the range between BS and MS (σ_{Sh} is the standard deviation of shadowing [43]). In other words, a shadow fading margin, which is usually predicted to be the 1–2% part of the log-normal probability density function (PDF), and a fast fading margin, which is typically predicted to be the 1–2% part of the Rayleigh or Rician PDF, may be taken into account simultaneously or separately in a link budget design depending on the propagation situation. This situation is often referred to as "fading margin overload" resulting in a very low-level received signal almost entirely covered in noise. The probability of such worst cases determines the event of how rapidly the signal level drops below the receiver's *noise floor level* (NFL). The probability of such an event was predicted in Reference 43 as a sum of the individual margin overload probabilities, the slow and the fast, when the error probability is close to 0.5, since the received signal is at the NFL. According to the three-step scenario, illustrated in Figure 10.25, the following practical algorithm of power budget design was proposed.

First step. Estimation of median path loss by using the well-known Hata model [6, 43] and deployment of a correction factor corresponding to the local antenna elevation, deduced from measurements, for example,

$$\bar{L} = L_{\text{LOS}} + L_{\text{NLOS}}. \tag{10.29}$$

Second step. Estimation of slow fade margin, using the characteristic slow fading variance of typically 6 to 9 dB [43], and assuming log-normal slow fading PDF with a 1–2% slow fading margin overload probability (see Fig. 10.25), one has a slow fading margin of $L_{\text{SF}} = 2\sigma_{\text{SF}} = 10 \div 15$ dB.

Third step. Assuming the Rician fast fading PDF, as more general PDF for multipath channel prediction, with the Rician parameter $K = 5$–10 [43] and a fast fading margin overload probability of 1% (see Fig. 10.25), one has a fast fading margin of $L_{\text{FF}} = 5$–7 dB.

Another concept that still uses the same statistical approach for obtaining the slow and fast fade margins was proposed in Reference 44. According to the proposed concept, the effect of slow fading or shadowing can be described as a difference between the median path loss and the maximum acceptable path loss, L_m. The median path loss can be predicted by any standard propagation model (see, e.g., References 2, 6, 42, and 43), which is depicted in Figure 10.26 by the continuous curve according to Reference 44.

The parameter L_m relates to the NFL of the concrete communication system. To obtain the shadow fade margin, one needs information on the PDF of the slow fading, which is defined in References 42–46 as a Gaussian process with a zero-mean Gaussian variable σ_{SF} and with a standard deviation of shadowing σ_{L}. Its PDF was described in Chapter 1; we rewrite it using a definition as:

$$PDF(x) = \frac{1}{\sigma_{\text{L}}\sqrt{2\pi}} \exp\left\{ -\frac{\sigma_{\text{SF}}^2}{2\sigma_{\text{L}}^2} \right\}. \tag{10.30}$$

Then the corresponding term of shadowing Equation (10.28) in dB for link budget design equals: $L_{\text{SF}} = 10 \log \sigma_{\text{SF}} \equiv \sigma_{\text{SF}}[\text{dB}]$. Now we can, as in References 43–45, introduce an error function to obtain the so-called CCDF, $\text{CCDF} \equiv Q(t) \equiv 1 - CDF(t)$, which describes the probability that the shadowing increases the median path loss by at least Z dB (Fig. 10.26):

$$Q(t) \equiv \Pr(\sigma_{\text{SF}} > Z) = \frac{1}{\sqrt{2\pi}} \int\limits_{x=t}^{\infty} \exp\left\{ -\frac{x^2}{2} \right\} dx, \tag{10.31}$$

where $t = Z / \sigma_{\text{L}}$. The CCDF is depicted in Figure 10.27 following References 42 and 43 versus the normalized parameter t. In this approach was assumed a

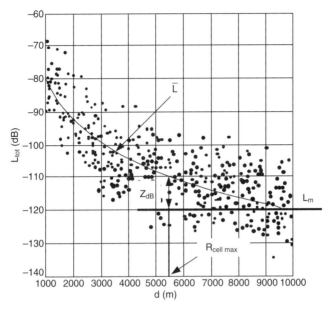

FIGURE 10.26. Median path loss and slow fading margin with respect to experimental dara versus distance d from the transmitter in meter (Source: Reference 44. Reprinted with permission of John Wiley & Sons).

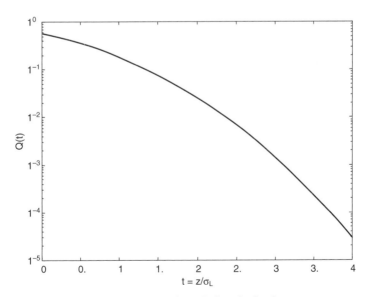

FIGURE 10.27. Gaussian CCDF versus the relative shadowing parameter normalized by standard deviation of slow fading.

priori that either a maximum acceptable path loss of the system compared to noise floor level is known, or the concrete percentage of successful communication at the fringe of coverage of test area is known. In both cases, the margin level Z can be obtained and, finally, for the concrete range of test area, the slow fading margin. To explain to the reader how to use the results of calculations according to References 43 and 44 that are presented in Figure 10.26 and Figure 10.27, let us explain an example illustrated in Figure 10.26.

For the case where the maximum acceptable path loss of the system is $L_m = 120$ dB and the median path loss at the range of the test area (or cell) is $\bar{L} = 110$ dB, the maximum fading margin is $Z = 10$ dB at the range $r = 5.5$ km. On the contrary, when the range of service and the maximum acceptable path loss are unknown, but the percentage of the successful coverage for each individual subscriber location inside the test side, as well as the standard deviation of slow fading are known, one can use the CCDF depicted in Figure 10.27. Now, if one considers the radio coverage of the area of service of about 90%, then we can find the value t for which the path loss is less than the maximum accepted path loss, L_m. It means that there is a probability of 90% to locate any user in the area of service. Thus, the shadowing effect (the worst case) can be found in the following manner: CCDF $= Q(t) = 100\% - 90\% = 10\%$ or $Q(t) = 10^{-1}$. From Figure 10.27, this occurs for $t = 1.28155$. Moreover, if the standard deviation of shadowing is also known, for example, $\sigma_L [dB] = 8$dB, we get that

$$L_{SF} \equiv Z[\text{dB}] = t\sigma_L = 1.28155 \cdot 8 = 10.2524 \text{ dB},$$

that is, the same result ($Z = 10$ dB) as was estimated when the range of test area and L_m were a priori known. Finally, if all parameters of shadowing and the system are known, one can easily obtain from Figure 10.27 the maximum range of the test area where full coverage is possible, using results shown in Figure 10.26.

Using the concept described in References 43–45, one can also estimate the *fast fade margin* (L_{FF}) using well-known Rician PDF distribution of such parameter (see also definitions introduced in Chapter 1). We will rewrite it using new notations as

$$\text{PDF}(L_{FF}) = \frac{2\sigma_{FF}}{(\text{rms})^2} \exp\left\{-\frac{\sigma_{FF}^2}{(\text{rms})^2}\right\} \cdot \exp(-K) \cdot I_0\left(\frac{2\sigma_{FF}}{\text{rms}}\sqrt{K}\right), \quad (10.32)$$

where, as in References 43–45, $rms = \sqrt{2} \cdot \sigma_{FF}$, σ_{FF} is the standard deviation of fast fading, K is the ratio of LOS component (deterministic part of the total signal) and NLOS component (random part of the total signal). Then the last term in the equation of link budget (10.28) can be defined as $L_{FF} = 101\log\sigma_{FF} \equiv \sigma_{FF}[\text{dB}]$. When parameters rms and K a priori are known, one can find L_{FF} by using results presented in Figure 10.28.

Now, if, for example, $K = 10$ and the fast fading probability equals 10^{-3}, it means the probability of the event that there is fast fading between subscribers

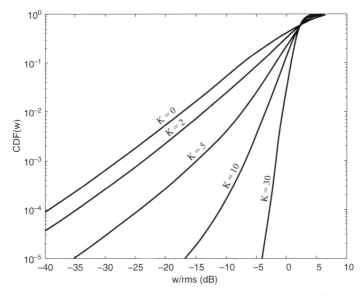

FIGURE 10.28. Rician CDF versus relative strength of the signal.

equals 0.1%, or that there are about 99.9% cases of successful communication links. If so, we have from the curve depicted in Figure 10.28, that for $K = 10$ we get $w = -10 \mathrm{llog}(L_{FF} / rms) = -L_{FF[dB]} + rms_{[dB]} = -7$ dB, that is, the fast fade margin L_{FF} is 7 dB below the rms. If rms is known, for example [44, 45], $rms = 5$ dB, then $L_{FF} = 5 + 7 = 12$ dB. According to the scenario of how to obtain the link power budget proposed in References 44 and 45, one has two variants of link-budget design prediction.

First Variant. Estimation of the *slow fade margin*, Z, deriving the median path loss, $\bar{L}$, by using well-known propagation models [42–46], and using the maximum acceptable path loss (L_m), as well as the range between concrete terminal antennas, the knowledge of which is done for the performed wireless system. Estimation of fast fade margin can be done, if the standard deviation and the Rician K-parameter are known for the concrete propagation channel and wireless system.

Second Variant. Estimation of the slow fade margin, Z, and the range between concrete terminal antennas using derivation of the median path loss, $\bar{L}$, through well-known propagation models [42–46], and using the standard deviation of slow fading, σ_L, as well as the percentage of the success communication for the concrete wireless system performed. In this case, using Figure 10.27 and the known value of σ_L, one can easily obtain the slow fade margin $L_{SF} \equiv Z = t\sigma_L$. As for the fast fade margin, L_{FF}, it can be obtained when the standard deviation of fast fading (σ_F) or rms, as well as the probability of the successful communication (i.e., the cumulative distributed function [CDF]) are known for the concrete wireless communication channels by using the results from Figure 10.26, Figure 10.27, and Figure 10.28.

10.2.2. Link Budget Design Based on the Stochastic Approach

Both approaches presented earlier need a priori information about the situation inside the propagation channel, that is, about the features of propagation environment and about the concrete wireless system. This means that designers of wireless networks need full information of the "success" (in percentages) of communication between subscribers and radio coverage of the area of service, as well as characteristics of the system, such as the standard deviation of slow- and fast-fading signal, the maximum acceptable path loss or noise floor level, and so forth. As was mentioned earlier, to have all these characteristics, one must obtain experimentally all features of the communication channel (environment) and have full information about the concrete wireless system. This way is unrealistic, because all these characteristics are not predictable simultaneously: either the concrete communication system parameters are known and designers need to carry out test experiments to find the characteristic features of the channel (environment), or conversely, having information about the terrain features, designers predict future wireless system for effective service of subscribers.

In the following section, we first propose an approach developed in Reference 27, which is based on information about the propagation characteristics of the channel obtained either by measurements or by using information obtained from topographic maps. At the same time, we have some information about the wireless networks. Namely, about the NFL or *maximum accepted path loss* of the wireless system.

Then, we will present another approach developed in Reference 29, which is based on the general stochastic multiparametric model, all main equations of which can be rearranged and simplified to the "straight-line" mathematical form (as was done in Chapter 6 for another model) for various scenarios occurring in the built-up terrain. In other words, we propose two approaches, which are based on the general stochastic model described earlier for rural, mixed residential, and built-up terrain.

The First Approach. We will present this approach for link budget design in the form of prediction algorithm.

First Step. At this step, we obtain the standard deviation of slow fading (σ_L) as a logarithm of ratio between the signal intensity with and without diffraction phenomena, which gives main influence on shadowing from building contours:

in the case of single diffraction

$$\sigma_L = 10\log\frac{\left[\left(\lambda d / 4\pi^3\right)+\left(z_2 - \bar{h}\right)^2\right]^{1/2}}{\left(z_2 - \bar{h}\right)}$$
(10.33)

in the case of multiple diffraction

$$\sigma_L = 10\log\frac{\langle I_1 \rangle + \langle I_2 \rangle}{\langle I_3 \rangle},$$
(10.34)

where $\langle I_1 \rangle$ is described by (5.76) and $\langle I_2 \rangle$ by (5.78) for urban and suburban environments. The term $\langle I_3 \rangle$ is described by the following expression:

$$\langle I_3 \rangle = \frac{\Gamma \lambda l_v}{8\pi\{\lambda^2 + [2\pi l_v\gamma_0(\bar{h} - z_1)]^2\}d^3}(z_2 - \bar{h}).$$
(10.35)

Finally, this allows us to obtain the fade margin using two options described in References 43 and 44:

- according to Reference 43, because of shadowing effect from building roofs and corners

$$\sigma_{SF} = 2\sigma_L \text{ and } L_{SF} = 10\log\sigma_{SF}$$
(10.36)

- according to Reference 44 using information about the maximum acceptable loss within the system or NFL, for the concrete range of test area, $d = constant$, we immediately obtain this value as $L_{SF} = L_m - \bar{L}$, where $\bar{L}$ is defined by (5.78)–(5.81) for each concrete terrain environment. In this case, obtaining σ_L and then, we finally (by using Fig. 10.28), can figure out (in percentages) how successful the communication link will be, without obtaining any shadowing effects.

$$t = L_{SF} / \sigma_L$$
(10.37)

Second Step. To obtain information about the *fast fading margin*, L_{FF}, we need, first of all, the knowledge of the Rician K-parameter as a ratio of coherent (LOS component) and incoherent (multipath component without diffraction) parts of the total signal intensity [27]; that is, $K = \langle I_{co} \rangle / \langle I_{inc} \rangle$. Here, for the mixed residential rural areas, $\langle I_{inc} \rangle$ and $\langle I_{co} \rangle$ are described by expressions (5.29) and (5.30). For urban and suburban areas, they are described by (5.76) or (5.78) and (5.79), respectively. At the same time, we obtained earlier (for each terrain type) the *rms* of the total signal intensity, $rms = \sqrt{\langle I_{total} \rangle}$, using for each case the corresponding Equation (5.80). Finally, we use, instead of the method proposed by Reference 27 and presented earlier with help of CDF (x) depicted in Figure 10.28, the corresponding equation of Rician distribution, which allows us to obtain σ_{FF} or L_{FF} using well-known equation:

$$L_{FF} = 10 \log \sigma_{FF} = 10 \log \left\{ \left[\int_0^\infty x^2 PDF(x)dx - \left(\int_0^\infty xPDF(x)dx \right)^2 \right]^{1/2} \right\}, \qquad (10.38)$$

where x is the random amplitude of the received signal. Using (10.38), following Reference 27, we get for σ_{FF}

$$\sigma_{FF} = \left[2 \cdot (rms) \cdot e^{-K} \right] \cdot \left\{ \frac{1}{2} e^K \int_0^\infty y^3 e^{-y^2} I_0 \left(2y\sqrt{K} \right) dy - \left[\int_0^\infty y^2 e^{-y^2} I_0 \left(2y\sqrt{K} \right) dy \right]^2 \right\}^{1/2}.$$

$$(10.39)$$

Now, the knowledge of corresponding *rms*, the ratio L_{FF} / rms, and K for the concrete range between subscribers or between base station and subscribers, allows us to obtain the outage probability function CDF(L_{FF} / rms). Conversely, if CDF(L_{FF} / rms) is known, using result of *rms* and K derivations, as well as the corresponding Rician CDF(w) distribution, depicted in Figure 10.28 according to References 44 and 45 versus the normalized magnitude of the fast fading, $w = L_{FF} / rms$, we can obtain the fast fade margin L_{FF}.

Finally, taking L_{FF} either from (10.39) or from Figure 10.28, and σ_L according to (10.33)–(10.35) and relations $\sigma_{SF} = 2\sigma_L$ ($L_{SF} = 10 \log \sigma_{SF}$) [27, 44, 45], we get the total power budget (in decibels) for the actual communication link design. Moreover, we simultaneously can strictly predict the percentages of the fast-fading effects (via CDF(L_{FF} / rms), see previous discussion) for the real situation within the urban communication channel.

The Second Approach. Another approach was proposed in Reference 29 based on the general equations of the path loss obtained for different terrain types, suburban and urban. Because these equations account for multiple diffraction and scattering, having the coherent and incoherent parts (see Eq. 5.76, Eq. 5.77, Eq. 5.78, Eq. 5.79, Eq. 5.80, and Eq. 5.81), they have been rearranged in the simple forms of "straight-line" equations, as was proposed earlier by another author (see References 2, 6, and 43–46). Depending on elevation of the BS antenna, with respect to building rooftops and the MS antenna, several scenarios occur in the built-up scene proposed in Reference 29.

Thus, for the *first quasi-LOS* scenario, when the BS antenna is above, and the MS antenna is below the rooftops (see Fig. 10.29a), the following equation for the total path loss (in dB) can be presented [29]:

$$L_1(r) = -32.4 - 20 \log f_{[MHz]} - 20 \log r_{[km]} - L_{fading} + (G_{BS} + G_{MS}), \qquad (10.40)$$

where following Reference 35 we can define L_{fading} as signal deviations with respect to average path loss. After straightforward derivation according to Reference 36, we get

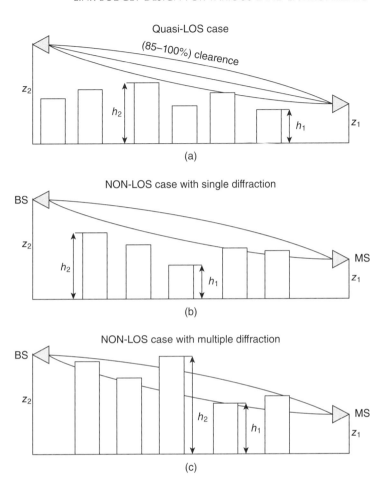

FIGURE 10.29. (a) BS antenna is higher than building rooftops; Quasi-LOS conditions with clearance exceeding 85%. (b) There are several buildings that obstruct the radio path; NLOS conditions. (c) The BS antenna is at the same level or lower than building rooftops; NLOS multipath conditions.

$$L_{fading} = 10\log\left[\gamma_0 r \frac{F(z_1, z_2)}{h_2 - h_1}\right] = 10\log\left[\gamma_0 r \frac{F(z_1, z_2)}{\Delta h}\right], \qquad (10.41)$$

where, h_1 and h_2 are the minimum and maximum heights of built-up terrain in meter, r is the range between the terminal antennas in km, z_2 and z_1 are the height of the BS and MS antenna, respectively. $\gamma_0 = 2\bar{L}\nu/\pi$ is the density of building contours in km^{-1}, $\bar{L}$ is the average length of buildings in km, ν is the number of buildings for square kilometer, G_{BS} and G_{MS} are the antenna gains for the BS and MS, respectively.

In urban areas, depending on the buildings density, γ_0 ranges between $5 \cdot 10^{-3}$ m^{-1} and $10 \cdot 10^{-3}$ m^{-1}. The function $F(z_1, z_2)$, in meters, describes the buildings' overlay profile surrounding two terminal antennas. It can be simply evaluated, following Reference 29, for two typical cases:

$$F(z_1, z_2) = \begin{cases} (h_1 - z_1) + \dfrac{\Delta h}{n+1}, & h_1 > z_1, z_2 > h_2 > z_1 \\ \dfrac{(h_2 - z_1)^{n+1}}{(n+1)(\Delta h)^{n+1}}, & h_1 < z_1, z_2 > h_2 > z_1 \end{cases}. \tag{10.42}$$

For urban areas with approximately equal amount of tall and small buildings $n = 1$; when the number of tall buildings is dominant (Manhattan-grid plan) $n = 0.1$–0.5; when the number of small buildings prevail (suburban and residential areas) $n = 5$–10. Thus, in most cities of Israel, $n = 1$ is a very precise value. In Lisbon, for example, this value varies from 0.89 to 1.17 (see previous discussion), that is, close to $n = 1$. Therefore, we propose to use $n = 1$ for areas that are not the same as Manhattan-grid scenario. In the case of $n = 1$, the average building height can be assumed as

$$\bar{h} = h_2 - \frac{n}{n+1}(h_2 - h_1) = \frac{(h_2 + h_1)}{2}.$$

In the *second scenario*, as shown in Figure 10.29b, in NLOS conditions, diffraction from roofs located close to the MS antenna is the source of shadowing and the slow fading phenomenon. Here, following Reference 29, we get

$$L_2(r) = -32.4 - 30\log f_{[MHz]} - 30\log r_{[km]} - L_{fading} + (G_{BS} + G_{MS}), \tag{10.43}$$

where in this specific case

$$L_{fading} = 10\log \frac{\gamma_0 l_v F^2(z_1, z_2)}{|\Gamma| \left[\dfrac{\lambda r}{4\pi^3} + (z_2 - \bar{h})^2 \right]^{1/2}}. \tag{10.44}$$

In (10.44), all the units of the parameters are in meters and have the same meaning as the parameters described earlier. Here also, $F^2(z_1, z_2) = (\bar{h} - z_1)^2$; and the parameter l_v is the walls' roughness, usually equals 1 m to 3 m; $|\Gamma|$ is the absolute value of reflection coefficient: $|\Gamma| = 0.4$ for glass, $|\Gamma| = 0.5$–0.6 for wood, $|\Gamma| = 0.7$–0.8 for stones, and $|\Gamma| = 0.9$ for concrete. The wavelength of radio wave has a wide range that varies from $\lambda = 0.05$ m to $\lambda = 0.53$ m and covers most of the modern wireless networks. Equation (10.43) with Equation (10.44) can be used for link budget design for various scenarios occurring in the built-up terrain where the BS antenna is higher than the buildings' overlay profile.

In the case of the *third scenario*, depicted in Figure 10.29c, diffraction from roofs of the buildings, located close to the MS antenna and BS antenna, are the sources of shadowing and the slow fading phenomena. Here we get [29]

$$L_3(r) = -41.3 - 30\log f_{[MHz]} - 30\log r_{[km]} - L_{fading} + (G_{BS} + G_{MS}), \qquad (10.45)$$

where

$$L_{fading} = 10\log \frac{\gamma_0^4 l_v^3 F^4(z_1, z_2)}{\lambda |\Gamma|^2 \left[\frac{\lambda r}{4\pi^3} + (z_2 - \bar{h})^2\right]}. \qquad (10.46)$$

All parameters presented in (10.46) are the same, as earlier, and are measured in meter. In this situation, the profile function can be presented following Reference 29 as

$$F(z_1, z_2) == \begin{cases} \left|(h_1 - z_1) + \dfrac{(\Delta h)^2 - (h_2 - z_2)^2}{2\Delta h}\right|, & h_1 > z_2, h_2 > h_1 > z_1 \\[3mm] \dfrac{(h_2 - z_1)^2 - (h_2 - z_2)^2}{2(\Delta h)}, & h_1 < z_2, h_2 > h_1 > z_1 \end{cases}. \qquad (10.47)$$

As in the second scenario, Equation (10.45), Equation (10.46), and Equation (10.47) can be used for link budget design for various scenarios of the built-up terrain, and for both BS and MS antennas lower than the rooftops.

The *fourth scenario* describes a situation when both antennas are located in direct visibility along a street at the range of $r = 100 - 1000$ m. The width of the street can vary from $a = 10$ m to $a = 30$ m, the waveband of the transmitter/receiver is in the wavelength range of $\lambda = 0.01 - 0.05$ m, and the parameter of brokenness can vary from 0.1 to 0.8 [29], which describes the distribution of buildings lining the street. The parameter of brokenness is defined by the ratio between average length of building $\bar{L}$ and the sum of average length of slits $\bar{l}$ and the average length of buildings, as shown in Figure 10.30, that is, $X = \bar{L}/(\bar{L} + \bar{l})$. For example, in the Manhattan-grid area of crossing avenues, the parameter of brokenness is $X = 0.5 - 0.8$. In such a scenario,

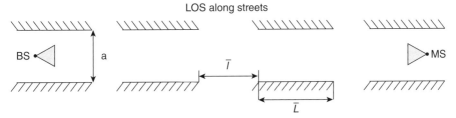

FIGURE 10.30. Both terminal antennas are within the street below rooftops.

the heights of the BS and MS antennas are lower or at the same height compared with average buildings' height. Here, following the waveguide model described in Chapter 5, we can present the total path loss (in dB) as

$$L_5 = -32.4 - 20 \log f_{[MHz]} - 17.8 \log r_{[km]} - 8.6 |\ln X| \frac{\lambda r}{2a^2}$$

$$+ 20 \log \frac{(1-X)}{(1+X)} + (G_{BS} + G_{MS}),$$

(10.48)

where r, a, and λ are in meter; all other parameters were determined earlier.

In the case of *fifth scenario*, as shown in Figure 10.31a, the BS antenna, with the height z_2, is higher than the maximum height of the building overlay profile, h_2, but the moving subscriber (MS) antenna with the height z_1 is located lower than h_2. Both of them are located within the street, as is seen from Figure 10.31b. In such a scenario, the guiding effects of channeling of radio waves along the street become weaker and the effect of building array randomly distributed at the terrain must be taken into account by using the proposed stochastic approach and combining it with the waveguide model, both described in Chapter 5.

However, situations presented in Figure 10.31a,b are more complicated compared with that described in Chapter 5, because here we must take into account the effects of buildings lining the street and placed within the layer denoted in Figure 10.31a by "1" in both parts of the street (due to symmetry of the problem, we depicted, in Fig. 10.31a only one side of the building layer). Then, for all other buildings located in area "2" (see Fig. 10.31b), we can use the stochastic description and Equation (10.40) for the *first situation* with BS antenna elevation and Equation (10.43) for the *second situation* with BS antenna elevation, where now instead of Equations (10.41) and (10.43), we should introduce an additional term. As was shown in Reference 29, this term, which accounts effects of buildings lining the street in area "1", cannot be used additively when we present all equations in decibels. Therefore, we can rewrite L_{fading} using the following equations:

A. For the *first situation*, accompanied by the buildings lining the street, we get

$$L_{fading} = 10 \log \left[\gamma_0 r \frac{F(z_1, z_2)}{h_2 - h_1} + \ln \left(\pi \frac{k l_v \bar{h}}{2a} \right) \right],$$

(10.49a)

where $\bar{h}$ is the average building height defined in meter, l_v is the parameter of roughness of buildings' walls (usually $l_v = 1$–3 m), and $k = 2\pi / \lambda$.

B. For the *second situation*, considering the buildings lining the street, we get

$$L_{fading} = 10 \log \left\{ \frac{\gamma_0^4 l_v^3 F^4(z_1, z_2)}{\lambda |\Gamma|^2 \left[\frac{\lambda r}{4\pi^3} + (z_2 - \bar{h})^2 \right]} + \ln \left(\pi \frac{k l_v \bar{h}}{2a} \right) \right\}.$$

(10.49b)

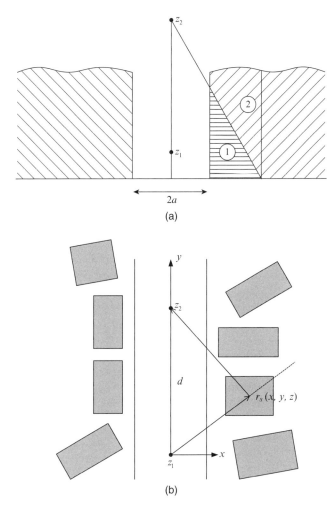

FIGURE 10.31. (a) The 2-D vertical plane of the fifth scenario, where the BS antenna is higher than rooftops. (b) The 2-D horizontal plane of the fifth scenario where both antennas are along the street.

As in earlier discussions, all parameters in (10.49) are in meter. Using Equations (10.49a) or (10.49b), we can finally introduce either in Equations (10.40) or (10.43), according to the corresponding scenario occurring in the urban scene.

10.2.3. Experimental Verification of Link Budget Accounting for Shadowing Phenomena

To account for the long-term or slow fading, the proposed stochastic approach and the corresponding Equation (10.34), Equation (10.35), and Equation

(10.36), are used to increase the accuracy of the proposed model by taking into account shadowing effects from building roofs for low antenna elevation. As was shown in experiments in Lisbon (see Section 10.1.4), this can significantly change results of the theoretical prediction of path loss for antennas that can be arranged deeply within the built-up layer. In this case, we need to estimate the additional signal attenuation due to shadowing and introduce in a link-budget equation not only mean value of path loss, but also the effects of slow fading, described by Equation (10.34), Equation (10.35), and Equation (10.36). Returning to Figure 10.21, we analyze a situation that occurs when the antenna is located at the height that corresponds to the angles greater than θ'. The procedure that estimates long-term (slow) fading phenomena for the propagation link budget design, using existing methods, was described earlier in Section 10.2.1. Taking into account the additional fading effects, using Equation (10.34), Equation (10.35), and Equation (10.36), for the median path loss obtained for tested area of Lisbon by using Equation (5.76), Equation (5.77), Equation (5.78), Equation (5.79), Equation (5.80), and Equation (5.81), we can significantly correct the results of total path loss prediction using the following equation:

$$L_{total} = 10\log\left[\lambda^2\left(\langle I_{co}\rangle + \langle I_{inc}\rangle\right)\right] + 10\log\sigma_{SF}. \tag{10.50}$$

For comparison with the experimental results presented in Figure 10.19a,b,c, we selected only the worst-case scenarios (i.e., with shadowing), where the difference between theoretical prediction and measurements exceeds 20–25 dB. These sites are presented in Figure 10.32a,b, where again, segments that describe the experimental data with asterisks correspond to theoretical calculations of mean pass loss according to (5.76) to (5.81).

Using the more general Equation (10.50), for the link budget design, with the effects of slow fading, we get the results depicted by diamond signs in Figure 10.32a,b. By inspection, it is seen that the additional excess attenuation, which we now add to the stochastic model, yields a better agreement with the experimental data for five of the six tested sites. The discrepancy of about 20–30 dB for these five points is reduced to about 3–5 dB, and the standard deviation does not exceed 5–6 dB.

10.2.4. Experimental Verification of Slow and Fast Fading

The 1970s and 1980s brought some pioneering experimental and theoretical investigations of spatial–temporal variations of signal strength or power in various built-up environments. It was shown that propagation process within the urban communication link is usually locally stationary in the time domain [14, 35, 40, 44–49]. The spatial variations of the signal passing such a channel have double natures, the large-scale and the small-scale [40, 44, 49], which we defined in Chapter 1 as slow and fast fading, respectively, according to well-known standard.

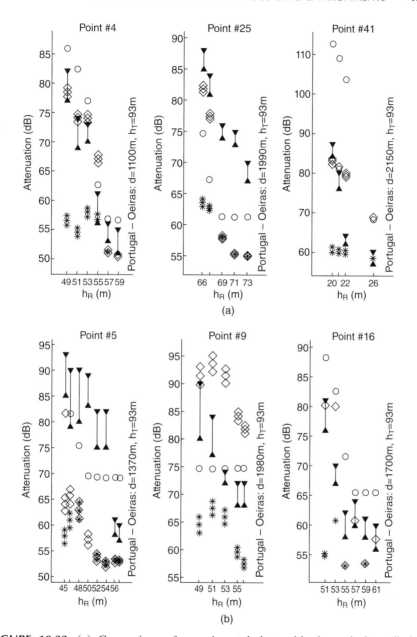

FIGURE 10.32. (a) Comparison of experimental data with theoretical prediction taking into account shadowing (denoted by diamonds). (b) The same as in panel a, but for other experimental sites.

Experimental Verification of Slow Fading. As was shown by numerous experimental observations of spatial and temporal fluctuations of radio signals in urban communication links, slow fading is observed mostly for higher elevated BS antenna and lower elevated MS antennas, compared to roofs of the buildings, and are caused by the shadowing effects of buildings surrounding both terminal antennas. Finally, deep "shadow" zones in communications are created along the radio path in NLOS (clutter) conditions between each BS and each MS antenna located within the area of service. A summary of experimental results, obtained in different cities of Israel, was presented in terms of the cross-correlation function of signal amplitude $D_i(d) = \langle [E_i(\mathbf{r}_1) - E_i(\mathbf{r}_2)] \rangle$ for three components of the field strength $(E_1 = E_x, E_2 = E_y, E_3 = E_z)$. These components were analyzed by measuring along the radio paths between the terminal antenna located at the point $\mathbf{r}_2$ and that at the point $\mathbf{r}_1$, that is versus $d = |\mathbf{r}_2 - \mathbf{r}_1|$. The result of this statistical analysis, shown in Figure 10.33, is direct evidence of the existence of such long-scale random variations of signal strength level along the radio path d.

These spatial signal variations have been obtained after signal processing of the average signal amplitudes (about 3800 magnitudes of signal strength has been investigated) with the scale of averaging of $\sim$8–10 wavelengths at the frequency band from 80 to 400 MHz, to exclude effects of fast fading due to random interference between multipath field components.

As can be seen from Fig. 10.33, there are two sharp maxima, which correspond to two scales of signal spatial variations, $L_1 = 15$–20 m and $L_2 = 80$–100 m,

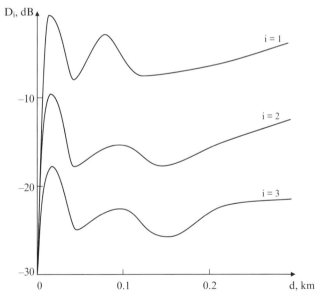

FIGURE 10.33. Statistical analysis of long-scale slow fading along the radio path d for three components of the electromagnetic field ($i = 1, 2, 3$).

respectively. As was shown experimentally [14, 44–49], the first scale can be related to the average length of gaps between buildings. It determines the so-called "light zones" observed within these gaps.

The second scale, according to References 14 and 44–49, can be related to the large "dark" zones that follow the "illuminated" zones, and can be explained only by the diffraction from buildings located along and across the radio path. Other experimental investigations carried out during the 1980s and the beginning of 1990s proved this principal result, which describes a nature of slow signal fading in urban communication links. In other words, all referred experimental works have shown that the effect of shadowing is a realistic cause for long-term variations of signal strength (or power) after its averaging with "window scale" exceeding several wavelengths during the statistical processing. Moreover, as was obtained both theoretically and experimentally [46–60], the statistical distribution of slow signal variations, measured in decibels (dB), are very close to log-normal distribution. As for this phenomenon, we described it in detail in Chapter 1 and also in this section using corresponding analytical presentation of the slow fading.

Experimental Verification of Rayleigh and Rician Laws for Fast Fading. As in Reference 14, we based on Rician statistical distribution because, as was mentioned in Chapter 1, the latter distribution is more general and covers the Rayleigh fast fading statistics. Therefore, it can be successfully used for the analysis of signal fast fading both for close and for open fixed radio links. At the same time, first Clarke [64] and Aulin [65], and later Ali Abdi with colleagues [66], have used Rayleigh statistics in their analysis of fast fading. To verify the accuracy of the models described for fast fading phenomena for built-up environments, special experiments were carried out [15–32, 47–52]. About 3800 magnitudes of signal strength were collected, and then divided at three separate groups depending on the type of radio path and antenna elevations, with respect to roofs of the buildings. The *first group* of signal outputs was measured at radio links, with the obstructive conditions for the low elevated terminal antennas where there is NLOS between the terminal antennas. The *second group* of signal outputs was obtained for radio links, where both LOS and NLOS conditions were observed between the terminal antennas. In this case, one antenna (usually the base station antenna) was higher than rooftops. The *third group* was obtained for radio links where only LOS conditions for radio signal were observed.

Thus, the measured output samples for the first group were differentiating as satisfying Rayleigh statistics, (about 800 from 1200 measured points). The measured output points for the second and the third groups were combined together, approximately 1000 points from 1200 samples for the second group, and 800–900 from 1300 sample points for the third group. The reason for not using all the samples will be explained later.

FIGURE 10.34. (a) Distribution of signal strength amplitude $r(t)$ versus the normalized parameter $x = r(t) / rms$; the histogram corresponds to experiments and the continuous and dashed curves describe the Rayleigh distribution and that from the stochastic approach, respectively. (b) Distribution of signal strength amplitude $r(t)$ versus the normalized parameter $x = r(t) / rms$; the histogram corresponds to experiments and the continuous and dashed curves describe the Rician distribution and that from stochastic approach, respectively. (c) Distribution of signal strength amplitude $r(t)$ versus the normalized parameter $x = r(t) / A$; histogram corresponds to experiments and the continuous curve describes general distribution (10.51).

→

   Statistical analysis of first group of outputs, which is approximately 1200 magnitudes of signal strength, is presented as histogram in Figure 10.34a versus the normalized parameter $x = r(t) / rms$. Where $r(t)$ is the local magnitude of signal envelope; the parameter rms is defined in Chapter 1.
   This histogram was constructed using approximately 800 points along the radio path at the ranges from 200 m to 1.5 km between the terminal antennas. The corresponding continuous curve presented in Figure 10.34a was derived from the Rayleigh distribution (see Chapter 1), with all parameters obtained from resulting statistical analysis of measured points. From Figure 10.34a, one can see that the experimentally obtained signal strength distribution is fully described by Rayleigh's law. The measured output points for the second and the third groups were combined together (approximately 1300 points for the second group and 700–800 points for the third group). Their statistical analyses were presented in the histogram in Figure 10.34b as a function of the normalized signal strength. The continuous curve in the figure describes calculations of the Rician distribution (see Chapter 1) of fast signal variations obtained by using the information about rms and K from statistical analysis of experimental data. As it is seen, a good explanation of experimental data was obtained by using the Rician's law. Results of the statistical analysis of numerous experiments and theoretical prediction, both presented in Figure 10.34a,b, allow us to conclude that the Rayleigh statistics can be used during analysis of closed urban communication links for low elevated mobile terminal antennas, that is, for a case of NLOS conditions in mobile communication. At the same time, the Rician law strictly describes fast fading effects for combined NLOS (close areas) and LOS (open areas) conditions where one of the antennas is higher than the rooftops.

Description of Multipath Effects by the Stochastic Approach. Now, we will explain why all the experimental samples in the statistics were not used. As was mentioned earlier, about 30–35% of more than 3800 measurement points cannot be described by the Rayleigh–Rician statistics. Thus, in the first group, only about 800 from the 1200 measured points could be analyzed by using the

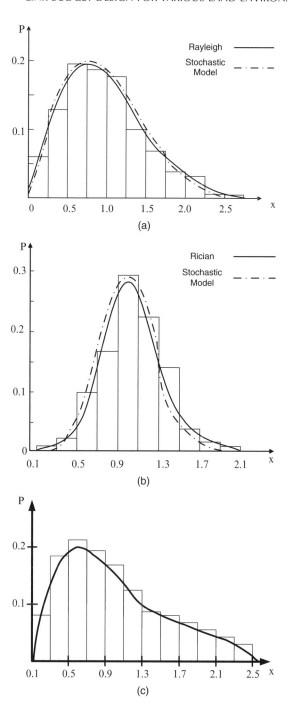

(a)

(b)

(c)

Rayleigh statistics description. The same situation occurred with the second group, where only about 1000 points of the 1200 satisfy the Rayleigh law. Finally, from the third group, only about 800–900 of the 1300 points could be analyzed by using Rician statistics. In all the points that are not described by the well-known Rayleigh–Rician statistics, we observe deeper signal strength fast variations.

In order to explain the mechanism of deep signal strength variations, we have used the unified stochastic approach fully described in Reference 30, taking into account the multiple reflection and scattering effects from buildings and other obstructions randomly distributed on the terrain according to Poisson's law. Such a model allows us to take into account the strength of the total field at the receiver because of additive effect of one-time to n-time scattered waves with independent strengths. As was obtained in References 29 and 30, in the zones close to the BS antenna ($d < 1$ km), the single scattered waves are predominant, whereas far from the BS antenna ($d > 2$ km) the two- and three-time scattered waves prevail. However, the field strength of the n-reflected waves exponentially attenuates with distance. Therefore, for microcell communication channels with ranges less than 2–3 km, only single and two-time scattered waves must be taken into account; this phenomenon was also mentioned in Chapter 5. Moreover the strengths of these waves are normally distributed with the zero-mean value and dispersion σ_1^2 (for single scattered waves) and σ_2^2 (for two-time scattered waves), and depend strongly on the characteristic features of the terrain. The main features are the density of the building ν/km^2, the average length of the building (or width, depending on the orientation to antenna direction) $\bar{L}$, and the contours density of the buildings γ_0 (see the corresponding expressions in Chapter 5). The average wave number also depends on the distance from the BS antenna d. Thus, following Equations (5.70) and (5.72) presented in Chapter 5, we can compute the average number of single and double scattered waves, as well as the probability to receiving single and double scattered waves at the MS antenna.

Following the earlier mentioned, we took into account in our computations of Equation (10.53) (see discussion that follows) for CDF the effects of independent single (the first term) and double (the second term) scattering, as well as their mutual influences on each other (the third term), that is,

$$CDF(r) = \frac{P_1(1-P_2)}{P_0}\frac{r}{\sigma_1^2}\exp\left\{-\frac{r^2}{2\sigma_1^2}\right\} + \frac{P_2(1-P_1)}{P_0}\frac{r}{\sigma_2^2}\exp\left\{-\frac{r^2}{2\sigma_2^2}\right\}$$
$$+ \frac{P_1P_2}{P_0}\frac{r}{(\sigma_1^2+\sigma_2^2)}\exp\left\{-\frac{r^2}{2(\sigma_1^2+\sigma_2^2)}\right\},$$
(10.51)

where $P_0 = 1-(1-P_1)(1-P_2) = 1-\exp\left[-\left(\bar{N}_1+\bar{N}_2\right)\right]$ is the probability of direct visibility (i.e., of LOS). In Figure 10.34(c), the corresponding histogram shows the cumulative distribution of signal strengths of these "anomalous" 1000–1100 samples is shown together with the CDF (the continuous curve) computed according to (10.51), taking into account the same experimental data as was done earlier for the Rayleigh and Rician statistics evaluation.

We have analyzed this more general distribution as a function of the normalized random signal strength envelope $x = r / \sqrt{\sigma_1^2 + \sigma_2^2}$ using the probabilities P_1 and P_2. These probabilities were computed for the terrain parameters $\gamma_0 = 10 \ km^{-1}$ and $d = 0.5$–0.8 km of an average city and for the ratio of $(\sigma_1/\sigma_2) \approx 12$–$13$ dB of losses between the single and double scattered waves [30].

It is clearly seen from Figure 10.34(c) that for d < 1 km, the obtained distribution, calculated according to (10.51), with all the parameters obtained from measurement, strongly differs from Rayleigh or Rician statistics, presented in Figure 10.34a,b. At the same time, using other 2500–2600 samples selected separately and depicted in Figure 10.34a,b, respectively, we computed (10.51) for each specific condition that occurred in the urban scene. Thus, for the histogram in Figure 10.34a, we took the same conditions of mean city with $\gamma_0 = 10 \ km^{-1}$, but with the NLOS conditions at the distance of 1.5 km from the BS antenna. For the histogram, depicted in figure 10.34b, we took the same conditions of mean city with $\gamma_0 = 10 \ km^{-1}$, but with the LOS and quasi-LOS conditions at the distance of 1.5 km from the BS antenna. The corresponding computations of general signal strength distribution (10.51) are shown by dashed curves in Figure 10.34a,b. It is obvious that general distribution (10.51) gives close solution both for Rician law (in LOS and quasi-LOS conditions) and for Rayleigh law (in NLOS conditions), at distances from BS antenna where effects of multipath become predominant. Therefore, with a great accuracy, we can use expression (10.51) to describe the fast fading phenomena for various situations occurring in multipath communication links for different elevations of base station antenna and subscriber antenna, with respect to obstructions surrounding them. We propose a new tool to investigate fast spatial and temporal signal strength fading by using the general signal strength distribution (10.54) based only on propagation multipath phenomena, using measurements data, and characteristic features of the built-up terrain, conditions of measurements, and antennas location, compared with height profile of the buildings.

10.3. CHARACTERIZATION OF MULTIPATH RADIO CHANNEL BY THE RICIAN FACTOR

In this section, we present the reader a simple physical explanation of multipath propagation effects that fit the Rician distribution. The K-factor of Rician distribution can also be a very useful tool for multiplicative noise description (see definitions in Chapter 1). We will show in Chapter 12 how to use this fading factor to predict information data stream parameters, such as capacity, spectral efficiency, and bit error rate, as well as the maximum radius of cell during cellular maps designed in different terrestrial environments.

According to the definition presented in Section 10.2, the Rician parameter can be defined according to stochastic model as the ratio of the coherent and incoherent component of the signal intensity, $K = \langle I_{co}\rangle/\langle I_{inc}\rangle$. Here the coherent component, $\langle I_{co}\rangle$, describes LOS conditions, and the incoherent component, $\langle I_{inc}\rangle$, describes the clutter or obstructive conditions in the propagation channel. As was also shown in References 67–69, the latter component determines the multiplicative noise inside the communication channel, and we can determine K as a ratio between the signal and the multiplicative noise, $K \equiv s/N_{mult}$. It has different physical meanings in different environments.

In fact, for the mixed residential area, using results obtained in Chapter 5, we get the expression for K-factor along the radio path d between two terminal antennas:

$$K = \frac{Ico}{Iinc}$$

$$= \frac{\exp\left\{-\gamma_0 \cdot d \dfrac{\bar{h} - z_1}{z_2 - z_1}\right\}\left[\dfrac{\sin\left(k \cdot z_1 z_2/d\right)}{2\pi d}\right]^2}{\dfrac{\Gamma}{8\pi} \cdot \dfrac{\lambda \cdot l_h}{\lambda^2 + [2\pi l_h \gamma_0]^2} \cdot \dfrac{\lambda \cdot l_v}{\lambda^2 + [2\pi l_v \gamma_0 (\bar{h} - z_1)]^2} \cdot \dfrac{\left[(\lambda d/4\pi^3)^2 + (z_2 - \bar{h})^2\right]^{1/2}}{d^3}} \cdot$$

(10.52)

For the urban and suburban environments following Chapter 5, we finally get

$$K = \frac{Ico}{Iinc_1 + Iinc_2}$$

$$= \frac{\exp\left\{-\gamma_0 d \dfrac{(z_1 - \bar{h})}{(z_2 - z_1)}\right\}\dfrac{\sin^2\left(k \cdot z_1 z_2/d\right)}{4\pi^2 d^2}}{\dfrac{\Gamma \cdot \lambda \cdot l_v \left[(\lambda d/4\pi^3) + (z_2 - \bar{h})^2\right]^{1/2}}{8\pi\left[\lambda^2 + (2\pi l_v \gamma_0 \cdot (z_1 - \bar{h}))^2\right]d^3} + \dfrac{\Gamma^2 \lambda^3 l_v^2 \left[(\lambda d/4\pi^3) + (z_2 - \bar{h})^2\right]}{24\pi^2\left[\lambda^2 + (2\pi \cdot l_v \cdot \gamma_0 \cdot (z_1 - \bar{h}))^2\right]^2 d^3}},$$

(10.53)

where, as in Chapters 5 and 8, $\gamma_0 = 2\bar{L}\nu/\pi\ km^{-1}$ is the density of building contours, ν is the density of buildings in the investigated area per square kilometer, and $\bar{L}$ is the average length (or width) of buildings in meters, l_v and l_h are the vertical and horizontal coherence scales of reflections from building walls in meters, and $\bar{h}$ is the average buildings' height in meters. Finally, $\Gamma \equiv |\Gamma|$ is the absolute value of the reflection coefficient, and z_2 and z_1 are the BS and MS antenna heights in meters.

For one of the mixed residential area in Poland described in Section 10.1.4, which was selected as an experimental site in References 23 and 26, we have

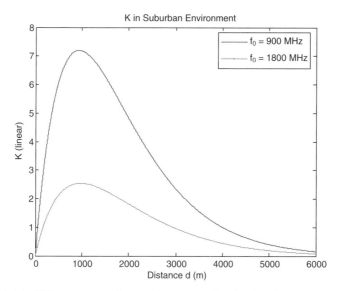

FIGURE 10.35. K-factor versus distance between BS and MS at $f_0 = 900$ and 1800 MHz.

obtained the following parameters of the terrain with houses and vegetation: $\bar{h} = 12$ m, $z_1 = 2$ m, $z_2 = 22$ m, $f_0 = 900$ MHz and $f_0 = 1800$ MHz, $\gamma_0 = 2 \cdot 10^{-3}$ m^{-1}, $d = (1\text{–}6000)$ m, $l_v = 1$ m, $l_h = 1$ m, $|\Gamma| = 0.6$. Results of numerical calculation are shown in Figure 10.35.

As follows from the results of this calculation, the ratio between a signal and the noise caused by fading can achieve a magnitude of 6–7 at the ranges of 1–2 km, that is, for microcell environment. Decrease of the LOS (dominant) component at far distances can be explained by the increase of the multipath component due to multiple scattering from obstructions, which are sufficient in far zones.

For the numerical analysis of the K-factor variations along radio path in urban area, we used the parameters obtained from one of experimental site described in References 24 and 25: $\bar{h} = 18.3$ m, $z_1 = 2$ m, $z_2 = 50$ m, $\gamma_0 = 1 \cdot 10^{-2}$ m^{-1}, $d = (1\text{–}1600)$ m, $l_v = 2$ m, $|\Gamma| = 0.8$. Results of calculations are shown in Figure 10.36 for 900 MHz and 1800 MHz. As follows from Figure 10.36, the results of simulations fully repeat the previous results obtained for mixed residential area; only the K-factor in urban environment can achieve magnitude of 15–18, the effect which depends on density of buildings and the height of the BS antenna. This effect also depends on the frequency of the system. Thus, when doubling the frequency, the magnitude of the K-factor is also increased more than twice (see Figure 10.35 and Figure 10.36). The obtained results are not in contradiction with those obtained in References 19 and 20, where with increase of the radiated frequency of the source, the coherent component becomes predominant, compared with the incoherent component of the total signal intensity.

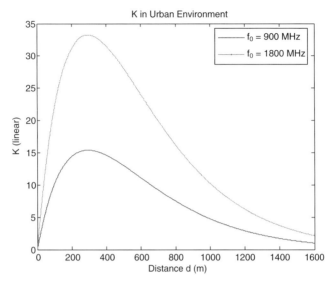

FIGURE 10.36. K-factor versus distance between BS and MS at $f_0 = 900$ and 1800 MHz.

As was mentioned in References 19 and 20, this result has a physically vivid explanation. Thus, with the increase of the radiated frequency, that is, the decrease of the wavelength with respect to the dimensions of rough structures and obstructions surrounding a transmitter, the effects of scattering (e.g., stochastic effects) becomes weaker, whereas the effects of reflection and diffraction from obstructions (e.g., deterministic effects) become predominant.

10.4. MAIN ALGORITHM OF RADIO COVERAGE (RADIO MAP) DESIGN

The physical–statistical multiparametric models proposed in Chapter 5 for various land environments are based on the stochastic approach and require, as initial data, statistical parameters of the terrain and environment such as the law of the distribution of the buildings at the rough terrain, their density, average length and height, as well as the locations of terminals, the transmitter and receiver, to obtain loss characteristics in various situations in the built-up areas. In addition to that, this approach provides an estimate of the total path loss and finally, allows us to obtain the radio coverage of the area of service for various built-up environments. In turn, we propose the following algorithm to obtain the radio coverage and the corresponding radio map. First of all, we need information about the terrain features, such as [15–20, 24–26, 32]:

- Terrain elevation data, digital terrain map, consisting of ground heights as grid Points $h_q(x, y)$.

- A clutter map, the ground cover of artificial and natural obstructions as a distribution of grid points ($h_0(x, y)$) for built-up areas. This is the overlay profile of the buildings $F(z_1, z_2)$; the average length or width of obstructions, $\bar{L}$ or $\bar{d}$; the average height of obstructions in the test area ($\bar{h}$); the obstructions density per square kilometer, ν.

- The effective antenna height, which is the antenna height plus a ground or obstruction height, if the antenna is assembled on a concrete obstruction: z_1 for the transmitter and z_2 for the receiver.

- The antenna pattern or directivity and its effective radiated power (ERP); the operating frequency, f.

According to data obtained earlier, we perform a prediction algorithm consisting several steps of analysis [19, 20, 24, 25]:

First Step. We introduce the built-up terrain elevation data for a three-dimensional radio path profile construction. As a result, there is a digital map (cover) with actual heights of obstructions in the computer memory.

Second Step. Using all parameters of built-up terrain and antennas, the transmitter, and receiver, the three-dimensional digital map is analyzed for estimations of the following parameters and features of the built-up terrain: the typical correlation scales of the obstacles, ℓ_v and ℓ_h, and the type of building material dominant in the test area to evaluate the absolute value of reflection coefficient in each "cell" of computation.

Finally, all these parameters allow us to obtain the density of building contours at the plane $z = 0$ (the ground level), $\gamma_0 = 2\langle L \rangle \nu / \pi$, and then the clearance conditions between the receiver and the transmitter, for example the average horizontal distance of the LOS $\langle \rho \rangle$: $\langle \rho \rangle = \gamma_0^{-1}$.

Third Step. The various factors obtained earlier must then be used for the computer program based on the 3-D multiparametric model for different types of irregular terrain. In the case of forested rural areas, the expressions (5.19) to (5.22) must be computed. In the case of mixed residential rural areas, the same expressions (5.80) and (5.81), combined with (5.29) and (5.30), must be used. In the case of built-up (urban and suburban) terrain, expressions (5.49a) or (5.49b) (straight crossing grid-street structure) or (5.76) to (5.79) (randomly distributed buildings), or combination of both of structures, as was shown in Chapter 8, must be used in final expressions (5.80) and (5.81).

At the same time, we should note that, if from a general topographic map of the area of service (or experimental site), the specific scenarios described in Section 10.2.4 can be performed, then we can use for each scenario the corresponding Equation (10.40), Equation (10.41), Equation (10.42), Equation (10.43), Equation (10.44), Equation (10.45), Equation (10.46), Equation (10.47), Equation (10.48), and Equation (10.49). As a final result, the signal path loss distribution can be obtained for both isotropic and directive antennas by introducing their effective radiation power (ERP) (see definitions proposed in Chapter 2).

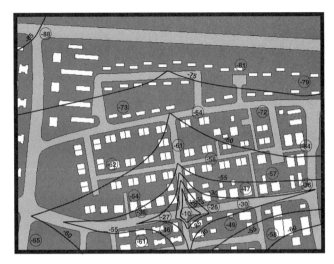

FIGURE 10.37. Radio coverage of experimental site with straight crossing streets according to Referenes 16 and 18; the curves are the results of computation of (5.49) or (10.48), and the circles are the measured data.

Fourth Step. The earlier data must be arranged in a form of a two-dimensional radio map, which describes the ground coverage of the radio signal for the built-up area being tested for service. The example of such radio coverage for one concrete experimental site with straight crossing streets described in References 16 and 18 is shown in Figure 10.37.

Here, the curves are the path loss obtained experimentally by direct measurements along various routes of moving antenna. The numbers inside circles are computed using Equation (5.49) according to crossing-waveguide model described in Chapter 5 or Equation (10.48). An agreement between theoretical prediction and measured data is clearly seen.

Another example of radio map can be presented for the Stockholm city, following results obtained in Section 10.1.4. On the basis of theoretical results, the radio map of the experimental site can be presented as 2-D and 3-D pictures, as shown in Figure 10.38. The vivid radio coverage of the service site of Stockholm was obtained to analyze propagation situation and possibility to communicate with the BS antenna for any subscriber's access point (from stp 1 to stp 10) located in the area of service.

According to the general algorithm presented earlier, we have designed radio coverage of Ramat-Gan market area in Israel, following only its topography map and building height profile presented in Figure 10.39.

It can be easily prototyped as a typical "downtown" area with tall buildings that are very dense. The first stage of the simulation was to process the physical data. We have measured the dimensions of each building near the antenna and calculated the average height of the buildings for every path stretched at 5°

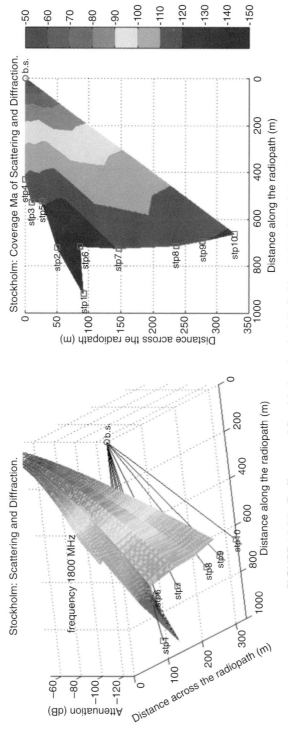

FIGURE 10.38. Radio map of Stockholm city: in 3-D (left) and 2-D (right) domains.

483

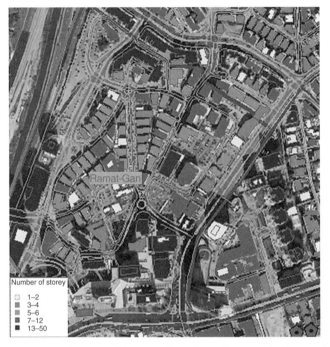

FIGURE 10.39. Topographic map of stock market area in Ramat-Gan.

TABLE 10.5. The Terrain and Antenna Parameters of Ramat-Gan Experimental Site

Sector (°)	γ_0 (gamma) (m^{-1})	z_2 (m)	d (m)	Δd (m)	Tilt (α) (°)	Eff (dB)	G dBi
90	10^{-3}	50	200	5	4	8.8	12.5
220	10^{-3}	50	1500	20	6	8.8	12.5
330	10^{-3}	50	500	10	4	8.8	12.5

intervals. We then estimated the number of buildings per square kilometer in the test-site area, their average length or width (related to the direction of the antenna main loop orientation), and the contour density of the buildings, γ_0. Finally, the 2-D map of building density has been performed as shown in Figure 10.40.

The antenna characteristics, such as "sector," tilt, height (z_2), effective power ("Eff") and gain ("G"), as well as the terrain parameters, such as γ_0, the distance d between BS and MS, and step Δd along the radio path between BS and MS used for this simulation are presented in Table 10.5.

We notice that in the first column, the values represent the angle of the sector with reference to $0°$ (north). The simulation running constants and variables are the following:

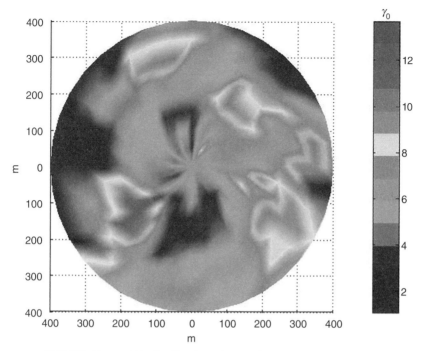

FIGURE 10.40. 2-D distribution of the building contours density.

$$z_1 = 2\,[\text{m}];\, l_v = 2\,[\text{m}];\, f_0 = 900\,[\text{MHz}];\, |\Gamma| = 0.7 - 0.8;\, \lambda = \frac{3 \cdot 10^8}{f_0} \approx 0.33\,[\text{m}].$$

The BS antenna was elevated above the built-up profile with the average building height of $\bar{h} = 18.3\,\text{m}$. Finally, using the stochastic model, we have created a radio map for the test-site area of –800 m to 800 m, depicted in Figure 10.41, where at the right side, the path loss (in dBm) is presented by colored segments corresponding to the obtained magnitudes of loss. As is seen from Figure 10.41, at the ranges close to BS antenna (100–300 m) the path loss is varied from –100 dBm to –120 dBm, whereas far from BS antenna (at 500–800 m) it varies from –150 dBm to –180 dBm.

The same computation process was done to simulate the premises of Ben-Gurion University in Israel, with one major difference. Here, we have divided the university into five major parts, as depicted in Figure 10.42, which are different in their buildings density.

This decision was made on the basis of the university's topographic plan and the values of building contours density γ_0: for section II, $\gamma_0 = 6 \cdot 10^{-3}\,\text{m}^{-1}$, for sections I, III and IV, $\gamma_0 = 3 \cdot 10^{-3}\,\text{m}^{-1}$, and for section V, $\gamma_0 = 4 \cdot 10^{-3}\,\text{m}^{-1}$. The heights of two antennas were 25 m and 22 m (compared to the average buildings height of $\bar{h} = 14.5\,\text{m}$).

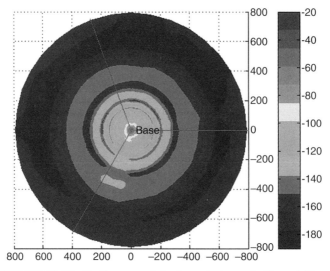

FIGURE 10.41. Radio map of stock market area in Ramat-Gan.

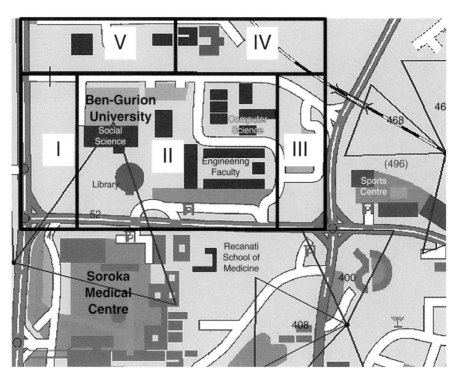

FIGURE 10.42. Topographic plan of Ben Gurion university campuses.

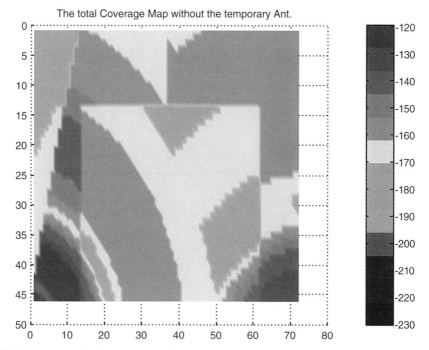

FIGURE 10.43. Radio coverage of two sectoral antennas, numbered as 52 and 400, and shown in Fig. 10.42.

Two-sectoral antennas with three sectors 52, 408, and 400, arranged only according to preliminary experiments, cover an area of the university and together give the pure coverage of radiated energy (see Fig. 10.43). Results presented in Figure 10.45 show that the two-sectoral antennas cannot fully cover the university's campuses, and more than 40% of the area of service are in shadow zones. Therefore, an effective service of subscribers cannot be achieved in more than 40% of the university area.

Therefore, we proposed another arrangement of three sectors, on the basis of our stochastic approach and the corresponding algorithm described in Chapters 5 and 8 and at the beginning of this section, according to which specific titles and heights of three sectoral antennas, and power distribution within each sector, were found according to a strict analysis of the topographic map of the university area. Now, using the same three more strictly assembled sectoral antennas according to the theoretical prediction, based only on topography of the tested area and the data of antennas, more than 90% of radio coverage was obtained (see Fig. 10.44) to allow each subscriber located in the university area a stable and effective radio service. We noticed that because the obtained data corresponds to the technical requirements of our customer, we cannot present here full information about derivations according to earlier proposed algorithm.

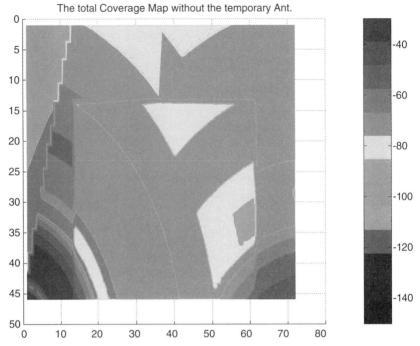

FIGURE 10.44. Radio coverage of three sectoral antennas, numbered 52, 400, and 408, according to the theoretical prediction of their new parameters.

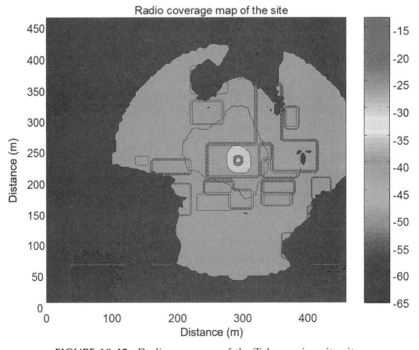

FIGURE 10.45. Radio coverage of the Tokyo university site.

The same algorithm was used for computations of the total path loss and the radio coverage performance of Tokyo University, the built-up terrain of which and the parameters of the terminal antennas were described in details in Chapter 5. The corresponding 2-D radio coverage (i.e., the radio map) of the university site areas when the transmitter antenna is located at the roof of the taller university campus, is shown in Figure 10.45. As was discussed in Reference 32, the proposed stochastic approach is a good predictor of radio coverage design (with accuracy of 85–90% with respect to the strict ray-tracing approach giving 90–95% of accuracy, but not continuously for each radio path, as is also discussed in Chapter 5).

All examples presented earlier show that using the proposed theoretical algorithm to predict link budget for various outdoor environments, each designer of the wireless networks can achieve the effective arrangement of antennas for full radio coverage of the tested environment with minimization of the path loss for each user located in the areas of service.

REFERENCES

1   Anderson, H.R., "A ray-tracing propagation model for digital broadcast systems in urban areas," *IEEE Trans. on Broadcasting*, Vol. 39, No. 3, 1993, pp. 309–317.

2   Bertoni, H.L., *Radio Propagation for Mobile Wireless Systems*, Prentice Hall PTR, New Jersey, 2000.

3   Bertoni, H.L., P. Pongsilamanee, C. Cheon, and G. Liang, "Sources and statistics of multipath arrival at elevated base station antenna," *Proc. IEEE Veh. Technol. Conference*, Vol. 1, 1999, pp. 581–585.

4   Liang, G. and H.L. Bertoni, "A new approach to 3D ray tracing for propagation prediction in cities," *IEEE Trans. Anten. Propagat.*, Vol. 46, 1998, pp. 853–863.

5   Lebherz, M., W. Weisbeck, and K. Krank, "A versatile wave propagation model for the VHF/UHF range considering three dimensional terrain," *IEEE Trans. Antennas Propagat.*, Vol. AP-40, 1992, pp. 1121–1131.

6   Lee, W.Y.C., *Mobile Communication Engineering*, McGraw Hill, New York, 1985.

7   Kouyoumjian, R.G. and P.H. Pathak, "A uniform geometrical theory of diffraction for an edge in a perfectly conducting surface," *Proc. IEEE*, Vol. 62, No. 9, 1974, pp. 1448–1469.

8   Vogler, L.E., "An attenuation function for multiple knife-edge diffraction," *Radio Sci.*, Vol. 19, No. 8, 1982, pp. 1541–1546.

9   Walfisch, J. and H.L. Bertoni, "A theoretical model of UHF propagation in urban environments," *IEEE Trans. Antennas Propagat.*, Vol. AP-38, 1988, pp. 1788-1796.

10 Bertoni, H.L., W. Honcharenko, L.R. Maciel, and H.H. Xia, "UHF propagation prediction for wireless personal communications," *Proc. IEEE*, Vol. 82, No. 9, 1994, pp. 1333–1359.

11 Rustako, A.J., Jr., N. Amitay, M.J. Owens, and R.S. Roman, "Radio propagation at microwave frequencies for line-of-sight microcellular mobile and personal communications," *IEEE Trans. Veh. Technol.*, Vol. 40, No. 2, 1991, pp. 203–210.

12 Tan , S.Y. and H.S. Tan, "Propagation model for microcellular communications applied to path loss measurements in Ottawa City streets," *IEEE Trans. Veh. Technol.*, Vol. 44, No. 3, 1995, pp. 313–317.

13 Tan, S.Y. and H.S. Tan, "UTD propagation model in an urban street scene for microcellular communications," *IEEE Trans. Electromag. Compat.*, Vol. 35, No. 4, 1993, pp. 423–428.

14 Ponomarev, G.A., A.N. Kulikov, and E.D. Telpukhovsky, *Propagation of Ultra-Short Waves in Urban Environments*, Rasko, Tomsk, 1991 (in Russian).

15 Blaunstein, N., "Average field attenuation in the non-regular impedance street waveguide," *IEEE Trans. Anten. and Propagat.*, Vol. 46, No. 12, 1998, pp. 1782–1789.

16 Blaunstein, N. and M. Levin, "VHF/UHF wave attenuation in a city with regularly spaced buildings," *Radio Sci.*, Vol. 31, No. 2, 1996, pp. 313–323.

17 Blaunstein, N. and M. Levin, "Loss characteristics prediction in urban environments with randomly distributed buildings," Proceedings of IEEE/URSI International Conference, Atlanta, USA, pp. 23–26, July 15–21, 1998.

18 Blaunstein, N., "Prediction of cellular characteristics for various urban environments," *IEEE Anten. Propagat. Magazine*, Vol. 41, No. 6, 1999, pp. 135–145.

19 Blaunstein, N., N. Yarkoni, and D. Katz, "Link budget performance for various outdoor scenarios taking into account fading phenomena," Proceedings of 13th Conference on Microwave Technologies, Prague, Czech Republic, pp. 56–59, September 26–28, 2005.

20 Blaunstein, N., N. Yarkoni, and D. Katz, "Innovative approach to radio propagation modeling in urban and suburban areas," Proceedings of International Conference on Electromagnetics in Advanced Applications (ICEAA '05), Torino, Italy, pp. 293–298, September 12–16, 2005.

21 Kovacs, I.Z., P.C.F. Eggers, and K. Olsen, "Radio channel characterization for forest environments in VHF and UHF frequency bands," Proceedings of Vehicular Technology Conference '99, Amsterdam, the Netherlands, pp. 1387–1391, September 1999.

22 Blaunstein, N., I.Z. Covacs, D. Katz, J.B. Andersen, et al., "Prediction of UHF path loss for forested environments," *Radio Science*, Vol. 38, No. 3, 2003, pp. 1059–1075.

23 N. Blaunstein, D. Katz, A. Freedman, I. Matityahu, "Prediction of loss characteristics in rural and residential areas with vegetation," Proceedings of International Conference on Electromagnetics in Advanced Applications, Torino, Italy, 2001, pp. 667–670.

24 Blaunstein, N., D. Katz, and D. Censor, "Loss characteristics in urban environment with different buildings' overlay profiles," *Proc. 2001 IEEE Anten. Propag. Int. Symp.*, Boston, Massachusetts, Vol. 2, 2001, pp. 170–173.

25 Blaunstein, N., D. Katz, D. Censor, et al., "Prediction of loss characteristics in built-up areas with various buildings' overlay profiles, *IEEE Anten. Propagat. Magazine*, Vol. 43, No. 6, 2001, pp. 181–191.

26 Blaunstein, N., D. Censor, D. Katz, et al., "Radio propagation in rural residential areas with vegetation," *J. Progress in Electromagnetic Research*, PIER, Vol. 40, 2005, pp. 131–153.

27 Blaunstein, N. and Y. Ben-Shimol, "Frequency dependence of pathloss characteristics and link budget design for various terrestrial communication links," *IEEE Trans. on Antennas and Propagat.*, Vol. 52, No. 10, 2004, pp. 2719–2729.

28 Blaunstein, N., "Distribution of angle-of-arrival and delay from array of buildings placed on rough terrain for various elevations of base station antenna," *Journal of Communications and Networks*, Vol. 2, No. 4, 2000, pp. 305–316.

29 Yarkoni, N., N. Blaunstein, and D. Katz, "Link budget and radio coverage design for various multipath urban communication links," *Radio Sci.*, Vol. 42, 2007, pp. 412–427.

30 Blaunstein, N., N. Yarkoni, and D. Katz, "Spatial and temporal distribution of the VHF/UHF radio waves in built-up land communication links," *IEEE Trans. Antennas and Propagation*, Vol. 54, No. 8, 2006, pp. 2345–2356.

31 Blaunstein, N. and Y. Ben-Shimol, "Spectral properties of signal fading and Doppler spectra distribution in urban communication mobile links," *Wireless Communic. and Mobile Computing*, Vol. 6, No. 1, 2006, pp. 113–126.

32 Katz, D., N. Blaunstein, M. Hayakawa, and Y.S. Kishiki, "Radio maps design in Tokyo city based on stochastic multi-parametric and deterministic ray tracing approaches," *J. Antennas and Propag. Magazine*, Vol. 51, No. 5, 2009, pp. 200–208.

33 D. O. Reudink and M. F. Wazowicz, "Some propagation experiments relating foliage loss and diffraction loss at X-band and UHF frequencies," *IEEE Trans. on Commun.*, Vol. COM-21, 1973, pp. 1198–1206.

34 Barton, F.A. and G.A. Wagner, "900 MHz and 450 MHz - mobile radio performance in urban hilly terrain," *Veh. Technol. Confer. Rec.*, Cleveland, Ohio, USA, June 1973, pp. 1–8.

35 Okumura, Y., E. Ohmori, T. Kawano, and K. Fukuda, "Field strength and its variability in the VHF and UHF land mobile radio service," *Review Elec. Commun. Lab.*, Vol. 16, No. 9–10, 1968, pp. 825–843.

36 Wells, P.J., "The attenuation of UHF radio signal by houses," *IEEE Trans. Vech. Technol.*, Vol. 26, No. 4, 1977, pp. 358–362.

37 Bramley, E.N. and S.M. Cherry, "Investigation of microwave scattering by tall buildings," *Proc. IRE*, Vol. 120, No. 8, 1973, pp. 833–842.

38 Allsebrook, K. and J.D. Parsons, "Mobile radio propagation in British cities at frequencies in the VHF and UHF bands," *IEEE Proc.*, Vol. 124, 1977, pp. 95–102.

39 Trubin, V.N., "Urban and suburban radio propagation characteristics in the VHF and UHF bands," *7th Int Symp. Electromagn. Compat.*, Wroclaw, Poland, Vol. 1, 1984, pp. 393–402.

40 Pedersen, K.I., P.E. Mogensen, and B.H. Fleury, "Power azimuth spectrum in outdoor environments," *IEE Electron. Lett.*, Vol. 33, No. 18, 1997, pp. 1583–1584.

41 Pedersen, K.I., P.E. Mogensen, and B.H. Fleury, "A stochastic model of the temporal and azimuthal dispersion seen at the base station in outdoor environments," *IEEE Trans. Veh. Technol.*, Vol. 49, No. 2, 2000, pp. 437–447.

42 Chan, G.K., "Propagation and coverage prediction for cellular radio systems," *IEEE Trans. Veh. Technol.*, Vol. 40, No. 5, 1991, pp. 665–670.

43 Steele, R., *Mobile Radio Communication*, IEEE Press, New York, 1995.

44   Harley, P., "Short distances attenuation measurements at 900 MHz and 1.8 GHz using low antenna heights for microcells," *IEEE J. Select. Areas Communic.*, Vol.7, No. 1, 1989, pp. 5–11.

45   Crosskipf, R., "Prediction of urban propagation loss," *IEEE Trans. Anten. Propagat.*, Vol. 42, No.5, 1994, pp. 658–665.

46   Chia, S.T.S., "Radiowave propagation and handover criteria for microcells," *British Telecom Tech. J.*, Vol. 8, No. 1, 1990, pp. 50–61.

47   Cox, D.C., "910 MHz urban mobile radio propagation: multipath characteristics in New York City," *IEEE Trans. Communic.*, Vol. 21, No. 11, 1973, pp. 1188–1194.

48   Erceg, V., M. Taylor, D. Li, and D.L. Schilling, "Urban/suburban out-of-sight propagation modeling," *IEEE Communication Magazine*, Vol. 19, No. 2, 1992, pp. 56–61.

49   Black, D.M. and D.O. Reudink, "Some characteristics of mobile radio propagation at 836 MHz in the Philadelphia area," *IEEE Trans. Vehicular Technol*, Vol. 21, No. 2, 1972, pp. 45–51.

50   Reudink, D.O., "Comparison of radio transmission at X-band frequencies in suburban and urban areas," *IEEE Trans. Anten. Propagat.*, Vol. 20, No. 4, 1972, pp. 470–473.

51   Ikegami, F. and S. Yoshida, "Analysis of multipath propagation structure in urban mobile radio environments," *IEEE Trans. Anten. Propagat.*, Vol. 28, No. 4, 1980, pp. 531–537.

52   Fumio, J. and J. Susumi, "Analysis of multipath propagation structure in urban mobile radio environments," *IEEE Trans Anten. Propagat.*, Vol. 28, No. 4, 1980, pp. 531–537.

53   Lee, W.C.Y. and Y.S. Yeh, "On the estimation of the second-order statistics of lognormal fading in mobile radio environment," *IEEE Trans. on Communic.*, Vol. 22, No. 6, 1974, pp. 869–873.

54   Lee, W.C.Y., "Estimate of local average power of a mobile radio signal," *IEEE Trans. Vehic. Technol.*, Vol. 34, No. 1, 1985, pp. 22–27.

55   Hata, M. and T. Nagatsu, "Mobile location using signal strength measurements in a cellular system," *IEEE Trans. Vehic. Technol.*, Vol. 29, No. 2, 1980, pp. 245–252.

56   Mockford, S., A.M.D. Turkmani, and J.D. Parsons, "Local mean signal variability in rural areas at 900 MHz," *Proc. of the 40th Vehic. Technol. Conference*, 1990, pp. 610–615.

57   Gudmunson, M., "Correlation model of shadow fading in mobile radio systems," *Electronic Letters*, Vol. 27, No. 23, 1991, pp. 2145–2146.

58   Goldsmith, A.J., L.J. Greenstein, and G.J. Foschini, "Error statistics of real-time power measurements in cellular channels with multipath and shadowing," *IEEE Trans. Vehic. Technol.*, Vol. 43, No. 3, 1994, pp. 439–446.

59   Holtzman, J.M. and A. Sampath, "Adaptive averaging methodology for handoffs in cellular systems," *IEEE Trans. Vehic. Technol.*, Vol. 44, No. 1, 1995, pp. 59–66.

60   Giancristofaro, D., "Correlation model for shadow fading in mobile radio channels," *Electronic Letters*, Vol. 32, No. 11, 1996, pp. 958–959.

61   Chockalingam, A., P. Dietrich, L. Milstein, and R. Rao, "Performance of closed-loop power control in DS-CDMA cellular systems," *IEEE Trans. Vehic. Technol.*, Vol. 47, No. 3, 1998, pp. 774–789.

62 Wong, D. and D.C. Cox, "Estimating local mean signal power level in a Rayleigh fading environment," *IEEE Trans. Vehic. Technol.*, Vol. 48, No. 3, 1999, pp. 956–959.

63 Greenstein, L.J., D.G. Michelson, and V. Erceg, "Moment-method estimation of the Ricean factor," *IEEE Communic. Letters*, Vol. 3, No. 6, 1999, pp. 175–176.

64 Clarke, R.H., "A statistical theory of propagation in the mobile radio environment," *BSTJ*, Vol. 47, No. 7, 1968, pp. 957–1000.

65 Aulin, T.A., "A modified model of the fading signal at a mobile radio channel," *IEEE Trans. Veh. Technol.*, Vol. 28, No. 3, 1979, pp. 182–203.

66 Tepedelenlioglu, C., A. Abdi, G.B. Giannakis, and M. Kaveh, "Estimation of Doppler spread and signal strength in mobile communications with applications to handoff and adaptive transmission," *J. Wireless Communicat. and Mobile Computing*, Vol. 1, No. 2, 2001, pp. 221–242.

67 Yarkoni, N. and N. Blaunstein, "Capacity and spectral efficiency of MIMO wireless systems in multipath urban environments with fading," *1st European Conference on Antennas and Propagation (EuCAP 2006)*, Nice, France, November 6–10, 2006, pp. 111–115.

68 Yarkoni, N. and N. Blaunstein, "Signal power distribution in the space, time delay and angle of arrival domains in indoor communication links," *URSI Symposium*, Ottawa, July 16-21, 2007, pp. 41–43.

69 Blaunstein, N. and N. Yarkoni, "Data stream parameters prediction in land and atmospheric MIMO wireless communication links with fading," *2nd European Conference on Antennas and Propagation* (EUCAP 2007), Edinburgh, November 2007, pp. 123–126.

CHAPTER ELEVEN

Cellular and Noncellular Communication Networks Design Based on Radio Propagation Phenomena

As was mentioned in Chapter 1, wireless systems are an enhancement of the traditional wireline telephone systems. However, unlike the wireline systems, they suffer from a poor grade of service (GOS) and low quality of service (QOS) because of low link reliability caused by propagation characteristics of the radio environment. Furthermore, most characteristics of wireless network design strongly depend on the sensitivity of receivers due to high service demand, on a low number of call resources due to limited frequency band of each network, and on degradation in the information data stream transmitting within each individual subscriber radio communication link. Important statistical characteristics that must be predicted are the path loss, slow, and fast fading. These allow designers of such networks to improve GOS and QOS characteristics, to create cellular maps of areas of service, and to optimize the information data stream within each radio communication channel.

In order to avoid measuring channel statistics for all operating environments and for all networks design, we introduced in Chapters 5 and 8 a unified stochastic approach for multipath radio channel description, based on real physical phenomena, such as multiple reflection, diffraction, and scattering from various nontransparent obstructions (trees, hills, houses, and buildings). In this chapter, we describe elements of wireless networks design via this

Radio Propagation and Adaptive Antennas for Wireless Communication Networks: Terrestrial, Atmospheric, and Ionospheric, Second Edition. Nathan Blaunstein and Christos G. Christodoulou.
© 2014 John Wiley & Sons, Inc. Published 2014 by John Wiley & Sons, Inc.

unified statistical approach, comparing it with existing standard methods of networks design.

Thus, in Section 11.1, we describe the unified concept of GOS design in multipath radio channels caused by fading phenomena. In Section 11.2, we discuss about strategies to create cellular maps, uniform, and nonuniform, for various built-up environments, referring the reader to the channel (frequency) assignment technique. Then, Section 11.3 describes the prediction mechanism of main parameters of information data stream, such as capacity, spectral efficiency, bit error rate (BER), level crossing rate (LCR), and average time duration versus Rician K-parameter that determines fading effects within the multipath radio channel.

11.1. GRADE OF SERVICE DESIGN OPERATING IN MULTIPATH FADING ENVIRONMENT

The classical GOS analysis, that is, the analysis of the probability of working without call congestion of wireless networks, uses the Erlang B model that is a private case of the birth and death (B&D) equations of the statistical traffic theory [1–5]. Also, the Erlang B model is usually used for a calculation of the total system capacity. This gives a calculation of the estimated GOS, which is the probability of a user trying to initiate a call and the call failing. However, this model is justified only in cases where all users have access to all resources of the system (a situation of full availability). In wireless systems, due to propagation limitation such as obstacles and wave fading, a user has limited access to system resources. This refers to a limited availability, and therefore, we cannot use the classical approach of this traditional model.

In wireless systems, a user gets service from one cell. The user is part of a group that is covered by one or more cells. If the user has optional access to more than one cell, the system has to allocate the user to one of these cells in order to achieve an optimal GOS—referring to load balancing. The decision rule of user allocation refers to the *load balancing algorithm*. Hence, the goal of the wireless system operator is to deploy a system such that the GOS will be maximized, while the number of cells serving a certain number of users is minimal. In fact, some studies that have been done to explore this type of the GOS problem have found a strict effect of the coverage overlapping between cells and the effect of the decision rules of user allocation [6–8].

In the following section, we will briefly explain the approach of how to define an effective methodology of the GOS estimation by calculating the fading and its statistical characteristics, according to the concept of link budget design described in Section 10.2 (Chapter 10). We compare the generalized GOS that takes into account the fading as well as the congestion probability due to high load, with the traditional GOS calculation. We also compare the results for different land area types (mixed residential and urban) and for different handoff decision rules, in order to show the destructive effects of slow fading due to shadowing, and those of fast fading due to multipath phenomena, in order to recommend an optimal deployment methodology.

11.1.1. The Concept

Until now, existing wireless systems considered only elements of the traditional deployment methodology that do not take into account the real propagation characteristics of the environment. The most elementary stage of a system deployment is the link budget design, which is the main parameter of wireless communication links (see Section 10.2, Chapter 10). It is based on propagation characteristics of the channel and describes the signal power distribution along the radio path between the base station (BS) antenna and the user antenna [9, 10]. The link budget is a balanced sheet of gains and losses [11]. Some of the link budget parameters such as antenna gain are fixed, and some are statistical such as fading (see Chapter 1). So, by calculating link budget, we evaluate the link performance. Section 10.2 (see Chapter 10) explains briefly link performance, considering the wave path loss and taking into account a fade margin, which is the average of the slow and fast fading due to shadowing and multipath phenomena in the entire area. As in Section 10.2, in our estimations of the GOS, we take into account two types of areas: *mixed residential areas*, as low buildings with trees around, and *urban areas*, as mostly high buildings in a dense built-up area. For each type of area, we may calculate the average path loss and estimate the statistics of slow and fast fading effects according to the corresponding algorithm presented in Section 10.2. The proposed model is adapted to use a known number of subscribers located in an area of service, which is assumed to be distributed equally over this area, with a known statistical distribution.

In our description, we concentrate on a model of two cells that are deployed in two ways: *contained*, when one cell's coverage area is fully contained in the other cell (see Fig. 11.1, case A) and *overlapped*, when two cells have a common area (see Fig. 11.1, case B).

If the user is in the common area, then it is allocated to one cell upon *predefined algorithms*:

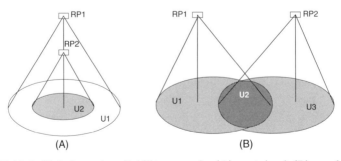

FIGURE 11.1. Unbalanced availability scenario: (A) contained, (B) overlapping.

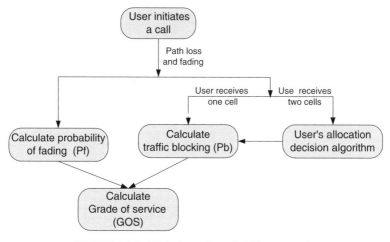

FIGURE 11.2. Unbalanced availability scenario.

(a) random assignment (denoted *random*)

(b) assign to site with higher signal strength (denoted *by_ss*)

(c) assign user to the less loaded cell (denoted *min-load*), investigated in detail in Reference 7.

For each user who initiates an outgoing call, it calculates its chance of not succeeding. All the steps of the corresponding algorithm are described by the block scheme shown in Figure 11.2.

11.1.2. Simulation Tests

The simulation includes two scenarios, as shown in Figure 11.1 according to Reference 8. In both cases, there are two radio ports (RPs); each with eight channels. In the contained scenario A, RP2 serves the user group U2, and RP1 serves both groups, U1 and U2. In the overlapped scenario B, RP1 serves group U1 and group U3, and RP2 serves U2 and U3; the coverage area by the group U3 is called the overlapped area.

In both scenarios, A and B, each user can initiate a call randomly, and if the signal level is high enough, it can request for a channel allocation. Here, it is clear that if the fading effect is high, that is, low signal is received, then there is degradation in service quality. If the user is in the overlapped area, a decision on which RP to allocate to it is taken by the following load balancing algorithms [7]:

- random (*random*)—the user selects randomly one RP
- best signal strength (*by_ss*)—the user selects the RP that is received with a highest signal strength

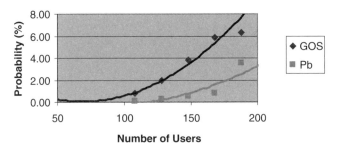

FIGURE 11.3. Pb is traffic blocking without fading effect; GOS is traffic blocking with fading effect.

- controlled load (*min_load*)—the user selects the RP that has the less occupied channels.

The goal of our simulations presented in the next section is to demonstrate the decrease of traffic as an effect of fading or a load balancing algorithm.

11.1.3. Traffic Computation in Wireless Networks with Fading

For each single scenario, overlapped or contained, and an algorithm, we have scattered various numbers of users who have call statistics of Poisson distribution, with an average arrival process of 1 hour and an average call time of 4 minutes, in order to measure the probability of dropped calls (call blocking). It is possible to measure the same results with and without considering the fading effect.

We now consider how the fading affects the results. In Figure 11.3, two graphs are presented showing the blocking rate (Pb) as a function of the number of users, as well as the traffic loss when considering the fading effects (GOS).

In this example, a system that has 2% blocking carried 180 users, while taking in account the fading, we get for a GOS of 2% with only 130 users. The fading therefore has an effect that reduces the system capacity by 50 users.

Results of Simulations. Our test and simulation results measured the three parameters:

- effect of fading on the average increase of the overlapping area
- increase of probability of call blocking when considering the fading effect
- effect of overlapping on the GOS.

In the numerical experiment, we checked the GOS for 128 users only. Figure 11.4 and Figure 11.5 depict the results achieved by computing the problem concerning the last two scenarios.

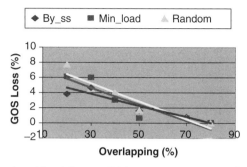

FIGURE 11.4. Mixed residential area: GOS versus overlapping (in %) for three load balancing algorithms.

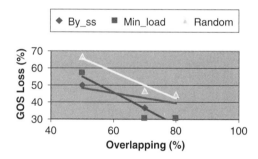

FIGURE 11.5. Urban area: GOS versus overlapping.

Mixed Residential Area. Figure 11.4 depicts a situation when GOS was computed, taking into account fading phenomena. Without fading, the probability of blocking was not more than 3%. For the same GOS of 5–7% fading effects can increase the overlapping area.

Urban Area. Figure 11.5 depicts a situation when GOS was computed, taking into account fading phenomena. Here, without accounting for the fading effect, we get no more than 5% of blocking probability, whereas in overlapping scenario in urban environments, fading has a tremendous effect on GOS (up to 20–30%), and in low overlapping, it is so large that unacceptable GOS was obtained (for overlapping of 50% it can achieve 50–70%).

Some Specific Examples. Finally, we will present some specific examples of a cellular system arranged in the mixed residential area to show the loading

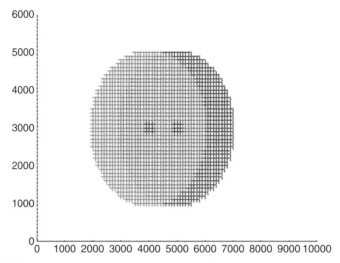

FIGURE 11.6. An example of two-cell propagation map in an overlapping case.

effect on each cell constructed according to the overlapping scenario. We use the following testing area configuration:

- two cells located at a distance of 1 km
- number of users is 128
- average arrival time is 60 minutes
- average holding time is 4 minutes
- decision rule is *random* with decision probability $P_1 = 35\%$.

After the corresponding simulation, we get the following overlapping map shown in Figure 11.6. The overlapping area (red area in the middle) is 53% of the entire service coverage area. Finally, we get two histograms that represent the load on each cell (see Fig. 11.7). These histograms show the percentage of time of occupancy for a certain number of channels. As it is seen, since we took the probability of decision of $P_1 = 35\%$, common users had higher probability to be in Cell #2, because we decided more time of occupancy of eight channels being in Cell #2 compared to Cell #1. We also get from this simulation the following results: the probability of fading (P_{ff}) is 0.6%. Thus, it is low enough for a high overlapping. At the same time, the probability of blocking (P_{bl}) from simulations was 1.8%. Finally, we get, according to Reference 8,

$$GOS = P_{bl}(1 - P_{sf} \cdot P_{ff}) + P_{sf} \cdot P_{ff}(1 - P_{bl}). \tag{11.1}$$

Substituting probabilities mentioned earlier, we get

$$GOS = 0.018 \cdot (1 - 0.006) + 0.006 \cdot (1 - 0.018) = 0.023 \text{ or } 2.3\%.$$

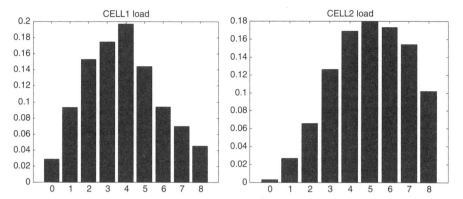

FIGURE 11.7. Traffic histograms of an overlapping case versus number of channels.

So, for the presented configuration of the system, the fading effect is of $2.3 - 1.8\% = 0.5\%$ on GOS, which is in agreement with results obtained for mixed residential areas shown in Figure 11.4.

11.2. PROPAGATION ASPECTS OF CELL PLANNING

In the previous section, we analyzed two scenarios of two cells assembling for allocation of subscribers in an area of service: overlapping scenario and contained scenario. In Chapter 3, we briefly considered what we mean by introducing the *cellular* maps' concept and presented the main parameters and characteristics of the cell. Now, we will introduce the reader to the strategy of *cell planning* and *cell splitting* depending on conditions of radio propagation in different built-up areas.

11.2.1. Advanced Methods of Cellular Map Design

To arrange the effective splitting of tested built-up area at cells, the designers need strict information about the law of signal power decay for the concrete situation in the site of consideration. Specifically, they need the strict link budget analysis of propagation situation within each communication channel, as well as full radio coverage of each subscriber located at line-of-sight (LOS) or non-line-of-sight (NLOS) conditions in areas of service, giving exact clearance between subscribers within each cell. On the basis of precise knowledge of the propagation phenomena inside the cellular communication channels, it is easy to optimize cellular characteristics, such as radius of a cell, reuse factor Q, co-channel interference parameter (C/I), and so on.

Definition of Radius of Cells via Propagation Models. As follows from Chapter 6, a better clearance between BS and the moving subscribers (MSs) in cluttered conditions may be reached only for LOS conditions (or direct visibility) between them. In this case, as follows from the two-ray model and the waveguide street model (see Chapter 5), the cell size (R_{cell}) cannot be larger than the break point range (r_B), at which the decay of the signal is changed from $\gamma = 2$ (as in free space propagation) to $\gamma = 4$ (propagation above flat terrain). If so, the law of signal decay between BS and each MS in the cell of radius $R_{cell} \leq r_B$ is R_{cell}^{-2}. Generally speaking, beyond the break point, the law of signal decays versus the range between terminal antennas, described by path loss slope parameter γ, depends on the concrete situation in the urban scene and may be proportional to $R^{-\gamma}$ with $\gamma > 2$ (for $\gamma = 4$–7, see discussions in Chapter 5). Such a distance dependence of signal decay law inside and outside the cell is shown in Figure 11.8a,b for two typical situations in regular cell splitting.

Thus, we can conclude that the best clearance between each BS and any subscriber inside the cell determines the minimum radius of the concrete cell. Wave propagation phenomena in urban environments with both antennas in NLOS (clutter) conditions were described earlier in Chapter 5 by using two physical–statistical models, street waveguide, and multiparametric stochastic. As follows from models described there, in rural and mixed residential areas with a rare building distribution, the path loss slope parameter γ describing

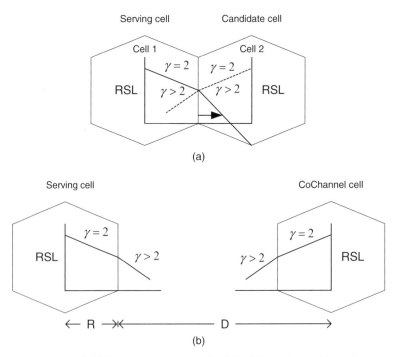

(a)

(b)

FIGURE 11.8. (a) LOS hand-off scenario. (b) LOS co-channel interference.

the received signal decay is changed from $\gamma = 2.5$ to $\gamma = 4.0$ (see Chapter 5). In other words, in such an area, field attenuation is faster than that in LOS conditions of free space.

From the multiparametric model described in Chapter 5, when both antennas are placed in a built-up area with a high density of irregularly distributed buildings, the law of signal power decay is changed with distance d between antennas. It changes from $\sim d^{-\gamma}$ ($\gamma = 2$) before the break point r_B (where the coherent part of total field is dominant) to $\sim d^{-\gamma}$ ($\gamma = 2.5$–4) beyond the break point r_B (where the incoherent part of total field is dominant). Therefore, it is easier to obtain the minimum cell size, R_{cell}, defined as the break point r_B, in a variety of cities with crossing-street grid plan. In fact, for urban areas with a rectangular grid plan of straight crossing streets, when the multislit street waveguide model (see Chapter 5) is successfully used [12–14], the cell size can be described by the following equation, where we also account for the effects of diffraction from building corners, as in Reference 13:

$$R_{cell} \equiv r_B = \frac{4 h_T h_R}{\lambda} \frac{(1 + \chi \, | \, R_n \, |)}{(1 - \chi \, | \, R_n \, |)} \frac{\left[1 + h_b \, / \, a + h_T h_R \, / \, a^2 \right]}{| \, R_n \, |^2 + | \, D_n \, |^2}, \qquad (11.2)$$

where all parameters in (11.2) are described in Chapter 5. As follows from (11.2), we need information about street geometry, such as the street width a, the average height of buildings h_b, and the mean gaps between buildings lining the street, that is, the parameter of brokenness χ. At the same time, we need information about both antennas' height, h_T and h_R, and the material of the building walls (which is determined by the absolute value of reflection coefficient R_n and diffraction coefficient D_n). Using these data, one can easily obtain the cell radius along the crossing streets.

In the case of built-up areas with nonregularly distributed buildings placed on rough terrain, consisting of hills, trees, and other obstructions located in residential zones, the cell size can be obtained by using the probabilistic approach presented in Chapter 5, according to the multiparametric stochastic model. As follows from this approach, the average distance of the direct visibility $\bar{\rho}$ between two arbitrary points, the source, and the observer, is described by the following equation:

$$R_{cell} \equiv \bar{\rho} = (\gamma_0 \gamma_{12})^{-1} \, (\text{km}), \qquad (11.3)$$

where γ_{12} is the dimensionless parameter describing the effects of the buildings' overlay profile with respect to both terminal antennas. For uniform distribution of buildings height, when $\gamma_{12} = 1$, (11.3) can be simplified as

$$R_{cell} \equiv \bar{\rho} = (\gamma_0)^{-1} = \frac{\pi}{2 \bar{L} \nu} \, \text{km}. \qquad (11.4)$$

Now, if one obtains the information about servicing area, that is, about the overlay profile of the buildings and parameters of building height pattern n and $\bar{h}$ (as was done for Lisbon, see Section 10.1.3, Chapter 10) and about length of the average buildings $\bar{L}$ (in km), as well as the density of buildings per square kilometer, it is easy to estimate the cell radius within the tested area by using Equation (11.3) or Equation (11.4).

Thus, by using Equation (11.2), Equation (11.3), and Equation (11.4) for specific scenarios occurring in the land communication environment, we can easily obtain the *minimum cell radius* for different built-up areas with various situations of the terminal antennas, BS and MS [15–18].

Definition of Radius of Cell via the K-factor. Taking into account fading phenomena, we can also estimate the maximum cell radius based on the definition of Rician K-factor and the corresponding evaluation of this parameter presented in Section 11.3. Thus, for the experiment carried out in mixed residential area and the results of the K-factor computation shown in Figure 10.23 for $f = 900\,\text{MHz}$, the criteria to use the maximum radius of cell can be determined by maximal value of K versus the distance. From the illustration presented in Figure 10.23, this radius can be estimated as about 1–1.2 km, whereas, for the same conditions, Equation (11.3) gives for the minimum cell radius a value of 570 m. The same computations made for urban area ("market site" of Ramat-Gan) according to an illustration presented in Figure 10.24 allow us to estimate the maximum cell radius of about 370–400 m. Whereas, for the same conditions using Equation (11.4), the estimated value of minimum radius was only 170–180 m. Real experiments carried out in both areas showed that the stable communication between BS and any user located in the area of service can be achieved for the mixed area up to 1 km, and for the tested urban area at about 400–550 m. We noticed, according to the experiment described in Section 10.4, that the tested area of Ramat-Gan is a high-dense area with tall buildings. These estimations show again that using the proposal stochastic approach with definitions of the minimum cell radius by (11.4) and of the maximum cell radius by K-parameter from (10.52) to (10.53) (Chapter 10), we can predict a priori the cell radius of the area where stable communication between users and BS antenna can be achieved.

Definition of Cell Radius via Maximum Acceptable Path Loss. Another method of cell radius estimation is to use information about total link budget in the channel accounting for both slow and fast fading. We discussed in Chapter 10 how to estimate this parameter. Now, using Equation (10.36) with Equation (10.33), Equation (10.34), and Equation (10.35) for slow fading, and Equation (10.39) with Equation (10.37) and Equation (10.38) for fast fading (see Chapter 10), on the basis of the results obtained from the stochastic model, we can estimate the maximum and minimum cell radius. Let us again use data from the experiment carried out in the Ramat-Gan experimental site described in Chapter 11. Following the same approach as was done for link budget design and K-factor estimation for this propagation scenario, we have found the following results. The attenuation loss according to slow fading is changed from 7.1 to 7.2 dB, which is approximately constant with the range between the BS and MS antennas. At the same time, the probability to obtain this loss is changed from about 0.4 at 100 m from the BS antenna to about 0.9 at 500 m from the BS.

The same computations for the attenuation loss due to fast fading give about 32 dB at the distance of 100 m from the BS antenna and about 42 dB at the distance of 500 m from the BS antenna, with the corresponding probability of 0.7 and about 1. Then, according to estimations made in Chapter 11 for link budget design in the Ramat-Gan experimental area, the total loss is changed from about 105 dB at the distance of 100 m to about 140 dB at the distance of 500 m. Taking into account the acceptable path loss of the system, which was used in the corresponding experiment which is 138 dB, we have found that the maximum radius of the cell cannot be more than 420 m. Therefore, according to these estimations, one can see that instead of using the equations for K-factor estimations presented in Section 10.3, which give the cell radius $R_{min} \approx 320$ m, we have $R_{max} \approx 420$ m accounting maximum acceptable signal loss with the corresponding probability of fast and slow fading phenomena. Of course, this radius is larger than the minimum one, but smaller or about the same as that of about 400–450 m obtained experimentally in the Ramat-Gan experimental area.

So, using information of the K-factor describing the fast fading phenomena, its probability, and the corresponding loss, or using the maximum accepted path loss, accounting for both kinds of fading, slow and fast, with their corresponding probabilities and loss, one can estimate more precisely the cell radius instead of using the simple Equation (11.3) and Equation (11.4), following the stochastic model. Using the more precise approach described earlier, we can increase the radius of a cell, which allows designers of cellular maps to increase the areas of service for each individual BS, and finally the number of users located in such areas, which increases both the quality and capacity of service.

Co-channel Interference Parameter Definition. The same detailed analysis based on propagation phenomena in different built-up areas can be done to determine the C/I defined earlier. According to the propagation situation in the urban scene for cell sites located beyond the break point range r_B, as shown in Figure 11.8b, the Equation (3.3) from Chapter 3 can be modified, taking into account the multipath phenomenon and obstructions that change the signal decay law from D^{-2} to $D^{-\gamma}$, $\gamma = 2 + \Delta\gamma$, $\Delta\gamma \geq 1$. Hence, instead of (3.3), we finally have:

$$\frac{C}{I} = 10 \log \left[\frac{1}{6} \left(\frac{D^{(2+\Delta\gamma)}}{R_{cell}^2} \right) \right]. \tag{11.5}$$

According to the concepts of cellular map construction presented earlier, the signal strength decay is weaker within each cell (with path loss slope parameter $\gamma = 2$) and corresponds to that in free space. At the same time, due to obstructions, the signal strength decay is stronger in regions outside the servicing cell and within the co-channel site (with path loss slope parameter $\gamma = 2 + \Delta\gamma, \Delta\gamma \geq 1$). Therefore, we can rewrite (11.5) in terms of the number of cells in cluster N, and of radius of the individual cell R_{cell}, by use of (3.2) (Chapter 3):

$$\frac{C}{I} = 10\log\left[\frac{N}{2}(3N)^{\frac{\Delta\gamma}{2}} R_{cell}^{\Delta\gamma}\right]. \tag{11.6}$$

Let us examine these equations for two typical cases described earlier. In the case of typical *crossing straight wide avenues*, for which according to multislit street waveguide model $\Delta\gamma = 2$ ($\gamma = 4$) (see Chapter 5), we get

$$\frac{C}{I} = 10\log\left[\frac{3}{2}(N)^2 R_{cell}^2\right]. \tag{11.7}$$

For the case of narrow streets, which is more realistic in cases of the urban scene, one can put in (11.6) $\Delta\gamma = 3 - 7$ ($\gamma = 5$–9). This is close to the exponential signal decay that follows from the street waveguide model [12–14].

For the case of propagation over *irregular built-up terrain*, following the probabilistic approach presented in Chapter 5, $\Delta\gamma = 1$ and the C/I ratio prediction equation

$$\frac{C}{I} = 10\log\left[\frac{N}{2}(3N)^{1/2} R_{cell}\right]. \tag{11.8}$$

From (11.6) to (11.8), the C/I ratio strongly depends on conditions of wave propagation within the urban communication channels (on path loss slope parameter $\gamma = 2 + \Delta\gamma, \Delta\gamma \geq 1$) and on the cellular map splitting strategy (on parameters N and R_{cell}). In fact, from the equations presented earlier, the C/I performance is enhanced if the cell radius R_{cell} is within the break point range and the reuse distance D is beyond this range. At the same time, stated differently, for a given C/I ratio, a channel can be reused more often, enhancing the cellular system capacity. This is an engineering subject that lies outside the scope and main goal of this book. The reader can find all these questions described in detail in References 19–22.

11.2.2. Strategy of Nonuniform Cellular Maps Design

In the previous section, we considered the classical strategy of how to split the radio map into regular cell lattice using hexagon cell configuration, with periodic constant frequency reuse pattern, that is, cluster layout concept by using

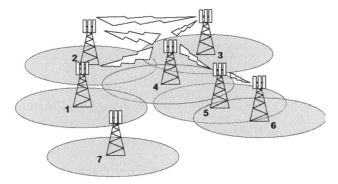

FIGURE 11.9. Nonhomogeneous cellular pattern.

the same frequencies at those cells, the ranges of which between their BS equal the reused distance D. This concept is good for uniform distribution of users/subscribers and calls, for uniform traffic (see Section 11.1). However, due to the demand to keep on increasing the network traffic capacity, designers of the most modern cellular networks started looking for new strategies and some alternative concepts to deal with inhomogeneous traffic. Moreover, in such modern approaches, the BS positioning within the cell pattern was stated to be nonuniform, with relocation in real time according to propagation conditions within the traffic hot spots, as shown in Figure 11.9. The corresponding approaches and models are fully discussed in References 23–32. Later, we will briefly show how to introduce a flexible concept of cell splitting and propagation effects in a new strategy of cellular maps design. Then, we will base on a specific experiment described in Reference 24, and use it for comparison with theoretical predicting model proposed in References 27–29. We will compare the obtained theoretical and experimental results with those obtained by using the proposed stochastic approach of radio propagation in different scenarios in the urban scene.

A Heuristic Approach for Channel Assignment Performance. Let us briefly introduce a modern heuristic algorithm developed in References 27–29 on how to obtain C/I ratio based on mathematical models proposed in References 23–26 and 30–32 by accounting for the Walfisch–Ikegami model (WIM) of signal power decay in an urban environment [33].

The frequency assignment problem was represented in References 27–29 in the form of the heuristic algorithm developed on the basis of cell configuration, which does not follow the classical hexagonal cell homogeneous concept with a periodic frequency reuse pattern. This algorithm is based on the binary constraints between a pair of the transmitters presented in References 27–29 that appear in the following form:

$$|f_i - f_j| > k, \quad k \geq 0, \tag{11.9}$$

where f_i and f_j are the frequencies assigned to transmitters i and j. In References 27–34, different configurations of the cellular pattern were analyzed for channel (frequency) assignment with applications to real nonregular, nonuniform radio networks, mobile, and stationary, considering

- cellular maps with different dimensions of cells
- cellular maps with irregular shapes of cells
- cellular maps with certain levels of intercell overlapping.

An example of nonuniform cell pattern distribution with different shapes and sizes of cells is presented in Reference 34 for a city in Germany [30], where the location of the BS, the cell shape, and dimensions of each cell were taken to be nonregular. It was then redesigned in References 27–29 by using of the new heuristic approach. In its redesigned form, this nonregular cell pattern is shown in Figure 11.10, according to References 27–30. For such a configuration of cells, we need to use the following equation:

$$\left(\frac{C}{I}\right)_i = \frac{R_i^{-4}}{\sum\limits_{j \in M_i} d_{ij}^{-4}}. \tag{11.10}$$

Here, following Reference 34, we take a simple two-ray propagation model with $\gamma = 4$ (see Chapter 5). Notice that all notations are changed here from those used in References 28, 30 and 34 to be unified with those used in this section. Here, R_i is a radius of cell i; M_i is the set of all the cells (excluding cell i) that uses the same bandwidths (channels) as cell i; d_{ij} is the worst-case distance between interfering cell j and cell i. The latter can be found as [34]

$$d_{ij} = \sqrt{(x_i - x_j)^2 + (y_i - y_j)^2} - R_i, \tag{11.11}$$

where (x_i, y_i) and (x_j, y_j) are the Cartesian coordinates of the BSs of cells i and j. Using the simplest propagation model in Reference 34, the co-channel interference constraint was obtained for C/I threshold $\alpha = 1/\beta = 18\,dB$:

$$\sum_{j \in M_i} \frac{d_{ij}^{-4}}{R_i^{-4}} \leq \beta. \tag{11.12}$$

In Reference 28, sufficient improvements of the model [34] were obtained by introducing adjacent channel interference $adj\_factor_k = -a(1 + \log_2 k)$ where k is the bandwidth separation (in number of channels) between the adjacent channel frequency and central frequency of the corresponding filter (see strict explanations in Reference 28). Typical value for a is $18\,dB$ (as $\alpha = 18\,dB$ in Reference 34), and for $k = 1$, an adjacent channel is attenuated by a factor equal to 0.015 [28].

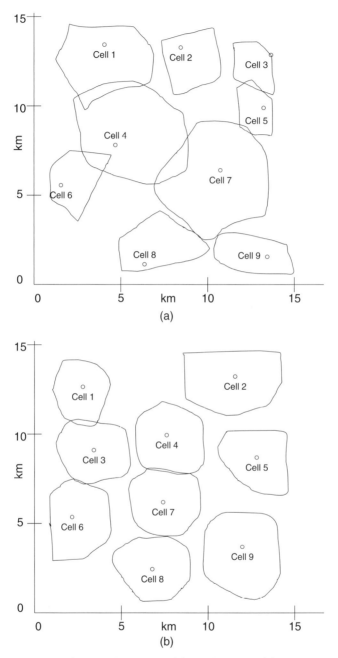

FIGURE 11.10. Nonuniform cell pattern: (a) original and (b) redesigned network configurations (Source: Reference 27. Reprinted with permission @2004 IEEE).

Channel Assignment Strategy Accounting for the Propagation Loss Law. Using such definitions, the co-channel interference constraint (11.11), accounting for the simple law of the received power attenuation $P_{Ri} \propto P_{Ti}d^{-4}$, can be rewritten as [28]

$$\sum_{j \in M_i} \frac{P_{Tj}d_{ij}^{-4}}{P_{Ti}R_i^{-4}} + \sum_{k=1}^{n} adj\_factor_k \sum_{j \in M_i} \frac{P_{Tk}d_{ik}^{-4}}{P_{Ti}R_i^{-4}} \leq \beta. \tag{11.13}$$

Using the two-ray model presentation (see Chapter 6), we can also express this constraint as [28]:

$$\sum_{j \in M_i} \frac{P_{Tj}L(d_{ij}, f)}{P_{Ti}L(R_i, f)} + \sum_{k=1}^{n} adj\_factor_k \sum_{j \in M_i} \frac{P_{Tk}L(d_{ik}, f)}{P_{Ti}L(R_i, f)} \leq \beta. \tag{11.14}$$

In Reference 28, the WIM with a slope attenuation parameter of $\gamma = 2.6$ (i.e., $P_{Ri} \propto P_{Ti}d^{-2.6}$), was also taken into consideration. Let us now compare results obtained according to Reference 28, using a simple two-ray model and WIM, with those obtained using the stochastic approach with slope attenuation parameter $\gamma = 3.0$ (i.e., $P_{Ri} \propto P_{Ti}d^{-3.0}$). To compare the effects of these three laws of propagation loss on strategy of frequency assignment, in computations, following References 27–29, we take into account the interference effects of first ($n = 1$) and second ($n = 2$) adjacent channels. During computations, we considered only co-channel interference ($n = 0$) and also considered a practical example of a 21-cell network, by varying the radii of the cells without changing the BS locations in order to produce three different configurations:

(a) nonoverlapping cells
(b) adjacent cells
(c) overlapping cells.

Finally, the channel assignment *span* and *order*, which guarantees a C/I of at least 18 dB in every point in an urban environment, was computed and presented in Figure 11.11a,b, and Figure 11.13a,b for *span* (a) and *order* (b) assignment. Here, three kinds of laws of path loss according to two-ray ($d^{-4.0}$), WIM ($d^{-2.6}$), and stochastic ($d^{-3.0}$) models for nonoverlapping (Fig. 11.11), adjacent (Fig. 11.12), and overlapping (Fig. 11.13) configurations of nonuniform cellular patterns are shown.

As it is clearly seen from the presented illustrations, the multiparametric model and the WIM give higher channel (frequency) assignments for the worst situations, with a configuration of cellular pattern planning, such as overlapped and adjacent, where the two-ray model is a weaker predictor. Consequently, for nonuniform and nonregular radio cellular networks, it is more realistic to use stochastic model (which also predicts a distance dependence of $d^{-2.5}$ in the presence of the diffraction phenomena, i.e., close to that for WIM $\sim d^{-2.6}$)

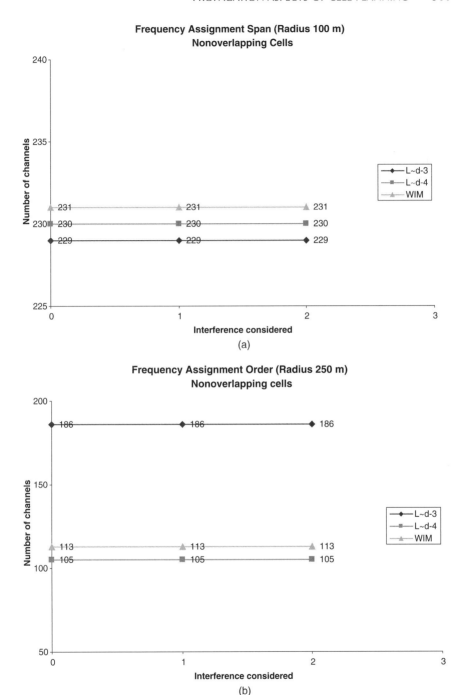

FIGURE 11.11. Frequency assignment for nonoverlapping cells (a) with radius 100 m; (b) 250 m.

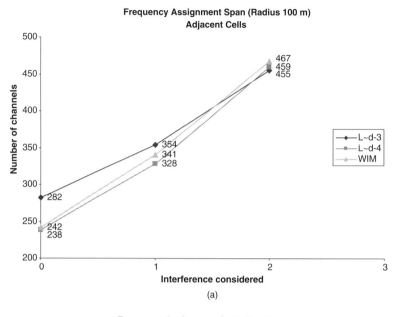

(a)

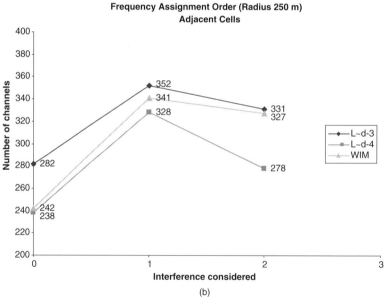

(b)

FIGURE 11.12. Frequency assignment for adjacent cells: (a) with radius 100 m; (b) 250 m.

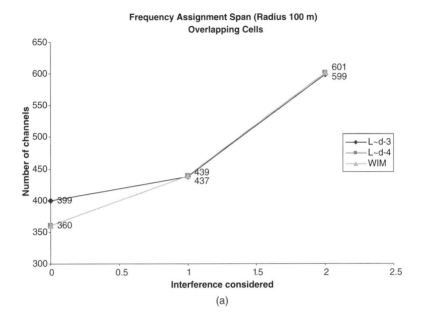

(a)

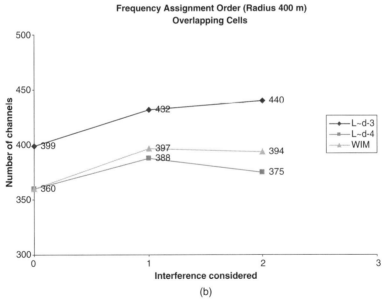

(b)

FIGURE 11.13. Frequency assignment for overlapping cells: (a) with radius 100 m; (b) 400 m.

compared with that predicted by the simple two-ray model usually used by other authors [30, 34].

So, we show again, as was done in earlier discussions of how to predict the efficiency and increase performance of cellular networks, that the strict description of propagation phenomena that occurred in specific urban radio communication channels allow designers to better and more precisely resolve both the BS location problem and the frequency assignment problem, which must be considered simultaneously. We may conclude that the receipt based on the heuristic model, which was discussed in References 27–29, is fully verified by existing experimental data described in Reference 30, as well as by other theoretical models [31–34]. It can be used for the purposes of nonuniform cellular maps design and channel (frequency) assignment and allocation within different land radio links by the use of the more realistic propagation models, the WIM (as COST-231 standard), and the stochastic multiparametric model. The results shown in Figure 11.11, Figure 11.12, and Figure 11.13 fully illustrate the actuality of these conclusions.

11.3. PREDICTION OF PARAMETERS OF INFORMATION DATA STREAM

In the literature for wireless communications, the term "capacity" has different meanings: determining the user capacity in cellular systems in users per channel, information stream capacity inside the communication channel in bits per second (bps), or considering data in bits per second per hertz (bps/Hz) per BS dealing with the spectral efficiency of the communication channel. Let us determine the main parameters of the information data stream sent through any wireless communication link.

11.3.1. Channel Capacity and Spectral Efficiency

According to standards utilized in information science, a *channel capacity*, denoted by C, is referred to as the maximum data rate of information in a channel of given bandwidth, which is measured in bps, whereas, the *spectral efficiency*, denoted by $\tilde{C} = C / B_w$, is considered as a measure in bps/Hz. Both these terms are used in the well-known Shannon–Hartley equation, which for one channel with the given signal-to-noise ratio (SNR) S/N_0B_w, where S is the signal power in $W = J/sec$, B_w is the channel bandwidth (in Hertz), and N_0 is the noise power spectral density in W/Hz, can be written as [35]

$$C = B_w \log_2\left(1 + \frac{S}{N_0 B_w}\right). \tag{11.15}$$

If we denote by $\tilde{C}$ the spectral efficiency, as the ratio C/B_w, we get instead of (11.15)

$$\tilde{C} = \log_2\left(1 + \frac{S}{N_0 B_w}\right). \tag{11.16}$$

These two equations for the capacity and spectral efficiency estimation are valid only for the channels with additive white Gaussian noise (AWGN channels), which is also called additive noise (see definitions in Chapter 1). In this case, the power of the additive noise equals $N_{add} = N_0 B_w$, which is simply defined in the literature as SNR. Usually, AWGN channels are called the ideal channels, and all practical radio channels are compared to the ideal channel by selecting detection error probability of 10^{-6} and finding SNR necessary to achieve it. Effects of interference can be regarded as another source of effective noise, which raises the noise level for calculating the error rate. In this case, we must introduce in (11.15) and (11.16) together with N_{add} the noise caused by interference N_{int}:

$$\tilde{C} = \log_2 \left(1 + \frac{S}{N_0 B_w + N_{int}}\right). \tag{11.17}$$

Earlier, we discussed the channels in which only white or Gaussian noise was taken into account. What will happen if there is additional noise, called multiplicative noise (see definitions in Chapter 1), which usually occurs in the wireless communication channel, land, atmospheric, and ionospheric, due to multipath fading phenomena?

In this case, on the basis of a unified algorithm of how to estimate fading effects described in Chapter 10, we can account all kinds of noise in the Shannon–Hartley Equation (11.15). To do that, we propose now a simple approach, which can be used only if LOS component is predominant with respect to the NLOS component, that is, when the Rician parameter $K > 1$ [36]. Taking into account fading phenomena, described by the Rician distribution with parameter $K > 1$ (see Chapters 1, 5, and 10), we can estimate the multiplicative noise by introducing its spectral density (N_{mult}) with its own frequency band (B_Ω) into (11.15) or (11.16):

$$\tilde{C} = \log_2 \left[1 + \frac{S}{N_0 B_w + N_{mult} B_\Omega}\right]. \tag{11.18}$$

This equation can be rewritten as

$$\tilde{C} = \log_2 \left(1 + \frac{S}{N_{add} + N_{mul}}\right) = \log_2 \left(1 + \left(\frac{N_{add}}{S} + \frac{N_{mul}}{S}\right)^{-1}\right), \tag{11.19}$$

where, according to our definitions introduced in Chapters 5 and 10, following the proposed stochastic approach, $S/N_{mult} = <I_{co}>/<I_{inc}>$. Using now the definition of K, introduced in these chapters, we get

$$\frac{S}{N_{mult}} = \frac{\langle I_{co} \rangle}{\langle I_{inc} \rangle} = K. \tag{11.20}$$

Combining together all these notations, we finally get capacity as function of the Rician K-factor:

$$C = B_w \log_2\left[1+\left(\frac{N_{add}}{S}+K^{-1}\right)^{-1}\right] = B_w \log_2\left(1+\frac{K \cdot SNR_{add}}{K+SNR_{add}}\right), \qquad (11.21)$$

where we denoted signal-to-additive-noise ratio as $SNR_{add} = S/N_{add}$.

The approximate Equation (11.21) can be used to estimate the capacity of the wireless communication channel with fading, additive, and multiplicative phenomena, only when the coherent component of the total signal power will exceed the incoherent component and a direct visibility for each subscriber is taken place. As follows from (11.22), a decrease of the K-factor in one order leads to decrease of the capacity maximum twice. To compensate this decrease caused by fading, the array antennas can be used in a multiple-input multiple-output (MIMO) communication link consisting of an M-element (or M-beam) antenna at one end and an N-element (or N-beam) antenna at the other (see Chapter 7, Fig. 7.1). We will analyze all parameters of MIMO system later. Here, we must notice only that to understand a problem of how noise characteristics caused by fading phenomena affect situation with information data rate and capacity within various land radio communication links, we must obtain the convenient equations that describe relations with the main parameters of data stream. Therefore, we can obtain the relation between the K-factor and a spectral efficiency $\tilde{C}$ using (11.21), that is,

$$K = \frac{SNR_{add}\left(2^{\frac{C}{B_w}}-1\right)}{SNR_{add}-\left(2^{\frac{C}{B_w}}-1\right)} = \frac{SNR_{add}\left(2^{\tilde{C}}-1\right)}{SNR_{add}-\left(2^{\tilde{C}}-1\right)}. \qquad (11.22)$$

This relation will be also used in order to derive BER versus capacity and spectral efficiency too, which we will present at the next section.

Now, we will discuss about bit error occurring in the multipath channels with fading, and will present its classical and advanced description.

11.3.2. Bit Error Rate in Wireless Communication Channel with Fading

The BER is another main parameter that determines the quality of the communication channel, that is, the relative quantity of the received data versus the sent data. BER is measured in the percentage of bits that have error relative to the total bits received in a transmission. First, we will consider in this section well-known classical description of data error rate and then we will give definitions of BER via the advanced statistical approaches.

As was mentioned in Chapter 1, multipath fading channels cause multiplicative variations in the transmitted signal $s(t)$ or in its envelope $r(t)$, which is defined as the *multiplicative* noise [19–22, 37–39].

Classical Description of Bit Error Rate. Let us now introduce a classical description of different channels with their specific "reaction" on the white (or additive) and multiplicative noise caused by the flat, slow, and fast fading (see definitions in Chapter 1).

Channel with the Gaussian Flat Fading. The ideal AWGN digital channel, where only additive (or "white") Gaussian noise ($n(t)$) is observed, the probability of information error (which defines BER) is described by error Q-function versus SNR (γ) as [40, 41]

$$P_e = \int_0^\infty P_e(\gamma)p(\gamma)d\gamma, \tag{11.23}$$

where $P_e(\gamma)$ is the probability of error for *any* digital modulation at a specific value of SNR and $p(\gamma)$ is a probability density function (PDF) due to the AWGN channel. The total transmitted signal after digital modulation in such a channel can be presented as [41]:

$$s(t) = Ag(t) + n(t). \tag{11.24}$$

This equation represents a complex baseband signal, as a sum of the modulated signal, $g(t)$, and the noise waveform, $n(t)$, where A is overall path loss, assumed not to vary in time. If so, both the real and imaginary parts of noise $n(t)$ are zero-mean, independent, real Gaussian processes, each with a standard deviation of σ_n [37, 38, 40, 41].

For digital signals, which consist of symbols with an individual energy E_s and a finite duration T_s we get

$$E_s = A^2 T_s / 2. \tag{11.25}$$

Similarly, if the noise is contained within a bandwidth $B = 1/T_s$, and has power spectral density $PSD \equiv N_0$, then the mean noise power [37, 38, 40, 41]

$$P_n = \frac{1}{2}\langle n(t)n^*(t)\rangle \equiv \sigma_n^2 = BN_0 = N_0/T_s. \tag{11.26}$$

The SNR at the input of the demodulator is then (see definitions in Reference 40 and 41)

$$\gamma \equiv SNR = \frac{\langle A^2 g^2(t)\rangle}{2P_n} = \frac{A^2\langle g^2(t)\rangle}{2\sigma_n^2} = \frac{E_s}{N_0} = \frac{A^2 T_s}{2N_0}. \tag{11.27}$$

It is common to express the error rate performance of a digital communication system in terms of this parameter, or in terms of the corresponding SNR per bit [40, 41]:

$$\gamma_b = \frac{\gamma}{m} = \frac{E_b}{N_0}, \tag{11.28}$$

where m is the number of bits per symbol. In References 40 and 41, it was determined

> The SNR is the key parameter in calculating the digital modulation system performance in any multipath communication channel with white noise.

This parameter determines the probability of information error, which can be estimated through the error Q-function as (see definition in Chapter 1)

$$P_e = Q\left(\sqrt{\frac{A^2 d^2}{2N_0}}\right) \equiv Q\left(\sqrt{\frac{d^2}{2N_0}}\right), \tag{11.29}$$

where d is the M-metric distance in the corresponding constellation diagram of geometric presentation of any digital modulated signal (see definitions in Reference 41).

In the example of binary phase shift keying (BPSK) digital modulation with 2-D metrics, $d = 2\sqrt{E_s}$ [41] and

$$P_r = Q\left(\sqrt{\frac{d^2}{2N_0}}\right) = Q\left(\sqrt{\frac{4E_b}{2N_0}}\right) = Q\left(\sqrt{2\gamma}\right). \tag{11.30}$$

The result of calculation of BER according to (11.30) is illustrated in Figure 11.14 versus SNR. The fast decrease in BER as SNR increases is the main

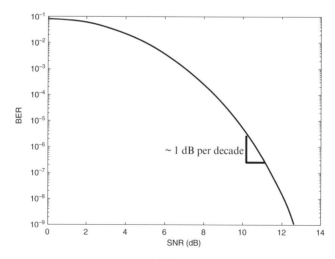

FIGURE 11.14. The BER function $Q\left(\sqrt{2\gamma}\right)$ versus SNR, γ, for the AWGN channel.

feature of an AWGN channel. This decrease is the fastest that could take place for digital modulation channels, so the AWGN channel is a best-case channel. For high SNRs, the BER decreases by a factor of 10 for every 1 dB increase in SNR.

Rayleigh Flat Fading Channel. Let us now consider the real case of narrowband flat fading, which describes the actual situation in the outdoor/outdoor communication channels and in which fading, slow and/or fast, affects all frequencies in the modulated signal equally, so it can be modeled as a single multiplicative process (see definitions in Chapter 1). In this channel, the fading process differs from the ideal AWGN channel considered earlier. We will show this for the worst case in an urban propagation channel, which is characterized by the Rayleigh fading, and will compare both kinds of channels, the ideal, and the worst one.

Since the fading varies with time, the SNR at the input of the receiver also varies with time. It is necessary, in contrast with the AWGN case, to obtain relations between the instantaneous SNR, γ, and the mean SNR, Γ, which we will define in this section. In the real multipath communication channel, the received signal $r(t)$ may be expressed as [37, 40, 41]

$$r(t) = A\alpha(t)g(t) + n(t), \tag{11.31}$$

where $\alpha(t)$ is the complex fading coefficient at time t. Equation (11.31) mathematically describes the narrowband fading multipath channel (see definitions in Chapter 1). If the fading is assumed constant over the transmitted symbol duration, then it is also constant over a symbol and is given:

$$\gamma(t) = \frac{A^2 |\alpha(t)|^2 \langle |g(t)|^2 \rangle}{2P_n} = \frac{A^2 |\alpha(t)|^2}{2P_n} \tag{11.32}$$

and then

$$\Gamma = \langle \gamma(t) \rangle. \tag{11.33}$$

Usually, we consider the fading that has the unit variance, and consider any change in mean signal power into the path loss, then,

$$\Gamma = \langle \gamma(t) \rangle = \frac{A^2}{2P_n}. \tag{11.34}$$

Two things are therefore needed in order to find the performance of a network consisting of the narrowband fading channel: the mean SNR and a description of the way the fading causes the instantaneous SNR to vary relative to this mean.

For this purpose, for a Rayleigh fading flat narrowband channel, we will rewrite (11.32) in the following manner according to Reference 41:

$$\gamma = \alpha^2 E_s / N_0. \tag{11.35}$$

For Rayleigh distribution, the PDF of error is now [37, 40, 41]

$$p(\gamma) = \frac{1}{\Gamma} \exp\left(-\frac{\gamma}{\Gamma}\right), \tag{11.36}$$

where

$$\Gamma = (E_s / N_0)\langle \alpha^2 \rangle \tag{11.37}$$

is the average value of the SNR. If so, for the cumulative distributed function (CDF) we get (see relation between PDF and CDF in Chapter 1)

$$\text{CDF}(\gamma) \equiv \Pr(\gamma < \gamma_s) = \int_0^{\gamma_s} p(\gamma)d\gamma = 1 - \exp\left(-\frac{\gamma_s}{\Gamma}\right). \tag{11.38}$$

Here γ_s is the constant threshold of SNR in the Rayleigh fading channel. The result of (11.38) is illustrated in Figure 11.15. This result can be used to calculate the required mean SNR above some threshold for an acceptable percentage of the time [37, 40, 41].

The same CDF from (11.38) can be successfully used to predict the error rate performance of digital modulation schemes in a Rayleigh channel by assuming that the SNR is constant over one symbol duration. In this case, the Rayleigh error rate performance can be predicted in the same manner, as was

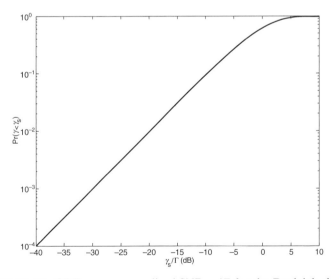

FIGURE 11.15. CDF versus normalized SNR, γ_s/Γ, for the Rayleigh channel.

done earlier for the AWGN channel case. We give, as an example, the BPSK modulated signals, using AWGN error rate performance from (11.30) and the Rayleigh statistics according to (11.36) for the instantaneous SNR (γ). Following (11.36), the average bit error probability P_e can be defined as [37, 40, 41]

$$P_{eBPSK} = \langle P_e(\gamma) \rangle = \int_0^\infty P_e(\gamma) p(\gamma) d\gamma$$

$$= \int_0^\infty Q(\sqrt{2\gamma}) \frac{1}{\Gamma} \exp\left(-\frac{\gamma}{\Gamma}\right) d\gamma = \frac{1}{2}\left[1 - \sqrt{\frac{\Gamma}{1+\Gamma}}\right], \qquad (11.39)$$

from which an approximation for large Γ, $P_e \approx 1/(4\Gamma)$, leads. This inverse proportionality is a characteristic of uncoded modulation in a Rayleigh flat channel, leading to a reduction of BER by one decade for every 10 dB increase in SNR. This contrasts sharply with approximately 1 dB per decade variation in the AWGN channel (see Fig. 11.14). To understand this better, we illustrate in Figure 11.16 how slow the BER changes for various narrowband digital modulations inside the Rayleigh multipath communication channel as a function of E_s/N_0, with respect to that for AWGN ideal channel. The same slow rate of BER reduction for other types of modulated signals, frequency shift keying (FSK), differential phase shift keying (DPSK), and phase shift keying (PSK), is observed compared to fast rate of a typical AWGN curve. We will not discuss in detail how different kinds of digital modulation influence BER performance; the reader can read about that subject in several other books, such as References 37, 38, 40, and 41.

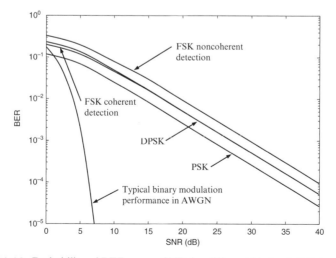

FIGURE 11.16. Probability of BER versus SNR for different kinds of digital modulation.

Rician Flat Fading Channel. To predict the actual reduction in BER for each situation occurring in the urban scene, one can use the Rician fading, described briefly in Chapter 1, where the Rician K-factor, as a ratio between LOS component and NLOS component of the total multipath signal, can be calculated theoretically using the proposed multiparametric model (see Section 10.3, Chapter 10). The Rician flat fading channel is usually used whenever one path field component exceeds or is at the same level with the other multipath components due to multiscattering, multireflection, and multidiffraction. The BER for any digital modulated narrowband signal and for a flat fading Rician channel, depending on K value, is intermediate between the AWGN and Rayleigh channel cases, as is shown in Figure 11.17. It is clear that the Rician channel behaves like AWGN channel in the limit as $K \to \infty$, and like Rayleigh multipath channel when $K = 0$.

In a real outdoor communication channel for the concrete parameter K, which usually varies from $K = 3$ to $K = 10$ for suburban and urban environments (see Section 10.3), all graphs presented in Figure 11.17 for Rayleigh fading must be transformed to those closer to the ideal case of AWGN channel.

Moreover, using any of wideband (spread spectrum) digital modulations, such as direct sequence spread spectrum modulation (DS-SS), and frequency and time hopping spectrum modulation (FH-SS and TH-SS) [37, 40, 41], as well as their combinations for the concrete scenario occurring in the urban communication channel (for the concrete K), one can significantly increase the performance of data flow regulation to reject significantly multipath fading (e.g., the multiplicative noise) and the co-channel interference between users, and, finally, to increase the performance of the multiple access in the wireless communication networks. In this section, we only mentioned the effects of fast

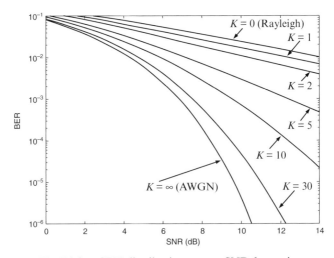

FIGURE 11.17. The Rician CDF distribution versus SNR for various parameters K.

fading on the BER performance, described by classical Gaussian, Rician, and Rayleigh distributions; other aspects of signal processing using different kinds of classical digital modulation can be found in References 19–22 and 37–41.

BER Description Using the Stochastic Approach. To estimate the BER parameter, the following equation is usually used [42]:

$$\text{BER} = \frac{1}{2}\int_0^\infty p(x)\,erfc\left(\frac{\text{SNR}}{2\sqrt{2}}x\right)dx. \tag{11.40}$$

Here $p(x)$ is the PDF and $erfc(\bullet)$ is the well-known error function [43] introduced also in Chapter 1. Now, using the BER definition (11.40), where $p(x)$ is Rician PDF with the standard deviation σ (see definitions in Chapter 1) and taking into account multiplicative noise, we finally get for BER the following general equations:

$$\text{BER}(K,\text{SNR},\sigma) = \frac{1}{2}\int_0^\infty \frac{x}{\sigma^2}\cdot e^{-\frac{x^2}{2\sigma^2}}\cdot e^{-K}\cdot I_0\left(\frac{x}{\sigma}\sqrt{2K}\right)$$

$$erfc\left(\frac{K\cdot\text{SNR}_{add}}{2\sqrt{2}\left(K+\text{SNR}_{add}\right)}x\right)dx. \tag{11.41}$$

Then, using (11.40), and the relation (11.22), we get the BER as a function of $\tilde{C}$:

$$\text{BER}(\tilde{C}) = \frac{1}{2}\int_0^\infty \frac{x}{\sigma^2}\cdot e^{-\frac{x^2}{2\sigma^2}}\cdot e^{-\frac{W_1(\tilde{C})}{W_2(\tilde{C})}}\cdot I_0\left(\frac{x}{\sigma}\sqrt{2\frac{W_1(\tilde{C})}{W_2(\tilde{C})}}\right)$$

$$erfc\left(\frac{\dfrac{W_1(\tilde{C})}{W_2(\tilde{C})}\cdot\text{SNR}_{add}}{2\sqrt{2}\left(\dfrac{W_1(\tilde{C})}{W_2(\tilde{C})}+\text{SNR}_{add}\right)}x\right)dx, \tag{11.42}$$

where $W_1(\tilde{C})=\text{SNR}_{add}\cdot\left(2^{\tilde{C}}-1\right)$ and $W_2(\tilde{C})=\text{SNR}_{add}-\left(2^{\tilde{C}}-1\right)$.

This is a practically important equation, which gives relation between the spectral efficiency of the multipath communication channel, caused by fading phenomena, and the probability of BER of the information data stream inside such a channel.

11.3.3. Analysis of Relations between Main Parameters of Data Stream

To show the reader how to minimize the BER and maximize the capacity or the spectral efficiency for different wireless networks (with the corresponding bandwidth of each channel), we created a special program based on earlier

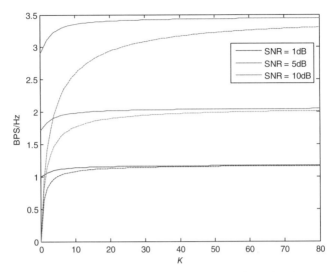

FIGURE 11.18. Spectral efficiency as a function of K-factor.

equations. First, we have analyzed the changes of the capacity as a function of K, that is, as a function of different conditions of the channel allocation, using Equation (11.18), Equation (11.19), Equation (11.20), and Equation (11.21). Results are shown in Figure 11.18. Here, we have examined the spectral efficiency for three different typical values of SNR_{add} which are shown in Figure 11.18. It is obvious that with the increase in SNR (from 1 dB to 10 dB), the spectral efficiency is increased by more than three times; this effect is more significant for the worst case of multipath fading channels (for $K < 5$).

Then, we made computations to estimate the capacity that can be used by various wireless networks with a given S/N_{add} ratio of 5 dB, using (11.21) in different bandwidths, corresponding to several well-used networks; such as the code division multiple access (CDMA) system with the bandwidth $B_w = 25$ MHz, the time division multiple access (TDMA) system ($B_w = 40$ MHz), the TDMA1 system (15 MHz), the TDMA2 system (5 MHz), and the Global System for Mobile (GSM), as combination of the TDMA and frequency division multiple access (FDMA) ($B_w = 12.5$ MHz). As it is seen from Figure 11.19, where the capacity is shown versus K-factor, the fading effect is dominant for $K < 5$, when the multipath component may play some significant role. This effect depends on parameter K, which characterized fading effects within the wireless channel. Furthermore, a network that uses larger bandwidth gives better throughput of data streams. Usually, the TDMA systems have bandwidth containing not more than 20 MHz, therefore, their capacity is larger than that for CDMA systems.

As is known from References 40, 41, 44 and 45, the TDMA and CDMA systems result in an increase in channel capacity over the standard FDMA

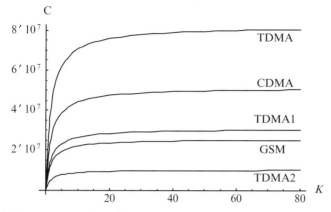

FIGURE 11.19. Capacity of various systems versus the K-factor.

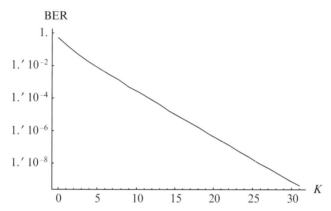

FIGURE 11.20. BER as a function of the K-factor.

system, allowing different time slots and different codes to be assigned to different users. Each increase is on the order of 5 to 10 [46–49].

Finally, we ran two simulations, the first one was BER as a function of K using Equation (11.41), and the second was BER as a function of spectral efficiency, using Equation (11.2). The parameters we used for computations are the following: $\sigma = 2$ and $SNR_{add} = S/N_{add} = 1\,dB$. The results of the computation are shown in Figure 11.20 and Figure 11.21, respectively. As seen from Figure 11.22 and Figure 11.23, with an increase of K-factor, that is, when LOS component becomes predominant, compared to NLOS multipath component of signal total intensity, the BER parameter is decreased essentially, from 10^{-2} for $K \approx 5$ to 10^{-6} for $K \approx 20$ (corresponding to communication channels with high BS antenna and quasi-LOS for any user located in the area of service). At the same time, as expected, the spectral efficiency is also increased

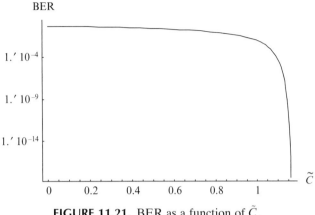

FIGURE 11.21. BER as a function of $\tilde{C}$.

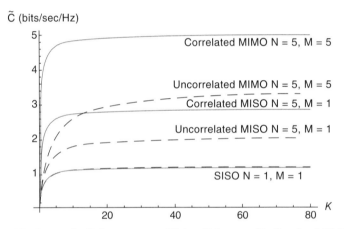

FIGURE 11.22. Spectral efficiency versus Rician K-factor of fading for MIMO, MISO and SISO antenna systems.

(see Fig. 11.23). Hence, with the increase of the spectral efficiency of the data stream simultaneously, sharp decrease of BER is also observed.

11.3.4. Data Stream Parameters in MIMO Communication Links

For the MIMO communication links consisting of the M-element (or M-beams) antenna at the one end and N-element (or N-beam) antenna at the other end, the link capacity and spectral efficiency depend on the degree of correlation between channels corresponding to each element or beam.

The AWGN Channel. First, we consider the MIMO channel consisting the white Gaussian noise, where the separate channels in the antenna array are

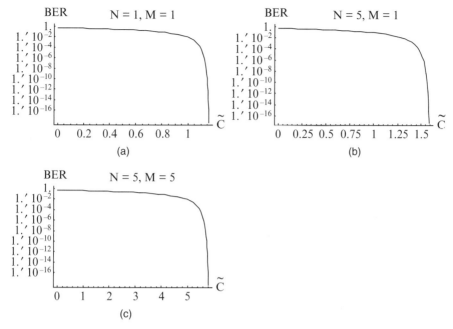

FIGURE 11.23. BER for correlated SISO (a), MISO (b), and MIMO (c) antenna systems versus the spectral efficiency.

uncorrelated. In this case, their capacity and spectral capacity may be added. Thus, for $M > N$, the following expression can be used in pure AGWN channel without multiplicative noise [35, 50, 51]:

$$\tilde{C}_{uncor} = N \cdot \log_2 \left(1 + \frac{M \cdot S}{N \cdot B_W \cdot N_0} \right). \tag{11.43}$$

We can interpret this equation as follows: the output signal power is divided equally between N channels, since they have the same gain as M. As the capacities of the channels are the same, we get a factor N outside the logarithm.

In the case of correlated antennas within the array, we have only one channel with gain MN [35, 50, 51] and

$$\tilde{C}_{corr} = \log_2 \left(1 + \frac{M \cdot N \cdot S}{B_W \cdot N_0} \right). \tag{11.44}$$

In the case of uncorrelated element antenna array, the channel capacity or spectral efficiency increases linearly with increase on number of channels N. While for correlated antenna elements, this increase is lesser according to logarithmic law. Nevertheless, in both cases, an adaptive array can significantly increase the data rate between the mobile and a BS in multiple spatial channels, as it was shown in Reference 43. For example, in an IS-136 system, an omni-antenna gives 48.6 kilobits per second (kbps) in a single 30 kHz channel, whereas the M spatially separated channels corresponded to M antenna elements at the BS can yield data rate of $\approx M$ 30 kHz [43]. Use of the additional techniques for space–time coding or space–time processing offer the potential to make practical spectral efficiency of many bps/Hz to each mobile user (in an earlier example of hundreds of kbps in a 30 kHz channel).

The Multipath Channel with Fading. Accounting now in (11.21) and (11.22) for the additional effects of the multiplicative noise, we finally get according to References 52 and 53 the following expressions instead of (11.43) and (11.44):

- for uncorrelated elements of the antennas

$$\tilde{C}_{uncor} = N \log_2 \left(1 + \frac{M}{N} \cdot \frac{K \cdot \mathrm{SNR}_{add}}{(K + \mathrm{SNR}_{add})} \right) \tag{11.45}$$

- for correlated elements of the antennas

$$\tilde{C}_{corr} = \log_2 \left(1 + MN \cdot \frac{K \cdot \mathrm{SNR}_{add}}{(K + \mathrm{SNR}_{add})} \right). \tag{11.46}$$

Now, comparing (11.21) with (11.45) or (11.46), we can obtain the compensation of multiplicative fading effects for link capacity in linear order of N for the adaptive antenna array with uncorrelated elements or in logarithmical order of NM for the adaptive antenna array with the correlated elements.

Next, we present, according to References 52 and 53, the equations for the K-factor and the BER accounting for the relations between the K-factor and the spectral efficiency $\tilde{C}$ described by (11.22) for the MIMO system, with the additive and multiplicative noises, by introducing in (11.22) the parameter $\tilde{C}/N$ instead of $\tilde{C}$ in exponent $2^{\tilde{C}}$. Finally we get instead of (11.22)

- for MIMO system with *uncorrelated* antenna N and M elements

$$K = \frac{\mathrm{SNR}_{add} \left[\frac{N}{M} \left(2^{\frac{\tilde{C}}{N}} - 1 \right) \right]}{\mathrm{SNR}_{add} - \left[\frac{N}{M} \left(2^{\frac{\tilde{C}}{N}} - 1 \right) \right]} \tag{11.47a}$$

- for the MIMO system with *correlated* antenna N and M elements

$$K = \frac{\mathrm{SNR}_{add} \left[\frac{1}{MN} \left(2^{\tilde{C}} - 1 \right) \right]}{\mathrm{SNR}_{add} - \left[\frac{1}{MN} \left(2^{\tilde{C}} - 1 \right) \right]}. \tag{11.47b}$$

Finally, according to References 52 and 53, we get for the BER the following expressions:

- for uncorrelated elements inside the MIMO system

$$
\mathrm{BER}\left(\tilde{C}\right)=\frac{1}{2}\int_0^\infty \frac{x}{\sigma^2}\cdot e^{-\frac{x^2}{2\sigma^2}}\cdot e^{-\frac{\mathrm{SNR}_{add}\frac{N}{M}\left(2^{\frac{\hat{C}}{N}}-1\right)}{\mathrm{SNR}_{add}-\frac{N}{M}\left(2^{\frac{\hat{C}}{N}}-1\right)}}\cdot I_0\left(\frac{x}{\sigma}\sqrt{2\frac{\mathrm{SNR}_{add}\frac{N}{M}\left(2^{\frac{\hat{C}}{N}}-1\right)}{\mathrm{SNR}_{add}-\frac{N}{M}\left(2^{\frac{\hat{C}}{N}}-1\right)}}\right)
$$

$$
\cdot erfc\left(\frac{\frac{N}{M}\left(2^{\frac{\hat{C}}{N}}-1\right)}{2\sqrt{2}}x\right)dx
$$

(11.48a)

- for *correlated* antenna elements inside the MIMO system

$$
\mathrm{BER}\left(\tilde{C}\right)=\frac{1}{2}\int_0^\infty \frac{x}{\sigma^2}\cdot e^{-\frac{x^2}{2\sigma^2}}\cdot e^{-\frac{\mathrm{SNR}_{add}\frac{1}{MN}\left(2^{\tilde{C}}-1\right)}{\mathrm{SNR}_{add}-\frac{1}{MN}\left(2^{\tilde{C}}-1\right)}}\cdot I_0\left(\frac{x}{\sigma}\sqrt{2\frac{\mathrm{SNR}_{add}\frac{1}{MN}\left(2^{\tilde{C}}-1\right)}{\mathrm{SNR}_{add}-\frac{1}{MN}\left(2^{\tilde{C}}-1\right)}}\right)
$$

$$
\cdot erfc\left(\frac{\frac{1}{MN}\left(2^{\tilde{C}}-1\right)}{2\sqrt{2}}x\right)dx.
$$

(11.48b)

The earlier equations can be successfully used for estimations of data stream parameters, such as the spectral efficiency and BER in various multibeam antenna channels with additive and multiplicative noises.

First, we will analyze the spectral efficiency for two cases, MIMO and multiple-input single output (MISO), of antenna arrangement concerning the capacity of data stream inside channel with multipath fading. As can be seen from Figure 11.22, the MIMO and MISO systems with the correlated antenna elements give better spectral efficiency compared to those on basis of the uncorrelated antenna elements, and therefore, have a higher capacity for a given bandwidth. Here, we present the single input single output (SISO) system only to compare systems and show difference between them, because in SISO system, the term correlated and uncorrelated are not relevant.

The presented illustrations in Figure 11.22 clearly show that as the *K*-factor increases, that is, when the LOS component exceeds the NLOS component in more than five times ($K \geq 5$), the increase of the spectral efficiency stabilizes, and K does not affect it anymore. At the same time, for a range of small values of K ($K \ll 5$), as K increases, the coherent part exceeds the incoherent part, and a sharp increase of the spectral efficiency is observed.

Analysis of the BER as a function of the spectral efficiency $\tilde{C}$ for two kinds of antenna systems, MISO and MIMO, with correlated and uncorrelated antenna elements, was made on the basis of derivations made in References 52 and 53. The results of numerical computations are shown in Figure 11.23a,b,c, and Figure 11.24a,b,c for correlated and uncorrelated cases, respectively. Here, the SISO system, which is not actual for the present discussions, is shown only as a simple case for comparison with more complicated antenna systems. As follows from the illustrations presented in Figure 11.23a,b,c, for the case of correlated antenna elements, with the increase of number of elements, the limit where the sharp decrease of BER is observed becomes larger, and the BER limits to zero. Thus, for $N = 1$ and $M = 1$ (SISO) we observe in Figure 11.23a a sharp decrease of BER with increase of $\tilde{C}$ beyond the unit; the same as shown in Figure 11.21. At the same time, for $N = 5$ and $M = 1$ (MISO) (see Fig. 11.23b), this sharp decrease is observed at $\tilde{C} \cong 2.5$. For $N = 5$ and $M = 5$ (MIMO) (see Fig.11.23c), this sharp decrease is observed at $\tilde{C} \cong 4.2$.

The same tendency is observed for uncorrelated case of antenna elements arrangement, but for smaller values of $\tilde{C}$ (see Fig. 11.24a,b,c). Therefore, we

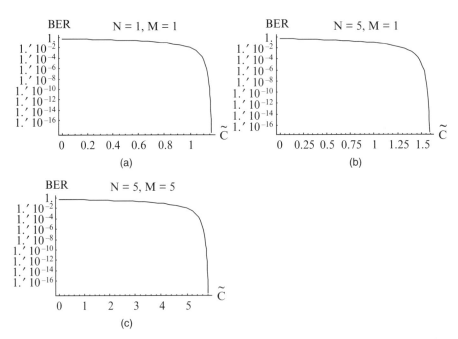

FIGURE 11.24. BER for uncorrelated MIMO and MISO antenna systems versus the spectral efficiency.

can conclude that the type of multielement antenna, with correlated or uncor-related elements, is not crucial parameter for the BER prediction in MIMO channels with fading. At the same time, the amount of elements at the BS and at the mobile user (N and M, respectively) is a significant and tremendously influenced element of BER prediction in the multipath MIMO channels with fading.

11.3.5. Estimation of BER via the Level Crossing Rate and Average Fade Duration

To estimate the capacity of the information data stream and the BER, general characteristics that determine the rate at which the input baseband signal, passing the radio communication link, falls below a given sensitivity of the receiver (defined by the level X in Chapter 1), and how long this signal remains below this level, should be analyzed. This is useful for relating the SNR during fading to the instantaneous BER that result. In other words, new parameters of fading phenomena, the LCR, and the average fade duration (AFD), allow the designers of digital wireless networks to determine the most likely number of signaling bits that may be lost during the fading [54–58], and finally, estimate BER for each scenario occurring in the land communication link.

Later, we will present an example of how to estimate the BER using knowl-edge of the LCR and AFD in the Rayleigh fading channel.

Rayleigh Fading Channel. The main equations describing LCR and AFD in the multipath channel, the fading of which can be determined by the Rayleigh PDF or CDF, were presented in Chapter 1. For the convenience, we repeat them again. Thus, the LCR can be presented as follows:

$$N_X = \sqrt{2\pi} f_m \zeta \exp\left(-\zeta^2\right) \qquad (11.49)$$

Here, $f_m = \nu/\lambda$ is the maximum Doppler frequency shift and $\zeta = X/\sqrt{2}\cdot\sigma \equiv X/rms$ is the value of specific level X, normalized to the local *rms* amplitude of fading envelope, where $rms = \sqrt{2}\cdot\sigma$ (see all definitions in Chapter 1).

The AFD can be found in the same manner according to definitions made in Chapter 1, that is, for the Rayleigh fading channel,

$$\langle\tau\rangle = \frac{\exp\left(\zeta^2\right)-1}{\sqrt{2\pi} f_m \zeta}. \qquad (11.50)$$

BER via AFD and LCR. We show a simple example of how to estimate LCR and AFD for the next estimation of BER. Let us suppose that via a digital radio channel, the information data is sent with the rate of $R = 500$ bps (i.e., kbps). Therefore, the time of one bit duration is $T_b = 1/R = 2$ ms. For $\zeta = 0.1$ and a maximum Doppler shift of $f_m = 20$ Hz (the velocity of the moving receiver [or transmitter] is 20 m/s for the radiated frequency of 1 GHz), we get from (11.50) that $<\tau> = 0.2$ m/s, which is 10 times less than the bit duration. So, only 10% of the bit time duration is affected by fast Rayleigh fading. At the same time, using the same data, we have found according to (11.49) that $N_X \approx 5$ crossings per second.

Therefore, the total number of bits in error per second are 5, and the BER is

$$\text{BER} = \frac{\text{number of bits per second}}{\text{number of bits in data per second}} = \frac{5}{500} = 10^{-2}.$$

This means that only 1 bit per 100 bits will be lost in such mobile digital communication link. It is obvious that the example presented earlier is actual only for the Rayleigh fast fading in the wireless channel, without accounting for the conditions of environment and the terrain features.

We must notice that in a more general case of the Rician channel, with fast fading in real situations occurring in the built-up scene, the more complicated computations should be done by using Equation (1.45), Equation (1.46), Equation (1.47), Equation (1.48), Equation (1.49), and Equation (1.50) from Chapter 1, where the information on the Rician K-factor of fast fading for each scenario (mixed residential, suburban, or urban) can be found using Equations (10.52) and (10.53) presented in Chapter 10, on the basis of the multiparametric stochastic approach. Using these equations, we developed a numerical code to evaluate BER that can be expected in the multipath Rician channels with fading. First, we return to Figure 10.35 and Figure 10.36 presented in Chapter 10, to obtain the data on the Rician K-factor distribution. Then, by using Equation (11.45), Equation (11.46), Equation (11.47), and Equation (11.48), we have found that for $K = 2$ BER $= 9 \cdot 10^{-3}$, for $K = 5$ BER $= 2 \cdot 10^{-3}$, and for $K = 10$ BER $= 8 \cdot 10^{-4}$. In our estimations of the BER, we used the predicted K-parameters usually observed experimentally in suburban and urban communication links, which vary from $K = 2$ to $K = 15$ [19–22, 37–39]. We noticed that the obtained results are not in contradiction with those presented by Figure 10.35 and Figure 10.36, which described the spatial distribution of K-factor, and Figure 11.17, which described the distribution of BER versus the Rician K-factor.

REFERENCES

1   Syski, R., *Introduction to Congestion Theory in Telephone Systems*, 2nd edition, McGraw-Hill, New York, 1986.

2   Nelson, R., *Stochastic Processes and Queuing Theory*, Springer-Verlag, Berlin, 1994.

3   Alanyali, M. and B. Hajek, "On simple algorithm for dynamic load balancing," *Proc. IEEE INFOCOM*, Vol. 1, 1995, pp. 230–238.

4   Yuan, W. et al., "Loading balancing in wireless networks," *Proc. GLOBECOM*, Vol. 23, 1997, pp. 1616–1620.

5 Papavassiliou, S. and L. Tassiulas, "Joint optimal channel base station and power assignment for wireless access," *IEEE/ACM Trans. Network.*, Vol. 4, 1996, pp. 857–872.

6 Freedman, A., A. Gil, R. Giladi, "An impact of unbalanced availability on GOS of wireless systems," *Wireless Personal Communications*, Vol. 20, 2002, pp. 21–40.

7 Blaunstein, N., A. Freedman, R. Giladi, and M. Levin, "Unified approach of GOS optimization for fixed wireless access," *IEEE Trans. on Veh. Technolog.*, Vol. 51, 2002, pp. 101–110.

8 Blaunstein, N. and R. Hassanov, "Grade of service design in wireless systems operating in multipath fading environments," *J. Business Briefing: Wireless Technology 2004*, 2004, pp. 121–124.

9 Lee, W.Y.C., *Mobile Communication Engineering*, McGraw Hill, New York, 1985.

10 Jakes, W.C., *Microwave Mobile Communications*, John Wiley & Sons, New York, 1974.

11 Blaunstein, N. and Y. Ben-Shimol, "Frequency dependence of path loss characteristics and link budget design for various terrestrial communication links," *IEEE Trans. on Antennas and Propagat.*, Vol. 52, No. 10, 2004, pp. 2719–2729.

12 Blaunstein, N. and M. Levin, "Propagation loss prediction in the urban environment with rectangular grid-plan streets," *Radio Sci.*, Vol. 32, No. 2, 1997, pp. 453–467.

13 Blaunstein, N., "Average field attenuation in the non-regular impedance street waveguide," *IEEE Trans. Anten. and Propagat.*, Vol. 46, No. 12, 1998, pp. 1782–1789.

14 Blaunstein, N., "Prediction of cellular characteristics for various urban environments," *IEEE Anten. Propagat. Magazine*, Vol. 41, No. 6, 1999, pp. 135–145.

15 Xia, H.H. and H.L. Bertoni, "Diffraction of cylindrical and plane waves by an array of absorbing half screens," *IEEE Trans. Antennas and Propagation*, Vol. 40, 1992, pp. 170–177.

16 Bertoni, H.L., W. Honcharenko, L.R. Maciel, and H.H. Xia, "UHF propagation prediction for wireless personal communications," *Proc. IEEE*, Vol. 82, No. 9, 1994, pp. 1333–1359.

17 Rustako, A.J. Jr., N. Amitay, M.J. Owens, and R.S. Roman, "Radio propagation at microwave frequencies for line-of-sight microcellular mobile and personal communications," *IEEE Trans. Veh. Technol.*, Vol. 40, No. 2, 1991, pp. 203–210.

18 Tan, S.Y. and H.S. Tan, "UTD propagation model in an urban street scene for microcellular communications," *IEEE Trans. Electromag. Compat.*, Vol. 35, No. 4, 1993, pp. 423–428.

19 Feuerstein, M.L. and T.S. Rappaport, *Wireless Personal Communication*, Artech House, Boston-London, 1992.

20 Lee, W.Y.C., *Mobile Cellular Telecommunications Systems*, McGraw-Hill, New York, 1989.

21 Linnartz, J.P., *Narrowband Land-Mobile Radio Networks*, Artech House, Boston-London, 1993.

22 Mehrotra, A., *Cellular Radio Performance Engineering*, Artech House, Boston-London, 1994.

23 Gamst, A. and E.G. Zinn, "Cellular radio network planning," *IEEE Aerosp. Electron. Syst. Magazine*, Vol. 1, No. 1, 1985\6, pp. 8–11.

24 Akl, R.G., M.V. Hegde, M. Naraghi-Pour, and P.S. Min, "Cell placement in CDMA network," in *Proc. of the IEEE Wireless Communic. and Networking Conf.*, Vol. 2, 1999, pp. 903–907.

25 Eisenblatter, A., A. Fugenschuh, T. Koch, A. Koster et al., "Modeling feasible network configurations for UMTB," ZIB, Berlin, *Tech. Rep.* 02–16, March 2002.

26 Hurley, S., "Planning effective cellular mobile radio networks," *IEEE Trans. Veh. Technol.*, Vol. 51, No. 2, 2002, pp. 48–56.

27 Santiago, R.Ch., and V. Lyandres, "A sequential algorithm for optimal base station location in a mobile radio network," in *Proc. of 2004 IEEE 15th Int. Symp. on Personal, Indoor and Mobile Radio Communic.*, Barcelona, Spain, September 5–8, 2004.

28 Santiago, R.Ch., A. Raymond, V. Lyandres, and V.Ya. Kontorovitch, "Effective base stations location and frequency assignment in mobile radio networks," in *Proc. of 2003 IEEE Int. Symp. on Electromagn. Compatibility*. Istanbul, Turkey, May 11–16, 2003.

29 Santiago, R.Ch., E. Gigi, and V. Lyandres, "An improved heuristic algorithm for frequency assignment in non-homogeneous cellular mobile networks," in *Proc. of 2004 IEEE 60th Veh. Technol. Conf.*, Los Angeles, California, September 26–29, 2004.

30 Tutschku, K. and P. Tran-Gia, "Spatial traffic estimation and characterization for mobile communication network design," *IEEE J. Select. Areas Communic.*, Vol. 16, No. 5, 1998, pp. 804–811.

31 Hurley, S., R.M. Whitaker, and D.H. Smith, "Channel assignment in cellular networks without channel separation constraints," in *Proc. of 2000 IEEE Veh. Technol. Conf.*, Boston, MA, September 24–28, 2000.

32 Hurley, S., "Automatic base station selection and configuration in mobile networks," in *Proc. of 2000 IEEE Veh. Technol. Conf.*, Boston, MA, September 24–28, 2000.

33 Lee, J.S. and L.E. Miller, *CDMA Systems Engineering Handbook*, Artech House, Boston, 1988.

34 Wu, J.-L.C. and L.-Y. Wey, "Channel assignment for cellular mobile networks with non-uniform cells," *IEE Proc. Communic.*, Vol. 145, 1998, pp. 451–456.

35 Andersen, J.B., "Antenna arrays in mobile communications," *IEEE Antenna Propagat. Magazine*, Vol. 42, No. 2, 2000, pp. 12–16.

36 Blaunstein, N., "Wireless Communication Systems," in *Handbook of Engineering Electromagnetics*, R. Bansal, ed., Marcel Dekker, New Jersey, 2004, pp. 417–481.

37 Saunders, S.R., *Antennas and Propagation for Wireless Communication Systems*, John Wiley & Sons, New York, 1999.

38 Rappaport, T.S., *Wireless Communications: Principles and Practice*, Prentice Hall, Englewood Cliffs, New Jersey, 1996.

39 Faruque, S., *Cellular Mobile Systems Engineering*, Artech House, Boston-London, 1994.

40 Stuber, G.L., *Principles of Mobile Communication*, Kluwer Academic Publishers, Boston, MA, 1996.

41 Proakis, J.G., *Digital Communications*, McGraw-Hill, New York, 1989.

42 Andrews, L.C., R.L. Philips, and C.Y. Hopen, *Laser Beam Scintillations with Applications*, SPIE Press, New York, 2001.

43　Alouini, M.-S., Simon, M.K., and Goldsmith, A.J., "Average BER performance of single and multi carrier DS-CSMA systems over generalized fading channels," *Wiley Journal on Wireless Systems and Mobile Computing*, Vol. 1, No. 1, 2001, pp. 93–110.

44　Jung, P., Z. Zvonar, and K. Kammerlander, eds., *GSM: Evolution Towards 3rd Generation*, Kluwer Academic Publishers, Boston, MA, 1998.

45　Prasad, R., *CDMA for Wireless Personal Communications*, Artech House, Boston-London, 1996.

46　Raith, K. and J. Uddenfeldt, "Capacity of digital cellular TDMA systems," *IEEE Trans. Veh. Technol.*, Vol. 40, No. 2, 1991, pp. 323–332.

47　Gilhousen, K.S., I.M. Jacobs, R. Padovani et al., "On the capacity of cellular CDMA system," *IEEE Trans. Veh. Technol.*, Vol. 40, No. 2, 1991, pp. 303–312.

48　Sivanand, S., "On adaptive arrays in mobile communication," in *Proc. IEEE Nat. Telesystems Conf.*, Atlanta, GA, 1993, pp. 55–58.

49　Balaban, P. and J. Salz, "Dual diversity combining and equalization in digital cellular mobile radio," *IEEE Trans. Veh. Technol.*, Vol. 40, No. 2, 1991, pp. 342–354.

50　Andersen, J.B., "Array gain and capacity for known random channels with multiple element arrays at both ends," *IEEE J. Select. Areas in Communic.*, Vol. 18, No. 11, 2000, pp. 2172–2178.

51　Andersen, J.B., "Role of antennas and propagation for the wireless systems beyond 2000," *J. Wireless Personal Communic.*, Vol. 17, 2001, pp. 303–310.

52　Yarkoni, N. and N. Blaunstein, "Capacity and spectral efficiency of MIMO wireless systems in multipath urban environments with fading," *1st European Conference on Antennas and Propagation* (*EuCAP 2006*), November 6–10, Nice, France, 2006, pp. 316–321.

53　Blaunstein, N. and N. Yarkoni, "Data stream parameters prediction in land and atmospheric MIMO wireless communication links with fading," *2nd European Conference on Antennas and Propagation* (EUCAP 2007), Edinburgh, November 2007, pp. 123–126.

54　Steele, R., *Mobile Radio Communication*, IEEE Press, New York, 1992.

55　Anderson, H.R., "A ray-tracing propagation model for digital broadcast systems in urban areas," *IEEE Trans. on Broadcasting*, Vol. 39, No. 3, 1993, pp. 309–317.

56　Bertoni, H.L. *Radio Propagation for Mobile Wireless Systems*, Prentice Hall PTR, New Jersey, 2000.

57　Liang, G. and H.L. Bertoni, "A new approach to 3D ray tracing for propagation prediction in cities," *IEEE Trans. Anten. Propagat.*, Vol. 46, 1998, pp. 853–863.

58　Lebherz, M., W. Weisbeck, and K. Krank, "A versatile wave propagation model for the VHF/UHF range considering three dimensional terrain," *IEEE Trans. Antennas Propagat.*, Vol. AP-40, 1992, pp. 1121–1131.

CHAPTER TWELVE

Effects of the Troposphere on Radio Propagation

Tropospheric effects involve interactions between radio waves and the lower layer of the Earth's atmosphere, embracing altitudes from the ground surface up to several tens of kilometers above the Earth, including effects of the gases composing the air and hydrometeors such as rain, clouds, fog, pollutions, as well as various turbulent structures created by turbulent wind streams, both in vertical and horizontal directions, gradient of temperature, moisture, and pressure in layered atmosphere at the near-the-earth altitudes.

12.1. MAIN PROPAGATION EFFECTS OF THE TROPOSPHERE AS A SPHERICAL LAYERED GASEOUS CONTINUUM

12.1.1. Model of the Troposphere and Main Tropospheric Processes

Troposhere is a region of the Earth's lower atmosphere that surrounds the Earth from the ground surface up to 10–20 km above the terrain, where it continuously spreads to the stratosphere (20–50 km), and then to the thermosphere, usually called ionosphere (50–400 km). The effects of the latter on radio propagation will be presented in the next chapter. Here, we will focus on the effects of the troposphere on radiowave propagation, starting with a definition of the troposphere as a natural layered air medium consisting of different gaseous, liquid, and crystal structures.

Radio Propagation and Adaptive Antennas for Wireless Communication Networks: Terrestrial, Atmospheric, and Ionospheric, Second Edition. Nathan Blaunstein and Christos G. Christodoulou. © 2014 John Wiley & Sons, Inc. Published 2014 by John Wiley & Sons, Inc.

The physical properties of the troposphere is characterized by the following main parameters such as *temperature, T* (in Kelvin), *pressure, p* (in millibars or in millimeters of Mercury), and *density, p* (in particles per cubic meter or centimeter). All these parameters significantly change with altitude, seasonal, and latitudinal variability, and strongly depend on weather [1–9].

Content of the Troposphere. The troposphere consists different kinds of gaseous, liquid, and crystal structures, including effects of gas molecules (atoms), aerosol, cloud, fog, rain, hail, dew, rime, glaze, and snow. Except for the first two components, the others are usually referred to as *hydrometeors* in past literature [10–24]. Furthermore, due to irregular and sporadic air streams and motions, such as irregular wind motions, the chaotic structures, defined as *atmospheric turbulences*, are also present in the troposphere [25–40].

Later, we will present a brief description of the various components that make up the troposphere [1–40].

Aerosol is a system of liquid or solid particles uniformly distributed in the atmosphere. Aerosol particles play an important role in the precipitation process, providing the nuclei upon which condensation and freezing take place. The particles participate in chemical processes and influence the electrical properties of the atmosphere. Actual aerosol particles range in diameter from a few nanometers to about a few micrometers. When smaller particles are in suspension, the system begins to acquire the properties of a real aerosol structure. For larger particles, the settling rate is usually so rapid that the system cannot properly be called a real aerosol. Nevertheless, the term is commonly employed, especially in the case of fog or cloud droplets and dust particles, which can have diameters of over $100\,\mu m$. In general, aerosols composed of particles larger than about $50\,\mu m$ are unstable, unless the air turbulence is extreme, as in a severe thunderstorm (see details in References 5 and 36).

Hydrometeors are any water or ice particles that have formed in the atmosphere or at the Earth's surface as a result of condensation or sublimation. Water or ice particles blown from the ground into the atmosphere are also classified as hydrometeors. Some well-known hydrometeors are rain, fog, snow, clouds, hail, dew, rime, glaze, blowing snow, and blowing spray.

(A) *Rain* is the precipitation of liquid water drops with diameters greater than 0.5 mm. The concentration of raindrops typically spreads from 100 to $1000\,m^{-3}$. Drizzle droplets usually are more numerous. Raindrops seldom have diameters larger than 4 mm, because as they increase in size, they break up. The concentration generally decreases as diameters increase, except when the rain is heavy. It does not reduce visibility as much as drizzle. Meteorologists classify rain according to its rate of fall. The hourly rates relating to light, moderate, and heavy rain corresponds to dimensions less than 2.5 mm, between 2.8 mm and 7.6 mm, and more than 7.6 mm, respectively. Less than 250 mm and more than

1500 mm/year represent approximate extremes of rainfall for all of the continents. Rainfall intensities greater than 30 mm in 5 minutes, 150 mm in 1 hour, or 500 mm/day are quite rare, but these intensities, on occasion, have been more than double for the respective duration (see details in References 6, 11–18, and 36). Later we will discuss the effects of rain on radio propagation.

(B) *Snow* is the solid form of water that crystallizes in the atmosphere and falls to the Earth, covering permanently or temporarily about 23% of the Earth's surface. At sea level, snow falls usually at higher latitudes, that is, above latitude 35°N and below 35°S. Close to the equator, snowfall occurs exclusively in mountain regions, at elevations of 4900 m and higher. The size and shape of the crystals depend mainly on the temperature and the amount of water vapor available as they develop. In colder and drier air, the particles remain smaller and compact. Frozen precipitation has been classified into seven forms of snow crystals and three types of particles: graupel, that is, granular snow pellets, (also called soft hail), sleet, (i.e., partly frozen ice pellets), and hail, for example, hard spheres of ice (see details in References 3 and 36).

(C) *Fog* is a cloud of small water droplets near ground level and sufficiently dense to reduce horizontal visibility to less than 1000 m. The word "fog" may also refer to clouds of smoke particles, ice particles, or mixtures of these components. Under similar conditions, but with visibility greater than 1000 m, the phenomenon is termed a mist or haze, depending on whether the obscurity is caused by water drops or solid particles. Fog is formed by the condensation of water vapor on condensation nuclei that are always present in natural air. This happens as soon as the relative humidity of the air exceeds saturation by a fraction of 1%. In highly polluted air, the nuclei may grow sufficiently to cause fog at humidities of 95% or less. Three processes can increase the relative humidity of the air: (1) cooling of the air by adiabatic expansion, (2) the mixing of two humid airstreams having different temperatures, and (3) the direct cooling of the air by radiation. According to the physical processes involved in the creation of fogs, there are different kinds of fogs that are usually observed: advection, radiation, inversion, and frontal. We do not enter deeply into the subject of their creation because this is a subject of meteorology, for which readers may refer to special literature [3, 7, 24, 36]. Here, we will only analyze their influence on radiowave propagation.

(D) *Clouds* have the dimensions, shape, structure, and texture that are influenced by the kind of air movements that result in their formation and growth and by the properties of the cloud particles. In settled weather, clouds are small and well scattered. Their horizontal and vertical dimensions are only a kilometer or two. In disturbed weather, they cover a large part of the sky, and individual clouds may tower as high as 10 km or more. Clouds often cease their growth only upon reaching the stable stratosphere, producing heavy showers, hail, and thunderstorms. Growing clouds are sustained by upward air currents, which may vary in strength from a few centimeters per second to several meters per second. Considerable growth of the cloud droplets with falling speeds of only about 1 cm/s, leads to their fall through the cloud, reaching the ground as drizzle or rain. Four principal classes are recognized when clouds are classified according to the kind of air motions that produce them: (1) layer clouds formed by the widespread regular ascent of air, (2) layer clouds formed by widespread irregular stirring or turbulence, (3) cumuliform clouds formed by penetrative convection, and (4) orographic clouds formed by ascent of air over hills and mountains. The reader who is interested to delve deeper into this subject can find information in References 7, 21, 24, and 36.

(E) *Atmospheric turbulence* is a chaotic structure generated by the irregular air movements in which the wind randomly varies in speed and direction. Turbulence is important because it churns and mixes the atmosphere and causes water vapor, smoke, and other substances, as well as energy, to become distributed at all elevations. Atmospheric turbulence near the Earth's surface differs from that at higher levels. Within a few hundred meters of the surface, turbulence has a marked diurnal variation, reaching a maximum about midday. When the sky is overcast, the low-level air temperature varies much less between day and night and turbulence remains nearly constant. At altitudes of several thousand meters or more, the frictional effect of the Earth's surface topography on the wind is greatly reduced and the small-scale turbulence, which is usually observed in the lower atmosphere, is absent.

Tropospheric Radio Phenomena. From investigations carried out in References 1, 2, 4, and 6, it follows that for clear gaseous atmosphere, if hydrometeors are absent, fading phenomena of radio waves can prevent an availability of 99.999% for the paths of 5 km and more, with the fade margin of 28 dB. However, there are refraction effects observed in the troposphere, which can significantly decrease the efficiency of satellite communication links (see Fig. 12.1 according to References 25–30).

Refraction occurs as the result of propagation effects of layered quasi-homogeneous layered structures of the troposphere that cause radio waves to propagate not along straight radio paths but to curve slightly toward the ground. This phenomenon will be described later.

Moreover, the troposphere consists of a mixture of particles having a wide range of sizes and characteristics, from molecules in atmospheric gases to different kinds of hydrometeors such as raindrops, drops of snow, hail, drops of fogs, clouds, and so forth. The main processes that cause total wave loss (in decibels) are absorption and scattering, that is, $L_{tot} = L_{abs} + L_{scat}$ [25–30].

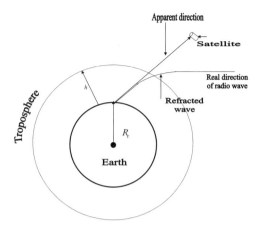

FIGURE 12.1. Displacement of the ray due to refraction.

Absorption (or *attenuation*) occurs as a result of conversion from radiowave energy to thermal energy within an attenuating particle, such as a gas molecule and different hydrometeors. We will consider this effect below for a gaseous layered atmosphere, as an air continuum, and for different kinds of hydrometeors.

Scattering occurs from the redirection of the radio waves into various directions so that only a fraction of the incident energy is transmitted onwards in the direction of the receiver [25–30]. This process is frequency dependent, since wavelengths that are long compared to the particles' size will be only weakly scattered. The main influencing mechanisms in radio links passing through the troposphere are hydrometeors, including raindrops, fog, snow, clouds, and so on. For such kinds of obstructions of radiowave energy, the scattering effects are only significant to systems operating below 10 GHz [35]. The absorption effects also rise with frequency of radio waves, although not so rapidly. We will discuss about the effect of scattering later.

To predict the effects of all such tropospheric structures on radiowave propagation through the atmosphere, we need some background knowledge about the concentration and size distribution of all kinds of structures, as well as their spatial and altitudinal distribution. We will briefly describe these questions later, considering the effects of each kind of atmospheric content separately.

12.1.2. Tropospheric Refraction

As a first step, we will consider the troposphere as a quasi-homogeneous gaseous layered medium, consisting of aerosol and molecules and atoms of gases. In other words, we consider the gaseous spherical medium around the Earth, the components of which are homogeneously distributed within the virtual layers along the height from the ground surface [32].

Refractive Index or Refractivity. The radio properties of the quasi-homogeneous layered troposphere are characterized by the refraction index n, related to the dielectric permittivity of the air (ε_r) as $n = \sqrt{\varepsilon_r}$. The refractive index n of the Earth's atmosphere is slightly greater than one, with a typical value at the Earth's surface of around 1.0003. Since the value is so close to unity, it is common to express the refractive index in N-units, usually called *refractivity* [1–3], which is the difference between the actual value of the refractive index and unit in parts per million:

$$N = (n-1) \cdot 10^6 \tag{12.1}$$

Thus, at the ground surface, the refractivity equals $N \equiv N_S \approx 315$ N-units. In a real atmosphere, refractivity N varies with gas pressure and temperature and with water vapor pressure in the atmosphere. The variations of temperature, pressure, and humidity from point to point within the troposphere cause the variations of the refractivity N, which can be calculated according to the semiempiric Debye equation [1–9]:

$$N = \frac{77.6}{T}(p_a + 4810 p_w / T), \tag{12.2}$$

where T is the absolute temperature in Kelvin (K), p_a is the atmospheric pressure in millibars (mb), and p_w is the water vapor pressure in millibars. There are seasonal and daily variations of the refractivity measured at the surface of the ground, N_0.

More important is the decrease of the refractive index with height. Usually, we can neglect the horizontal variations of N and consider the troposphere as a quasi-homogeneous spherically layered medium. If so, the dominant variation of N is vertical, with height above the Earth's surface: N reduces toward zero (n becomes close to unity) as the height is increased. The variation is approximately exponential within the first few tens of kilometers of the Earth's atmosphere; that is, this region is called the *troposphere* [1, 2]:

$$N = N_S \exp\left\{-\frac{h}{H}\right\}, \tag{12.3}$$

where h is the height above sea level, and $N_S \approx 315$ and $H = 7.35$ km are standard reference values; H is defined as the height scale of the standard atmosphere. Equation (12.3) is called the standard exponential model of the troposphere.

Tropospheric Refraction. This refractive index variation with height causes the phase velocity of radio waves to be slightly slower closer to the Earth's surface, such that the ray paths are not straight but tend to curve slightly toward the ground. In other words, the elevation angle α_1 of the initial ray at any arbitrary

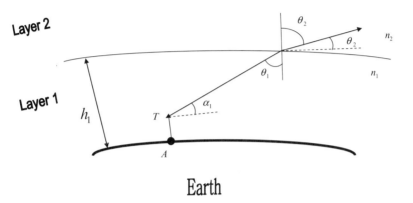

FIGURE 12.2. Refraction caused by the layered atmosphere.

point (see Fig. 12.2) is changed after refraction at angle α_2. The same situation will be at the next virtual layer of atmosphere with a different refractive index n. Finally, the ray launched from the Earth's surface propagates over the curve, whose radius of curvature (ρ) at any point, is given in terms of the rate of change of n with height [1, 2, 30]:

$$\rho = -\left(\frac{\cos\alpha_1}{n}\frac{dn}{dh}\right)^{-1}. \tag{12.4}$$

As a result, a ray passing the troposphere, instead of the apparent direction, propagates in a direction far from that to the satellite. The resulting ray curvature is illustrated in Figure 12.1. The gradient of the refractivity is given by

$$g(h) = dN/dh.$$

Usually, it is assumed [1–9] that near the Earth's surface, this gradient varies exponentially as

$$g_s(h) = -0.04\exp(-0.136h), \text{km}^{-1}. \tag{12.5}$$

Linear Approximation. According to (12.5), the gradient depends nonlinearly with height. However, as a first approximation, we can use a linear model, setting the gradient as a constant equal to its value at $h = 0$: $g = g(0)$. This occurs for small heights, when the standard atmosphere in (12.3) can be

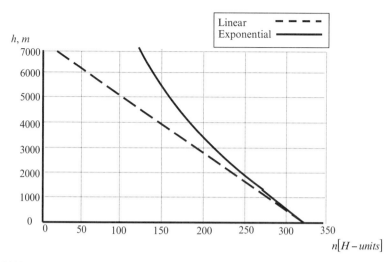

FIGURE 12.3. Linear and exponential height dependence of the refractive index.

approximated as linear, as shown in Figure 12.3, according to the following equation [1–9, 30]:

$$N \approx N_S - \frac{N_S}{H} h. \qquad (12.6)$$

The refractivity thus has nearly a constant gradient of about -43 N-units/km. If so, the curvature of the ray trajectory is constant (this follows from (12.3) for $dn/dh = \text{const.}$). A common way to take this factor into account is to introduce, instead of the actual Earth's radius, the effective Earth's radius:

$$R_{\text{eff}} = K_e R_e \qquad (12.7)$$

where $R_e = 6375\,\text{km}$, and K_e is the Earth radius factor. As was shown in References 1 and 2, the large values of the K-factor facilitate the propagation over long paths, and small values may cause obstruction fading. In order to predict such fading, the statistics of the low values of the K-factor have to be known. However, since the instantaneous behavior of the K-factor differs at various points along a given path, an effective K-factor for the path (K_e) should be considered. In general, K_e represents a spatial average, and the distribution of K_e shows less variability than that derived from point-to-point meteorological measurements. The variability decreases with increasing distance. The effective factor is given by [1, 2, 30]

$$K_e = \frac{1}{R_e \dfrac{dn}{dh} + 1} \cong \frac{R_e}{1 - R_e / \rho}. \qquad (12.8)$$

As the variation of refractive index is mostly vertical, rays launched and received with relatively high elevation angles, usually used in fixed satellite communication links, will be mostly unaffected. But for the near horizontal rays, where

$$\rho \approx -10^6/g,$$ (12.9)

we obtain

$$R_{eff} = K_e R_e,$$ (12.10)

where now the effective earth radius factor is

$$K_e = \left(1 + 10^{-6} g R_e\right)^{-1}.$$ (12.11a)

Another form of this relation reads

$$K_e = \frac{0.157}{0.157 + g}.$$ (12.11b)

For the standard atmosphere and in limits of a linear model ($g = -3.925 \cdot 10^{-2}$ 1/km), one can immediately obtain from (12.11b) $K_e = 4/3$, so the effective radius from (12.10) is about 8500 km. Although the linear model leads to an excessive ray bending at high altitudes, this is not so important in our calculations, because the critical part of the trajectory is located near the ground antenna.

Statistical Approximation. Usually, it is assumed that for the radio path with length d greater than about 20 km, the standard deviation of the effective gradient (g_e) tends to normal distribution, with the mean value g_0 as in standard atmosphere and *rms* deviation (see definitions of statistical parameters and distribution functions in Chapter 1):

$$\sigma_e \approx \frac{\sigma_0}{\sqrt{1 + d/d_0}},$$ (12.12)

where $d_0 \approx 13.5$ km for the European climate conditions. Estimations show that for a radio path of length $d = 150$ km, we have $\sigma_e \approx 0.3\sigma_0$, and for $d = 350$ km, we have $\sigma_e \approx 0.2\sigma_0$. The reasonable estimate of σ_0 is $\sigma_0 \sim 0.04$.

In Chapter 1, the Gaussian probability density function (PDF) was introduced, the cumulative distributed function (CDF) $F(x)$ of which can be presented by the error function (*erf*) in the following manner [30, 33, 34]:

$$F(x) = \frac{1}{2}\left[1 + \mathrm{erf}\left(\frac{x - m}{\sigma\sqrt{2}}\right)\right],$$ (12.13)

where the error function is defined as

$$\mathrm{erf}(x) = \frac{2}{\sqrt{\pi}} \int_0^x dt e^{-t^2}.$$

Then, the characteristic Q-function of the normal distribution is given by

$$Q = \sqrt{2}\,\mathrm{erfinv}(2t - 1), \tag{12.14}$$

where $erfinv(x)$ is the inverse error function, and t is the time availability expressed in relative units (if t is in percentage, there is a need to divide this value by 100%). Thus, for the 95% time availability we get $Q = 1.64$ (see References 30, 33, and 34) and

$$g_e \approx g_0 + 1.64\sigma_e. \tag{12.15}$$

Therefore, $g_e \approx -0.020$ ($K_e \approx 1.14$) for $d = 150\,\mathrm{km}$ and $g_e \approx -0.027$ ($K_e \approx 1.21$) for $d = 350\,\mathrm{km}$. For the 99% time availability we get $Q = 2.33$ (see References 30, 33, and 34) and

$$g_e \approx g_0 + 2.33\sigma_e. \tag{12.16}$$

That leads to a $g_e \approx -0.012$ ($K_e \approx 1.08$) for $d = 150\,\mathrm{km}$ and $g_e \approx -0.021$ ($K_e \approx 1.15$) for $d = 350\,\mathrm{km}$. We can see that for the real model of the spherical layered troposphere, the median value of K_e differs from four-thirds, which follows from the linear model of the reflectivity profile.

Approximation Based on the Geometric Optics Approach. This approximation is valid if each element of the curved radiowave trajectory in the layered atmosphere, as well as the spatial dimensions of the gradient of refractive index, is larger than the wavelength. Then, according to Reference 41, for any angle θ between the radius vector **r** of some point at the trajectory of the radio wave and the direction of the wave line at this point, defined by unit vector $\mathbf{l}_0$ (see Fig. 12.4), we can present the equation of the wave line in the following form:

$$r \cdot n(r) \cdot \sin\theta = const. \tag{12.17}$$

Expression (12.17) is the main equation for analysis of the wave trajectory refraction in the spherical–symmetric media.

If, for example, the receiver is located at point A at the ground surface, as shown in Figure 12.4, and the transmitter is at point B outside the atmospheric layer, then from (12.17), based on the law of conservation of energy along the wave trajectory [41], we get

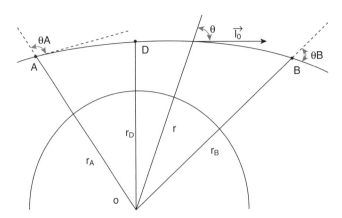

FIGURE 12.4. Geometry of the refraction problem.

$$R_c \cdot n_A \cdot \sin \theta_A = r \cdot n(r) \cdot \sin \theta = n_B \cdot \sin \theta_B. \qquad (12.18)$$

All parameters presented in Equation (12.18) are clearly seen from Figure 12.4. From Equation (12.18), it is easy to obtain the angle θ that describes the effect of refraction for any point at the wave (e.g., ray) trajectory, that is,

$$\tan \theta = \frac{\sin \theta}{\cos \theta} = \frac{\sin \theta_A}{\left[\left(\dfrac{rn}{R_c n_A} \right)^2 - \sin^2 \theta_A \right]^{1/2}} . \qquad (12.19)$$

We are interested in estimating the deviations of the refractive angle ξ along the ray trajectory that we can define by using a more complicated geometrical model [41] shown in Figure 12.5a. The angle ξ is the angle between the true direction from the receiver located at point A, and the tangent to the wave line at point B of the transmitter location. As can be seen from Figure 12.5a, the value of ξ is equal to the angular difference $\theta_0 - \theta_b$, that is, ξ shows how much the zenith angle of the beam transmission line θ_b differs from the actual zenith angle of direction to the transmitter antenna defined by θ_0 (see Fig. 12.5a). Next, we denote the refractive index within the tropospheric layer as $n(h_1)$, and the refractive index in an immediate proximity above this layer as $n(h_2)$. Let us apply the refraction law in the spherically symmetric medium, that is, Equation (12.16) rewritten according to a new geometry shown in Figure 12.5a:

$$n(h_1)(R_e + h_1)\sin \theta_1 = n(h_2)(R_e + h_2)\sin \theta_2, \qquad (12.20)$$

where, as earlier, R_e is the radius of the Earth, $\theta_{1,2}$ are angles between the direction of the ray transmission line and the radius vector $\mathbf{r}_{1,2}$ (see Fig. 12.5a).

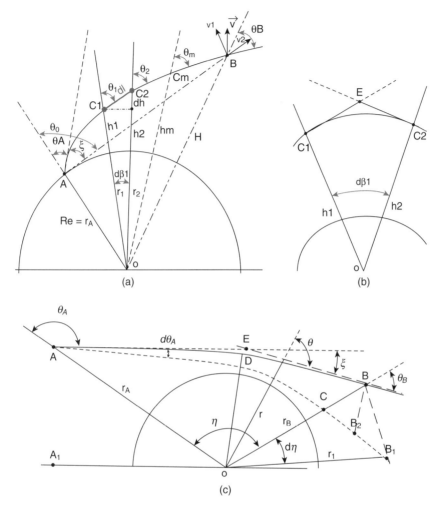

FIGURE 12.5. (a) Geometry of the problem to find the refractive angle deviation ξ according to [41]. (b) Geometry of the local points, C_1 and C_2, at the ray trajectory to obtain the refraction parameter deviations along the radio path. (c) Geometry of the problem to find the angle η between vectors $\mathbf{r}_A$ and $\mathbf{r}_B$.

The expression (12.20) is true at any tropospheric altitude h. Therefore, from (12.20) we get

$$n(h)(R_e + h)\sin\theta = n_0(R_e)\sin\theta_b, \tag{12.21}$$

where n_0 and θ_b are the refraction index and the zenith angle of the ray transmission line at the Earth's surface, that is, at point B on Figure 12.5a. Let us now derive the expression for the ray curvature radius in the spherically symmetric medium, that is, find $R_0 = dl/d\xi$, where $dl = C_1C_2$ is an element of the beam line length, and $d\xi$ is a change in the refraction angle at points C_1 and C_2. From Figure 12.5b, it follows

$$d\xi = d\beta_1 + d\theta \tag{12.22}$$

$$dh = dl \cdot \cos\theta \tag{12.23}$$

$$d\beta_1 = (R_c + h)^{-1}\tan\phi \cdot dh, \tag{12.24}$$

where $d\beta_1$ is the angle between vectors $\mathbf{r}_1$ and $\mathbf{r}_2$, $d\theta = \theta_2 - \theta_1$, $dh = h_2 - h_1$. Equation (12.22), Equation (12.23), and Equation (12.24) follow from the quadrangle OC_1EC_2 (see Fig. 12.5b). Taking into account these equations, we get the following expression for ray curvature radius:

$$\rho_0 = \frac{R_e + h}{\sin\theta + (R_e + h)\dfrac{d\theta}{dh}\cos\theta}, \tag{12.25}$$

which is more complicated with respect to that obtained earlier by using the classical approach (see Eq. 12.4). On the one hand, by using the relations between the refractive index and all other geometrical parameters presented earlier, we finally get

$$\rho_0 = -\frac{n}{\dfrac{dn}{dh}\sin\theta}, \tag{12.26}$$

where n, θ, and ρ_0 depend on altitude h. On the other hand, according to Reference 41, the angle θ is determined by an integral:

$$\xi = \int_0^H \left(\frac{d\beta_1}{dh} + \frac{d\theta}{dh}\right)dh, \tag{12.27}$$

where H is an altitude of the radiowave source. Making use of (12.22), (12.23), and (12.24), and differentiating (12.22) to find $d\theta/dh$, we can rewrite (12.27) as

$$\xi = -\int_0^H \left(\frac{1}{n}\frac{dn}{dh}tg\,\theta \cdot dh\right), \tag{12.28}$$

or expressing $\tan\theta$ through $\sin\theta_0$, we finally obtain an equation for deviations of the angle of refraction in the spherically symmetric layered troposphere with the upper boundary of the layer, $H < 40\,\text{km}$

$$\xi_t = -\sin\theta_b \int_0^\infty \left[\left(\frac{n}{n_0}\right)^2\left(1 + \frac{h}{R_e}\right)^2 - \sin^2\theta_c\right]^{-1/2}\frac{1}{n}\frac{dn}{dh}dh. \tag{12.29}$$

Analyzing (12.29), let us take into account that h/R_e is much less than unity, and the refraction index differs slightly from unit and decreases with an increase in altitude, according to (12.4). Let us transform the subintegral function in the denominator of the expression (12.29) by neglecting the terms h/R_e and N^2, which are much less than the unity. Finally, we get the following equation:

$$n\left[\left(\frac{n}{n_0}\right)^2\left(1+\frac{h}{R_c}\right)^2 - \sin^2\theta_b\right]^{1/2} = \left[\cos\theta_b + \left(\frac{2h}{R_c} + 3N - N_0\right)\right]^{1/2}. \tag{12.30}$$

A further reduction of (12.30) is possible at $\theta_b < 80°$, when the terms in the round brackets are significantly less than $\cos^2\theta$. In this case, the right side of expression (12.30) equals $\cos\theta_b$. Then, from (12.30) we obtain the simple equation for the refraction angle deviations in the troposphere as

$$\xi_t = N_0\tan\theta_b \approx N_0\tan\theta_0. \tag{12.31}$$

In this expression, $\theta_b = \theta_0 - \xi = \theta_0$ is taken into account. A more detailed analysis [41] showed that the relationship (12.31) is true only for the zenith angles $\theta_0 < 80°$. It is also evident that the refraction angle deviations depend on a shape of a specific profile of the refractivity $N(h)$. That is why, the difference of $N(h)$ in the actual troposphere with the exponential law (12.3) does not manifest itself on the value of angle ξ_t, if $\theta_0 < 80°$. With this, the refraction angle is determined only by the near-surface value of N_0 and the zenith angle θ_0. In a small sector of angles θ_0 from $80°$ to $90°$, ξ depends on a type of the function $N(h)$. In Table 12.1, ξ_t values are presented for different angles θ_0 and for average meteorological conditions, computed for $N_0 = 3.29\cdot10^{-4}$. As can be

TABLE 12.1. Typical Data of the Reduced Refraction Coefficient N at Different Altitudes h and Refraction Angle Deviations, ξ_t, for Different Values of the Zenith Angle θ_0

h (km)	$N \times 10^6$	θ_0 (°)	ξ_t''(s)
0	318	0	0
2	239	10	11.4
4	190	20	23.6
6	152	30	37.5
8	121	40	54.6
10	94	50	77.5
12	76	60	112
14	59	65	139
16	45	70	177
18	33	74	223
20	23	80	354
22	15	84	561

seen from the data presented in Table 12.1, the angle ξ_t depends on an altitude of a radiowave source, H. For this reason, the refraction angle is given for the source located at altitude $H = 200$ km. The results of calculations of the troposphere refraction angle deviations were compared with the experimental data taken from Reference 41. Thus, for $N_0 = 3.29 \cdot 10^{-4}$ and for $\theta_0 \approx 82°$, the measured refractive angle deviation equals $\xi_t = 470''$, whereas from (12.31) we get $\xi_t = 480''$. Hence, the simple Equation (12.31) gives satisfactory fit of experimental data even at large zenith angles θ_0.

12.1.3. Changes of Radio Wave Characteristics caused by Refraction

We should expect some significant effects of the refraction phenomenon on the radiowave propagation which we will consider later separately.

Changes of the Radiowave Path in the Atmosphere. Let us examine the effect of the troposphere on an apparent increase in range. As shown in Figure 12.4 or Figure 12.5, there is an increase of radiowave path, from direct to curved, caused by refraction. With the relationship between the reduced refraction factor and altitude h taken as an exponential relationship [1, 2]

$$N = N_0 \exp\{-b_1 h\} \tag{12.32}$$

with, according to References 1 and 2,

$$b_1 = -\frac{1}{10} \ln\left(\frac{9.2 \cdot 10^{-5}}{N_0}\right), \tag{12.33}$$

and by combining with (12.31), we obtain

$$\Delta L_t = N_0 b_1^{-1} \cos^{-1} \theta_0. \tag{12.34}$$

From Equation (12.34), it follows that $\Delta L_t = N_0 b_1^{-1}$ for the vertical direction. Also, according to (12.33), the parameter b_1 is expressed through N_0, that is why ΔL_t is determined by the near-the-ground value of the reduced factor of radio waves refraction (N_0) and the zenith angle θ_0. The correction on radiowave lag in the troposphere for midlatitudes at $\theta_0 = 0$ and for $b_1 = 0.126$/km equals 2.3–2.4 m in winter, and 2.4–2.5 m in summer. At larger zenith angles, the ΔL_t value increases; for example, at $\theta_0 = 80°$, ΔL_t reaches the value of 16–21 m. We should note that Equation (12.37) is not true for $\theta_0 << 80°$. When determining ΔL_t based on the near-the-ground value of pressure, P_a, and the integral moisture, W_a, it is possible to use the empirical equation [1, 2]

$$\Delta L_t = 0.228 \cdot P_a + 6.3 \cdot W_a. \tag{12.35}$$

According to (12.35), ΔL_t can be represented as a sum of two components: one component corresponding to an atmosphere with no humidity and one related to atmosphere with humidity. The first component gives the main contribution to ΔL_t; for $\theta_0 = 0$ it is equal to 2.25–2.35 m. This component can be determined based on the "near-the-ground" values of pressure and temperature, with an error of about 4 cm. The main variation of ΔL_t is caused by variability in atmospheric humidity; for midlatitudes, in winter, at $\theta_0 = 0$, it is equal to 3–5 cm, and in summer, it varies in the range 11–23 cm. With a satisfactory accuracy, it can be taken that this component is proportional to the integral moisture in the atmosphere, that is, to W_a.

Attenuation of Radio Waves caused by Refraction. According to Reference 41, the weakness of radio waves after refraction, can be presented in a general form (based on the geometry shown in Fig. 12.6 below) as

$$X = \frac{\left(r_A^2 + r_B^2 - 2r_A r_B \cos\eta\right)\cdot n_B \cdot p}{r_A \cdot r_B \cdot \sin\eta\left[\left(r_A^2 - p^2\right)\right]^{1/2} + \left(r_B^2 - p^2\right)^{1/2} - \dfrac{d\xi}{dp}\left(r_A^2 - p^2\right)^{1/2}\left(r_B^2 - p^2\right)^{1/2}}.$$

(12.36a)

Here, p is the "straight distance to the target," η is the angle between vectors $\mathbf{r}_A$ and $\mathbf{r}_B$; all other geometrical parameters are shown in Figure 12.5c.

We can simplify this general equation for the case of $n_A = n_B \approx 1$. In this case, we get for the practical applications the following equation:

$$X = \frac{p\cdot\left(r_A^2 + r_B^2 - 2r_A r_B \cos\eta\right)}{r_A \cdot r_B \cdot \sin\eta\left[\left(r_A^2 - p^2\right)\right]^{1/2} + \left(r_B^2 - p^2\right)^{1/2} - \dfrac{d\xi}{dp}\left(r_A^2 - p^2\right)^{1/2}\left(r_B^2 - p^2\right)^{1/2}}.$$

(12.36b)

Derivations of Equation (12.36b) for the angle $\eta = 30°, 45°, 60°$ give for the radiowave path loss $L = -10\log X = -8.1\,\text{dB}, -5.72\,\text{dB}, -3.34\,\text{dB}$, respectively.

12.1.4. Wave Attenuation by Atmospheric Gaseous Structures

Let us consider the wave attenuation caused by atmospheric gas. Then, in the next sections, we will consider all effects of hydrometeors, as the most important factor in determining communication system reliability. The molecular absorption is due primarily to atmospheric water vapor and oxygen. Although for frequencies around 1–20 GHz, this kind of attenuation is not large, it takes place as a permanent factor. The absorption in the atmosphere over a path length r is given by [4, 5, 30]

$$A = \int_0^r dr\gamma(r), \text{dB}$$

(12.37)

where $\gamma(r)$ is the specific attenuation consisting of two components:

$$\gamma(r) = \gamma_o(r) + \gamma_w(r), [\text{dB/km}] \tag{12.38}$$

where $\gamma_o(r)$ and $\gamma_w(r)$ are the contributions of oxygen and water vapor, respectively.

Using these equations, one can find that the attenuation due to water vapor dominates, and for typical European or North American summer weather conditions, the specific attenuation does not exceed 0.02 dB/km at sea level. This corresponds to the maximal attenuation of 7 dB for a horizontal path length of 350 km. Under summer conditions, the absorption due to oxygen does not exceed typically $8 \cdot 10^{-3}$ dB/km, which corresponds to 2.8 dB for the maximal distance. In the winter, the oxygen contribution to the specific attenuation does not exceed 10^{-2} dB/km. The total attenuation at sea level due to atmospheric gases can be estimated as 0.025 dB/km. However, for slant paths, the total attenuation does not exceed the value of 1 dB, but for the 99% level of probability, it may be estimated as 2 dB.

As was mentioned earlier, gaseous molecules in the atmosphere may absorb energy from radio waves passing through them, thereby causing attenuation. Thus, nonpolar molecules, such as oxygen (O_2) may also absorb wave energy due to the existence of magnetic moments. Moreover, the increase of absorption is observed with an increase of wave frequency [5, 9, 30]. But here, several resonance peaks of absorption, each corresponding to different modes of molecule vibration, the lateral, the longitudinal, and so forth, are occurring. The main resonance peaks of H_2O are around 22.3, 183.3, and 323.8 GHz, and for O_2 are around 60 GHz, covering a complex set of closely spaced peaks that prevent the use of the band 57–64 GHz for practical satellite communication (see Fig. 12.6). The specific attenuation in decibels per kilometer for water vapor (γ_w) and for oxygen (γ_o) is given in Figure 12.6 according to References 5 and 9 for a standard set of atmospheric conditions. The total atmospheric attenuation L_a for a particular path is then found by integrating the total specific attenuation over the total path length r_T in the atmosphere [3–5, 9, 30]:

$$L_a = \int_0^{r_T} \gamma_a(l)dl = \int_0^{r_T} [\gamma_w(l) + \gamma_o(l)]dl \, [\text{dB}]. \tag{12.39}$$

This integration calculated for the total zenith ($\theta = 90°$) attenuation carried out in References 5 and 9 is presented in Figure 12.7 by assuming an exponential decrease in gas density with height. The attenuation for an inclined path with an elevation angle $\theta > 10°$ can then be found from the zenith attenuation L_z as [5, 9, 30]

$$L_a = \frac{L_z}{\sin \theta}. \tag{12.40}$$

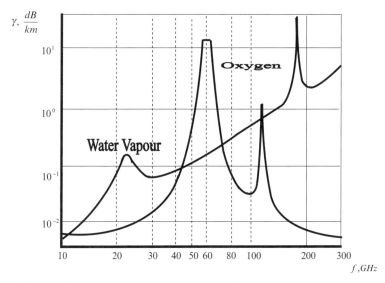

FIGURE 12.6. Attenuation versus frequency for water vapor and oxydent.

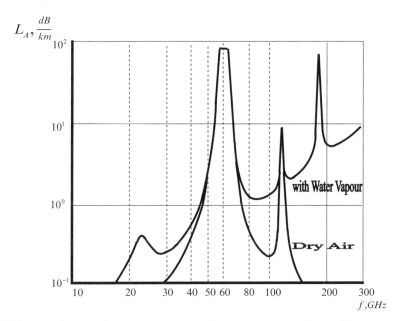

FIGURE 12.7. Attenuation versus frequency for gaseous atmosphere with water vapor and dry air.

We must note that atmospheric attenuation results in an effective upper frequency limit for mobile-satellite communications.

12.1.5. Scattering in the Troposphere by Gaseous Structures

Pure scattering occurs if there is no absorption of the radiation in the process, and hence, no loss of energy but only a redistribution of it [5, 9, 30]. Most of the scattering encountered in the atmosphere is essentially pure and is discussed in this section.

The attenuation due to scattering of the radio wave depends upon the pattern (or main lobe) of the receiving antenna. If the antenna pattern is very large, the wave field arriving at the receiver consists of only a few dozen scattered components and mainly of the direct radiated component.

Theoretically, scattering can be treated using three separate approaches depending on the wavelength and the size of the particles causing the scattering. These approaches are Rayleigh scattering, Mie scattering, and nonselective scattering.

Rayleigh Scattering applies when the radiation wavelength is *smaller* than the particle size. The volume scattering coefficient for Rayleigh scattering can be expressed as [25, 27]

$$\sigma = \frac{\left(4\pi^2 N V^2/\lambda^4\right)\left(n^2 - n_0^2\right)^2}{\left(n^2 + 2n_0^2\right)^2}, \tag{12.41}$$

where N is a number of particles per unit volume, (cm^3); V is the volume of scattering particles, (cm^3); λ is a wavelength of radiation, (cm); n_0 is the refractive index of the atmosphere in which molecules (atoms) of gases are suspended as particles; and n is a refractive index of scattering particles. For spherical water droplets in air, (12.41) becomes

$$\sigma = 0.827 \frac{N \sigma_S^{\,3}}{\lambda^4}, \tag{12.42}$$

where σ_S is the cross-sectional area of the scattering droplet. The expression (12.42) must be integrated over the range of λ and σ_S encountered in any given circumstance. As long as the particle diameter $2\sqrt{\sigma_S/\pi}$ is very small compared to λ, the same scattering can be experienced from a large number of small particles or a small number of large particles, assuming the product $N\sigma_S^3$ is the same.

At conditions of standard temperature and pressure, the scattering coefficient is $\sigma_\lambda = 1.07 \times 10^{-3}\,\lambda^{-4.05}$/km ($\lambda$ in micrometers), σ_λ is the scattering coefficient for wavelength, (cm).

Mie Scattering is applicable when the particle size is *comparable* to the radiation wavelength. The Mie scattering area coefficient is defined as the ratio of the incident wave front that is affected by the particle to the cross-sectional area of the particle itself. The scattering coefficient σ can be obtained from References 25 and 26:

$$\sigma = NK\pi a^2, \tag{12.43}$$

where the value of K rises from 0 to nearly 4 and asymptotically approaches the value 2 for large droplets. For the almost universal condition, in which there is a continuous size distribution in the particles, we have from Reference 25

$$\sigma_\lambda = \pi \int_{a_1}^{a_2} N(a) K(a, n) a^2 d, \tag{12.44}$$

where $N(a)$ is a number of particles per cubic centimeter in the interval da, (cm³); $K(a,n)$ is the scattering area coefficient; a is the radius of spherical particle, (cm); and n is an index of refraction of particle. Many authors present a detailed treatment of scattering theory [25–27] for a wide variety of particle composition, size, and shape. The Mie scattering area coefficient is given in References 25–27.

12.1.6. Propagation Clearance

The maximal distance r_0 of line-of-sight (LOS) propagation in nonrefractive atmosphere and spherical earth surface is given by [31]

$$r_0 = \sqrt{2R_{\mathrm{eff}}} \left(\sqrt{h_1} + \sqrt{h_2} \right) = 3.57 \left(\sqrt{h_1} + \sqrt{h_2} \right), \tag{12.45}$$

where h_1 and h_2 are the heights of the antennas, (m). When refraction is taken into account, we have [31]

$$r_0 = 3.57\sqrt{K_e} \left(\sqrt{h_1} + \sqrt{h_2} \right) [\mathrm{km}]. \tag{12.46}$$

In practice, the term associated with the height of the ground-based terminal antenna ($h_1 < 10\,\mathrm{m}$) can be neglected with respect to the height h_2 of the air vehicle antenna. Then, the latter equation may be easily inverted to obtain the minimal altitude of the object (air vehicle antenna), which is visible for a given distance d:

$$h_{2\min} = 0.0785 d^2 / K_e \ [\mathrm{m}]. \tag{12.47}$$

Although this dependence looks like a parabolic function, its behavior differs from pure parabolic because of the presence of K_e, which is the function of the distance.

Let us now present some examples on how refraction affects the range of direct visibility (LOS conditions) between two antennas within the tropospheric radio link. Thus, for the vehicle antenna at height of $h_2 = 2\,km$, we have that $r_0 = 178\,km$ for 95% availability, and $r_0 = 173\,km$ for 99% level. Similarly, for $h_2 = 6\,km$, we have $r_0 = 312\,km$ (availability is 95%) and 304 km (availability is 99%). On the contrary, knowledge of the range between antennas allows us to obtain the minimal height of air vehicle, from which LOS conditions are valid. Thus, the range of 350 km will be covered, according to (12.11), only for heights h_2 larger than 8 km.

Tropospheric radio paths are classified as *open* (corresponds to the same LOS conditions as in terrestrial links described in Chapter 6), *semiopen*, and *closed* (corresponds to the same non-line-of-sight [NLOS] conditions in terrestrial links). As was shown in Chapter 4, wave propagation takes place within the first Fresnel zone (ellipsoid) around the ray connecting the terminal/vehicle antennas. We will state this concept also for tropospheric radio links. Thus, the radius of the first Fresnel zone, at a point between the transmitter and the receiver antennas, is determined by the following equation (we use notations introduced in Chapter 5):

$$\ell_F(d_1, d_2) = \sqrt{\lambda d_1 d_2/(d_1 + d_2)}, \tag{12.48}$$

where d_1 and d_2 are the distances to the antennas at the point where the ellipsoid radius is calculated. The maximum value is achieved in the middle of the path ($d_1 = d_2 = L/2$) and is equal to the first Fresnel zone radius (see Fig. 12.8):

$$\ell_F = \sqrt{\lambda L/2}. \tag{12.49}$$

For example, for a frequency of 15 GHz and a maximal distance of 150 km, $\ell_F \approx 39\,m$, and for a frequency of 13 GHz and a distance of 350 km, $\ell_F \approx 64\,m$.

Using the Fresnel zone concept, we can determine all kinds of tropospheric radio links. Thus, we will state that for $f = 15\,GHz$ and a radio path of $d_{max} = 150\,km$, $\ell_F \approx 39\,m$, and for $f = 13\,GHz$ and a radio path of $d_{max} = 350\,km$, $\ell_F \approx 64\,m$.

In open (or within the horizon) paths, there are no obstacles located between the antennas (see Fig. 12.8). In this case, the propagation of the wave is similar to that in free space (see Chapter 4), taking into account only the attenuation due to atmospheric gases and hydrometeors.

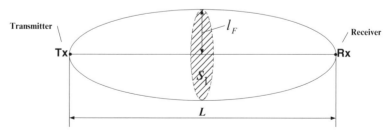

FIGURE 12.8. The line-of-sight (LOS) conditions; S_1 is the area of cross section of Fresnel's zone with radius l_F.

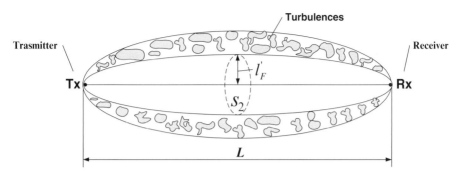

FIGURE 12.9. The non-line-of-sight (NLOS) conditions; S_2 is the area of cross section of the part of the Fresnel's zone with radius l'_F.

For semiopen (near the radio horizon) paths, the obstacles cover a part of the ellipsoidal cross-section and the effects of the obstacles can be important (see Fig. 12.9). However, the size of the first Fresnel zone is very small compared to the variability of the air vehicle antenna height, and this intermediate case is of much less importance.

For closed paths, including the hilly or mountainous terrain described in Chapter 4, the wave attenuation, due to diffraction from such obstructions and due to the effects in the troposphere, may exceed 300 to 350 dB, and the propagation is possible only by using the so-called troposcatter mechanism (see Section 12.3).

12.2. EFFECTS OF HYDROMETEORS ON RADIO PROPAGATION IN THE TROPOSPHERE

12.2.1. Effects of Rain

The attenuation of radio waves caused by rain increases with the number of raindrops along the radio path, the size of the drops, and the length of the path through the rain.

Statistical–Analytical Models. If such parameters of rain, such as the density and size of the drops are constant, then according to Reference 30, the signal power P_r at the receiver decreases exponentially with radio path r, through the rain, with the parameter of power attenuation in e^{-1} times, α, that is,

$$P_r = P_r(0)\exp\{-\alpha r\}. \tag{12.50}$$

Expressing (12.50) logarithmic scale gives

$$L = 10\log\frac{P_t}{P_r} = 4.343\alpha r. \tag{12.51}$$

Another way to estimate the total loss via the specific attenuation in decibels per meter was shown by Saunders in Reference 30. He defined this factor as

$$\gamma = \frac{L}{r} = 4.343\alpha, \tag{12.52}$$

where now the power attenuation factor α can be expressed through the integral effects of the one-dimensional (1-D) diameter D of the drops, defined by $N(D)$, and the effective cross-section of the frequency-dependent signal power attenuation by rain drops, $C(D)$ (dB/m), as

$$\alpha = \int_{D=0}^{\infty} N(D) \cdot C(D) dD. \tag{12.53}$$

As was mentioned in References 4, 6, and 30, in real tropospheric situations, the drop size distribution $N(D)$ is not constant and one must account for the range dependence of the specific attenuation, that is, the range dependence, $\gamma = (r)$, and integrate it over the whole radio path length r_R to find the total path loss:

$$L = \int_{0}^{r_R} \gamma(r) dr. \tag{12.54}$$

To resolve Equation (12.57), a special mathematical procedure was proposed in Reference 11 that accounts for the drop size distribution. This procedure yields an expression for $N(D)$ as

$$N(D) = N_0 \exp\left\{-\frac{D}{D_m}\right\}, \tag{12.55}$$

where $N_0 = 8\cdot10^3\,\mathrm{m}^{-2}\,\mathrm{mm}^{-1}$ is a constant parameter [11], and D_m is the parameter that depends on the rainfall rate R, measured above the ground surface in millimeters per hour, as

$$D_m = 0.122 \cdot R^{0.21}\mathrm{mm}. \tag{12.56}$$

As for the attenuation cross-section $C(D)$ from (12.53), it can be found using the so-called Rayleigh approximation that is valid for lower frequencies, when the average drop size is small compared to the radio wavelength. In this case, only absorption inside the drop occurs and the Rayleigh approximation is valid, giving a very simple expression for $C(D)$, that is,

$$C(D) \propto \frac{D^3}{\lambda}. \tag{12.57}$$

Attenuation caused by rain increases more slowly with frequencies approaching a constant value known as the *optical limit*. Near this limit, scattering forms a significant part of attenuation that can be described using the *Mie* scattering theory discussed earlier.

In practical situations, an empirical model is usually used, where $\gamma(r)$ is assumed to depend only on rainfall rate R and wave frequency. Then according to References 4, 6, 8, and 9, we can obtain

$$\gamma(f, R) = a(f)R^{b(f)}, \tag{12.58}$$

where γ has units dB/km; $a(f)$ and $b(f)$ depend on frequency (GHz). For 15–70 GHz frequency band, the attenuation coefficients, $a(f)$ and $b(f)$, can be approximated by [6, 9]

$$a(f) = 10^{1.203 \log(f) - 2.290}$$
$$b(f) = 1.703 - 0.493 \log(f). \tag{12.59}$$

There is another frequency dependence of the parameters a and b, proposed in Reference 8 for practical applications in satellite communications, where usually, the frequency band is divided into several frequency bands. For the subband of 8.5–25 GHz we have

$$a(f) = 4.21 \cdot 10^{-5} \cdot f^{2.42}$$
$$b(f) = 1.41 \cdot f^{-0.0779} \tag{12.60a}$$

for the subband of 25–54 GHz

$$a(f) = 4.21 \cdot 10^{-5} \cdot f^{2.42}$$
$$b(f) = 2.63 \cdot f^{-0.272} \tag{12.60b}$$

and for the subband of 54–100 GHz

$$a(f) = 4.09 \cdot 10^{-2} \cdot f^{0.699}$$
$$b(f) = 2.63 \cdot f^{-0.272}. \tag{12.60c}$$

In Reference 6 and 9, it was shown that for a ground vehicle antenna elevation angle θ smaller than $90°$, it is necessary to account for the variation in the rain in the horizontal direction. This allows us to focus on the finite size of rain clouds, that is, on the areas called the *rain areas* (or *cells*) (see Fig. 12.10). In this case of finite rain sizes, the path length is reduced by using a reduction factor s. If so, the rain attenuation is [6, 15]

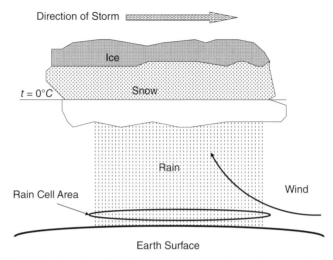

FIGURE 12.10. A rain cell with respect to snow, ice, and storm phenomena.

$$L = \gamma s r_R = a(f) R^{b(f)} s r_R. \tag{12.61}$$

Also, rain varies in time over various scales: *seasonal, annual,* and *diurnal.* All of these temporal variations are usually accounted by using (12.61) to predict the rain attenuation ($L_{0.01}$) that must not exceed 0.01% of the time during an average year.

Now, using the semiempirical approach presented in References 6, 11, 15, and 30, we can finally obtain the total path loss caused by rain as

$$L_{total} = L_{FS} + L, \tag{12.62}$$

where L is defined by Equations (12.61) and (12.62), and L_{FS} is a path loss in free space (see Chapter 4).

The field of rainfall rate is inhomogeneous in space and time. Rain observations by weather radar show small intervals of higher rain rate imbedded in longer periods of lighter rain. Also, such observations show small area of higher rain rate imbedded in large regions of lighter rain [12–18]. The geometrical characteristics of rain cells depend on the rain intensity and climate conditions, seasonal, annual, and diurnal, which are related to the coordinates of the region.

The cell diameter appears to have an exponential probability distribution of the form [12–18]

$$P(D) = \exp(-D/D_0), \tag{12.63}$$

where D_0 is the mean diameter of the cell and is a function of the peak rainfall rate R_{peak}. For Europe and in the United States, the mean diameter D_0 decreases

TABLE 12.2. Rain Rate R for Different Rain Cells

R, (mm/h)	100	50	25	20	10	5
Cell scale (km)	3	4	6	7	10	20

slightly with increasing R_{peak} when $R_{peak} < 100$ mm/h. This relationship appears to obey a power law:

$$D_0 = aR_{peak}^{-b}, R_{peak} > 10\ R_e = 6375\ \text{mm/h}. \tag{12.64}$$

Values for the coefficient a ranging from 2 to 4, and the coefficient b from 0.08 to 0.25 have been reported. An example of the correlation between rainfall rate and the typical cell scale is given in Table 12.2, following the results obtained in Reference 30 according to the observations made in References 12–18.

The most difficult parameter in attenuation modeling is the *spatial distribution* of rain. As the precipitations are characterized by variations in both the horizontal and vertical directions, a correction factor is required in the modeling path lengths of rain attenuation. Here, we use a simple prediction method on the basis of the so-called effective path length to take into account the nonuniform profile of rain intensity along a given path. The effective path length L_r is the length of a hypothetical path obtained from radio data, dividing the total attenuation by the specific attenuation exceeded for the same percentage of time. The transmission loss due to attenuation by rain is then given by

$$A_r = \gamma_r L_r. \tag{12.65}$$

The effective path length L_r can be estimated, according to the empirical model [10, 11] as

$$L_r = \frac{L_s}{1 + 0.0286 L_h R^{0.15}}, \tag{12.66}$$

where, neglecting the ray bending

$$L_h = d, L_s = \sqrt{d^2 + h_2^2}, h_2 < h_r \tag{12.67a}$$

$$L_h = d\,h_r/h_2, L_s = h_r\sqrt{1 + d^2/h_2^2}, h_2 > h_r \tag{12.67b}$$

d is the horizontal component of the distance between antennas, h_r is the average rain height (approximately 3 km for European weather conditions), and h_2 is the height of the air vehicle antenna. An example of the corresponding calculations of rain attenuation (in dB) for the frequency of 15 GHz and

TABLE 12.3. Effective Path Length L_r and the Rain Attenuation F_r for Time Availability of 99.0% and 99.9% for $f = 15\,$GHz and the Antenna Height $h_2 = 2\,$km

Distance (km)	50	100	150
L_r, (km) (99.0% level)	22.6	26.6	28.2
L_r, (km) (99.9% level)	18.5	21.1	22.1
F_r, (dB) (99.0% level)	1.7	2.0	2.1
F_r, (dB) (99.9% level)	12.8	14.6	15.3

TABLE 12.4. Effective Path Length L_r and the Rain Attenuation F_r for Time Availability of 99.0% and 99.9% for $f = 13\,$GHz and the Antenna Height $h_2 = 6\,$km

Distance (km)	100	200	300
L_r, (km) (99.0% level)	19.9	24.9	27.1
L_r, (km) (99.9% level)	16.8	20.1	21.8
F_r, (dB) (99.0% level)	1.3	1.6	1.8
F_r, (dB) (99.9% level)	8.5	9.8	10.6

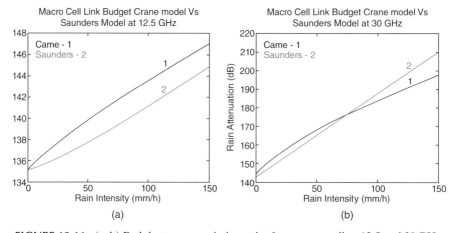

FIGURE 12.11. (a, b) Path loss versus rain intensity for a macrocell at 12.5 and 30 GHz.

antenna height of $h_2 = 2\,$km are presented in Table 12.3, and for the frequency of 13 GHz and antenna height of $h_2 = 6\,$km in Table 12.4.

Comparison of path loss, caused by specific rain attenuation versus the rain intensity, (in mm/h), obtained from Saunders' statistical model [30] and the Crane empirical model [14] at frequencies of 12.5 and 30 GHz, are shown in Figure 12.11a,b for a macrocell area ($r_R \geq 10\,km$), and Figure 12.12a,b for microcell areas ($r_R \leq 2$–$3\,km$). From the comparison shown in Figure 12.11a and Figure 12.12a, it follows that there is a good match between the Saunders

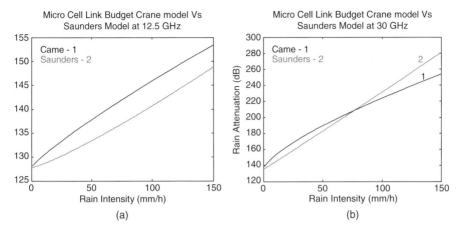

FIGURE 12.12. (a, b) Path loss versus rain intensity for a macrocell at 12.5 and 30 GHz.

and the Crane model (with the average deviation of 2 dB for macrocell areas and an average deviation of 5 dB for microcell areas) for all rain intensities at 12.5 GHz. The same tendency appears in the results shown in Figure 12.11b and Figure 12.12b, with the average deviation of 4 dB for macrocell areas and an average deviation of 7 dB for microcell areas for rain intensity in the range of 10–100 mm/h. At the same time, there is no good match between the Saunders model and the Crane model, from which we got the average deviation of 10 dB for macrocell areas and the average deviation of 20 dB for microcell areas, for rain intensity in the range of 100–150 mm/h. As the rain intensity increases, this difference becomes even more pronounced.

There is a significant difference in rain attenuation between macro and microcells. The path loss caused by rain attenuation reaches 270 dB, at 30 GHz, for microcell areas versus the 200 dB attenuation that occurs in the macrocell areas. All these results are very important for designers of land-satellite link performance, because in the radio path through a microcell area containing intensive rain, there is much more signal attenuation observed than in radio paths through macrocell areas, where the areas of intensive rain cover only few percentages of the total radio path.

12.2.2. Effects of Clouds and Fog

In cloud models, which will be described later, a distinction between *cloud cover* and *sky cover* must be explained. Sky cover is an observer's view of the cover of the sky dome, whereas cloud cover can be used to describe areas that are smaller or larger than the floor space of the sky dome.

Attenuation and Path Loss. It follows from numerous observations that in clouds and fog, the drops are always smaller that 0.1 mm, and the theory for the small size scatterers is applicable [7, 20–24, 36]. This gives for the attenuation coefficient

$$\gamma_c \approx 0.438 c(t) q / \lambda^2, \, dB/km, \tag{12.68}$$

where λ is the wavelength measured in centimeters, and q is the water content measured in gram per cubic meter. For the visibility of 600 m, 120 m, and 30 m, the water content in fog or cloud is 0.032 g/m³, 0.32 g/m³, and 2.3 g/m³, respectively. The calculations show that the attenuation, in a moderately strong fog or cloud, does not exceed the attenuation due to rain with a rainfall rate of 6 mm/h.

Due to lack of data, a semiheuristic approach is presented here. Specifically, we assume that the thickness of the cloud layer is $w_c = 1 \, km$, and the lower boundary of the layer is located at a height of $h_c = 2 \, km$. The water content of clouds has a yearly percentage of [7]

$$P(q > x) = p_c \exp\left(-0.56\sqrt{x} - 4.8x\right)\%, \tag{12.69}$$

where p_c is the probability of cloudy weather (%). Neglecting ray bending, we have for the length of the path within the cloud layer

$$L_c = 0, \, h_2 \le h_c$$
$$L_c = \sqrt{d^2 + h_2^2} \, (1 - h_c / h_2), \, h_c < h_2 < h_c + w_c \tag{12.70}$$
$$L_c = w_c / \sin\theta, \, h_2 \ge h_c + w_c$$

where

$$\theta = \arctan(h_2 / d)$$

Here, h_2 is the air vehicle antenna height. Although the attenuation in clouds is less than in rain, the percentage of clouds can be much more essential than that of the rain events. Thus, the additional path loss due to clouds can be estimated as 2 dB and 5 dB for 350 km path and $h_2 = 6 \, km$, and for the time availability of 95% and 99%, respectively.

Depolarization Effects. The polarization of a radio wave changes when passing through the cloud, as an anisotropic medium. As was shown in Reference 30, a purely vertical polarized wave may obtain an additional horizontal component, or the right-hand circularly polarized (RHCP) wave may obtain an additional left-hand circularly polarized (LHCP) component. The extent of this depolarization may be measured by the cross-polarization discrimination (XPD) and cross-polarization isolation (XPI) terms, which can be presented schematically in Figure 12.13 according to Reference 30.

Essentially, the XPD term expresses how much of a signal in a given polarization is transformed into the opposite polarization caused by cloud, while

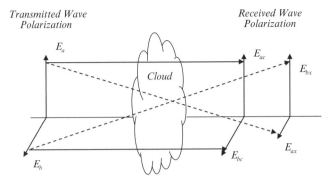

Transmitted Wave Polarization   *Received Wave Polarization*

E_a E_{ac} E_{bx}

Cloud

E_b E_{bc} E_{ax}

FIGURE 12.13. Depolarization of the transmitted wave caused by a cloud.

the XPI term shows how much two signals of opposite polarizations, transmitted simultaneously, will interfere with each other at the receiver. As was shown in literature (see, e.g., Reference 30 and the References section herein), depolarization factors are strongly correlated with attenuation and path loss described earlier. Thus, to predict XPD directly, one must, first of all, obtain the effects of attenuation and path loss. These parameters are related logarithmically [30]: XPD $= a - b\log L$, where L can be evaluated from (12.70), a and b are constants: $a = 35.8$ and $b = 13$. This equation can be an accurate empirical predictor for frequencies below 10 GHz. Rain and other hydrometeors, as well as tropospheric scintillations caused by turbulence, can be the additional source of signal depolarization in land-satellite communication links (see Chapter 14).

12.3. EFFECTS OF TROPOSPHERIC TURBULENCES ON RADIO PROPAGATION

As a result of turbulent flows caused in the turbulent structure (sometimes called *eddies*) of the wind in the troposphere, the horizontal layers of equal refractive indices become mixed, leading to rapid refractive index variations over small distances. These are the *small-scale* variations that appear over short time intervals, and yield *rapid* refractive index variations. Let us first consider the main characteristics and parameters of atmospheric turbulence and then discuss the tropospheric scintillations mentioned earlier, as well as the effects of multiple scattering due to the irregular structure of the troposphere.

12.3.1. Main Characteristics and Parameters of Atmospheric Turbulence

Atmospheric turbulence is a chaotic phenomenon created by random temperature, wind magnitude variation, and direction variation in the propagation medium. This chaotic behavior results in index-of-refraction fluctuations. The turbulence spectrum is divided into three regions by two scale sizes [42–45]:

- the outer scale of turbulence: L_0
- the inner scale (or micro scale) of turbulence: l_0.

These values vary according to atmosphere conditions, distance from the ground, and other factors. The inner scale l_0 is assumed to lie in the range of 1 mm to 30 mm. Near ground, it is typically observed to be around 3 to 10 mm, but generally increases to several centimeters with increasing altitude h. A vertical profile for the inner scale is not known. The outer scale L_0, near ground, is usually taken to be roughly Kh, where K is a constant on the order of unity. Thus, L_0 is usually either equal to the height from the ground (when the turbulent cell is close to the ground) or in the range of 10 m to 100 m or more. Vertical profile models for the outer scale have been developed based on measurements, but different models predict very different results. Let l be the size of turbulence eddies, $k < 2\pi/\lambda$ is a wave number, and λ is a wavelength. Then one can divide turbulences at the three regions:

$$\text{Input range: } L_0 < l, \quad k < \frac{2\pi}{L_0}$$

$$\text{Inertial range: } l_0 < l < L_0, \quad \frac{2\pi}{L_0} < k < \frac{2\pi}{l_0} \tag{12.71}$$

$$\text{Dissipation range: } l < l_0, \quad \frac{2\pi}{l_0} < k.$$

These three regions induce strong, moderate, and weak spatial and temporal variations, respectively, of signal amplitude and phase, referred to in the literature as *scintillations*.

Now, as the troposphere is a random medium, these variations of the index-of-refraction (or turbulences) are random by nature and can be described only (with meaning of a stochastic process) by the PDF and CDF defined in Chapter 1, or by the corresponding spectral distribution functions.

The main goal of studying radiowave propagation through a turbulent atmosphere is the identification of a tractable PDF and CDF or the corresponding spectra of the irradiance under all irradiance fluctuation conditions. Obtaining an accurate mathematical model for a PDF and a CDF of the randomly fading irradiance signal will enable the link planner to predict the reliability of a radio communication system operating in such an environment. In addition, it is beneficial if the free parameters of that PDF and CDF can be tied directly to atmospheric parameters.

Energy Cascade Theory of Turbulence. The Kolmogorov energy cascade theory of turbulence is based on the division of three types of processes defined in (12.71) by two scale sizes: inner l_0 and outer L_0. As was mentioned earlier, the value of L_0 and l_0 may vary widely.

In the input range of (12.71), the large-scale atmospheric characteristic, such as wind, forms the turbulent eddies. Here, the thermal and kinetic energy of the atmosphere is the input to the turbulence system. The process is, in general, anisotropic and varying, depending on climatic conditions. In the inertial range of (12.71), the eddies formed in the input range are unstable and fragmented into smaller regions. These break up as well, continuing in this manner and causing energy to be distributed from small to large turbulence wave numbers. There is very little energy loss in this process. Most cases of microwave propagation are affected predominantly by this region of the wave number spectrum [23–26].

As for the dissipation range of (12.71), here, the energy in the turbulence, which was transferred through the inertial subrange, is dissipated through viscous friction by very small eddies. Kolmogorov's cascade theory is presented schematically in Figure 12.14 according to References 44 and 46–48.

Turbulence Power Spectrum. Results from theoretical models of scintillation depend strongly on the assumed model for the spatial power spectrum of refractive index fluctuations. If we ignore the outer-scale effects, which are usually not important in scintillation studies, the commonly used spectral models are all special cases of

$$\Phi_n(K) = \kappa(\alpha) C_n^2 K^{-\alpha-2} f(\kappa l_0), \tag{12.72}$$

where κ is the magnitude of the spatial wave number, α is a power-law index, K is a dimensionless factor, C_n^2 is the index-of-refraction structure parameter (will be described later separately because of its importance in scintillation studies), and l_0 is the inner scale (the reader can find all detailed information

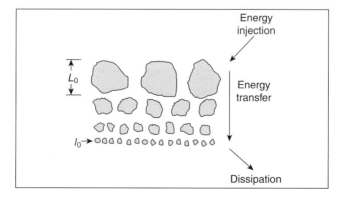

FIGURE 12.14. Kolmogorov cascade theory of turbulence.

about such parameters in References 42–50). The normalization factor K is set so that the index-of-refraction structure function is $C_n^2 r^\alpha$ for r larger than l_0. Kolmogorov turbulence theory predicts that α is near five-thirds, in which case, K is $\sim 1/30$. The information mentioned earlier allows us to rewrite (12.72) in the following manner:

$$\Phi_n(\kappa) = 0.033 C_n^2 \kappa^{-11/3} f(\kappa l_0),\tag{12.73}$$

where $f(\kappa l_0)$ is a factor that describes inner-scale modifications of the basic power-law form.

For example, the *Kolmogorov spectrum* is characterized by $f(\kappa lo) = 1$, whereas $f(\kappa lo) = \exp(-[\kappa l_0/5.92]^2)$ in the case of the *Tatarskii spectrum*; the latter sometimes called the *traditional* spectrum [43, 44]. However, neither of these spectrum models can be used to describe the spectrum outside of the inertial range. They both show the correct behavior (in terms of fitting experimental results) only in the inertial range. The Tatarskii spectrum has been shown to be inaccurate by as much as 50% for predicting the irradiance variance for the strong-focusing regime in optical propagation experiments and by as much as 40% for weak fluctuations. A more accurate model for scintillation studies is provided by the *Hill spectrum*, or by an analytic approximation that is given by the *modified atmospheric spectrum*. Let us briefly discuss these models.

Kolmogorov Spectrum. For statistically homogeneous turbulences, the related structure function exhibits the asymptotic behavior of the form [42–50]:

$$D_n(R) = \begin{cases} C_n^2 R^{2/3}, & l_0 < R < L_0 \\ C_n^2 l_0^{-4/3} R^2, & R < l_0 \end{cases},\tag{12.74}$$

where R is an eddy size.

On the basis of the earlier two-thirds power-law expression, it can be deduced, and the associated power spectral density for refractive index fluctuations can be described by the following expression:

$$\Phi_{nk}(\kappa) = 0.033 C_n^2 \kappa^{-11/3}, \quad \frac{2\pi}{L_0} < \kappa < \frac{2\pi}{l_0}.\tag{12.75}$$

This is the well-known Kolmogorov spectrum, which was calculated and shown in normalized form, $\bar{\Phi}_{nk}(\kappa) = \Phi_{nk}(\kappa)/0.033 C_n^2$, in Figure 12.15 for *inertial* and *dissipation* ranges.

Tatarskii Spectrum. The Kolmogorov's spectrum is theoretically valid only in the inertial subrange. The use of this spectrum is justified only within that subrange or over all wave numbers if the outer scale is assumed to be infinite and the inner scale negligibly small. Other spectrum models have been pro-

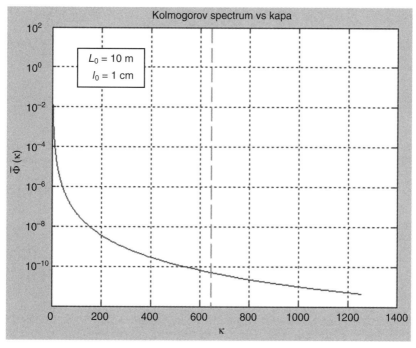

FIGURE 12.15. Kolmogorov normalized spectrum, shown for the inertial and dissipation ranges; the green vertical line indicates $k = 2\pi/l_0$.

posed for calculations when inner-scale and/or outer-scale effects cannot be ignored. In order to extend the power-law spectrum (12.75) into the dissipation range (where $\kappa > 2\pi/l_0$), a truncation of the spectrum at high wave numbers is required. Tatarskii suggested doing that by modulating the Kolmogorov spectrum model (12.75) by a Gaussian function, which led to the Tatarskii spectrum (or the traditional spectrum):

$$\Phi_{nk}(\kappa) = 0.033 C_n^2 \kappa^{-11/3} \exp\left(-\kappa^2 / \kappa_m^2\right), \quad \kappa > \frac{2\pi}{L_0}, \tag{12.76}$$

where $\kappa_m = 5.92/l_0$. This is used to express the composite spectrum, in the region other than the input *range*. Figure 12.16 shows the Tatarskii model spectral behavior, calculated and presented in the same normalized form, as in Figure 12.15. It can be seen from the graph that the power spectrum is "truncated" in high wave numbers relative to the Kolmogorov spectrum, specifically above $k = 2\pi/l_0$. Below this value, the two spectrums are almost identical.

Von Kármán Spectrum. For mathematical convenience, we may assume that the turbulence spectrum is statistically homogeneous and isotropic over all

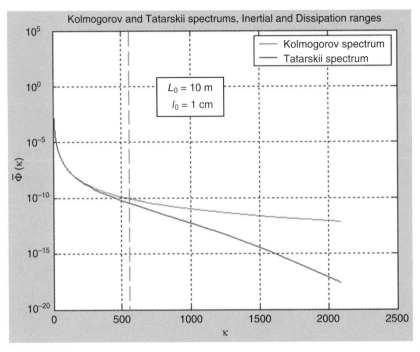

FIGURE 12.16. Tatarskii normalized spectrum, shown with Kolmogorov spectrum for inertial and dissipation range; the green vertical line indicates $k = 2\pi/l_0$.

wave numbers. A spectral model that is often used in this case, one that combines the three regions defined by (12.73), is the Von Kármán spectrum:

$$\Phi_{nk}(\kappa) = 0.033 C_n^2 \left(\kappa^2 + \frac{1}{L_0^2}\right)^{-11/6} \exp\left(-\kappa^2/\kappa_m^2\right), \quad 0 \le \kappa < \infty, \qquad (12.77)$$

where $\kappa_m = 5.92/l_0$. Note that even though the last equation describes the *entire* spectrum, its value in the input range must be considered only approximate, because it is general anisotropic and depends on how the energy is introduced into the turbulence. This model, unlike the previous models, does not have a singularity at $\kappa = 0$. Therefore, the Von Kármán spectrum is almost identical to the Tatarskii spectrum, except for a difference in small values of wave number. The Von Kármán spectrum was calculated and shown in Figure 12.17. Figure 12.18 focuses on the small wave numbers to emphasize the difference between the Tatraskii and Von Kármán models in that region. It is clearly seen from these illustrations that although the Tatraskii spectrum "explodes" near the origin, the Von Kármán spectrum inclination is suppressed in that region. For other regions, the two spectrums are almost identical.

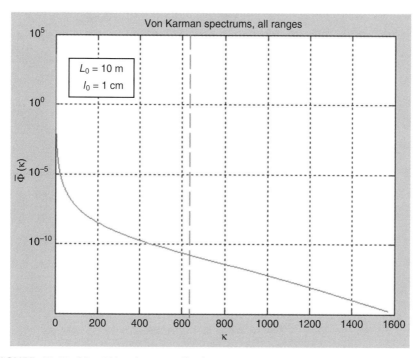

FIGURE 12.17. Von Kármán normalized spectrum, shown for all ranges; the green vertical line indicates $k = 2\pi/l_0$.

Modified Atmospheric Spectrum. The last models of turbulent spectra, defined by (12.75), (12.76), and (12.77), are commonly used in theoretical studies of radiowave propagation because they are relatively traceable models. Strictly speaking, however, these spectrum models have the correct behavior only in the inertial range: that is, the mathematical form that permits the use of these models outside the inertial range is based on mathematical convenience, and not because of any physical meaning. The Tatarskii spectrum has been shown to be inaccurate by as much as 50% in predicting the irradiance variance for the strong-focusing regime in optical propagation experiments and by as much as 40% for weak fluctuations. Hill (see details in References 46–48) developed a numerical spectral model with a high wave number rise that accurately fits the experimental data. However, as it is described in terms of a second-order differential equation that must be solved numerically, the Hill spectrum cannot be used in analytic developments. An analytic approximation to the Hill spectrum, which offers the same tractability as the Von Kármán model (12.77), was developed by Andrews with colleagues [46–48]. This approximation, commonly called the *modified atmospheric spectrum* (or just *modified spectrum*), is given by References 45–48, and it is valid for wave numbers in the range $0 \le \kappa < \infty$:

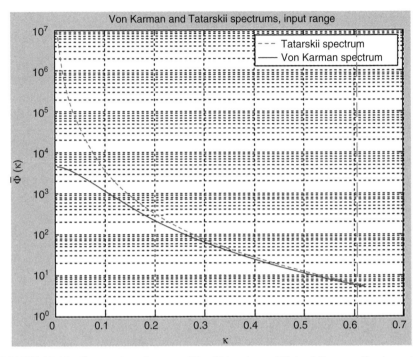

FIGURE 12.18. Comparison between Von Kármán and Tatraskii normalized spectra, shown for the input range; the gray vertical line near 0.6 indicates $k = 2\pi/L_0$.

$$\Phi_n(\kappa) = 0.033 C_n^2 [1 + 1.802(\kappa/\kappa_l) - 0.254(\kappa/\kappa_l)^{7/6}] \frac{exp(-\kappa^2/\kappa_l^2)}{(\kappa^2 + 1/L_0^2)^{11/6}}, \quad (12.78)$$

where $\kappa_l = 3.3/l_0$. Numerical comparison of results based on the earlier equation and the Hill spectrum reveal differences no larger than 6% but generally within 1–2% of each other. A comparison between the normalized modified atmospheric spectrum and the Von Kármán spectrum, $\overline{\Phi}_{nk}(\kappa) = \Phi_{nk}(\kappa)/0.033 C_n^2$, can be seen in Figure 12.19. The whole modified spectrum, on a log-log scale, is shown in Figure 12.20 computed according to expression (12.78) [46–48]. The modified model, which is based on Hill's numerical spectral model, provides good agreement with experimental results.

The Refractive Index Structure Parameter. As was mentioned earlier, any turbulence in the atmosphere can be characterized by three parameters: the inner scale l_0, the outer scale L_0, and the structure parameter of refractive index

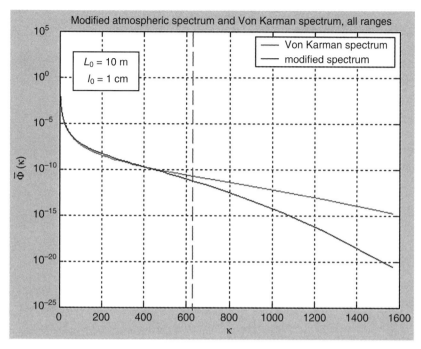

FIGURE 12.19. Comparison between the modified and Von Kármán normalized spectra; the green vertical line indicates $k = 2\pi/l_0$.

fluctuation C_ε^2. The refractive index structure parameter, as the measure of the "strength" or "power" of the turbulent structure, is considered the most critical parameter along the propagation path in characterizing the effects of atmospheric turbulence. It was defined earlier as a refractive index structure parameter C_n^2 (in radio propagation, it also is denoted by C_ε^2, accounting relationship between the refractive index n and permittivity ε). Values of C_n^2 near the ground, in warm climates, generally vary between 10^{-14} m$^{-2/3}$ to 10^{-12} m$^{-2/3}$. Various near-ground experiments carried out over different daytime hours during winter time showed that with an increase of altitude h, the structure parameter C_n^2 decreases to an altitude of 3 to 5 km. It then increases to some maximum near 10 km, after which, it rapidly decreases with increasing altitude. Usually, to obtain relationships between the structure parameter C_n^2 and the atmospheric refractive index fluctuations δn, we assume stationary, homogeneity, and isotropism of atmospheric refractive index fluctuations. However, as was mentioned in previous sections, the refractive index n is a complicated function of various meteorological parameters. For example, for

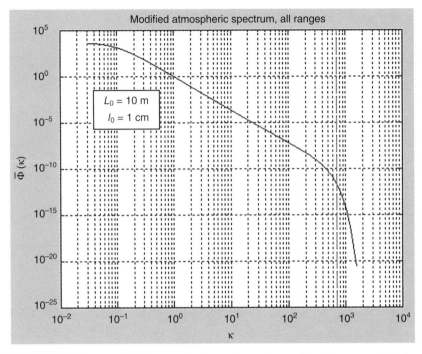

FIGURE 12.20. The modified normalized spectrum, shown for all ranges at the logarithmic f scale; the green vertical lines indicate $k = 2\pi/l_0$ and $k = 2\pi/L_0$.

above sea atmosphere, the value of the refractive index n can be presented as [42]

$$n \approx 1 + \frac{77p}{T}\left[1 + \frac{7.53 \cdot 10^3}{\lambda^2} - 7733\frac{q}{T}\right], \tag{12.79}$$

where p is the air pressure, (mb); T is the temperature (K); q is a specific humidity, $(g/(mg)/m^3)$, and λ is a wavelength. A simple approximation of relationship between the refractive index fluctuations and the structure parameter C_n^2 is given by [44]

$$C_n^2 \approx \frac{\langle(\delta n)^2\rangle}{x^{2/3}}, \tag{12.80}$$

where x represents the distance between antennas, the transmitter, and the receiver. If so, the structure parameter C_n^2 also varies according to variations of meteorological parameters. So, for marine atmosphere [44],

$$C_n^2 \approx \left(79 \cdot 10^{-6}\frac{p}{T^2}\right)^2 (C_T^2 + 0.113C_{Tq} + 0.003C_q^2), \tag{12.81}$$

where C_T^2 and C_q^2 are the air temperature and water vapor structure coefficients, respectively; C_{Tq} is the combined temperature–water vapor structure coefficient or covariance. The C_T^2 term in (12.81), which is the mean-square statistical average of the difference in temperature ΔT between two points along the radio path separated by a distance x, is given by

$$C_T^2 = \langle (\Delta T)^2 \rangle x^{2/3}. \tag{12.82}$$

The structure parameter can thus be written as [42, 44]

$$C_n^2 = \left(79 \cdot 10^{-6} \frac{p}{T^2} \right)^2 C_T^2. \tag{12.83}$$

In daytime and near the ground surface (at the height of several meters), the value of C_n^2 can range from $10^{-16}\,\mathrm{m}^{-2/3}$ to $10^{-12}\,\mathrm{m}^{-2/3}$, with changes of magnitude in only 1 minute.

12.3.2. Tropospheric Scintillations

Waves travel through tropospheric layers that are characterized by rapid variations of index, therefore, these waves undergo fast changes in amplitude and phase in a random way. This effect is called *dry tropospheric scintillation*. Rain is another source of tropospheric scintillation, which is called *wet*; it leads to a wet component of scintillation, which tends to be slower than the dry effects. The phase and amplitude fluctuations occur both in the space and time domains. Moreover, this phenomenon is strongly frequency dependent: the shorter wavelengths lead to more severe fluctuations of signal amplitude and phase, resulting from a given scale size. The scale size can be determined by experimental monitoring the scintillation of a signal on two nearby paths and by examining the cross-correlation between the scintillations on the paths. If the effects are closely correlated, then the scale size is large compared with the path spacing [45].

Additional investigations have shown that the distribution of the signal fluctuations (in decibels) is approximately a Gaussian distribution, whose standard deviation is the intensity [42–45].

Scintillation Index. A wave propagating through a random medium, such as the atmosphere, will experience irradiance fluctuations, called scintillations, even over relatively short propagation paths. Scintillation is defined as [42–45]

$$\sigma_I^2 = \frac{\langle I^2 \rangle - \langle I \rangle^2}{\langle I \rangle^2} = \frac{\langle I^2 \rangle}{\langle I \rangle^2} - 1. \tag{12.84}$$

This is caused almost exclusively by small temperature variations in the random medium, resulting in index-of-refraction fluctuations (i.e., turbulent structures).

In (12.84), the quantity I denotes irradiance (or intensity) of the radio wave and the angle brackets denote an ensemble average or equivalently, a long-time average. In weak fluctuation regimes, defined as those regimes for which the scintillation index is less than one [42–45], derived expressions for the scintillation index show that it is proportional to *Rytov variance*:

$$\sigma_1^2 = 1.23 C_n^2 k^{7/6} x^{11/6}. \tag{12.85}$$

Here, C_n^2 is the index-of-refraction structure parameter, k is the radiowave number, and x is the propagation path length between transmitter and receiver. The Rytov variance represents the scintillation index of an unbounded plane wave in the case of its weak fluctuations, but is otherwise considered a measure of the turbulence strength when extended to strong-fluctuation regimes by increasing either C_n^2 or the path length x, or both. It is shown in References 42–45 that the scintillation index increases with increasing values of the Rytov variance until it reaches a maximum value greater than unity in the regime characterized by random focusing, because the focusing caused by large-scale inhomogeneities achieves its strongest effect. With increasing path length or inhomogeneity strength, multiple scattering weakens the focusing effect, and the fluctuations slowly begin to decrease, saturating at a level for which the scintillation index approaches the value of one from above. Qualitatively, saturation occurs because multiple scattering causes the wave to become increasingly less coherent in the process of wave propagation through random media.

Signal Intensity Scintillations in the Turbulent Atmosphere. Early investigations concerning the propagation of unbounded plane waves and spherical waves through random media obtained results limited by weak fluctuations [45]. To explain the weak-fluctuation theory, three new parameters of the problem must be introduced instead of the inner and outer scales of turbulences described earlier. They are (a) the coherence scale, $l_1 \equiv l_{co} \sim 1/\rho_0$, which describes the effect of coherence between two neighboring points (see Fig. 12.21); (b) the first Fresnel zone scale, $l_2 \equiv \ell_F \sim \sqrt{x/k}$, as was mentioned in Chapter 4, which describes the clearance of the propagation link (see Fig. 12.8); and (c) the scattering disk scale, $l_3 \sim x/\rho_0 k$, which models the turbulent structure (see Fig. 12.22), where L is the length of the radio path.

On the basis of such definitions, Tatarskii [44] predicted that the correlation length of the irradiance fluctuations is of the order of the first Fresnel zone $\ell_F \sim \sqrt{L/k}$ (see Fig. 12.8). However, measurements of the irradiance covariance function under strong fluctuation conditions showed that the correlation length decreases with increasing values of the Rytov variance σ_1^2 and that a large residual correlation tail emerges at large separation distances. That is, in the strong-fluctuation regime, the spatial coherence radius ρ_0 of the wave determines the correlation length of irradiance fluctuations, and the scattering disk characterizes the width of the residual tail: $x/\rho_0 k$.

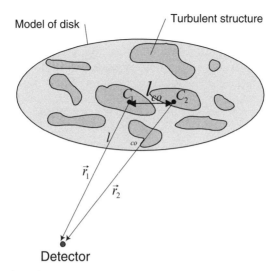

FIGURE 12.21. The coherency between two neighboring points, described by close-to-unit correlation coefficient, i.e., $0 << \rho(\mathbf{r_1}, \mathbf{r_2}) \leq 1$.

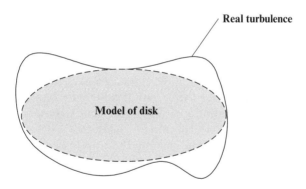

FIGURE 12.22. The area of scattering modeled by a disk.

In References 46–48, the theory developed in References 43 and 44 was modified for strong fluctuations and showed why the smallest scales of irradiance fluctuations persist into the saturation regime. The basic qualitative arguments presented in these works are still valid. Kolmogorov theory assumes that turbulent eddies range in size from a macroscale to a microscale, forming a continuum of decreasing eddy sizes.

The largest eddy-cell size, smaller than that at which turbulent energy is injected into a region, defines an effective outer scale of turbulence L_0, which near the ground is roughly comparable with the height of the observation point above ground. An effective inner scale of turbulence l_0 is associated with the smallest cell size before energy is dissipated into heat.

Here, we will present briefly modifications of the Rytov method obtained in References 46–48 to develop a relatively simple model for irradiance fluctuations, that is, applicable to moderate-to-strong fluctuation regimes. In References 46–48, the following basic observations and assumptions have been stated:

- atmospheric turbulence as it pertains to a propagating wave is statistically inhomogeneous
- the received irradiance of a wave can be modeled as a modulation process in which small-scale (diffractive) fluctuations are multiplicatively modulated by large-scale (refractive) fluctuations
- small-scale processes and large-scale processes are statistically independent
- the Rytov method for signal intensity scintillation is valid even in the saturation regime, with the introduction of a spatial frequency filter to account properly for the loss of spatial coherence of the wave in strong-fluctuation regimes
- the geometrical optics method can be applied to large-scale irradiance fluctuations.

These observations and assumptions are based on recognizing that the distribution of refractive power among the turbulent eddy cells of a random medium is described by an inverse power of the physical size of the cell. Thus, the large turbulent cells act as *refractive lenses* with focal lengths typically on the order of hundreds of meters or more, creating the so-called *focusing effect* or *refractive scattering* (see Fig. 12.23a). This kind of scattering is defined by the coherent component of the total signal passing the troposphere. The smallest cells have the weakest refractive power and the largest cells the strongest. As a coherent wave begins to propagate into a random atmosphere, the wave is scattered by the smallest of the turbulent cells (on the order of millimeters), creating the so-called *defocusing effect* or *diffractive scattering* (see Fig. 12.23b). This kind of scattering is defined by the incoherent component of the total signal. Thus, they act as defocusing lenses, decreasing the amplitude of the wave by a significant amount, even for short propagation distances. The diffractive scattering spreads the wave as it propagates. Refractive and diffractive scattering processes are compound mechanisms, and the total scattering process acts like a modulation of small-scale fluctuations by large-scale fluctuations. Schematically, such a situation is sketched in Figure 12.24, containing both components of the total field.

Small-scale contributions to scintillation are associated with turbulent cells smaller than the Fresnel zone $\sqrt{x/k}$ or the coherence radius ρ_0, whichever is smaller. Large-scale fluctuations in the irradiance are generated by turbulent

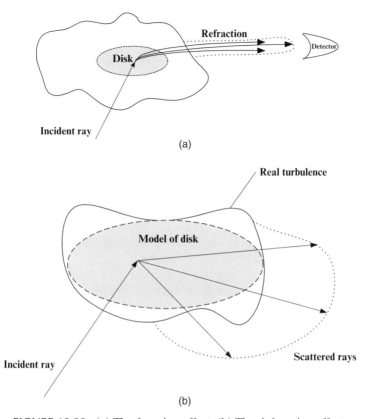

FIGURE 12.23. (a) The focusing effect. (b) The defocusing effect.

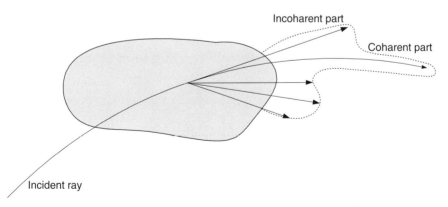

FIGURE 12.24. The total field pattern consisting of the coherent part (I_{co}) and incoherent part (I_{inc}).

cells larger than that of the first Fresnel zone or the scattering disk $x/k\rho_0$, whichever is larger, and can be described by the method of geometrical optics. Under strong-fluctuation conditions, spatial cells having size between those of the coherence radius and the scattering disk contribute little to scintillation. Hence, because of the loss of spatial coherence, only the very largest cells nearer the transmitter have focusing effect on the illumination of small diffractive cells near the receiver. Eventually, even these large cells cannot focus or defocus. When this loss of coherence happens, the illumination of the small cells is (statistically) evenly distributed and the fluctuations of the propagating wave are due to random interference of a large number of diffraction scattering of the small eddy cells.

To finish this paragraph, let us consider an example of the scintillation index of a plane radio wave that has propagated a distance x through unbounded turbulent atmosphere. First, we assume the inner scale of turbulence is zero, so it is not taken into account. In this case, we consider three states of atmospheric turbulence: weak, moderate, and strong.

(A) *Weak fluctuations.* Under weak-fluctuation theory and the Rytov method, the scintillation index can be expressed in the form

$$\sigma_I^2 = \exp(\sigma_{\ln I}^2) - 1 \cong \sigma_{\ln I}^2, \quad \sigma_I^2 \ll 1, \tag{12.86}$$

where $\sigma_{\ln I}^2$ is the log-irradiance variance defined under the Rytov approximation by

$$\sigma_{\ln I}^2 = 8\pi^2 k^2 \int_0^L \int_0^\infty \kappa \Phi_n(\kappa) \left[1 - \cos\left(\frac{\kappa^2 z}{k}\right)\right] d\kappa\, dz$$

$$= 1.06\sigma_1^2 \int_0^1 \int_0^\infty \eta^{-11/6} (1 - \cos\eta\xi)\, d\eta\, d\xi. \tag{12.87}$$

In the last step, we have assumed a conventional Kolmogorov spectrum and introduced the nondimensional quantities, $\eta = x\kappa^2/k$ and $\xi = z/x$. Performing the integration earlier, we obtain the result

$$\sigma_I^2 \cong \sigma_1^2 = 0.847(x)^{5/6}, \quad \sigma_1^2 \ll 1 \tag{12.88}$$

$$x/k\rho_0^2 = 1.22(\sigma_1^2)^{6/5}. \tag{12.89}$$

(B) *Moderate fluctuations.* At the other extreme, the asymptotic behavior of the scintillation index in the *saturation regime* is described by

$$\sigma_I^2 \cong 1 + \frac{0.86}{\sigma_1^{4/5}} = 1 + 0.919\left(\frac{k\rho_0^2}{x}\right)^{1/3}, \quad \sigma_1^2 \gg 1. \tag{12.90}$$

The resulting log-irradiance scintillation is

$$\sigma_{\ln I}^2 = 8\pi^2 k^2 \int_0^L \int_0^\infty \kappa \Phi_n(\kappa) G_x(\kappa) \left[1 - \cos\left(\kappa^2 z/k\right)\right] d\kappa dz \cong$$

$$\cong 1.06\sigma_1^2 \left(\frac{L}{k}\right)^{7/6} \int_0^1 \xi^2 \int_0^\infty \kappa^{4/3} \exp\left(-\frac{\kappa^2}{\kappa_x^2}\right) d\eta d\xi \cong 0.15\sigma_1^2 \eta_{1x}^{7/6},$$

(12.91)

where $\eta_x = x\kappa_x^2 / k$.

(C) *Strong fluctuations.* In the case of strong turbulence regime, the scintillation index for a plane wave in the absence of inner scale is given by

$$\sigma_I^2 = \exp\left[\frac{0.54\sigma_1^2}{\left(1 + 1.22\sigma_1^{12/5}\right)^{7/6}} + \frac{0.509\sigma_1^2}{\left(1 + 0.69\sigma_1^{12/5}\right)^{5/6}}\right] - 1, \quad 0 \le \sigma_1^2 < \infty. \quad (12.92)$$

An example of signal intensity scintillation index computation according to (12.91) from 1 GHz to 50 GHz versus the refractive index structure parameter that varies from 10^{-13} to 10^{-11}, for the distance $x = 10$ km, and the inner scale $l_0 = 0$ mm, is shown in Figure 12.25. It can be seen that the scintillation index for any $C_n^2 = const.$ (denoted, e.g., for $C_n^2 = 10^{-12}$ by the vertical line) becomes twice as strong as the frequency increases from 20 GHz to 50 GHz. This result is very important for predicting the fast fading of the signal within land aircraft and land-satellite radio communication links passing through the turbulent

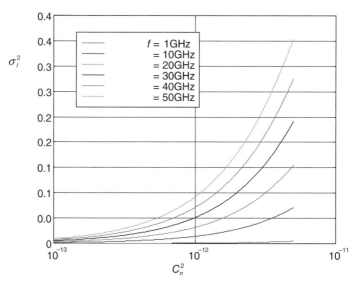

FIGURE 12.25. Index of signal intensity scintillations versus the intensity of refractive index scintillations for different frequencies from 1 GHz to 50 GHz.

troposphere and operating at frequencies in the L/X-band (i.e., more than 1 GHz).

When inner-scale effects become important ($l_0 \neq 0$), the atmospheric power spectrum is more strictly described by a modified spectrum with high wave number rise, that is, the traditional Tatarskii spectrum [44].

Under weak irradiance fluctuations, in the case of an unbounded plane wave, the scintillation index based on the modified spectrum is described for $\sigma_I^2 < 1$ by

$$\sigma_I^2(x) \cong 3.86\sigma_1^2 \left\{ \left(1+Q_l^{-2}\right)^{11/12} \left[\sin\left(\frac{11}{6}\tan^{-1}Q_l\right) + \right.\right.$$
$$+ \frac{1.507}{\left(1+Q_l^2\right)^{1/4}}\sin\left(\frac{3}{4}\tan^{-1}Q_l\right) \tag{12.93}$$
$$\left.\left. - \frac{0.273}{\left(1+Q_l^2\right)^{7/24}} \times \sin\left(\frac{5}{4}\tan^{-1}Q_l\right)\right] - 3.5Q_l^{-5/6}\right\}.$$

Here, $Q_l = 10.89x / kl_0^2$ is a nondimensional inner-scale parameter. Asymptotic expressions for the scintillation index in the saturation regime, based on the modified atmospheric spectrum, are

$$\sigma_I^2 \cong 1 + \frac{2.39}{\left(\sigma_{1I}^2 Q_l^{7/6}\right)^{1/6}}, \quad \sigma_I^2 Q_l^{7/6} >> 100. \tag{12.94}$$

Under general conditions, the size of the inner scale l_0 relative to the Fresnel zone $\sqrt{x/k}$ is an important consideration. For example, in weak irradiance fluctuations associated with short propagation paths, the inner scale may be of similar size to or larger than the width of the Fresnel zone; hence, there will be little contribution to scintillation from eddy cells smaller than the inner scale. On the contrary, over longer propagation path lengths, the inner scale can be much smaller than the Fresnel zone. In this latter situation, size of the cells is the same as the inner scale and smaller size contributes mostly to small-scale scintillation. Large-scale scintillation is dominated by cells with size larger than x/kl_0.

Eventually, however, the coherence radius becomes smaller than the inner scale, and small-scale scintillation depends less and less on cell size of about the same as the inner scale in the saturation regime. Large-scale scintillation, which continues to depend on the inner scale, begins to diminish in the saturation regime, as only those cells larger than the scattering disk are strong enough to still cause focusing effects. The scintillation index for a plane wave in the presence of finite inner scale is

$$\sigma_I^2 = \exp\left[\sigma_{\ln x}^2(l_0) + \frac{0.509\sigma_1^2}{\left(1+0.69\sigma_1^{12/5}\right)^{5/6}}\right] - 1, \tag{12.95}$$

TABLE 12.5. Scintillation Index σ_I^2 for Radiated Frequency of 25 GHz

$1 \cdot 10^{-13}$	$5 \cdot 10^{-13}$	$1 \cdot 10^{-12}$	$5 \cdot 10^{-12}$	C_n^2
~0.0028	~0.014	~0.027	~0.134	σ_I^2 for 0 mm inner scale
~0.0027	~0.013	~0.024	~0.091	σ_I^2 for 1 mm inner scale

TABLE 12.6. Scintillation Index σ_I^2 for Radiated Frequency of 50 GHz

$1 \cdot 10^{-13}$	$5 \cdot 10^{-13}$	$1 \cdot 10^{-12}$	$5 \cdot 10^{-12}$	C_n^2
~0.0093	~0.046	~0.092	~0.404	σ_I^2 for 0 mm inner scale
~0.0088	~0.038	~0.068	~0.241	σ_I^2 for 1 mm inner scale

where $x/k\rho_0^2 = 1.02\sigma_1^2 Q_l^{1/6}$ is in the presence of the inner scale. Results of the scintillation index according to (12.94), versus the refractive index parameter C_n^2, is shown in Figure 12.26a,b for $f = 10$ and 20 GHz, respectively. The results are for $x = 10$ km and for the inner scale l_0 changed from 1 mm to 7 mm. It is seen that for $C_n^2 = const$ (denoted, e.g., as $C_n^2 = 10^{-12}$ by the vertical dotted line), the scintillation index does not vary significantly with an increase of the inner scale of the initial turbulence, becoming in any way smaller than the values for the case of zero-order inner-scale model, described by (12.91). These results are summarized in Table 12.5 and Table 12.6 for inner scale $l_0 = 0$ mm and 1 mm, and for frequencies of 25 GHz and 50 GHz, respectively.

12.3.3. Effects of Troposheric Turbulences on Signal Fading

The fast fading of the signal at open paths is caused mainly by multipath propagation and turbulent fluctuations of the refractive index. Some very interesting ideas were proposed in References 49 and 50, which will be presented briefly later. As it is known, the fluctuations of the signal intensity due to turbulence arc distributed log-normally. For the Kolmogorov model, the normalized standard deviation of this distribution can be presented in terms of C_ε^2 instead of Rytov's Equation (12.85) presented in terms of C_n^2

$$\sigma^2 = 0.12 C_\varepsilon^2 k^{7/6} d^{11/6}, \tag{12.96}$$

where $k = 2\pi/\lambda$ is the wave number, and C_ε^2 is the structure constant of the turbulence averaged over the path. In the atmosphere, the structure constant

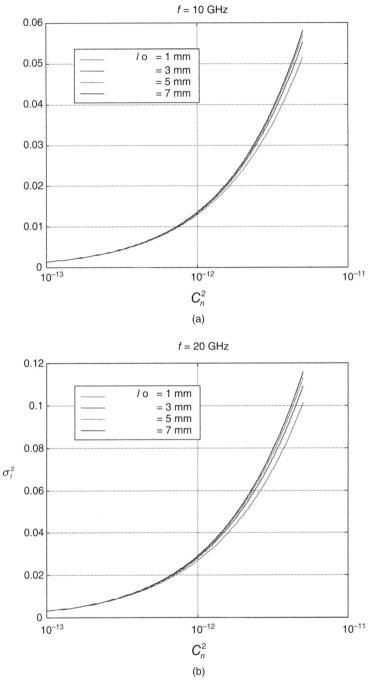

FIGURE 12.26. (a) Index of signal intensity scintillations versus the intensity of refractive index scintillations for frequency of 10 GHz and for different inner scales of turbulence ranging from 1 mm to 7 mm. (b) The same, as in Fig. 14.26(a), but for frequency of 20 GHz.

C_ε^2 may vary within at least four orders of magnitude, from 10^{-15} to $10^{-10}\,m^{-2/3}$. As the path-averaged statistics of these variations is unknown, the margin related to this kind of fading may be estimated only heuristically. The normalized temporal correlation function was obtained in [49, 50]

$$K(\tau) = \frac{1}{\sin(\pi/12)} \left[(1+\alpha^4/4)^{11/12} \sin\left(\frac{\pi}{12} + \frac{11}{6}\arctan\frac{\alpha^2}{2}\right) - \frac{11}{6}\left(\frac{\alpha}{\sqrt{2}}\right)^{5/3} \right], \quad (12.97)$$

where $\alpha = \tau/\tau_0$, $\tau_0 = \sqrt{d/k}\,/v$, and v is the projection of the vehicle velocity to the plane that is perpendicular to the path. The correlation time τ_c defined as $K(\tau_c) = 0.5$, can be estimated as $\tau_c \approx 0.62\tau_0$. The spectrum of the intensity fluctuations is [49, 50]

$$S(\omega) = \beta^2 w(\omega)/\omega \qquad (12.98)$$

is calculated by using the notion of the normalized spectral density

$$w(\omega) = 4\omega \int_0^\infty d\tau \cos(\omega\tau) K(\tau), \qquad (12.99)$$

which at high and low frequencies is given, respectively by

$$w(\Omega) = 12.0\,\Omega^{-5/3}, \Omega \geq 5 \qquad (12.100a)$$

and at low frequencies can be approximated as

$$w(\Omega) = 3.47\,\Omega\exp\left[-0.44\,\Omega^{\varphi(\Omega)}\right], \Omega \leq 5 \qquad (12.100b)$$

where

$$\varphi(\Omega) = 1.47 - 0.054\,\Omega, \qquad (12.101)$$

and $\Omega = \tau_0\omega$ is the dimensionless frequency. The normalized density $w(\Omega)$ has the maximal value of about 2.30 at $\Omega_m \approx 1.60$, and therefore, $\omega_m \approx 1.60/\tau_0$ [49, 50]. The typical values for both τ_0 and ω_m are shown in Table 12.7 for the frequency 15 GHz and vehicle velocity $v = 50$ m/s, and in Table 12.8 for the frequency 13 GHz and vehicle velocity $v = 350$ m/s, calculated according to

TABLE 12.7. Characteristic Time and Frequency versus the Radio Path for $f = 15$ GHz and $v = 50$ m/s

Distance (km)	50	100	150
τ_0 (s)	0.25	0.36	0.44
ω_m (Hz)	6.4	4.4	3.6

TABLE 12.8. Characteristic Time and Frequency versus the Radio Path for $f = 13\,GHz$ and $v = 350\,m/s$

Distance (km)	100	200	300
τ_0 (s)	0.055	0.077	0.095
ω_m (Hz)	29.1	20.8	16.8

TABLE 12.9. Link Budget in the Troposphere for $f = 15\,GHz$, $d = 150\,km$, $h_2 = 2\,km$

Time availability, (%)	95	99
Basic transmission loss (dB)	159	159
TX antenna gain (dB)	G_1	G_1
RX antenna gain (dB)	G_2	G_2
Molecular absorption (dB)	1	2
Hydrometeors (dB)	1	3
Refraction phenomena (dB)	4	5
Fast fading (turbulence) (dB)	1	2
Fast fading (multipath) (dB)	3	5
Diffuse scattering (dB)	2	4
Total (dB)	$171 - G_1 - G_2$	$182 - G_1 - G_2$

the earlier equations. The phase fluctuations have normal distribution with dispersion

$$\sigma_s^2 = 0.075 C_\varepsilon^2 k^2 ds_0^{-5/3}, \qquad (12.102)$$

where $s_0 \sim 2\pi/L_0$ and L_0 is the outer scale of the turbulent spectrum depending on the height and equal approximately to 10–100 m. Estimations carried out according to References 49 and 50 showed that the phase fluctuations caused by turbulence are negligible under typical atmospheric conditions and even for extremely strong turbulence.

12.4. LINK BUDGET DESIGN FOR TROPOSPHERIC COMMUNICATION LINKS

Let us summarize what was mentioned earlier by introducing some examples of link budget calculations in dB for the communication link between a ground-based antenna, defined by its gain G_1, and an air vehicle (helicopter, aircraft, or satellite) antenna, defined by its gain G_2. We do not present these parameters because they are different for different types of antennas and can be easily computed using special equations [51–53]. In the examples presented in Table 12.9 and Table 12.10, we considered different conditions of radio propagation

TABLE 12.10. Link Budget in the Troposphere for $f = 15\,$GHz, $d = 150\,$km, $h_2 = 4\,$km

Time availability (%)	95	99
Basic transmission loss (dB)	159	159
TX antenna gain (dB)	G_1	G_1
RX antenna gain (dB)	G_2	G_2
Molecular absorption (dB)	1	2
Hydrometeors (dB)	2	7
Refraction phenomena (dB)	5	8
Fast fading (turbulence) (dB)	1	2
Fast fading (multipath) (dB)	3	5
Diffuse scattering (dB)	2	4
Total (dB)	$173 - G_1 - G_2$	$187 - G_1 - G_2$

by introducing the time availability, which varied from 95% to 99%, assuming that the existence of fast fading effects described by Rayleigh statistics are from 1% to 5%. In these tables, we summarize all effects of hydrometeors, as well as fast fading caused by atmospheric turbulences and multipath phenomena caused by diffuse scattering.

REFERENCES

1 *International Telecommunication Union, ITU-R Recommendation*, "The radio refractive index: its formula and refractivity data," pp. 453–456, Geneva, 1997.

2 *International Telecommunication Union, ITU-R Recommendation*, "Effects of tropospheric refraction on radiowave propagation," pp. 834–842, Geneva, 1997.

3 *International Telecommunication Union, ITU-R Recommendation*, "Attenuation by hydrometeors, in precipitation, and other atmospheric particles," Vol. 5, pp. 721–723, Geneva, 1990.

4 Gordon, G.D. and W.L. Morgan, *Principles of Communications Satellites*, John Wiley & Sons, New York, 1993.

5 *International Telecommunication Union, ITU-R Recommendation*, "Attenuation by atmospheric gases," pp. 676–683, Geneva, 1997.

6 *International Telecommunication Union, ITU-R Recommendation*, "Specific attenuation model for rain for use in prediction methods," pp. 838–840, Geneva, 1992.

7 *International Telecommunication Union, ITU-R Recommendation*, "Attenuation due to clouds and fog", pp. 840–842, Geneva, 1997.

8 International Telecommunication Union, *ITU-R Recommendation* P530–7, "Propagation data and prediction methods required to design of terrestrial line-of-sight systems," Geneva, 1997, pp. 378–385.

9 Sadiku, M.N.O., "Satellite Communication Systems," in *Handbook: Engineering Electromagnetics Applications*, R. Bansal, ed., CRC, Taylor and Francis, New York, 2006, pp. 99–120.

10   Pruppacher, H.R. and R.L. Pitter, "A semi-empirical determination of the shape of cloud and rain drops," *J Atmos. Sci.*, Vol. 28, 1971, pp. 86–94.

11   Marshall, J.S. and W.M.K. Palmer, "The distribution of raindrops below 10 GHz," *NASA Reference Publication* 1108, 1983.

12   Joss, J. and A. Waldvogel, "Raindrops size distributions and sampling size error," *J. Atmos. Sci.*, Vol. 26, 1969, pp. 566–569.

13   McMorrow, D.J. and A.R. Davis, "Stochastic model for deriving instantaneous precipitation rate distributions," *J. Applied Meteorology*, Vol. 16, 1977, pp. 757–774.

14   Crane, R. K., "Prediction of attenuation by rain," *IEEE Trans. Commun.* Vol. 28, 1980, pp. 1717–1733.

15   Hogg, D.C. and T.S. Chu, "The role of rain in satellite communications," *Proc. IEEE*, Vol. 63, No. 9, 1975, pp. 1308–1331.

16   Bussey, H.E., "Microwave attenuation statistics estimated from rainfall and water vapor statistics," *Proc. IRE*, Vol. 38, 1950, pp. 781–783.

17   Stutzman, W.L. and K.M. Yon, "A simple rain attenuation model for earth space radio link operating at 10–33 GHz," *Radio Sci.*, Vol. 21, No. 1, 1986, pp. 65–72.

18   Tattelman, P. and K. Scharr, "A model for estimating 1-minute rainfall rates," *J. Climate Applied Meteorology*, Vol. 22, 1983, pp. 1575–1580.

19   Sekhon, R.S. and R.C. Srivastave, "Doppler radar observations of drop-size distributions in a thunderstorm," *J. Atmos. Sci.*, Vol. 28, 1971, pp. 983–984.

20   Slingo, A., "A GSM parameterization for the shortwave radiative properties of water clouds," *J. Atmos. Sci.*, Vol. 46, 1989, pp. 1419–1427.

21   Chou, M.D., "Parameterizations for cloud overlapping and shortwave single scattering properties for use in general circulation and cloud ensemble models," *J. Climate*, Vol. 11, 1998, pp. 202–214.

22   Ray, P.S., "Broadband complex refractive indices of ice and water," *Applied Optics*, Vol. 11, 1972, pp. 1836–1844.

23   Rossow, W.B., L.C. Garder, and A.A. Lacis, "Global, seasonal cloud variations from satellite radiance measurements. Part I: Sensitivity of analysis," *J. Climate*, Vol. 2, 1989, pp. 419–458.

24   Liou, K.N., *Radiation and Cloud Processes in the Atmosphere*, Oxford University Press, Oxford, 1992.

25   Deirmendjian, D., *Electromagnetic Scattering on Spherical Polydispersions*, American Elsevier, New York, 1969.

26   Nussenzveig, H.M. and W.J. Wiscombe, "Efficiency factors in Mie scattering," *Phys. Rev. Let.*, Vol. 45, 1980, pp. 1490–1494.

27   Zhang, W., "Scattering of radiowaves by melting layer of precipitation in backward and forward directions," *IEEE Trans. Anten. Propag.*, Vol. 42, 1994, pp. 347–356.

28   Zhang, W., S.I. Karhu, and E.T. Salonen, "Prediction of radiowave attenuations due to a melting layer of precipitation," *IEEE Trans. Anten. Propag.*, Vol. 42, 1994, pp. 492–500.

29 Zhang, W., J.K. Tervonen, and E.T. Salonen, "Backward and forward scattering by the melting layer composed of spheroidal hydrometeors at 5–100 GHz," *IEEE Trans. Anten. Propag.*, Vol. 44, 1996, pp. 1208–1219.

30 Saunders, S.R., *Antennas and Propagation for Wireless Communication Systems*, John Wiley & Sons, New York, 1999.

31 Hufnagle, R.E., "Line-of-sight wave propagation though the turbulent atmosphere," *Proc. IEEE*, Vol. 56, 1966, pp. 1301–1314.

32 Brehovskih, L.M., "Reflection of plane wave from layered inhomogeneous media," *Journal of Technical Physics*, Vol. 19, 1949, pp. 1126–1135.

33 Blaunstein, N. and Ch. Christodoulou, *Radio Propagation and Adaptive Antennas for Wireless Communication Links*, 1st edn, Wiley InterScience, New Jersey, 2007.

34 Rappaport, T.S., *Wireless Communications: Principles and Practice*, IEEE Press, New York, 1996.

35 Flock, W.L., "Propagation effects on satellite systems at frequencies below 10 GHz," *NASA Reference Publication* 1108, 1983.

36 Bean, B.R. and E.J. Dutton, *Radio Meteorology*, Dover, New York, 1966.

37 Crane, R.K., "A review of transhorizon propagation phenomena," *Radio Sci.*, Vol. 16, 1981, pp. 649–669

38 Krasuk, N.P., V.L. Koblov, and V.N. Krasuk, *Effects of Troposphere and Terrain on Radar Performance*, Radio Press, Moscow, 1988.

39 Janaswamy, R., "A curvilinear coordinate based split-step parabolic equation method for propagation predictions over terrain," *IEEE Trans. Anten. Propagat.*, Vol. 46, No. 7, 1998, pp. 1089–1097.

40 Allnutt, J.A., *Satellite-to-Ground Radiowave Propagation*, IEEE Press, New York, 1989.

41 Yakovlev, O.I., V.P. Yakubov, V.P. Uruadov, and A.G. Pavel'ev, *Propagation of Radio Waves*, URSS Publisher, Moscow, 2009.

42 Ishimaru, A., *Wave Propagation and Scattering in Random Media*, Academic Press, New York, 1978.

43 Rytov, S.M., Yu.A. Kravtsov, and V.I. Tatarskii, *Principles of Statistical Radiophysics*, Springer, Berlin, 1988.

44 Tatarski, V.I., *Wave Propagation in a Turbulent Medium*, McGraw-Hill, New York, 1961.

45 Doluhanov, M.P., *Propagation of Radio Waves*, Nauka, Moscow, 1972.

46 Andrews, L.C. and R.L. Phillips, *Laser Propagation Through Random Media*, Society of Photo-Optical Instrumentation Engineers, Bellingham, WA, 1998.

47 Andrews, L.C., R.L. Phillips, C.Y. Hopen, and M.A. Al-Habash, "Theory of optical scintillations," *J. Optical Society of America*, Vol. 16, 1999, pp. 1417–1429.

48 Andrews, L.C., R.L. Phillips, and C.Y. Hopen, *Laser Beam Scintillation With Applications*, International Society for Optical Engineering (SPIE), Bellingham, WA, 2001.

49 Samelsohn, G.M., "Effect of inhomogeneities' evolution on time correlation and power spectrum of intensity fluctuations of the wave propagating in a turbulent medium," *Sov. J. Commun. Technol. Electron.*, Vol. 38, 1993, pp. 207–212.

50 Samelsohn, G.M. and B.Ya. Frezinskii, *Propagation of Millimeter and Optical Waves in a Turbulent Atmosphere*, Telecommunication University Press, St. Petersburg, 1992.

51 Bello, P.A., "A troposcatter channel model," *IEEE Trans. Commun.*, Vol. 17, 1969, pp. 130–137.

52 Howell, R.G., R.L. Stuckey, and J.W. Harris, "The BT Laboratories slant-pass measurement complex", *BT Tech. J.*, Vol. 10, 1992, pp. 9–21.

53 Belloul, B., S.R. Saunders, and B.G. Evans, "Prediction of scintillation intensity from sky-noise temperature in earth-satellite links," *Electronics Letter*, Vol. 34, 1998, pp. 1023–1024.

CHAPTER THIRTEEN

Ionospheric Radio Propagation

The effects of the ionosphere on radio propagation are very important in radio communication between the terrestrial antennas and air vehicles (stationary or moving) or satellites. The problem of wave propagation and scattering in the ionosphere has become increasingly important in recent years: the ionosphere, atmosphere, and the Earth's background environment all play a significant role in determining the service level and quality of the land-satellite or satellite-satellite communication channel.

In recent decades, the increasing demand is observed in mobile-satellite networks designed to provide global radio coverage using constellations of low and medium Earth orbit satellites, which are now in operation. Such systems form regions, called megacells (see definitions in References 1–3), consisting of a group of co-channel cells, and clusters of spot beams from each satellite, which move rapidly across the Earth's surface. Signals are typically received by a moving or stationary vehicle at very high elevation angles. Only the local environmental features, ionospheric, atmospheric, and terrestrial, which are very close to a specific radio path, contribute significantly to the propagation process. Therefore, the performance of predicting models of fading phenomena, slow and fast, for ionospheric communication links has the same importance as for terrestrial links (described in Chapter 5) and tropospheric links (described in Chapter 12). This is because the same propagation effects, such as multiray reflection, diffraction, and scattering of radio waves, occur in three types of over-the-Earth communication links: land, atmospheric, and ionospheric.

Radio Propagation and Adaptive Antennas for Wireless Communication Networks: Terrestrial, Atmospheric, and Ionospheric, Second Edition. Nathan Blaunstein and Christos G. Christodoulou.
© 2014 John Wiley & Sons, Inc. Published 2014 by John Wiley & Sons, Inc.

However, as in ionospheric links, the slow fading tends to occur on similar distance scales as in a fast one, and hence, they cannot be easily separated, as was done for the land communication links described in Chapter 4. Unlike land communication channels, the predictions of ionospheric communication channels tend to be highly statistical in nature, because coverage across very wide areas must be included in consideration, while still accounting for the large variations due to the local environmental features.

In Section 13.1, we briefly present information about the ionosphere as a continuous medium consisting of plasma and describe the common effects of ionospheric plasma on radio propagation, following the description of ionospheric effects in References 1–12, as well as on the authors' computations. Then, in Section 13.2, we discuss the effects of large-scale and small-scale ionospheric plasma inhomogeneities [13–28] and illustrate the main results of signal amplitude and phase variations, that is, the fast fading, resulting from the inhomogeneous structure of the ionosphere, on the basis of numerical computations carried out by the authors according to proposed ionospheric models [29–41]. Finally, we consider the effects of an inhomogeneous ionosphere on radio propagation at long distances [42–52].

13.1. MAIN IONOSPHERIC EFFECTS ON RADIO PROPAGATION

13.1.1. Parameters and Processes Affecting Radio Propagation in the Ionosphere

The ionosphere is a region of ionized plasma (i.e., ionized gas composed of neutral atoms and molecules on one side and charged particles, electrons, and ions on the other), which surrounds the Earth at a distance ranging from 50 km to 500–600 km, where it continuously extends to the magnetosphere (600–2000 km) [4, 5]. The ions and electrons are created in the ionosphere by the Sun's electromagnetic radiation, solar wind, and cosmic rays that are the sources of atoms and molecule ionization. As the solar radiation penetrates deeper into the Earth's atmosphere at zenith, the ionosphere extends closest to the Earth around the equator and is more intense on the daylight side. Figure 13.1 shows the separation of the ionosphere into four distinct layers during the day: D layer that covers 50–90 km, E layer that covers 90–140 km, $F1$ layer, and $F2$ layer, located at altitudes of 140–250 km and above 250 km, respectively. During night time, these four layers are continuously transformed into the E and F layers, as the D layer does not exist in the night time because of the absence of solar radiation.

The Content of the Ionosphere. For the neutral component of the ionospheric plasma, in conditions of hydrostatical equilibrium and isothermical case, when $T_m(z) = $ const, the change of neutral particle concentration can be described by the following barometric equation [4]:

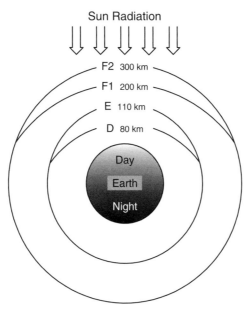

FIGURE 13.1. Presentation of ionospheric layers around the Earth.

$$N_m = N_{m0} \exp\left\{-\frac{(z - z_0)}{H_m}\right\}, \tag{13.1}$$

where N_m is the total plasma concentration; N_{m0} is the same, but for the altitude of $z \le z_0$; $H_m = T_m/M_m g$ is the height of the homogeneous atmosphere; $M_m = (1/N_m)\sum_\alpha m_\alpha N_\alpha$ is the summary content of neutral particles in the higher ionosphere; α is the specification of the neutral particles; g is the acceleration of free falls; and T_m is the gas temperature expressed in energetic units. Equation (13.1) is valid in the ionosphere of the Earth at altitudes of $z_e = z_0$, where $z_e \cong 1000$–2000 km (called exosphere) [4, 6, 7]. In such altitudes, components of the neutral atmospheric gas with specification α leave the atmosphere as their kinetic energy exceeds the potential energy of the Earth's gravitation field and the free path length of neutral particles (λ_α) exceeds the height of the homogeneous atmosphere, H_m (see definitions presented earlier). Furthermore, in the ionosphere, the temperature of neutral molecules and atoms T_m is increased with height. All these factors lead to deviations from the barometric Equation (13.1). As for the neutral content of the ionosphere, at altitudes less than 100 km, the atmosphere is fully presented by molecules of nitrogen (N_2) and oxygen (O_2). At such heights, concentration of other components of the neutral gas (such as He, O, H_2, NO, etc.) is very small and depends on the transport process due to turbulence in the neutral atmosphere, as well as on solar and wind processes in the atmosphere.

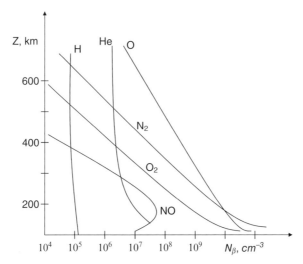

FIGURE 13.2. Height profile of the main neutral components of the ionosphere.

At altitudes of more than 100 km, atmospheric turbulence is absent; the corresponding limit of $z \geq z_T \approx 100$ km is called the turbopause [4, 6, 7]. In contrast, for up to $z_e \cong 1000$–2000 km, Equation (13.1) is still valid. In this range, the main components are atoms of nitrogen and oxygen created by the chemical processes of dissociations of the corresponding molecules: $N_2 \rightarrow N + N$ and $O_2 \rightarrow O + O$. At altitudes more than 500–600 km, in the magnetosphere, the concentrations of helium (He) and hydrogen (H) increase rapidly. Finally, at altitudes of the exosphere of more than 1000–1500 km, atoms of hydrogen become predominant [4, 6, 7] (see Fig. 13.2, where the height distribution of all neutral components of the ionosphere is summarized).

The ion content of ionospheric plasma is changed widely depending on the latitude (Φ) of the Earth. Thus, at middle latitudes ($|\Phi| < 55°$), the plasma is not changed drastically, but at high latitudes ($|\Phi| \geq 55°$–60°), a full concentration of ions decreases rapidly with height. In the middle latitude ionosphere, at the height of the D layer, the main components of plasma are ions NO^+ and O_2^+, and at the heights of E and F layers are ions of oxygen O^+. At altitudes of more than 300 km, plasma also contains components of He^+, H^+, N^+, N_2^+, which with the main ions mentioned earlier determine the total plasma content in the ionosphere. The structure of the ion content of plasma in the polar (high-latitude) ionosphere is more complicated [4, 6, 7].

The key parameter that has an effect on radio communications is the total electron and ions concentration N measured in free electrons number per cubic meter, because the ionospheric plasma is quasi-neutral and in each of its region, with dimensions larger than the Debye radius (definition can be found in References 5 and 7–10), the concentration of electrons (N_e) is equal to the total concentration of various ions (N_i), that is, $N_e \approx N_i = N$. The varia-

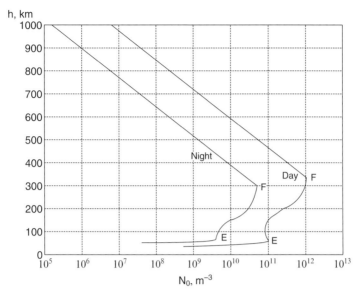

FIGURE 13.3. Ionospheric plasma density versus height of the ionosphere above the ground surface.

tion of N with height in the ionosphere for a typical day and night is shown in Figure 13.3, extracted from References 9 and 10. It must be noted that the total content of charged particles, electrons, and ions depend on the processes that create the structure of the ionosphere such as the solar and cosmic rays radiation, wave radiation, and photochemistry processes, which fully determine the ionization–recombination balance in the ionosphere. The degree of ionization of the ionospheric plasma is determined by the ratio of concentration of the plasma particles N and that of the neutral molecules and atoms N_m. In the lower ionosphere ($z < 100$–120 km), the degree of plasma ionization is very small ($N/N_m \sim 10^{-8}$–10^{-3}) and increases with height h (for $h > 300$ km $N/N_m \sim 10^{-4}$–10^{-3}). At the exospheric heights of $h > 1000$ km, the degree of ionization is $N/N_m \to 1$, that is, plasma becomes fully ionized [4–10]. The temperature of electrons (T_e) and ions (T_i) strongly depend on day time and the increase in height. Thus, at lower ionospheric heights, the temperature of electrons is close to the temperature of ions, that is, $T_e \approx T_e$, whereas at the upper ionosphere ($z > 200$–250 km), the temperature of electrons can exceed by 1.5–2 times the temperature of ions [7, 9]. More detailed information about the structure and properties of the ionosphere, midlatitude, and high latitude, can be found in References 4–7.

Main Characteristics of the Ionospheric Plasma. The characteristics, which are functions of the main parameters of the plasma such as concentration, content, and temperature, are described next. These include the frequency of interactions (collisions) of charged particles with neutral molecules and atoms, interactions between charged particles, the free path length between interactions, and the pertinent coefficients of diffusion and drift. All these characteristics determine the dynamic processes, diffusion, and drift of plasma in ambient electrical magnetic fields. They occur in the ionosphere and cause the formation of large-scale and small-scale plasma irregularities, which are the main "sources" of fading of radio signals in the ionosphere.

Next, we present the various equations used to evaluate the effects of plasma inhomogeneities on radio signal amplitude and phase scintillation. First, we note that the main characteristics are the frequencies of collisions of electrons and ions with neutral particles and between charged particles. Approximate expressions for these characteristics are fully presented in References 4 and 6–10, from which the frequency of electron-neutral collisions is given as

$$\nu_{em} = 1.23 \cdot 10^{-7} N_m T_e^{5/6}, \tag{13.2a}$$

where $N_m = N_{N_2} + N_{O_2}$, all other parameters were defined earlier. The estimation of frequency of electron–ion interactions for different ions (N_i) and electron temperature (T_e) can be achieved using the following approximation [4, 6–10]:

$$\nu_{ei} = \frac{5.5 \cdot N_i}{T_e^{3/2}} \ln \frac{220 T_e}{(N_i)^{1/3}}. \tag{13.2b}$$

Effective frequency of ion-neutral collisions can be presented as [4, 6–8]

$$\nu_{im} = \beta_{im}^0 N_m (T_i + T_m)^{1/2}, \tag{13.2c}$$

where the coefficient β_{im}^0 accounts for the difference in mass of different molecules, the effects of nonelastic interaction between particles, polarization, and recharge of molecules (see details in References 4 and 6–8). Furthermore, to analyze the effects of the ambient magnetic field on charged particles, the hydromagnetic frequencies of electrons (ω_H) and ions (Ω_H) are also introduced to estimate the effects of charged particles magnetization. As was investigated in References 9 and 10, at lower altitudes of the ionosphere ($z < 120$–150 km), $\omega_H > \nu_{em}$ and $\Omega_H << \nu_{im}$. In this case, electrons are magnetized and ions are not. Hence, the degree of magnetization of ionospheric plasma, which is defined by $\Omega_H \omega_H >> \nu_{im} \nu_{em}$, is very small, whereas in the upper ionosphere, at altitudes more than 150 km, $\omega_H >> \nu_{em}$ and $\Omega_H > \nu_{im}$. In this case, both electrons and ions, and hence, the ionospheric plasma, are fully magnetized.

For actual ionospheric applications in radio communications and radars, another parameter of ionization is usually used [8–12, 46–48], $p = \nu_{ei}/\nu_{em}$, instead of the earlier defined parameter N/N_m. As was mentioned in References 8 and 9, the ionospheric plasma has mainly low temperature (200 K at

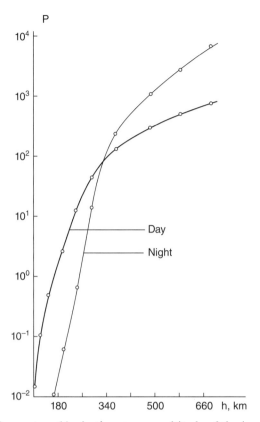

FIGURE 13.4. Parameter of ionization p versus altitude of the ionosphere (in km).

80–90 km to 2500–2800 K at 500–600 km). In such conditions, the cross-sections of collisions between electrons and ions are much greater than the cross-sections of collisions of electrons with neutral particles. So, the peculiarity of the real ionospheric plasma consists of the fact that even for $N/N_m << 1$, the parameter $p = \nu_{ei}/\nu_{em}$ becomes larger than 1. Therefore, electron–ion interactions are an essential part of the transport processes in the ionospheric plasma. Figure 13.4 presents the parameter of ionization p for diurnal and nocturnal middle latitude ionosphere at altitudes of 90–700 km. It is clear that this parameter can reach large values, though in the ordinary sense, the ionospheric plasma is weakly ionized. Another argument for using this parameter as a degree of plasma ionization is based on the fact that at altitudes of 80–500 km, which mostly affects radio propagation, $N_m > N_e$ and $N_m > N_i$ provides the condition of stability of a full plasma pressure necessary for the transport process of diffusion and drift. Under such circumstances, different disturbances of neutral molecules or atoms and their movement do not influence the diffusion of electrons and ions in the ionospheric plasma [8–10]. That is why we

TABLE 13.1. Phase Spectra S(k) per unit Mean Square Fluctuation of Phase Together with the Corresponding Autocorrelation Function $\rho(x)$ (Inner Scale l_0 Is Zero Relative to Outer Scale L_0) (Rearranged from Reference 41)

p	$S(k)$	$\rho(x)$
2	$\dfrac{4L_o}{1+k^2 L_O^2}$	$\exp\left(-\dfrac{x}{L_o}\right)$
3	$\dfrac{2\pi L_o}{(1+k^2 L_O^2)^{3/2}}$	$\dfrac{x}{L_o} K_1\left(\dfrac{x}{L_o}\right)$
4	$\dfrac{8L_o}{(1+k^2 L_O^2)^2}$	$\left(1+\dfrac{x}{L_o}\right)\exp\left(-\dfrac{x}{L_o}\right)$
5	$\dfrac{3\pi L_o}{(1+k^2 L_O^2)^{5/2}}$	$\dfrac{x}{L_o}\left[K_1\left(\dfrac{x}{L_o}\right)+\dfrac{1}{2}\dfrac{x}{L_o}K_0\left(\dfrac{x}{L_o}\right)\right]$

evaluate the degree of ionization of the ionospheric plasma by a parameter of ionization p shown in Figure 13.4 and presented in Table 13.1 for the daily and nocturnal ionosphere. The increase in the degree of ionization in the ionosphere, observed in Figure 13.4, has a very simple physical explanation: with an increase in altitude, the effective values of the collisional frequencies of charged particles with neutral particles drop, as defined by Equations (13.2a) and (13.2c), whereas the frequency of electron–ion interactions increases, as defined by Equation (13.2b).

That is, because with an increase in height, the number of neutral plasma components decreases rapidly, while the total number of ions in the ionosphere increases. It must be noted that from $z > 500$ km, the growth of the parameter p is reduced with an increase in height, while the degree of plasma magnetization (as was mentioned earlier) is monotonically increasing with height.

Ionization–Recombination Balance in the Ionosphere. Changes of plasma components, electrons, and ions in real time is determined by the equation of ionization balance:

$$\frac{\partial N}{\partial t} = q_i - \alpha N^2. \tag{13.3}$$

This is actual if the processes of ionization and recombination are predominant with respect to the transport processes. The latter, diffusion, thermodiffusion, and drift, are predominant, starting from 120 to 150 km, that is, from the E layer of the ionosphere. So, Equation (13.3) is correct only in the D and lower E layers of the ionosphere. In (13.3), $q_i = q_{iO} + q_{iO_2} + q_{iN_2}$ is the total intensity of ionization that accounts for the intensities of oxygen atom and oxygen and nitrogen molecules. α is a coefficient of dissociative recombination. The dissociative recombination is predominant in the ionosphere and is described in the interaction of molecular ion M^+, with electron e^- accompanied by dissociation of a molecular ion on two excited neutral atoms M^* [4–9]. For example, for the molecular ion of oxygen, the dissociative recombination gives [8, 9]

$$O_2^+ + e^- \rightarrow O^* + O^*. \qquad (13.4)$$

Therefore, during the process of dissociative recombination, an amount of electrons involved in this process during 1 second in the area of 1 cm^2, can be presented by Equation (13.3), as αN^2. The coefficient of dissociative recombination can be presented as [8–10]

$$\alpha = \alpha_1 \frac{N_{NO^+}}{N} + \alpha_2 \frac{N_{O_2^+}}{N}. \qquad (13.5a)$$

Here, α_1 and α_1 are coefficients of dissociative recombination of ions NO^+ and O_2^+, respectively, which can be presented in the following form [8, 9]:

$$\alpha_1 \approx 5 \cdot 10^{-7} \left(\frac{300}{T_e} \right)^{1.2} \qquad (13.5b)$$

$$\alpha_2 \approx 2.2 \cdot 10^{-7} \left(\frac{300}{T_e} \right)^{0.7}. \qquad (13.5c)$$

According to (13.5a), (13.5b), and (13.5c), coefficients of dissociative recombination are decreased with the growth of electron temperature. So, in the upper E layer and above, the concentration of electrons (i.e., plasma) is increased with an increase in temperature. For the detailed investigations of ionization–recombination balance in the ionosphere, we refer the reader to References 1–9. We will only note here that all presented equations characterize only a local balance of ionization, because they were obtained without accounting for transport processes, which are actual and important at altitudes above the E layer of the ionosphere.

Transport Processes in the Ionosphere. During the investigation of effects of ionosphere on radio propagation, it is important to analyze the transport processes in the ionosphere, such as diffusion, thermodiffusion, and drift of ionospheric irregularities, large-scale and/or small-scale, connected with the turbulent structure of the atmosphere at lower ionospheric altitudes [27], as well as with numerous instabilities of ionospheric plasma at the upper ionosphere [8, 9]. To understand the main mechanisms of each process, it is important to compare the initial dimensions of the irregularities created in the plasma and the characteristic scales of charged particles in plasma, which we briefly present following References 4–10. Thus, according to the empirical presentation of ionospheric models [4–10], for noon ionospheric conditions and low or moderate solar activity, the mean free path of charged particles can be estimated as: for electrons λ_e is of about 30 cm to 10 km, for ions is about 0.5 cm to 2 km for $z = 90$–500 km. Radius of electron magnetization (called the *Larmor radius* [4–10]), ρ_{He}, is changed in the range of 1 to 5 cm, and that for ions, ρ_{Hii}, is from 0.4 m to 7.5 m at altitudes of 90–500 km.

At the same time, the Debye radius, which defines the plasma quasi-neutrality (see description provided earlier), is changed in the range of 0.5–1 cm at altitudes under consideration. The comparison of inhomogeneities of scales from several meters to several tens of kilometers, which under our consideration (see discussion that follows) with earlier parameters allows us to state that later, we can use the theory of magnetohydrodynamics for all components of the ionospheric plasma: electrons, ions, and neutral particles, and estimate the influence of plasma irregularities of a wide range of scales on radio propagation. This subject will be considered in the next sections.

The qualitative analysis carried out earlier has shown that all characteristic scales of plasma are less than the dimensions of plasma irregularities. Moreover, at altitudes beyond the E layer ($h > 160$–200 km), the transfer processes, diffusion, thermodiffusion, and drift prevail compared with the chemical processes of ionization and recombination. Therefore, for the description of transfer processes beyond altitudes of the upper E layer, we can use a system of magnetohydrodynamic equations for all components of the ionospheric plasma, electrons, and ions, to describe the evolution of the irregular structures of ionospheric plasma of various nature.

The real coefficients of diffusion vary with altitude of the ionosphere non-monotonically [21–26]. Moreover, it was shown that components of the conductivity tensor have their maxima at altitudes of 100, 150, 250, and 300 km, respectively. The same tendency of nonmonotonically height dependence with local maxima was observed for other transport coefficients [21–26]. Hence, apart from the macroscale regular distribution of plasma in the ionosphere, there exists the naturally created nonregular distributions of plasma concentration, the so-called macroscale plasma irregularities, with dimensions from hundreds of kilometers (large-scale disturbances) to tens and few meters (moderate- and small-scale, respectively).

That is why, during the analysis of processes of formation and evolution of plasma inhomogeneities in the ionosphere and their influence on radio propagation, it is necessary to account for the whole spectrum of physical–chemical processes, which form both large-scale regular, moderate-scale, and small-scale irregular structures in the ionospheric plasma [27–37], the effects of which we will investigate in the following section.

13.1.2. Main Effects of Radio Propagation Through the Ionosphere

Now we will discuss the common effects of the ionosphere as a continuous plasma layered medium on radiowave propagation.

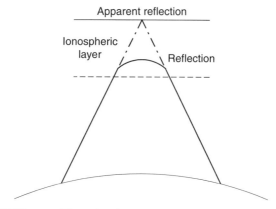

FIGURE 13.5. The reflection phenomenon in the ionosphere.

Refraction. The plasma content of the ionosphere changes the effective refractive index encountered by radio waves transmitted from the Earth, changing their direction by changing the phase velocity. Depending on certain conditions, which are determined by wave frequency, elevation angle of a ground-based or an air-based vehicle antenna, and electron/ion content, the radio wave may fail to escape from the Earth and may appear to be reflected back to Earth. This process is actually due to refraction (see Fig. 13.5). The same effects are described in Chapter 12 for the troposphere as a gaseous quasi-homogeneous continuum. The refractive index n_r of an ordinary radio wave depends on both the plasma density N (in m^{-3}) and the wave frequency f (in MHz) according to References 1–3, 7, and 11:

$$n_r^2 = 1 - \frac{f_c^2}{f^2},$$ (13.6a)

where f_c (in MHz) is the critical frequency of plasma at the given height, defined by the following expression [1–3, 7, 11]:

$$f_c = 8.9788\sqrt{N}\ [\text{Hz}].$$ (13.6b)

Apparent reflection from the ionosphere back to the Earth, as shown in Figure 13.5 [1–3], can occur whenever the wave frequency is below this critical frequency f_c, from which it follows that the "working" frequencies for satellite communications must be above this critical frequency f_c. The greatest critical frequency, usually observed in the ionosphere, does not exceed 12 MHz. This is the other extreme of an overall atmospheric "window" which is bound at the high-frequency end by atmospheric absorption at hundreds of gigahertz, as is shown in Chapter 12. A number of ionospheric effects for radio waves with frequencies above 12 MHz, which are very important in land-satellite communications, will be considered briefly according to the corresponding References 1–3 and 7.

Absorption of Radio Waves. In the absence of local inhomogeneities of the ionospheric plasma, the radio wave passing through the ionosphere as a homogeneous plasma continuum is absorbed due to pair interactions between electron and ion components of plasma. In such a situation, the intensity of radio wave is determined as [7, 11, 36]

$$I = I_0 \exp\left\{-2\frac{\omega}{c}\int \kappa ds\right\}, \tag{13.7}$$

where I_0 is the intensity of the incident radio wave, κ is the coefficient of absorption, ω is the angular frequency of the incident wave $\omega = 2\pi f$, c is the velocity of light in free space $c = 3\cdot10^8$ m/s, and integration in (13.9) occurs along the wave trajectory s.

For weakly magnetized plasma, when $\omega >> \omega_{He}$ (ω_{He} is the hydrofrequency of plasma electrons), and for high radio frequencies, when $\omega >> \omega_{pe}$ (ω_{pe} is the plasma frequency, $\omega_{pe} = (e^2 N/\varepsilon_0 m_e)^{1/2}$, e and m_e are the charge and mass of electron, respectively, ε_0 is the average dielectric parameter of the ambient ionospheric plasma), the coefficient of absorption can be presented in the following form:

$$\kappa = \frac{1}{2}\frac{\omega_{pe}^2(\nu_{em}+\nu_{ei})}{\omega[\omega^2+(\nu_{em}+\nu_{ei})^2]}. \tag{13.8}$$

Here, ν_{em} and ν_{ei} are the frequencies of interactions of plasma electrons with neutral molecules and atoms, and with ions, respectively, defined by Equation (13.2a), Equation (12.2b), and Equation (13.2c) [8]. To estimate losses due to absorption, that is, the part of energy of the radio wave that is absorbed in the quasi-regular layers of the ionosphere, from D to F, a special value of absorption in decibels (dB) is usually introduced [7–12]:

$$A_\omega = 10\log\frac{I_0}{I} \approx \frac{4.3}{c}\int \frac{\omega_{pe}^2(\nu_{em}+\nu_{ei})}{[\omega^2+(\nu_{em}+\nu_{ei})^2]}ds. \tag{13.9}$$

The expected value of absorption at the ionospheric radio links is estimated usually by using the measured radiometric absorption along radio traces at the fixed frequency and by experimental knowledge of frequency dependence of absorption determined by (13.9) [8–10].

Ionospheric Scintillation. There is a wind present in the ionosphere, just as in the troposphere, which causes rapid variations in the local electron density, particularly close to sunset. These density variations cause changes in the refraction of the radio wave in the Earth-satellite channel and hence, in signal levels. Portions of the ionosphere then act like lenses, cause focusing, defocusing, and divergence of the wave, and hence, lead to signal level variations, that is, signal scintillation (see Section 12.2).

To summarize the main propagation effects through the ionosphere as a plasma continuum, we must follow the results presented earlier. For frequencies beyond the range of 20 to 50 GHz, which are usually employed for Earth-satellite communication links, the effects of Faraday rotation are negligible (about a dozen of degrees), the propagation delay is very small (a dozen nanoseconds), and the radio frequency dispersion is very weak (a dozen picoseconds per one megahertz), so we can omit them from our computations. As for the attenuation caused by absorption and refraction, and signal amplitude and phase scintillations (i.e. fading), these effects are strongly dependent on the nonregular features of the ionosphere, usually referred to as inhomogeneities or irregularities [1–12].

13.1.3. Effects of Refraction

Let us now consider the effects of radiowave refraction in the ionosphere, just as was considered in Chapter 12 for the spherically symmetric layered troposphere. The radiowave refraction coefficient depends on the frequency of propagation and ionospheric altitude. The simplified refraction coefficient of plasma for high frequencies is determined by [35]

$$N = -\chi \cdot N_e \cdot f^{-2}, \tag{13.10}$$

where $\chi = 40.4$ if the electron density N_e is expressed in m^{-3}, and f is expressed in Hz. From (13.10) it follows that N is a negative value, and the relationship $N(h)$ repeats the altitude profile of the electron density of the ionosphere— $N_e(h)$. It is essential that this refraction coefficient decreases when the frequency increases as f^{-2}.

Effects of Refractive Angle Deviations. Next, we use Equations (13.10) and (12.29) from Chapter 12, based on the geometry shown in Figure 12.5a, to obtain the expression for the ionosphere refraction angle as

$$\xi_i = -\chi f^{-2} R_c \sin\theta_b \int_0^H \frac{\frac{1}{n}\frac{dN_e(h)}{dh} dh}{\left[n^2(R_c+h)^2 - R_c^2 \sin\theta_b\right]^{1/2}}. \tag{13.11}$$

Here R_c is the Earth's radius, H is the altitude of the transmitter antenna, and $N_e(h)$ is the altitude profile of the electron density, an example of which for middle-latitude ionosphere is shown in Figure 13.3. From Equation (13.11), it follows that ξ_i decreases with an increase in frequency as f^{-2}, and is determined, basically, by the vertical gradient of the electron density. Difficulty to get specific data on ionospheric refraction is connected with the fact that it is

necessary to know a varying altitude profile of the electron density and, especially, the values of the gradient dN_e/dh for each ionospheric latitude, from the
polar cap to the equator. The integral in the expression (13.11) can be divided,
following Reference 35, into two parts corresponding to the refraction in the
lower part of ionosphere, where electron density increases with an increase in
altitude, and in the upper part, where electron density decreases with an
increase in h (see Fig. 12.5, Chapter 12):

$$\xi_i = -\chi \cdot f^{-2} R_c \sin \theta_b (I_1 + I_2), \tag{13.12}$$

where for the lower ionosphere

$$I_1 = \int_0^{b_m} \frac{\dfrac{dN_c}{dh} dh}{\left[(R_c + h)^2 - R_c^2 \sin^2 \theta_b \right]^{1/2}}, \tag{13.13a}$$

and for the upper ionosphere

$$I_2 = \int_{h_m}^{H} \frac{\dfrac{dN_c}{dh} dh}{\left[(R_c + h)^2 - R_c^2 \sin^2 \theta_b \right]^{1/2}}. \tag{13.13b}$$

Here, h_m is the altitude of the main ionospheric maximum. In the expression
(13.13), it was suggested that the index of refraction n is equal to unity. This
approximation is possible if $\theta_b < 80°$ and $f > 40$ MHz. A detailed analysis of
Equation (13.12) with expressions (13.13a) and (13.13b) shows the following.
If the altitude of the radio transmitter is equal or less than 400 km ($H < 400$ km),
then the first integral I_1 becomes the main contribution to refraction. It means
in this case that the beam transmission line bends toward the Earth. If the
altitude of the radio waves source is more than 400 km, it is necessary, as well,
to take into account the second integral I_2. In the upper part of the ionosphere,
the gradient dN/dh has the opposite sign. Therefore, in this part of the ionosphere, the beam bends toward the opposite to the Earth side.

The ionosphere refraction angle ξ_i for $H > 1000$ km is smaller by 30% than
in the case when the altitude H is approximately equal to 400 km. It is essential
to mention that in the ionosphere, the beam bends toward the Earth despite
the fact that the refraction angle is the sum of these two components, I_1 and
I_2. It is necessary also to keep in mind that the ionospheric refraction angle
goes through slow variations caused by daily variations in the $N_e(h)$ relationship and through irregular quick fluctuations caused by statistical inhomogeneities in electron density. Let us compare refraction angle deviations in the
troposphere, ξ_t, and the ionosphere, ξ_i. Thus, at frequency $f = 100$ MHz, angles
ξ_t (see Chapter 12) and ξ_i are approximately the same, but at $f > 1000$ MHz,
$\xi_i << \xi_t$, and this means that the ionospheric refraction can be neglected for
high frequencies with respect to the tropospheric refraction.

Bending of the Radiowave Trajectory. Let us consider the influence of the ionosphere on the bending of the radio beam trajectory, denoted here as ΔL_i. In this case, following Reference 35, we can write

$$\Delta L = \int_0^H N(h)(R_c + h)\left[R_c + h)^2 - R_c^2 \sin^2 \theta_b\right]^{-1/2} dh, \qquad (13.14)$$

which leads to

$$\Delta L_i = \chi f^{-2} \int_0^H N_c(h)(R_c + h_m)\left[(R_c + h_m)^2 - (R_c^2 \sin^2 \theta_b)\right]^{-1/2} dh. \qquad (13.15)$$

From this relationship, we obtain for the vertical radio beam

$$\Delta L_i = \chi f^{-2} \int_0^H N_c(h) dh. \qquad (13.16)$$

Expression (13.15) can be simplified, if we will use the angle θ_m introduced in Chapter 12, and Equation (13.16), according to Reference 35. Finally, we get

$$\Delta L_i = \chi f^{-2} I_c \cos^{-1} \theta_m, \qquad (13.17)$$

where

$$I_c = \int_0^H N_c(h) dh \qquad (13.18)$$

is an integral electron density for the vertical beam. From Chapter 12, based on the geometrical optics approximation, $\cos \theta_m$ can be determined by the following expression:

$$\cos \theta_m = (R_c + h)^{-1}\left[(R_c + h)^2 + R_c^2 \sin^2 \theta_b\right]^{1/2} dh. \qquad (13.19)$$

From Equation (13.17), it follows that ΔL_i is proportional to the integral electron density and decreases with an increase in frequency as f^{-2}. Let us consider the results of experimental determinations of the integral electron density I_c, described in Reference 35 and, on this basis, find ΔL_i. Values of I_c were determined by receiving satellite signals at two frequencies. Measurements, which were carried out during the period of low solar activity, revealed that the integral electron density has a clearly defined quotidian variability: in summer at midnight, I_c is equal to Equation $(4–8) \times 10^{16} \text{ m}^{-3}$, and for given values of the integral electron density, ΔL_i will have the midnight values of 18–36 m and the midday values of 45–135 m. With the medium and high solar activity of the Sun in summer, at night, and predawn hours, I_c has the values of Equation $(8–12) \times 10^{16} \text{ m}^{-3}$, and at midday, a significantly higher integral electron density, reaching Equation (3–5) $\times 10^{17}$ m^{-3}, is observed. Under these conditions, for $\theta_0 = 0$ and the given wave length, ΔL_i will be in the range of 36–54 m at night, and variations within the limits of 130–230 m will be observed at noon. In winter, the quotidian variability I_c is expressed especially strongly: with a medium activity of the Sun during night and predawn hours, this value is usually equal to Equation (2–5) $\times 10^{16}$ m^{-3}, and at noon it is equal to Equation (2–4) $\times 10^{17}$ m^{-3}. This corresponds to ΔL_i variations for the vertical radio beam in the range of 9–22 m at night and of 90–180 m at noon. The values of radio-wave lag in the ionosphere (ΔL_i) can be expressed also versus the wave length. Thus, at wavelengths close to 10 cm, ΔL_i of the vertical radio beam can vary within the limits of 0.2–2.3 m. Moreover, it was found that the influence of the ionosphere on ΔL_i can be ignored at wavelengths of $\lambda < 3$ cm. We should mention here that because of the dynamic nature of the ionosphere, the ΔL_i variations are strong and theoretical calculations of the bending of radio waves in the ionosphere are of low accuracy. That is why, the two-frequency exclusion method is used, practically, for the ΔL_i estimations in navigational satellite measurements (for more details we refer the reader to the reference presented in Reference 35).

13.2. EFFECTS OF THE INHOMOGENEOUS IONOSPHERE ON RADIO PROPAGATION

Let us now introduce the reader to some very important "thin" effects on radio propagation that occur in the ionospheric inhomogeneous plasma medium consisting of different kinds of irregulatities in a wide range of scales: from small to large [7–12].

The ionosphere varies randomly in time and space, such that the amplitude and phase of propagating waves may similarly fluctuate randomly in these domains. The inhomogeneity of the ionosphere is an important factor in determining VHF/X-band wave propagation conditions (see References 1–3, 7, 11, and 12). As a result, interest in satellite communications has stimulated investigations of ionospheric properties, in particular, the analysis of the spatial–temporal distribution of ionospheric irregularities [8–10].

Many experiments are carried out using ground facilities (radars and iono-sounds) [13–20]. The methods of active modification of the ionosphere [21–27] and direct satellite measurements [28, 29] show that in the normal ionosphere, there exists a wide spectrum of irregular inhomogeneities, which cause various radio physical effects, such as interference, scattering, diffraction, and refraction of radio waves passing through the ionosphere, as well as variations of the incident angles of reflected waves caused by refraction (multirays effect), and so on [30–37]. When waves are propagated through an irregular medium, small- to large-angle scattering causes a phenomenon that is known as scintil-lation. We must note that the same phenomenon was found in the troposphere (see Section 12.3). All these effects result in amplitude and phase fluctuations of radio signals near the ground surface, changes in the duration and shape of radio waves, and finally, a decrease in signal-to-noise ratio (SNR or S/N).

The influence of inhomogeneities with various scales (large-scale to small-scale) on the effectiveness of the satellite terrestrial communication will be discussed later in Sections 13.2.1, 13.2.2, and 13.2.3. It should be emphasized that because of the nature of the problem, it is necessary to employ various approximations to obtain useful results and hence, approximation techniques applicable to a variety of different plasma inhomogeneities, are presented.

13.2.1. Propagation Effects of Large-Scale Inhomogeneities

In a spherical-symmetric homogeneous ionosphere, radio waves propagate in a plane of a *great circle* [1–7]. In the presence of large-scale inhomogeneities (with the horizontal scale L larger than the radius of the first Fresnel zone $d_F = (\lambda R)^{1/2}$, where λ is the wavelength and R is the distance from the ground facilities to the inhomogeneous area of the ionosphere, that is, $L \gg d_F$, the radio waves can change their direction from this plane.

Main Equations. For the description of radiowave propagation in an anisotropic medium, such as plasma, written in the Cartesian coordinate system with the origin in the center of the Earth and for a radiation frequency exceeding the plasma hydromagnetic frequency ($\omega_0 \gg \omega_{He}$), we can present the wave equation as (see Chapter 4)

$$\Delta E + k_0^2 \varepsilon E = 0, \tag{13.20}$$

where E is the component of electric field of the radio wave along the radius vector $\mathbf{r}$, $k_0 = 2\pi/\lambda$, $\varepsilon = \varepsilon(\mathbf{r})$ is the complex relative dielectric permittivity of the ionospheric plasma, $R = |\mathbf{r}|$, ω_0 is the wave frequency, ω_{He} is the electron hydromagnetic frequency, and c is the speed of light. Neglecting absorption and taking the case when the frequency ω_0 is more than the plasma frequency ω_{pe}, the dielectric permittivity can be presented as

$$\varepsilon = 1 - \frac{\omega_{pe}^2}{\omega_0^2}. \tag{13.21}$$

Clearly, the plasma frequency is a function of the spatial coordinates. Let us present the field E as [7, 11, 12]

$$E = E(\mathbf{r}) \exp[i\Phi(\mathbf{r})], \tag{13.22}$$

where E is the amplitude and $\Phi(\mathbf{r})$ is the phase of the radio wave. The wave vector is

$$\mathbf{k} = \nabla\Phi(\mathbf{r}). \tag{13.23}$$

For small variations of the amplitude at distances comparable with the wave length, from (13.21), (13.22), and (13.23), one obtains

$$k^2 = k_0^2 \varepsilon(\mathbf{r}). \tag{13.24}$$

Method of Characteristics. After differentiating Equation (13.20) and knowing the relationship $\nabla\cdot\mathbf{k} = \nabla\nabla\Phi(\mathbf{r}) = 0$, the equation for the wave vector k was found to be [11, 12]

$$(\mathbf{k}\cdot\nabla\cdot\mathbf{k}) = (1/2)k_0\nabla\varepsilon(\mathbf{r}). \tag{13.25}$$

The solution of Equation (13.25) can be defined using the method of characteristics used in References 11 and 12. This method transforms (13.25) into a characteristic equation, which in the spherical coordinate system $\{r, \theta, \varphi\}$ can be presented for all the components of wave vector $\mathbf{k} = \{k_r, k_\varphi, k_\theta\}$ as

$$\frac{dr}{k_r} = r\frac{d\theta}{k_\theta} = r\sin\theta\frac{d\varphi}{k_\varphi} = \frac{dk_r}{\dfrac{k_0^2}{2r^2}\dfrac{\partial(r^2\varepsilon)}{\partial r} - \dfrac{k_r^2}{r}} = \frac{dk_\theta}{\dfrac{k_0^2\cos\theta}{r} - \dfrac{k_0^2}{2r}\dfrac{\partial\varepsilon}{\partial\theta} - \dfrac{k_r k_\theta}{r}}$$
$$= \frac{dk_\phi}{\dfrac{k_0^2}{2r\sin\theta}\dfrac{\partial\varepsilon}{\partial\varphi} - \dfrac{k_\theta k_\varphi}{r}\cot\theta - \dfrac{k_r k_\varphi}{r}}. \tag{13.26}$$

From expression (13.26), it follows that

$$\frac{\partial r}{\partial\varphi} = \frac{rk_r}{k_\theta}, \quad \frac{\partial\varphi}{\partial\theta} = \sin\theta\frac{k_\varphi}{k_\theta}. \tag{13.27}$$

After differentiation of (13.26) over the θ, we finally get

$$\frac{\partial^2\varphi}{\partial\theta^2} + 2\cot\theta\frac{\partial\varphi}{\partial\theta} + \frac{1}{2}\sin 2\theta\left(\frac{\partial\varphi}{\partial\theta}\right)^3 = \frac{k_0^2}{2k_0^2\sin^2\theta}\frac{\partial\varepsilon}{\partial\varphi} - \frac{1}{2}k_0^2\frac{\partial\varphi}{\partial\theta}\frac{\partial\varepsilon}{\partial\theta}. \tag{13.28}$$

If inhomogeneities do not exist, that is, $\partial\varepsilon/\partial\varphi = \partial\varepsilon/\partial\theta = 0$, then from (13.28), the integral becomes

$$\tan\theta\sin(\varphi - \varphi_0) = C_0, \tag{13.29}$$

where φ_0 and C_0 are the integration constants obtained. Equation (13.29) for the selected C_0 and φ_0 describes a circle with its center at the origin of the coordinate system. Thus, in References 11 and 12, a rule is followed, that is

In the homogeneously layered ionosphere (independent of θ and φ), the ray trajectory lies in the plane of a great circle.

However, in the presence of large-scale irregularities the derivative $d\varphi/d\theta = 0$, but $d\varepsilon/d\theta \ll 1$. This enables terms proportional to $(d\varphi/d\theta)^3$ and $(d\varphi/d\theta)$ $(\partial\varepsilon/\partial\theta)$ to be neglected. Thus, the ray trajectory in a weak inhomogeneous plasma is defined by the equation

$$\frac{\partial^2\varphi}{\partial\theta^2} + 2\cot\theta\frac{\partial\varphi}{\partial\theta} = \frac{1}{2}\sin^2\theta\frac{\partial\varepsilon}{\partial\varphi}. \tag{13.30}$$

As was shown in References 11 and 12, analyzing Equation (13.30) results in

The existence of large-scale inhomogeneities causes the ray trajectory to deviate from the plane of the great circle.

The same effect, defined as refraction in the troposphere, is discussed in Section 13.2.

The Curved Smooth Screen Model. The problem of multimode reflection from ionospheric layers can be resolved by using Equation (13.25). However, there exists a simpler method based on the model of a curved smooth screen moving with respect to the observer. This model is correct if $\omega_{pe} \gg \omega_0$, where ω_{pe} is a critical frequency of plasma for each ionospheric layer, and $L \gg d_F$. The geometric optic approximation is applied in this method. The simplest model of a curved mirror is the smooth screen with a sinusoidal shape, which moves with a constant speed without changing shape. Let us consider a rectangular coordinate system $\{X, Y, Z\}$ with Z-axis in the vertical direction and the base on the ground surface. The receiver and transmitter are at the origin of the system, the altitude Z is a function of time t and coordinate X, $Z = Z(X,t)$, as is shown in Figure 13.6. The coordinates of the reflected points are defined from the following condition:

$$Z(X,t)\frac{\partial Z(X,t)}{\partial X} = -X. \tag{13.31}$$

If ionospheric inhomogeneities have the shape of a wave with amplitude δ, wavelength Λ, and period T, then

$$Z(X,t) = Z_0[1 + \delta\sin 2\pi(X/\Lambda - t/T)], \tag{13.32}$$

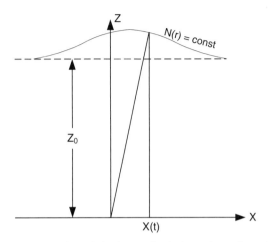

FIGURE 13.6. Modeling of the ionospheric layer by a sinusoidal layer.

and from (13.31), it follows that

$$(2\pi Z_0^2 \delta / \Lambda) \cos 2\pi (X / \Lambda - t / T)[1 + \delta \sin 2\pi (X / \Lambda - t / T)] = -X. \quad (13.33)$$

Here, Z_0 is the average height of the reflected level. For relatively small oscillations of the reflected level ($\delta \ll 1$), ordinates X of the reflected points are found from the solution of the transcendental equation [11, 12]:

$$(2\pi Z_0^2 \delta / \Lambda) \cos 2\pi (X / \Lambda - t / T) = -X. \quad (13.34)$$

From (13.34), it follows that the number of solutions depends on the position of the screen and on the amplitude δ. As was shown in References 11 and 12, the solution is unique if

$$2\pi Z_0^2 \delta / \Lambda < 1. \quad (13.35)$$

The inequality in (13.35) represents the condition of a single ray reflection from a smooth surface. If condition (13.35) is not valid, then there are several rays that contribute to the interference effect at the receiving point.

From Equations (13.31) and (13.32), we can calculate the incident angle θ' of the radio wave at the receiving point:

$$\cot \theta' = X / Z = (2\pi Z_0 \delta / \Lambda) \cos 2\pi (X / \Lambda - t / T). \quad (13.36)$$

The maximum ray reflection from the vertical axis, which is a half width of the angle spectrum γ equals

$$\gamma = 2\pi Z_0 \delta / \Lambda. \quad (13.37)$$

Amplitude and Phase of Reflected Waves. If A_0 is the amplitude of the signal reflected from the level of equal plasma density, then the amplitude A of a radio wave reflected from the smooth surface at the point X is [11, 12]

$$A(X,t) = \frac{A_0}{(1 - Z_0/\rho)},$$
(13.38a)

where $\rho = [1 + (dZ/dX)2]^{3/2}/(d^2Z/dX^2)$ is the radius of curvature of the screen at the point (X, Z). Relationship (13.38a) describes the focusing and defocusing effects of the reflected radio waves caused by large-scale inhomogeneities. The phase at the receiving point is

$$\Phi(X,t) = 2\omega R/c + \pi/4,$$
(13.38b)

where R is the optical path length from the reflector to the receiving point, and factor $\pi/4$ corresponds to the change of wave phase when reflection from the ionosphere takes place. When oblique propagation of radio waves takes place for large distances, the ray is reflected from an ionospheric surface of about 10,000 km². Such a surface contains many inhomogeneities, with about 20–100 km scales, on which radio wave scattering takes place. In the geometrical optics approximation, the wave scattering due to ionospheric density fluctuations is the same as the scattering from the rough surface (see Chapter 5). If now the radio wave falls on the bottom boundary of the ionosphere under the angle θ_0, then the angle $\theta = \theta(X, Y, Z)$ will be changed according to Snell's law:

$$[\varepsilon(X, Y, Z)]^{1/2} \sin\theta(X, Y, Z) = \sin\theta_0$$
(13.39)

during the field penetration to the upper altitudes. At the turning point $Z = Z_m(X,Y)$ the angle $\theta = \pi/2$ and reflection takes place. The height of a turning point is defined as a minimum root $Z = Z_m$ of the following equation [11, 12]:

$$1 - \varepsilon(X, Y, Z) = \cos^2\theta_0.$$
(13.40a)

If $N_0(Z)$ is an average concentration in the ionospheric level and $N_1(Z)$ is a perturbed density caused by plasma inhomogeneities, then the right part of (13.40a) can be rewritten as

$$1 - \varepsilon(X, Y, Z) = 1 - \tilde{\varepsilon}_0 - \tilde{\varepsilon}_1,$$
(13.40b)

where

$$\tilde{\varepsilon}_0 = e^2 N_0(Z)/m_e\omega^2\varepsilon_0$$
(13.41a)

and

$$\tilde{\varepsilon}_1 = e^2 N_1(Z) / m_e \omega^2 \varepsilon_0. \qquad (13.41b)$$

In the absence of inhomogeneities, the height of the reflection point (turning point) Z_{0m} is a function of coordinates X and Y, and the thickness of the reflected layer is defined by the vertical oscillations of equal electron density level (i.e., by a mean square deviation of the turning point from the height Z_{0m}). Thus, even weak large-scale inhomogeneities also increase the thickness of the reflecting layer. The radiowave trajectory due to horizontal changes of screen heights has a complex oscillatory character.

If the frequency of the ionospheric layer is f_0, radiowave reflection can occur for frequencies $f < f_0 \cos ec\,\theta_0$, and radiowave penetration for frequencies $f > f_0 \cos ec\,\theta_0$. The first constraint shows the possibility to reflect and scatter radio waves with frequencies $f < f_0 \cos ec\,\theta_0$. The second condition shows the possibility of land-satellite radiowave communication for frequencies more than the maximum useful frequency $f_{MUF} \equiv f_0 \cos ec\,\theta_0$.

Radio Waves with Frequency $\omega > \omega_{pe}$. Now, some effects of large-scale inhomogeneities for radio waves with frequency $\omega > \omega_{pe}$ are presented, where ω_{pe} is the plasma frequency of the ionosperic layer defined earlier.

Let us suppose that the wave propagates vertically down and passes through the layer with inhomogeneous density [11, 12]. After passing the layer at the height Z_0, the phase Φ of the wave will be a function of the horizontal coordinate X (see Fig. 13.7, extracted from References 11 and 12):

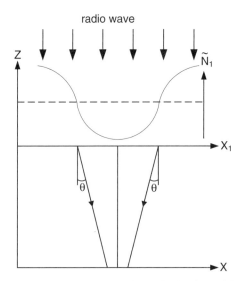

FIGURE 13.7. Penetration of radio waves through the sinusoidal ionospheric layer.

$$\Phi(X_1) = \frac{\omega}{c} R(X_1). \tag{13.42}$$

It can be seen that inhomogeneities are stretched along the Z-axis and N_1 does not depend on Z. For radio frequencies with $\omega > \omega_{pe}$, $\varepsilon_0 \sim 1$ and

$$R(X_1) \sim \frac{e^2}{m_e \omega^2 \varepsilon_0} N_1(X_1), \tag{13.43}$$

where

$$N_1(X_1) = \int N_1(X_1, Z) dz. \tag{13.44}$$

A ray passing through the layer at the point X_1 changes its trajectory from the vertical axis by the angle θ

$$\theta = \frac{c}{\omega} \frac{\partial \Phi}{\partial X_1} = \frac{\partial R_1}{\partial X_1} \tag{13.45}$$

and comes to the Earth's surface at the point

$$X = X_1 + Z_0 \tan \theta. \tag{13.46}$$

If the wave amplitude at the height Z_0 is equal to $A(Z_0)$, then from the law of energy conservation in the ray tube with scale dX, it follows that

$$|A(Z_0)|^2 dX_1 = |A(Z=0)|^2 dX. \tag{13.47}$$

From Equations (13.46) and (13.47), the wave amplitude at the Earth's surface is found to be

$$|A(Z=0)| = |A(Z_0)| \left| \frac{dX_1}{dX} \right|^{1/2} = |A(Z_0)| \cdot \left| 1 + Z_0 \frac{d^2 R_1}{dX^2} \right|^{-1}. \tag{13.48}$$

If $d^2 R_1/dX^2 > 0$, then $A(0) < A(Z_0)$, and there is a *defocusing* effect; whereas, when $d^2 R_1/dX^2 < 0$, then $A(0) > A(Z_0)$, and there is a *focusing* effect. The same effects occur in the troposphere (see Chapter 12). The phase of the wave passing through the layer at point X_1 and reaching point X on the Earth's surface (Fig. 13.7) is

$$\Phi(X, X_1) = \Phi_0 + \frac{\omega}{c} Z_0 [\sec \theta(X_1) - 1], \tag{13.49}$$

where Φ_0 is the wave phase in the absence of inhomogeneities. The rays come from different points at point X on the layer boundary. The difference of ray phases between points X_1 and X_1', observed at point X of the Earth's surface is [11, 12]

$$\Delta\Phi = \Phi(X, X_1) - \Phi(X, X_1') = \frac{\omega}{c} Z_0 \left[\sec\theta(X_1) - \sec\theta(X_1') \right]. \quad (13.50)$$

If rays coming to point X at the Earth's surface have the phase difference $\Delta\Phi \geq \pi/2$, then the essential interference picture is recorded. Calculations show that inhomogeneities in the ionospheric F region ($Z = 200$–300 km) cause an interference pattern for waves with frequency $f < 40$ MHz and only for the relative changes of ionospheric plasma density $N_1/\tilde{N}_0 > 0.002$, where, according to Reference 11, $\tilde{N}_0 = \int_{Z_0}^{\infty} N_0 dZ$ is the plasma electrons content in the entire ionospheric layer. Thus, large-scale inhomogeneities exist when the horizontal scale of the irregularity is larger than the width d_F of the first Fresnel zone. The phenomena of large-scale inhomogeneities are important for different and varied ranges of radiowave frequencies, during which radio waves are reflected from the ionospheric layer with large-scale inhomogeneities. In this situation, a number of radiophysical effects occur:

- a deviation of the radio wave's direction from the great circle plane
- an increase in the vertical size of the layer forming the reflected signal
- a change of propagation of the radio wave to "multimode"
- a complication of pattern and additional modulation of radiowave amplitude due to the arrival of several rays at the receiving point with different phases.

13.2.2. Propagation Effects of Small-Scale Inhomogeneities

Radiowave diffraction effects due to small-scale inhomogeneities that are usually contained in the ionospheric E and F layers are examined next. The influence of small-scale inhomogeneities results mainly in phenomena like diffraction and scattering.

Perturbation Method. For inhomogeneities with characteristic scale of $l << d_F$, (where $d_F = [\lambda R]^{1/2}$ and l represents the scale of inhomogeneity), and with $l < \lambda$, the geometrical optic approximation is not valid, and we must use the method of perturbation for Equation (13.25). As before, the dielectric permittivity ε in (13.25) is presented as a sum of $\tilde{\varepsilon}_0 + \tilde{\varepsilon}_1$, where $\tilde{\varepsilon}_0$ and $\tilde{\varepsilon}_1$ can be defined by Equations (13.41a) and (13.41b), respectively. Later, we will summarize some of the general results mentioned in References 11, 12, and 29–41.

The height $Z = Z_0$ is the height for which reflection takes place ($\tilde{\varepsilon}_0 = 0$). For the heights $Z_0 - D < Z < Z_0$ a model of a linear layer with thickness D can be used. As a result, the dielectric permittivity can be rewritten in the following form [11, 12]:

$$\tilde{\varepsilon}_0(Z) = -\frac{Z - Z_0'}{D}. \tag{13.51}$$

The field change E in the horizontal homogeneous layer of the ionosphere with $\tilde{\varepsilon}_0$ defined from (13.51) and for the normal wave incidence on the layer, can be described by the equation [11, 12]

$$\frac{\partial^2 E}{\partial Z^2} - \left[k_0^2 \frac{Z - Z_0'}{D}\right] E = 0. \tag{13.52}$$

This equation has an exact solution which can be expressed by the Airy functions, U and V:

$$U(-\tau) = \frac{1}{\pi^{1/2}} \int_0^{+\infty} \sin(x^3/3 - x\tau) dx \tag{13.53a}$$

$$V(-\tau) = \frac{1}{\pi^{1/2}} \int_0^{+\infty} \cos(x^3/3 - x\tau) dx, \tag{13.53b}$$

where for $\tau = k_0^{2/3} Z / D^{1/3}$. Small-scale inhomogeneities have a weak influence on wave propagation. If small-scale inhomogeneities of plasma density exist in the ionospheric layer, then the field changes are described by the equation

$$\Delta E + k_0(\tilde{\varepsilon}_0 + \tilde{\varepsilon}_1) E = 0, \tag{13.54}$$

which is solved by the method of perturbations, that is, the Airy function $V(-\tau)$ can be assumed as a standard function described in the following form:

$$E = E_0 \exp\{ik_0 g\} V(-k_0^{2/3}\Phi), \tag{13.55}$$

where E_0 is the amplitude of the incident wave on the boundary of the layer; g and Φ are the logarithm of wave amplitude and the wave phase, respectively. By putting expression (13.55) in (13.54), the following system is obtained

$$2\nabla g \nabla \Phi + k_0^{-1}\nabla^2 \Phi = 0$$
$$-k_0^{-1}\Delta g + (\nabla g)^2 - \Phi(\nabla \Phi)^2 + \tilde{\varepsilon}_0 + \tilde{\varepsilon}_1 = 0. \tag{13.56}$$

Let us suppose that k_0^{-1} and ε_1 are small, and present functions g and Φ as perturbation sums:

$$g = g_0 + g_1 + \cdots, \quad \Phi = \Phi_0 + \Phi_1 + \cdots. \tag{13.57}$$

Then for the first and second order approximation of (13.56), the wave phase can be given as [11, 12]

$$\Phi_0 = 2k_0 X_0 = 2k_0 \int_Z^{Z_0'} (\tilde{\varepsilon}_0)^{1/2} dZ \tag{13.58a}$$

$$\Phi_1 = 2k_0 X_1 = 2k_0 \int_Z^{Z_0'} \tilde{\varepsilon}_1 / (\tilde{\varepsilon}_0)^{1/2} dZ. \tag{13.58b}$$

Parameter Φ_0 defines the phase of a non-perturbed wave and Φ_1 defines the disturbance of phase in the ionospheric layer with small-scale inhomogeneities. The function g_1 in expansion (13.57) can be expressed as [11, 12]

$$g_1 = (1/2k_0) \ln[|E(Z)|/E_0] = -(1/2k_0) \int_Z^{Z_0'} [\Delta_\perp X_1 / (\tilde{\varepsilon}_0)^{1/2}] dZ, \tag{13.59}$$

and it describes the changes of the signal level in the process of scattering, and $\Delta_{\perp;} = \partial^2/\partial X^2 + \partial^2/\partial Y^2$.

The Cross-correlation Function of Phase Disturbances. Using expressions (13.53), (13.54), (13.55), (13.56), (13.57), and (13.58), we can find the cross-correlation function of phase disturbance Φ of the radio wave reflected from a layer of thickness D:

$$\Gamma_\Phi(\xi, \eta) = k_0 \int_{Z_0'-D}^{Z'} dZ_1 \int_{Z_0'-D}^{Z_0'} dZ_2 \langle \tilde{\varepsilon}_1(X_1, Y_1, Z_1) \tilde{\varepsilon}_1(X_2, Y_2, Z_2) \rangle$$

$$= 2k_0^2 \int_0^D \Gamma_\varepsilon(\xi, \eta, \zeta) d\zeta \int_{Z_0'-D-\zeta/2}^{Z_0'-\zeta/2} D/[(Z-Z_0') - \zeta^2/4] dZ, \tag{13.60a}$$

where $\xi = X_1 - X_2, \eta = Y_1 - Y_2, \zeta = Z_1 - Z_2$. The second integral in the earlier equation equals

$$\int_{Z_0'-D-\zeta/2}^{Z_0'-\zeta/2} D/[(Z-Z_0') - \zeta^2/4] dZ = D\{\ln[D - \zeta/2 + (D^2 - \zeta^2/4)^{1/2} - \ln(\zeta/2)\}$$

$$\approx D \ln(4D/\zeta). \tag{13.60b}$$

For the following calculations, we consider that the shape of the inhomogeneities of plasma density $\delta N \equiv N_1 < N_0$, to be distributed according to the Gaussian law inside an inhomogeneous ionospheric layer at the height Z_0, as [21–27]:

$$\Gamma_\varepsilon(\xi, \eta, \varsigma) = \Gamma_\varepsilon(\xi, \eta) \exp\left\{\frac{\varsigma^2}{l^2}\right\}, \tag{13.61}$$

where l is the characteristic scale of $|N_1|^2$ changing along the Z-axis. If

$$\Gamma_\varepsilon(0, 0, 0) = \langle \tilde{\varepsilon}_1^2 \rangle = \frac{\omega_{pe}^4}{\omega^4} \left\langle \left|\frac{N_1}{N_0}\right|^2 \right\rangle \tag{13.62}$$

does not depend on Z, then the maximum of the cross-correlation function of the phase fluctuations is

$$\Gamma_\Phi(0, 0) = \langle \Phi_1^2 \rangle = \pi^{1/2} k_0^2 l D \langle \tilde{\varepsilon}_1^2 \rangle \ln\left(\frac{8D}{l} + \frac{C}{2}\right), \tag{13.63}$$

where C is the Euler constant [11]. Expression (13.63) takes into account that $\langle N_1^2 \rangle$ is the same for all ionosphere altitudes. If relative fluctuations of plasma density $\langle |N_1/N_0|^2 \rangle$ depend on the height, then,

$$\langle \tilde{\varepsilon}_1^2 \rangle = (1 - \tilde{\varepsilon}_0)^2 \left\langle \left|\frac{N_1}{N_0}\right|^2 \right\rangle \tag{13.64}$$

is the function of Z. In this case

$$\langle \Phi_1^2 \rangle = \pi^{1/2} k_0^2 l D \left\langle \left|\frac{N_1}{N_0}\right|^2 \right\rangle \ln\left(\frac{8D}{l} + C - \frac{3}{2}\right). \tag{13.65}$$

From Equation (13.65), it is seen that the main contribution to the phase fluctuations of the radio wave are determined by the inhomogeneities placed near the reflected level Z_0', where $\tilde{\varepsilon}_0 \sim 0$. If the following form of spectrum of plasma density fluctuations $U_N(K)$ or of dielectric permittivity fluctuations $U_\varepsilon(K)$ is used, then

$$U_N(K) \sim U_\varepsilon(K) = M_\varepsilon \left[1 + (K_X^2 + K_Y^2 + K_Z^2) L_0^2 / 4\pi^2\right]^{-p/2}, \tag{13.66}$$

where L_0 is an external (outer) scale of inhomogeneities and M_ε is determined by the condition $\langle \tilde{\varepsilon}_1^2 \rangle = \int U_\varepsilon(K) dK_X dK_Y dK_Z$, then the correlation function of phase fluctuations can be presented as

$$\Gamma_\Phi(\xi, \eta) = 2k_0^2 D M_\varepsilon \int\limits_{Z_0'-D}^{Z_0'} dK_X dK_Y dK_Z \int\limits_{Z_0'-D-\varsigma/2}^{Z_0'-\varsigma/2} dZ \exp\{i(\xi K_X + \eta K_Y + \varsigma K_Z)\}$$

$$\times \left\{[1 + (K_X^2 + K_Y^2 + K_Z^2) L_0^2 / 4\pi^2]^{-p/2} / Z - Z_0' - \varsigma^2/4^{1/2}\right\}. \tag{13.67}$$

In Reference 23, it was pointed out that Equation (13.67) cannot be analytically integrated. Moreover, for the case $L_0 << D$, calculations of the spectrum of phase fluctuations Φ_1 have shown that spectrum U_{Φ_1} does not reproduce spectrum U_N.

The Thin Screen (Kirchhoff) Approximation. Now, using the equations presented earlier, the question of phase and amplitude fluctuations for the case of low-orbit satellite communication can be investigated in more detail, using the thin screen approximation method [33, 40–41] presented schematically in Figure 13.8. Let us suppose that the satellite trajectory and the receiving point on the Earth's surface are in the magnetic meridian plane. To simplify the problem, it is assumed that geomagnetic field lines are vertical at point O on the Earth's surface (see Fig. 13.9). The z-axis is directed along the magnetic field lines and the x-axis lies in the meridian plane. The case $\omega_0 >> \omega_{pe}$, a real case in satellite mobile communication, is now discussed. Due to diffraction and scattering effects at the small-scale inhomogeneities, radio waves from a satellite located at point P have a stochastic modulation of phase after passing through the ionospheric layer having thickness L.

For the case of waves from VHF to X-band, that is, for $\lambda << l$, where l is the scale of the inhomogeneity, that is, from a few meters up to a few centimeters, and for weak inhomogeneities ($\langle \Phi_1^2 \rangle << 1$), the angle of scattering of

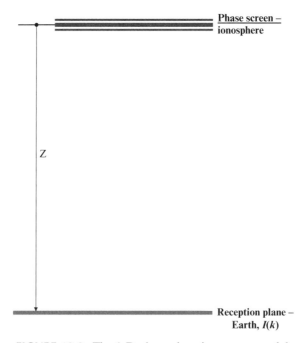

FIGURE 13.8. The 1-D phase changing screen model.

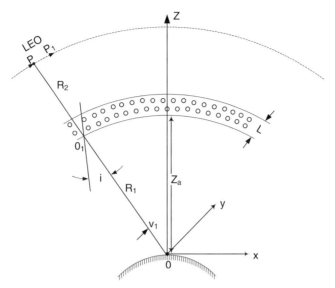

FIGURE 13.9. Geometry of a link LEO-satellite communication with the terminal antenna placed at the Earth's surface.

radio waves $\Phi_s \sim (\lambda/l)\langle\Phi_1^2\rangle$ is small. In the coordinate system $\{x',y',z'\}$ with the z'-axis directed along the ray OP, the phase fluctuations Φ_1 on the bottom boundary of the layer are (see Fig. 13.9)

$$\Phi_1(x', y') = k_0 L \sec i \int\limits_{Z_a}^{Z_a+L\sec i} \varepsilon((x', y', z')dz'. \tag{13.68}$$

Then, a cross-correlation function Γ_ϕ can be presented as

$$\Gamma_\phi(\xi, \eta) = k_0^2 L \sec i \int\limits_{-L\sec i}^{L\sec i} \Gamma_\varepsilon(\xi', \eta', \zeta')d\zeta', \tag{13.69}$$

where $\xi' = x_1' - x_2'$, $\eta' = y_1' - y_2'$, $\zeta' = z_1' - z_2'$. Using Kirchhoff's diffraction equations, we can calculate the wave field strength at the receiving point following Reference 33:

$$E_r = iE_0 \frac{\exp[-ik_0(R_1 + R_2)]}{\lambda R_1 R_2} \int \exp\{i\Phi_1(x', y')\}\exp\left\{-i\pi\frac{x'^2 + y'^2}{\lambda\tilde{R}}\right\}dx'dy', \tag{13.70}$$

where E_0 is the amplitude of wave at a transmitting point P, $\tilde{R} = R_1 R_2 /(R_1 + R_2)$.

If Φ_1 is distributed according to $<\exp(i\Phi_1)> = \exp(-\langle\Phi_1^2\rangle/2)$, we can obtain the average field $\langle|E_r|\rangle$ at the receiving point:

$$\langle|E_r|\rangle = \frac{E_0}{R}\exp\left(-\frac{\langle\Phi_1^2\rangle}{2}\right). \tag{13.71}$$

The average field is attenuated because part of the wave energy is transformed into the noncoherent component of the field. When the satellite moves from point P to point P_1 with coordinates $\{x', y', z'\}$ and $R' = R_1' + R_2'$ (see Fig. 13.9), the received field equals

$$E_r = iE_0\frac{\exp[-ik_0(R_1' + R_2')]}{\lambda R_1 R_2}\int\exp\{i\Phi_1(x'', y'')\}$$

$$\times\exp\left\{-i\pi\frac{\left[\left(x'' - \frac{R_2}{\tilde{R}}x_1'\right)^2 + y'^2\right]}{\lambda\tilde{R}}\right\}dx''dy''. \tag{13.72}$$

Here, the difference between R and R' is taken into account only in phase multiplication. Thus, the cross-correlation function of the received field can be presented as

$$\Gamma_E = \langle|(E_r - <E_r>)(E_{r1} - <E_{r1}>)|\rangle$$

$$= \frac{E_0^2}{R^2}\left\{\frac{1}{\lambda^2\tilde{R}^2}\int\langle\exp[i\Phi_1(x', y') - i\Phi_1(x'', y'')]\rangle\right.$$

$$\times\exp\left[-\frac{i\pi}{\lambda\tilde{R}}\left(x'^2 - \left(x'' - \frac{R_2}{\tilde{R}}x_1'\right)^2 + y'^2 - y''^2\right)\right]dx'dy'dx''dy'' - \exp(-\langle\Phi_1^2\rangle)\right\}. \tag{13.73}$$

Using the relationship $\rho_{\Phi1}(\xi', \eta') = \Gamma_{\Phi1}/\langle\Phi_1^2\rangle$, we can calculate the cross-correlation function of the wave field [36, 37]:

$$\Gamma_E = \frac{E_0^2}{R^2}\left\{\exp\left[-\langle\Phi_1^2\rangle\left(1 - \rho_{\Phi1}\left(\frac{R_1}{R}x_1'\right)\right)\right] - \exp(-\langle\Phi_1^2\rangle)\right\}. \tag{13.74}$$

The parameter $R_1 x_1'/R = x_0'$ is the coordinate of ray OP along the x'-axis at height Z_0 (see Fig. 13.9). It can be found from the speed V_0 of point O when the satellite moves as

$$x_0' = |V_0|t \tag{13.75}$$

Here, a coherent function that gives us the correlation function of the field in a specific plane is defined. By using this function, it is possible to obtain later the intensity of field fluctuations, $I(r)$.

Spectrum of Amplitude Fluctuations. Expression (13.74) enables the relationship between the cross-correlation function of phase fluctuations on the bottom boundary of the ionospheric layer and the correlation function of signal amplitude fluctuations, $\Gamma_E(t)$. The latter can be found for weak phase fluctuations ($\langle \Phi_1^2 \rangle < 1$) as

$$\Gamma_E(t) = \frac{E_0^2}{R^2} \langle \Phi_1^2 \rangle \rho_{\Phi 1}(x_0'). \tag{13.76}$$

The time fluctuations of the field amplitude are found from the spatial phase fluctuations on the bottom boundary of the ionospheric layer with small-scale inhomogeneities. Taking into account relationship (13.76), we obtain

$$\Gamma_E(\xi, \eta) = (E_0^2/R^2)k_0^2 L \sec i \int_{-L \sec i}^{L \sec i} \Gamma_\varepsilon(x_0', \zeta') d\zeta'. \tag{13.77}$$

The Fourier transform of Equation (13.77) gives the spectrum of amplitude oscillations:

$$U_g(f) \sim \int U_\varepsilon(K_{x'}, K_{y'}, K_{z'}) dK_{y'}. \tag{13.78}$$

Now, at point O, the angle i_1 is the angle between the magnetic field and ray OP (see Fig. 13.9). In the coordinate system (x_0, ζ), with the ζ-axis along the geomagnetic field and with the base at point O, there are two cases:

1. For the case $i_1 = 0$, when the ray is parallel to the magnetic field $\mathbf{B}_0$ lines, the disturbance of plasma density is averaged and the field correlation function Γ_E is determined [36, 37] by the scales of the inhomogeneities perpendicular to $\mathbf{B}_0$.
2. In the other critical case when $i_1 = 90°$, the density of oscillations are averaged in the direction transverse to the magnetic field and the spectrum Γ_E is expressed from the spectrum of the scales of the inhomogeneities elongated along $\mathbf{B}_0$.

One should note that the earlier expressions were obtained for the small-scale inhomogeneities ($l << d_F$). In real cases of satellite experiments, a wide spectrum of scales were observed—from centimeters up to kilometers. For a more general two-dimensional (2-D) case, the spectrum of plasma density $U_N(K)$ of inhomogeneities was calculated by Shkarofsky [38]. At the same time, the spectrum of amplitude scintillations, measured in satellite experiments, shows that for the spatial frequencies $L_0^{-1} << K_\perp << l_0^{-1}$ the spectrum of plasma density $U_N(K)$ can be presented in a simpler manner in one-dimensional (1-D) case as [38]

$$U_N(K_\perp) = \frac{\langle N_1^2 \rangle \Gamma\left(\dfrac{p-1}{2}\right)}{\pi \Gamma\left(\dfrac{p-2}{2}\right)} \frac{K_0^{p-2}}{(K_0^2 + K_\perp^2)^{(p-2)/2}} \approx K_\perp^{1-p}, \qquad (13.79)$$

where l_0, $L_0 = 2\pi/K_0$ are the inner and outer scale of inhomogeneities, respectively; $K_{\perp;} = 2\pi/l_{\perp;}$ is the scale of inhomogeneity perpendicular to the geomagnetic field $\mathbf{B}_0$, and $\Gamma(w)$ is the gamma function.

It was also obtained from satellite experiments that in the direction parallel to $\mathbf{B}_0$ for the longitudinal scales $l_\parallel > d_F$, the spectrum of inhomogeneities $U_N(K_\parallel)$ is Gaussian

$$U_N(K_\parallel) \sim \exp\{-K_\parallel^2 l_\parallel^2\}. \qquad (13.80)$$

The transformation from spectrum (13.79) to spectrum (13.80) was observed in satellite experiments at the angles i_1' between the radio ray and the geomagnetic field $\mathbf{B}_0$.

$$i_1' \sim \tan^{-1}\left(\frac{l_\perp}{l_\parallel}\right). \qquad (13.81)$$

In many experiments, it was shown that ionospheric inhomogeneities of the F region, which produce radio scintillations, are stretched along the geomagnetic field ($l_{\perp;}/l_\parallel << 1$) and the angle i_1 has a value of only several degrees. Thus, the spectrum of field amplitude fluctuations is determined by the spectrum of transversal scales of inhomogeneities.

It should also be noted that for the metric and decimetric wave band, the case $l_0 << d_F << l_{\perp;}$ occurs and more general equations than (13.79) and (13.80), such as (13.66), must be used. But analytical calculation for the more general case of anisotropic inhomogeneities and of oblique incidence of radio waves on the ionospheric layer, is a very complicated mathematical problem. The analytical result was obtained only for the case when the wave was incident on the ionospheric layer perpendicular to its surface and when the magnetic field lines were also perpendicular to the ionosphere surface. In this case, one can present the spectra of signal logarithmic amplitude g, $Ug(\mathbf{K}_\perp)$, and signal phase S_1, $U_{S_1}(\mathbf{K}_\perp)$ fluctuations as

$$U_g(\mathbf{K}_\perp) = k_0^2 L U_\varepsilon(\mathbf{K}_\perp) \sin^2\left(\frac{K_\perp^2 d_F^2}{2}\right) \qquad (13.82a)$$

$$U_{S_1}(\mathbf{K}_\perp) = k_0^2 L U_\varepsilon(\mathbf{K}_\perp) \cos^2\left(\frac{K_\perp^2 d_F^2}{2}\right). \qquad (13.82b)$$

Expressions (13.82a,b) show that for a wide spectrum of inhomogeneities, the amplitude and phase spectra are determined by the spectra $U_\varepsilon(\mathbf{K}_{1:})$ of inhomogeneities and by "filter" functions $\cos^2\left(K_\perp^2 d_F^2/2\right)$ and $\sin^2\left(K_\perp^2 d_F^2/2\right)$.

The Scintillation Index. In estimating the fading of radio signals passing through the ionosphere, usually we use a scintillation index, denoted by σ_I^2 [29–41], in the same manner as in Chapter 12. This scintillation index determines the "strength" or "power" of inhomogeneities inside the ionospheric layer and can be defined as the dispersion of the radiowave intensity I fluctuations. We rewrite its equation for our future discussions as

$$\sigma_I^2 = \frac{\langle I^2\rangle - \langle I\rangle^2}{\langle I\rangle^2} \equiv \frac{\langle I^2\rangle}{\langle I\rangle^2} - 1. \tag{13.83}$$

For weak scintillations, this index was obtained in References 29 and 39–41 as

$$\sigma_I^2 = 4\langle g^2\rangle = 4\int U_g(K_\perp)dK_x dK_y. \tag{13.84}$$

Here again, g is the logarithm for the amplitude of the radio signal. The earlier equation allows us to conclude that if small-scale inhomogeneities exist, then two conditions apply:

- The inhomogeneity scale l is smaller than the first Fresnel zone, d_F
- l is larger than the wavelength λ.

The influence of small-scale inhomogeneities is mainly manifested in phenomena like diffraction and scattering [29–41]. However, the effect of small-scale inhomogeneities on wave propagation in the ionosphere is not well recognized. Although a lot of literature exists about large-scale inhomogeneities, the role of small-scale inhomogeneities seems to be less understood. In the next section, we will focus mainly on the small-scale inhomogeneities effects, we will analyze them, and we will generalize some theoretical results. Calculations of phase fluctuations $\left[\langle \Phi_1^2\rangle\right]^{1/2}$ on the bottom boundary of the ionospheric layer with inhomogeneities for various radio frequencies and for different plasma density fluctuations $\left[\langle N_1^2/N_0^2\rangle\right]^{1/2}$ will be given in Section 13.2.3. The 1-D spectra of plasma density disturbances $U_N(K_x) \sim K_x^{-p+2}$ for various extinction parameters in exponent, $p' = p - 2$, will also be given. Furthermore, the index of scintillation σ_I^2 as a function of phase fluctuations $\left[\langle \Phi_1^2\rangle\right]^{1/2}$ and of parameter p' will be analyzed, because it characterizes the power of ionospheric inhomogeneities.

13.2.3. Scattering Phenomena Caused by Small-Scale Inhomogeneities

As was mentioned in the previous section, when radio waves are propagating through an irregular ionosphere, small-angle scattering causes what is known as scintillation of the signal strength or intensity. In such phenomena, a distinction can be drawn between diffractive scattering from small-scale irregularities and refractive scattering from large-scale irregularities. The same phenomena are observed in the troposphere, caused by large-scale and small-scale turbulent gaseous structures (see Chapter 12). We put the same question as was done in the previous chapter on how we can separate these effects as well as the inhomogeneities that caused them. For a given location of the transmitting and receiving antennas, a Fresnel scale (d_F) is the parameter which can give the corresponding separation. This also depends on the wavelength and the coordinate of locations of the source and the observer. In such an assumption, diffractive scattering is caused by irregularities whose scale is less than the Fresnel scale. Diffractive scattering of electromagnetic waves by a scintillation medium is described in References 34 and 41. Refractive scattering involves irregularities whose scale is greater than the local Fresnel scale [31].

In order to present the effects of refractive and diffractive scattering, the thin phase changing screen model (see Fig. 13.8) of the scintillation medium was introduced by Booker with coauthors [31, 34, 40, 41]. Such a model replaces weak multiple scattering by strong single scattering in a way that enables us to understand the relation between "diffractive" and "refractive" scattering in scintillation phenomena, which, as was shown in References 49 and 50, strongly depend on wavelength and on the angle sensitivity of the transmitter and receiver antennas.

Main Parameters of the Problem. The phase changing screen model has the following characteristic parameters:

(a) A phase changing screen, representing the ionospheric F region, with ionospheric plasma inhomogeneities characterized by the following parameters:
   - mean square fluctuation of phase $[(\Delta\Phi)^2] \equiv \langle \Phi_1^2 \rangle$
   - outer scale L_0
   - inner scale l_0.

(b) A reception plane, representing the surface of the Earth (see Fig. 13.8).

The spectrum of intensity fluctuation is created and obtained in the reception plane. At the same time, to determine the physical processes that accompany radiowave propagation through the inhomogeneous ionosphere, we must compare the outer scale of the irregular ionospheric region with the Fresnel scale defined as $d_F = (\lambda Z/2\pi)^{1/2}$. Z is the distance from the screen to the reception plane and λ is the wavelength. In References 31, 34, 40, and 41, it was assumed that the RMS fluctuation of phase $[\langle \Phi_1^2 \rangle]^{1/2}$ is large compared to 1 radian.

Moreover, in the 1-D case of ionospheric layer presented in Figure 13.8, as follows from Equation (13.60a), the power spectrum of phase fluctuations $S(k)$ is proportional to k^{-p}, when $k >> (1/L_0)$ and p is referred to as the *spectral index*.

For a practical scintillation medium, parameter p is defined as the spectral index that is observed in any measurements of phase fluctuations along a straight line. For the ionosphere, it is also the spectral index that is observed when the source is at the satellite moving above the ionosphere on a straight line. Usually, in literature, the spectral index p is determined as one integer greater than that observed when measurements of the average refractive index $\langle n \rangle$ are made along a straight line in the medium with scintillations; that is, $p = \langle n \rangle + 1$ [29–41].

At the same time, the spectral index p is one integer less than that obtained by analyzing phase fluctuations made over an area rather than along a line. Observed values of the spectral index p range from about 2 to 4, with values between about 2.5 and 3.5 being most common [31, 34–37, 40, 41]. The smaller values of p are found when the scintillation phenomenon is strong.

Let us also define the expression $[(\Delta\Phi)^2]S(k)$ as the power spectrum of phase fluctuations, where $[(\Delta\Phi)^2]$ is the mean square fluctuation of phase, and $S(k)$ is the phase spectra. The corresponding autocorrelation function $\rho(x)$ is obtained by an inverse Fourier transformation of $S(k)$ (see Section 12.2.2). Table 13.1, Table 13.2, and Table 13.3 present the values of $S(k)$ and the corresponding autocorrelation functions $\rho(x)$ obtained in Reference 41.

Now, to differentiate the effects of "refractive" scattering from large-scale irregularities and "diffraction" scattering from small-scale irregularities, and to

TABLE 13.2. Autocorrelation Function for an Outer Scale L_0 and an Inner Scale l_0 (Rearranged from Reference 41)

P'	$\rho(x)$
2	$$\dfrac{\exp\left(-\dfrac{(x^2 + l_0^2)^{1/2}}{L_o}\right)}{\exp\left(-\dfrac{l_0}{L_o}\right)}$$
3	$$\dfrac{\dfrac{(x^2 + l_0^2)^{1/2}}{L_o} K_1\left(\dfrac{(x^2 + l_0^2)^{1/2}}{L_o}\right)}{\dfrac{l_0}{L_o} K_1\left(\dfrac{l_0}{L_o}\right)}$$
4	$$\dfrac{\left[1 + \dfrac{(x^2 + l_0^2)^{1/2}}{L_0}\right]\exp\left(-\dfrac{(x^2 + l_0^2)^{1/2}}{L_0}\right)}{\left(1 + \dfrac{l_0}{L_0}\right)\exp\left(-\dfrac{l_0}{L_0}\right)}$$
5	$$\dfrac{\dfrac{(x^2 + l_0^2)^{1/2}}{L_o}\left[K_1\left(\dfrac{(x^2 + l_0^2)^{1/2}}{L_o}\right) + \dfrac{1}{2}\dfrac{(x^2 + l_0^2)^{1/2}}{L_o} K_0\left(\dfrac{(x^2 + l_0^2)^{1/2}}{L_o}\right)\right]}{\dfrac{l_0}{L_o}\left[K_1\left(\dfrac{l_0}{L_o}\right) + \dfrac{1}{2}\dfrac{l_0}{L_o} K_0\left(\dfrac{l_0}{L_o}\right)\right]}$$

TABLE 13.3. Phase Spectra per unit Mean Square Fluctuation of Phase for an Outer Scale L_0 and an Inner Scale l_0 (Rearranged from Reference 41)

P'	$S(k)$
2	$$4l_0 \exp\left(\frac{l_0}{L_o}\right) \frac{K_1\left(\frac{l_0}{L_o}(1+k^2L_o^2)^{1/2}\right)}{(1+k^2L_o^2)^{1/2}}$$
3	$$\frac{2\pi l_0^2}{L_o K_1\left(\frac{l_0}{L_o}\right)} \frac{1+\left(\frac{l_0}{L_o}(1+k^2L_o^2)^{1/2}\right)}{\left(\frac{l_0}{L_o}(1+k^2L_o^2)^{1/2}\right)^3} \exp\left(-\frac{l_0}{L_o}(1+k^2L_o^2)^{1/2}\right)$$
4	$$\frac{8l_0^4}{L_o^3\left(1+\frac{l_0}{L_o}\right)\exp\left(-\frac{l_0}{L_o}\right)} \frac{K_1\left(\frac{l_0}{L_o}(1+k^2L_o^2)^{1/2}\right)+\frac{1}{2}\left(\frac{l_0}{L_o}(1+k^2L_o^2)^{1/2}\right)K_0\left(\frac{l_0}{L_o}(1+k^2L_o^2)^{1/2}\right)}{\left(\frac{l_0}{L_o}(1+k^2L_o^2)^{1/2}\right)^3}$$
5	$$\frac{3\pi l_0^4}{L_o^3\left(K_1\left(\frac{l_0}{L_o}\right)+\frac{l_0}{L_o}K_0\left(\frac{l_0}{L_o}\right)\right)} \frac{1+\left(\frac{l_0}{L_o}(1+k^2L_o^2)^{1/2}\right)+\frac{1}{3}\left(\frac{l_0}{L_o}(1+k^2L_o^2)^{1/2}\right)^2}{\left(\frac{l_0}{L_o}(1+k^2L_o^2)^{1/2}\right)^5}$$ $$\exp\left(-\frac{l_0}{L_o}(1+k^2L_o^2)^{1/2}\right)$$

analyze the significant roles in the physical processes, in addition to the earlier introduced outer scale L_0, the inner scale l_0, and the Fresnel scale d_F, new parameters are introduced [31, 34, 41]. They are the lens scale l_L, the focal scale l_F, and the peak scale l_P.

The *lens scale* (l_L) is defined as the size of the inhomogeneity in the phase changing screen [31, 34, 41]. An array of optical foci is produced in a plane parallel to the screen at distance Z. These foci lie in the reception plane at distance Z if

$$Z = \frac{l_L^2}{(\lambda/(2\pi))[2(\Delta\Phi)^2]^{1/2}} \tag{13.85a}$$

from which we get

$$l_L = \left(\frac{\lambda Z}{2\pi}\right)^{1/2}[2(\Delta\Phi)^2]^{1/4}. \tag{13.85b}$$

Lens action occurs in the screen, producing focal action in the reception plane. As follows from (13.85b), the lens scale l_L can be defined through the Fresnel scale:

$$l_L = d_F[2(\Delta\Phi)^2]^{1/4}. \tag{13.86}$$

Hence, the irregularities with scale of l_L defined by (13.85b) or (13.86) give at the Earth's surface a focusing effect of a radio wave passing the ionosphere and thus, this scale is defined as a lens scale [41].

The *focal scale* (l_F) is defined as the width of the average focal spots at the reception plane by the scattering of radio waves from the medium- and large-scale ionospheric irregularities. It determines the degree of fluctuations of the radio signal amplitude at the Earth's surface after being scattered. The focal scale, which varies with the spectral index, is associated with the arrival at the reception plane of an angular spectrum of waves that are approximately cophased within an angle of about $\pm\lambda/(2\pi l_F)$ of the norm. It is determined as [31, 34, 41]

$$l_F = l_L[2(\Delta\Phi)^2]^{1/2}. \tag{13.87}$$

It is seen from (13.87) that for a given scale in the screen, the larger the mean square fluctuation of phase, the closer is the focal plane to the screen and the sharper are the foci.

The *peak scale* (l_P) represents the peak in the intensity spectrum $I(k)$, such that the angular spatial frequency $K = l_P^{-1}$ gives the low frequency edge of the peak in the intensity spectrum in the reception plane. When "refractive" scattering dominates, the reciprocals of the scales l_P and l_F give the lower and upper roll-off angular spatial frequencies K for the intensity spectrum in the reception plane, whether this is a focal plane or not. The intensity spectrum therefore extends roughly from the angular spatial frequency $K = l_P^{-1}$ to the angular spatial frequency $K = l_F^{-1}$. The focal scale and the peak scale have a geometric mean equal to the Fresnel scale:

$$l_P = d_F^2 / l_F. \tag{13.88}$$

Table 13.4 represents the focal scale and the peak scale obtained in Reference 31 for polynomial spectrum of irregularities with $p' = p - 2 = 2 - 5$, as a

TABLE 13.4. The Focal Scale l_F and the Peak Scale l_P for Large L_0/d_F and $(\Delta\Phi)^2$ [31, 41]

	Spectral Index p'				
	2.5	3.0	3.5	4.0	4.5
l_F	$\dfrac{L_0}{\left[2(\overline{\Delta\Phi})^2\right]^{2/3}}$	$\dfrac{L_0}{\left[2(\overline{\Delta\Phi})^2 \ln(\overline{\Delta\Phi})^2\right]^{1/2}}$	$\dfrac{L_0}{\left[4(\overline{\Delta\Phi})^2\right]^{1/2}}$	$\dfrac{L_0}{\left[2(\overline{\Delta\Phi})^2\right]^{1/2}}$	$\dfrac{L_0}{\left[(\overline{\Delta\Phi})^2\right]^{1/2}}$
l_P	$\dfrac{d_F^2}{L_0}\left[2(\overline{\Delta\Phi})^2\right]^{2/3}$	$\dfrac{d_F^2}{L_0}\left[(\overline{\Delta\Phi})^2 \ln(\overline{\Delta\Phi})^2\right]^{1/2}$	$\dfrac{d_F^2}{L_0}\left[4(\overline{\Delta\Phi})^2\right]^{1/2}$	$\dfrac{d_F^2}{L_0}\left[2(\overline{\Delta\Phi})^2\right]^{1/2}$	$\dfrac{d_F^2}{L_0}\left[(\overline{\Delta\Phi})^2\right]^{1/2}$

function of the outer scale L_0 and the mean square fluctuation of phase $(\Delta\Phi)^2$, for $(\Delta\Phi)^2 > 1$. As shown in the results presented in Table 13.4, for $(\Delta\Phi)^2 > 1$ spectrum of amplitude fluctuations of radio signals, within a range of spatial frequencies, $l_F^{-1} < K < l_0^{-1}$, and "diffractive" scattering is mainly due to small-scale irregularities and can describe the corresponding spectrum of ionospheric inhomogeneities. At the same time, at frequencies $K < l_F^{-1}$, the scattering is related to scattering from medium and large inhomogeneities with dimensions $l > d_F$. That also describes the focusing properties of the ionospheric plasma. Therefore, determination of the spectrum of ionospheric inhomogeneities using measurements of the amplitude fluctuations can be done only when $l_F \gg l_0$.

What does it mean in the practical world of satellite communications? Results obtained earlier show that the effects of "diffractive" scattering occur mostly for medium-scale and small-scale irregularities because they become significant for high frequencies beyond the UHF/X-band. Moreover, large-scale irregularities cause the "refractive" scattering. This effect was also described earlier in Section 13.2.2. In Reference 41, it was shown that "refractive" scattering at large-scale inhomogeneities is predominant with respect to "diffraction" scattering from small-scale inhomogeneities, if $l_F < l_0$. This effect is also stronger if the power spectrum parameter p' is higher (see Table 13.4). The same effect is observed with an increase of signal phase fluctuations. Thus, for $(\Delta\Phi)^2 \equiv \langle\Phi_1^2\rangle \geq 10^3$, the spectrum of amplitude variations, $U_g(K)$ defined by (13.82a) for all p' is determined by the "refractive" scattering, which gives the so-called focusing effects.

Signal Intensity Fluctuations. Using the selected appropriate autocorrelation function $\rho(x)$ (from Table 13.2 and Table 13.3), two functions are formulated that each depend on x and k. The first is $f(x,k)$, which is obtained by substituting $\rho(x)$ into the following equation:

$$f(x, k) = 2\rho(x) - \rho(x - kd_F^2) - \rho(x + kd_F^2). \tag{13.89}$$

The second function is $g(x,k)$, which is obtained by substituting $f(x,k)$ into the following equation:

$$g(x, k) = \exp\{-[(\Delta\Phi)^2][f(0, k) - f(x, k)]\} - \exp\{-[(\Delta\Phi)^2]f(0, k)\}. \tag{13.90}$$

The intensity spectrum in the reception plane is then [34, 41]

$$I(k) = 4 \int_0^\infty g(x, k)\cos(kx)dx. \tag{13.91}$$

Using now Equation (13.91), we can obtain the square of the scintillation index, σ_I

$$\sigma_I = \frac{1}{2\pi} \int\limits_0^\infty I(k)\,dk. \tag{13.92}$$

For weak scattering $((\Delta\Phi)^2 = 10^{-1}, 1)$ the general expression for the spectrum of signal intensity fluctuation (13.91) can be reduced to [41]

$$I(k) = 4\langle(\Delta\Phi)^2\rangle S(k)\sin^2\left(\frac{1}{2}k^2 d_F^2\right), \tag{13.93}$$

where $S(k)$ has the value shown in Table 13.3 if the concrete inner scale l_0 is taken into account and in Table 13.2 if $l_0 = 0$.

We mentioned that the Fresnel oscillation associated with the $\sin^2(0.5k^2 d_F^2)$ term in the equation is depicted for the main lobe and the first side lobe. For the remaining lobes, only the average value is considered (corresponding to replacement of $\sin^2[0.5k^2 d_F^2]$ by 0.5). For a spectral index of $p' = 2$, and using Table 13.2, a substitute of $S(k) = 4L_0/(1 + k^2 L_0^2)$ leads to the following equation for $I(k)$:

$$I(k) = 16\langle(\Delta\Phi)^2\rangle \frac{L_0}{1 + k^2 L_0^2}\sin^2\left(\frac{1}{2}k^2 d_F^2\right). \tag{13.94a}$$

The same substitute of $S(k)$ for $p' = 3$ leads to an intensity fluctuation of

$$I(k) = 8\pi\langle(\Delta\Phi)^2\rangle \frac{L_0}{(1 + k^2 L_0^2)^{3/2}}\sin^2\left(\frac{1}{2}k^2 d_F^2\right). \tag{13.94b}$$

Finally, for $p' = 4$ we get

$$I(k) = 32\langle(\Delta\Phi)^2\rangle \frac{L_0}{(1 + k^2 L_0^2)^2}\sin^2\left(\frac{1}{2}k^2 d_F^2\right). \tag{13.94c}$$

Therefore, substituting in (13.92) different $I(k)$ values from (13.94) leads to the following equations for σ_I:

- for $P' = 2$

$$\sigma_I^2 = \frac{2\sqrt{2}}{\sqrt{\pi} L_0} d_F \langle(\Delta\Phi)^2\rangle \tag{13.95a}$$

- for $P' = 3$

$$\sigma_I^2 = \frac{\pi}{2 L_0^2} d_F^2 \langle(\Delta\Phi)^2\rangle \tag{13.95b}$$

- for $P' = 4$

$$\sigma_I^2 = \frac{8\sqrt{2}}{3\sqrt{\pi}L_0^3}d_F^3\left\langle(\Delta\Phi)^2\right\rangle.$$ (13.95c)

Computation of the intensity spectrum has been performed in accordance with Equations (13.91) and (13.92) for strong fluctuations, and Equation (13.94a), Equation (13.94b), Equation (13.94c), Equation (13.95a), Equation (13.95b), and Equation (13.95c) for weak fluctuations for an outer scale of $L_0 = 10d_F$ and an inner scale of $L_i = 10^{-2}d_F$ for the spectral index $p' = 3, 4, 5$. Figure 13.10

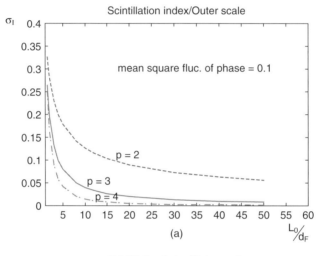

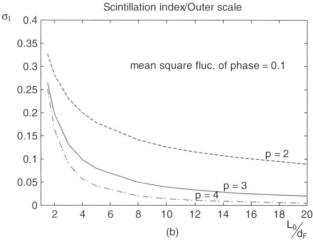

FIGURE 13.10. (a) Illustration for various values of the spectral index, the scintillation index σ_I as a function of the outer scale L_0, $1.5\,d_F \le L_0 \le 50\,d_F$, $\langle\Delta\Phi^2\rangle = 0.1$. (b) Illustration for various values of the spectral index, the scintillation index σ_I as a function of the outer scale L_0, $1.5\,d_F \le L_0 \le 20\,d_F$, $\langle\Delta\Phi^2\rangle = 0.1$.

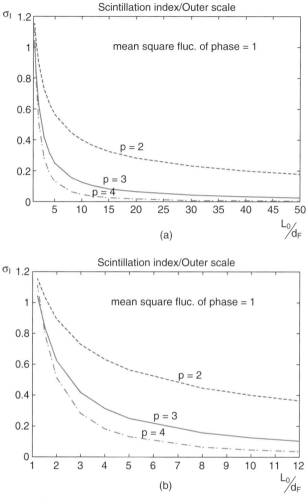

FIGURE 13.11. (a) Illustration for various values of the spectral index, the scintillation index σ_I as a function of the outer scale L_0, $1.2\,d_F \le L_0 \le 50\,d_F$, $\langle \Delta\Phi^2 \rangle = 0.1$. (b) Illustration for various values of the spectral index, the scintillation index σ_I as a function of the outer scale L_0, $1.2\,d_F \le L_0 \le 20\,d_F$, $\langle \Delta\Phi^2 \rangle = 0.1$.

and Figure 13.11 represent the scintillation index calculated numerically according to (13.94) and (13.95) for weak signal phase fluctuations and according to (13.91) and (13.92) for strong signal phase fluctuations, respectively. The scintillation is plotted as a function of the square mean deviations of signal phase for various parameters of 1-D spectrum $p' = p - 2$ and different scales of ionospheric irregularities. It is seen that for $p' = 2$ ($p = 4$) the scintillation index with an increase of phase fluctuations approaches the limit of one. For higher spectral indices ($p' > 2$), σ_I exceeds the unit value, which explains the focusing properties of the ionospheric layer consisting of various irregularities for strong variations in the signal phase after passing the ionosphere.

Signal Phase Fluctuations. We once again model the ionospheric F region as a slab of mean ionization density N with a uniform mean square fractional fluctuation of ionization density $(\Delta N/N)^2$ (see Fig. 13.8), with the thickness D and an outer scale L_0. We assume that on the Earth's plane, we receive radiation of wavelength λ from a distant point source at zenith angle χ.

In our computation, to illustrate results obtained in References 31, 34, and 41, we shall take the outer scale equal to the scale height H of the F region, and we shall also take the thickness of the F region to be H. In this case, the mean square fluctuation of signal phase experienced on passage through the F region may then be expressed as

$$\langle (\Delta\Phi)^2 \rangle \equiv \langle \Phi_1^2 \rangle = 4r_e^2 N^2 \left\langle \left(\frac{N_1}{N_0} \right)^2 \right\rangle \lambda^2 H^2 \sec\chi. \qquad (13.96)$$

Here, all parameters were defined earlier; r_e is the radius of the electron. For numerical computations we use $H = 100$ km and $N_0 = 10^{12}$ m^{-3}. The curves in Figure 13.12a,b, extracted from Reference 36, present the RMS fluctuation of phase as a function of the RMS, fractional fluctuation of ionization density for a series of frequencies running from 32 MHz to 60 GHz. The figures illustrate the ionospheric propagation of various wave frequencies at zenith angles of 10, 45, 60, 80°, respectively. Both axes, vertical and horizontal, are plotted logarithmically.

It is seen that for a given frequency, an increase of ionization density causes an increase in phase fluctuations. Furthermore, for a given ionization density, when we use high frequencies for the satellite communication channel (from UHF to X-band and higher), we can see a decrease in phase fluctuations to values appropriate for weak scattering conditions. Finally, for a given ionization density, when the zenith angle χ becomes larger, the effect of phase fluctuations becomes stronger. In fact, for a zenith angle of 60°, the phase fluctuation experienced in the passage of a 32 MHz wave through the F region with a fractional ionization density $\langle (N_1/N_0)^2 \rangle$ of 10^{-2} is about 750 radians. But when the zenith angle is 80°, and for the same frequency and ionization density, we obtain phase fluctuations of ~1270 radians. In order to obtain 750 radians, we need the ionization density $\langle (N_1/N_0)^2 \rangle$ to be ~3.5×10^{-3} m^{-3}.

To understand the role of the satellite position with respect to observer at the Earth's plane, additional analysis of the RMS fluctuation of phase as a function of the zenith angle was done and is shown for a series of frequencies running from 32 MHz to 60 GHz in Figure 13.13a,b. The figures illustrate the ionospheric propagation of various wave frequencies at ionization densities of 100%, 80%, 30%, and 1%. Again, both the vertical and horizontal axes are plotted logarithmically. As we mentioned earlier, in the following figures, we can see that for a given frequency, an increase of zenith angle causes an

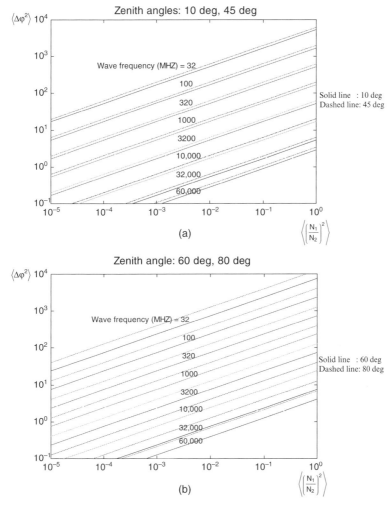

FIGURE 13.12. (a) The RMS fluctuation of phase versus the fractional fluctuation of ionization density for different frequencies at zenith angles of $10°, 45°$. Outer scale is equal to layer thickness $L = 100$ km. Mean ionization density is 10^{12} m^{-3}. (b) The RMS fluctuation of phase versus the fractional fluctuation of ionization density for different frequencies at zenith angles of $60°, 80°$. Outer scale is equal to layer thickness $L = 100$ km. Mean ionization density is 10^{12} m^{-3}.

increase in phase fluctuations. Furthermore, for a given zenith angle, when we use high frequencies for the communication channel (more than 1 GHz), we can see a decrease in phase fluctuations to values appropriate for weak scattering. The same features, as in Figure 13.12a,b, are clearly seen from illustrations of Figure 13.13a,b.

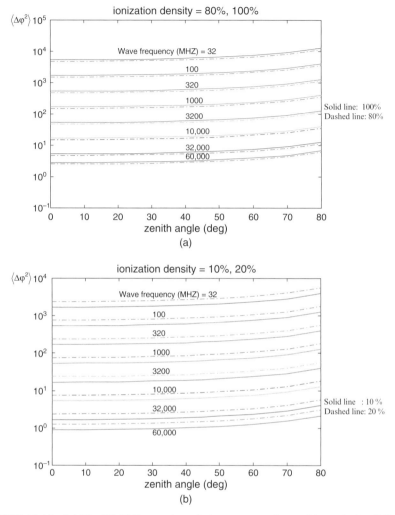

FIGURE 13.13. (a) The RMS fluctuation of phase versus the zenith angle for different frequencies at ionization densities of 80% and 100%. Outer scale is equal to layer thickness $L = 100$ km. Mean ionization density is 10^{12}m^{-3}. (b) The RMS fluctuation of phase versus the zenith angle for different frequencies at ionization densities of 10% and 20%. Outer scale is equal to layer thickness $L = 100$ km. Mean ionization density is 10^{12}m^{-3}.

Frequency Dependence of Signal Intensity Fluctuation Spectrum. Earlier, we evaluated the expressions of the signal intensity fluctuations defined by (13.91). On the basis of this expression, we present in Figure 13.14 the normalized intensity fluctuation as a function of the normalized plasma density deviations $\langle (N_1/N_0)^2 \rangle$ for various values of radiated frequency,

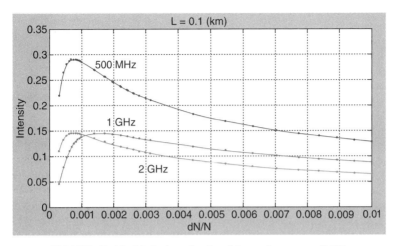

FIGURE 13.14. Variations in signal intensity versus $\delta N/N$.

500 MHz, 1 GHz, and 2 GHz. The computations were done in Reference 50 based on the measurements carried out in global positioning system (GPS) experiments and described in References 51 and 52. It is seen that for a given frequency, the behavior of the intensity fluctuation is nonlinear and decreases with an increase in plasma density fluctuations. Furthermore, for a given plasma density fluctuation, a decrease in the intensity fluctuation is observed with a decrease of frequency from 2 GHz to 500 MHz.

REFERENCES

1 Evans, B.G., *Satellite Communication Systems*, IEE, London, 1999.

2 Pattan, B., *Satellite-Based Cellular Communications*, McGraw-Hill, New York, 1998.

3 Saunders, S.R., *Antennas and Propagation for Wireless Communication Systems*, John Wiley & Sons, New York, 1999.

4 Ratcliffe, J.A., *Physics of the Upper Atmosphere*, Academic Press, New York, 1960.

5 Ginzburg, V.L., *Propagation of Electromagnetic Waves in Plasma*, "Science," Moscow, 1964.

6 Ratcliffe, J.A., *Introduction in Physics of Ionosphere and Magnetosphere*, Academic Press, New York, 1973.

7 Alpert, Ya.L., *Propagation of Electromagnetic Waves and Ionosphere*, "Science," Moscow, 1972.

8 Gurevich, A.V. and A.B. Shvartzburg, *Non-linear Theory of Radiowave Propagation in the Ionosphere*, "Science," Moscow, 1973.

9 Philipp, N.D., N.Sh. Blaunstein, L.M. Erukhimov, V.A. Ivanov, and V.P. Uriadov, *Modern Methods of Investigation of Dynamic Processes in the Ionosphere*, "Shtiintsa," Kishinev, Moldova, 1991.

10 Philipp, N.D., V.N. Oraevsky, N.Sh. Blaunstein, and Yu.Ya. Ruzhin, *Evolution of Artificial Plasma Inhomogeneities in the Earth's Ionosphere*, "Shtiintsa," Kishinev, Moldova, 1986.

11 Gurevich, A.V. and E.E. Tsedilina, *Extremely-Long-Range Propagation of Short Waves*, "Science," Moscow, 1979.

12 Gel'berg, M.G., *Inhomogeneities of High-Latitude Ionosphere*, "Science," Novosibirsk, USSR, 1986.

13 Banks, P., "Collision frequencies and energy transfer," *Planet. Space Sci.*, Vol. 14, 1969, pp. 1085–1122.

14 Belikovich, V.V., E.A. Benediktov, M.A. Itkina et al., "Frequency dependence of anomalous absorption of cosmic radio radiation in the ionosphere in areas of polar cups," *Geomagn. and Aeronomy*, Vol. 9, 1969, pp. 485–490.

15 Taubenheim, J. and V. Hense, "The contribution of the ionospheric region to cosmic noise absorption," *Annales Geophys.*, Vol. 22, 1966, pp. 320–322.

16 Mehta, N.C. and N.D. D'Angelo, "Cosmic noise absorption by E-region plasma waves," *J. Geophys. Res.*, Vol. 85, 1980, pp. 1779–1782.

17 Titheridge, J.E., "The diffraction of satellite signals by isolated ionospheric irregularities," *J. Atmosph. Terrest. Phys.*, Vol. 33. 1974, pp. 47–69.

18 Leadabrand, R.L., A.G. Larson, and J.C. Hodges, "Preliminary results on the wavelength dependence and aspect sensitivity of radar echoes between 50 and 3000 MHz," *J. Geophys. Res.*, Vol. 72, 1967, pp. 3877–3887.

19 Reinisch, B.W. and X. Huang, "Automatic calculation of electron density profiles from digital ionograms: 3. Processing of bottomside ionograms", *Radio Sci.*, Vol. 18, 1983, pp. 477–492.

20 Huang, X. and B.W. Reinisch, "Vertical electron content from ionograms in real time," *Radio Sci.*, Vol. 36, No. 2, 2001, pp. 335–342.

21 Blaunstein, N.Sh. and Ye.Ye. Tsedelina, "Spreading of strongly elongated inhomogeneities in the upper ionosphere," *Geomagn. and Aeronomy*, Vol. 24, No. 3, 1984, pp. 340–344.

22 Blaunstein, N.Sh. and Ye.Ye. Tsedilina, "Effects of the initial dimensions on the nature of the diffusion spreading of inhomogeneities in a quasi-uniform ionosphere," *Geomagn. and Aeronomy*, Vol. 25, No. 1, 1985, pp. 39–44.

23 Blaunstein, N. "Diffusion spreading of middle-latitude ionospheric plasma irregularities," *Annales Geophysicae*, Vol. 13, 1995, pp. 617–626.

24 Blaunstein, N., "The character of drift spreading of artificial plasma clouds in the middle-latitude ionosphere," *J. Geophys. Res.*, Vol. 101, 1996, pp. 2321–2331.

25 Blaunstein, N., "Changes of the electron concentration profile during local heating of the ionospheric plasma," *J. Atmosph. Terr. Phys.*, Vol. 58, No. 12, 1996, pp. 1345–1354.

26 Blaunstein, N. "Evolution of a stratified plasma structure induced by local heating of the ionosphere," *J. Atmos. and Terr. Phys.*, Vol. 59, No. 3, 1997, pp. 351–361.

27 Belikovich V.V., E.A. Benediktov, A.V. Tolmacheva, and N.V. Bakhmet'eva, *Ionospheric Research by Means of Artificial Periodic Irregularities*, Copernicus, Katlenburg-Lindau, Germany, 2002.

28 Reinisch, B.W., D.M. Haines, K. Bibl et al., "The radio plasma imager investigation on the image spacecraft." *Space Sci. Rev.*, Vol. 91, 2000, pp. 319–359.

29 Wernik, A.W. and C.H. Liu, "Application of refractive scintillation theory to iono-spheric irregularities studies", *Artificial Satellites*, Vol. 10, 1975, pp. 37–58.

30 Erukhimov, L.M. and V.A. Rizhkov, "Study of focusing ionospheric irregularities by methods of radio-astronomy at frequencies of 13–54 MHz," *Geomagn. and Aeronomy*, Vol. 5, 1971, pp. 693–697.

31 Crain, C.M., H.G. Booker, and S.A. Fergusson, "Use of refractive scattering to explain SHF scintillations," *Radio Sci.*, Vol. 14, 1974, pp. 125–134.

32 Briggs, B.H. and I.A. Parkin, "On variation of radio star and satellite scintillations with zenith angle," *J. Atmos. Terrest. Phys.*, Vol. 25, 1963, pp. 339–365.

33 Ga'lit, T.A., V.D. Gusev, L.M. Erukhimov, P.I. Shpiro, "About the spectrum of phase fluctuations during sounding of the ionosphere," *Radiophysics, Izv. Vuzov*, Vol. 26, 1983, pp. 795–801.

34 Booker, H.G., S.A. Rateliffe, and D.H. Shinn, "Diffraction from an irregular screen with applications to ionospheric problems," *Philos. Trans. Royal Soc. London*, Ser. A, Vol. 242, 1950, pp. 579–607.

35 Yakovlev, O.I., V.P. Yakubov, V.P. Uryadov, and A.G. Pavel'ev, *Radio Waves Propa-gation*, Lenand, Moscow, 2009.

36 Blaunstein, N., S. Pulinets, and Y. Cohen, "Computation of the main parameters of radio signals in the land-satellite channels during propagation through the per-turbed ionosphere," *Geomagn. and Aeronomy*, Vol. 53, No. 2, 2012, pp. 1–13.

37 Erukhimov, L.M., G.P. Komrakov, and V.L. Frolov, "About the spectrum of the artificial small-scale ionospheric turbulence," *Geomagnetism and Aeronomy*, Vol. 20, 1980, pp. 1112–1114.

38 Shkarovsky, I.P., "Generalized turbulence space-correlation and wave-number spectrum function pairs," *Can. J. Phys.*, Vol. 46, 1968, pp. 524–528.

39 Rino, C.L. and E.J. Fremouw, "The angle dependence of single scattered wave-fields," *J. Atmos. Terrestr. Phys.*, Vol. 39, 1977, pp. 859–868.

40 Booker, H.G., "Application of refractive scintillation theory to radio transmission through the ionosphere and the solar wind and to reflection from a rough ocean," *J. Atmos. Terrestr. Phys.*, Vol. 43, 1981, pp. 1215–1233.

41 Booker, H.G. and A.G. Majidi, "Theory of refractive scattering in scintillation phenomena," *J. Atmos. Terrestr. Phys.*, Vol. 43, 1981, pp. 1199–1214.

42 Burke, W.J., A Rubin, N. Maynard et al., "Ionospheric disturbances observed by DMSP at mid to low latitudes during magnetic storm in June 4–6, 1991," *J. Geophys. Res.*, Vol. 105, No. 18, 2000, pp. 391–398.

43 Mishin, E.V., W.J. Burke, Su. Basu, Sa. Basu et al., "Stormtime ionospheric irregu-larities in SAPS-related troughs: Causes of GPS scintillations at mid latitudes," *AGU Fall Meeting*, SH52A, Colorado, 2003.

44 Booker, H.G. and W.E. Gordon, "A theory of radio scattering in the ionosphere," *Proc. IRE*, Vol. 38, 1950, pp. 400–412.

45 Booker, H.G., "A theory of scattering by non-isotropic irregularities with applica-tions to radar reflection from the Aurora," *J. Atmos. Terr. Phys.*, Vol. 8, 1956, pp. 204–221.

46 Philipp, N.D., "Power of H_E-scatter signals," *Radiophysics*, Vol. 22, No. 4, 1979, pp. 407–411.

47　Philipp, N.D. and N.Sh. Blaunstein, "Effect of the geomagnetic field on the diffusion of ionospheric inhomogeneities," *Geomagnetism and Aeronomy*, Vol. 18, 1978, pp. 423–427.

48　Philipp, N.D. and N.Sh. Blaunstein, "Drift of ionospheric inhomogeneities in the presence of the geomagnetic field," *Izv. Vuzov, Radiophysics*, Vol. 18, 1978, pp. 1409–1417.

49　Leadabrand, R.L., A.G. Larson, and J.C. Hodges, "Preliminary results on the wavelength dependence and aspect sensitivity of radar echoes between 50 and 3000 MHz," *J. Geophys. Res.*, Vol. 72, 1967, pp. 3877–3887.

50　Blaunstein, N. and E. Plohotniuc, *Ionosphere and Applied Aspects of Radio Communication and Radar*, CRC Press, New York, 2008.

51　Mishin, E. and N. Blaunstein, "Irregularities within subauroral polarization stream-related troughs and GPS radio interference at midlatitudes," *Geophys. Monograph Series, AGU*, Vol. 181, pp. 291–296, 10.1029V181GM26, August, 2009.

52　Blaunstein, N., "Modeling of radio propagation in the land-satellite link through the stormtime ionosphere," in *Proc. of Int. Conf. on Mathematical and Informational Technologies*, Kopaonik, Serbia; Budva, Montenegro, August 27–September 5, 2009, pp. 71–74.

CHAPTER FOURTEEN

Land–Satellite Communication Links

14.1. OBJECTIVE

The Space Age started on October 4, 1957, when the first artificial satellite, Sputnik 1, was placed in orbit by the Soviet Union. Before October 1957, the term satellite referred to essentially a small body that revolved around a larger astronomical object. Thus, all the moons circling the planets of the solar system were called *satellites*. Today, these bodies are specifically called *natural satellites*, and any artificial object that revolves around a larger astronomical object along the elliptical or circular orbits is called an *artificial satellite*.

The main task of an artificial satellite is to allow stable wireless communication among users located over the Earth [1, 2]. During the last four decades, satellites have been used for wireless communication, weather forecasting, navigation, military observation, and other purposes. In this book, as was mentioned in Chapter 1, we deal only with the aspects of radio communication between various terminals, personal, stationary or moving, portable telephone or ground-based facilities, which will secure a satisfactory quality of service for each subscriber located within such a channel (see Fig. 14.1).

More than 40 years ago, J.R. Pierce and R. Kompfner [1, 2], discussed a number of alternatives, problems, and potential solutions for transoceanic communication by means of satellites, addressing the following subjects:

Radio Propagation and Adaptive Antennas for Wireless Communication Networks: Terrestrial, Atmospheric, and Ionospheric, Second Edition. Nathan Blaunstein and Christos G. Christodoulou.
© 2014 John Wiley & Sons, Inc. Published 2014 by John Wiley & Sons, Inc.

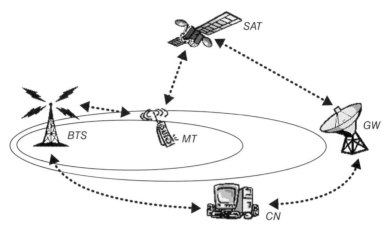

FIGURE 14.1. Satellite-land service terminals.

- alternative satellite repeater schemes; orbits, mutual visibility, and distances
- number of satellites
- path loss calculations
- modulation systems
- unknowns in satellite communications.

Surprisingly, after 40 years, the fundamental ideas have not changed much. Generally, there are differentiations between satellite orbits. They are called "far" geostationary earth orbit (GEO) and low earth orbit (LEO). Between these two orbits, there is medium earth orbit (MEO). In terms of transponders or repeaters, Pierce and Kompfner mentioned *active* and *passive*.

Signal propagation in land-satellite communication (LSC) systems for the last several decades has become an essential consideration, especially when high-rate data services are involved [1–19]. Path conditions may yield harmful impairments, severely degrading the system performance and availability. As far as urban or suburban built-up areas are concerned, the extent of the influence is mainly restricted to the roadside obstacles because the satellite is positioned at relatively high elevation angles in most practical situations. The advantages of satellite communication compared to local land are [1–4]:

1. large capacity; the total transmission capacity can exceed 1 *Gbit per second*
2. distance insensitive cost
3. wideband service that allows transmission of high bit rate multimedia information: video, audio, voice, digital, and so on
4. broadcast transmission capability that allows distribution of any information from one point to multipoint users located on the Earth.

TABLE 14.1. A Comparison of the LEO, MEO, and GEO Orbits [3]

	LEO	MEO	GEO
Altitude	700–1000 km	10,354 km	35,786
Orbital period	100 minutes	6 hours	Stationary
Number of satellite for global coverage	48–66	10–12	3
Space segment cost	Highest	Lowest	Medium
Satellite lifetime (years)	Imperceptible	Imperceptible	Long
Propagation delay call handover	Frequent	Infrequent	None

The GEO is a circular orbit in the equatorial plan. The angular velocity of the satellite is the same as that of the Earth. Therefore, the satellite seems to remain stationary in the sky. The LEO and MEO orbits are lower altitude orbits. They are particularly suitable for the personal communications systems because of their low path loss, fading effects, and low propagation delays. A comparative analysis of the main parameters of the LEO, MEO, and GEO orbits is shown in Table 14.1 after analysis of several tables presented in Reference 3.

Let us briefly consider advantages and disadvantages of each kind of satellite following References 1–4.

(a) *GEO satellites*

Advantages:

1. The satellite appears to be fixed (immovable) when viewed from the Earth, which means no tracking is required for each Earth station antenna.

2. About 40% of the Earth's surface is in view from the satellite.

Disadvantages:

1. High power loss of about 200 dB over the radio path.

2. Large signal delay of 238–284 ms.

3. Polar region of the Earth with latitudes more than $81°$ are not covered by GEO satellite.

(b) *LEO and MEO satellite*

Advantages:

1. Much smaller path loss compared to GEO satellites.

2. Lower signal delay (about several tens ms).

3. The reduction in range provides a large decrease in path loss resulting in much smaller receiving antennas.

4. The reduction in range provides a significant reduction in propagation delay, making voice conversation more pleasing to the user and increasing most data communication protocols.

TABLE 14.2. Frequency Allocations for Satellite Communications [4]

Band	Uplink (GHz)	Downlink (GHz)
C	6	4
X	8.2	7.5
Ku	14	12
Ka	30	20
S	40	20
Q	44	21
L (mobile)	1.525 to 1.559	1.626 to 1.660

Disadvantages:

1. Short period of satellite observation, both visual and radio.
2. Doppler Effects are significant because of the high speed of the satellite.
3. Many satellites are required to establish continuous transmission and full radio coverage of subscribers.

The notion of "free space" has long passed for satellite communications. For minimizing satellite-to-satellite interference and sharing the limited frequency spectrum among the ever increasing number of operational or proposed satellites, regulations become a main part of channel design. Coordination of frequency and orbit is necessary to ensure a noninterfering satellite operation.

In the same manner, as was defined for land communication links (see Chapters 1, 5, and 6), one can define the transmitted path form an Earth terminal antenna to the satellite as *uplink*, and the transmitted path from the satellite to an Earth's station as *downlink*. Table 14.2 presents the "working frequencies" that are commonly used both in uplink and downlink LSCs.

LSC systems enable users of portable computers or mobile phones to communicate with one another from any two points worldwide. Signal propagation for such systems has become an essential consideration. Path conditions may indeed cause harmful impairments that severely corrupt the system availability and performance. Hence, propagation considerations are very important for successful operation. Most satellites employ fixed, not mobile terminals, as in LSC systems. Satellite-mobile links operate with low signal margins and obstructions due to overpasses and vegetation cause outages and reduce communications quality.

Therefore, to design successful wireless LSC links, stationary or mobile, it is very important to predict all propagation phenomena occurring in such links to give a satisfactory physical explanation of the main parameters of the channel, such as path loss, slow, and fast fading, and finally, to develop a link budget. Moreover, in land satellite communications, we must divide the channel into three parts. The upper channel that covers ionospheric radio propagation was described in Chapter 13. Table 14.2 shows the frequency bands that are used for satellite communications. Fading phenomena are not very significant if we deal with regular ionosphere without any fluctuations caused by magnetic storms, disturbances on the Sun, cosmic rays, and so on. Taking into account such effects, for the frequency band (see Table 14.2), we can say that the fast fading phenomena effects change the link budget only within several decibels. The same situation is observed in the turbulent troposphere (see Chapter 12), as the middle part of the land-satellite channel, at frequencies in C/Q-band (see Table 14.2), taking into account the effects of hydrometeors (rain, snow, smoke etc.). Here, all features together give a cumulative effect, in terms of fast fading, of about few decibels in the total path loss (see estimations in Chapter 12).

The main effect in the link budget and the total path loss comes from the lower part of the land-satellite channel, where effects of the terrain profile, which cause shadowing (or slow fading) become more dominant. In a land subchannel, *local shadowing effects*, caused by multiple diffractions from numerous wedges and corners of several obstructions, become predominant and can significantly corrupt information sent from a ground-based terminal to a satellite and vice versa. Furthermore, in LSC typical land built-up scenarios, the line-of-sight (LOS) path between the satellite and the land terminal (stationary or mobile) can be affected by multipath mechanisms arising from reflection on rough ground surface and wall surfaces, multiple scattering from trees, and obstacles. As was mentioned by Saunders [2], in such very complicated environments accounting for high speed satellite movements, it is very complicated to differentiate slow fading and fast fading effects, as was done for land communication links (see Chapters 6 and 11). They must be accounted together.

Therefore, later, we will show the reader how to take into account the effects of the terrain profile and multiple diffraction and scattering effects in the fading description and for the link budget design within the LSC channel. All these elements will allow us to obtain the radio coverage of service areas. Finally, we will present a completed algorithm for the megacell radio map construction for different areas of the Earth, depending on the propagation situation within each LSC channel, the subscriber position, and satellite elevation above the horizon, with respect to the desired subscriber location.

In this chapter, we will analyze two main concepts on how to account for the terrain effects on LSCs. The first is based on *statistical models*, and the second is based on *physical–statistical* models. To unify these models and to use them together in our analysis, we assume that the radio signal is moving within a channel only between two states: *good* and *bad*, as is shown in Figure 14.2.

A *good state* occurs when the LOS component is predominant, whereas the *bad state* occurs when the LOS component is absent and only non-line-of-sight (NLOS) components are present because of shadowing and/or multipath

FIGURE 14.2. Markov's chain.

phenomena. This scheme is called the Markov's chain [5, 19]. The statistical models are based on a transfer from bad to good states and vice versa. We should note that in the next section, we will present only these statistical models, which more precisely predict the link budget in the LSC links, and with which we will compare the proposed physical–statistical models. At the same time, the physical statistical models are based on the classical aspects of radio propagation over the terrain and on the statistical description of obstructions placed randomly on the rough terrain. As will be shown later, such models can also be adapted to use the Markov's chain approach, as a basic aspect of pure statistical models. So, despite the fact that some researchers separate statistical and physical–statistical models and show them separately, we will show how to unify these approaches. Finally, prediction of fading viability in different satellite megacell networks will be presented, as a proof of the proposed physical–statistical approach based on multiparametric stochastic model.

14.2. TYPE OF SIGNALS IN LAND SATELLITE COMMUNICATION LINKS

Propagation between a satellite and a mobile receiver can be classified as either *nonshadowed*, when the mobile or stationary subscriber has an unobstructed LOS path to the satellite, or *shadowed*, when the LOS path to the satellite is obstructed by a feature placed at the terrain, whether natural or manmade.

The nonshadowed signal received at the mobile receiver is composed of three components: direct, specular, and diffuse (see Fig. 14.3). Propagation measurements indicate that a significant fraction of the total energy arrives at the receiver by way of a direct path.

The remaining power is received by the specular ground reflected path and the many random scattering paths that form the diffuse signal component.

The shadowed signal occurs when the signal fade is caused primarily by scattering and absorption from both branches and foliage where the attenuation path length is the interval within the first few Fresnel zones intersected by the canopies or building roofs.

Measured results indicate that shadowing is the most dominant factor determining slow signal fading. Its effect depends on the signal path length through the obstruction, type of obstruction, elevation angle, direction of traveler, and carrier frequency. The shadowing is more severe at low elevation angles where the projected shadow of the obstacle is high. The effect of shad-

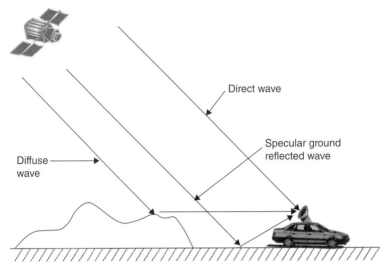

FIGURE 14.3. LOS propagation in land-satellite communications.

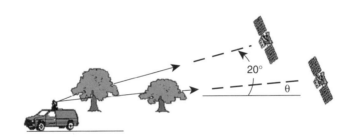

FIGURE 14.4. Shadowing caused by trees for low-elevated satellites.

owing due to diffraction from buildings is clear to understand, based on the results discussed in Chapter 6. The effect of shadowing induced by vegetation is more complicated and depends on how frequently trees are distributed in the surrounding area, the path length trough the trees, and the density of branches and foliage. Some typical scenarios of shadowing that occur due to tree canopies are depicted in Figure 14.4. At frequencies lower than 1 GHz, trees are virtually transparent to the signal. For higher frequencies, trees are regarded as ideal edge refractors in order to estimate the amount of signal attenuation. The shadowed signal received at the mobile receiver is composed of two components: a shadowed *direct* component and a *diffuse* component.

The shadowed direct component is generated when the LOS signal from the satellite passes through roadside vegetation and is attenuated and scattered by leaves, branches, and limbs of vegetation. The attenuation of the direct component depends on the path length through the vegetation. Scattering by vegetation generates a random forward scattered field that interferes with the direct component, causing it to fade and to lose its phase coherency. Thus, the shadowed direct component can be modeled as the sum of an attenuated LOS signal and a random forward scattered field [6]:

$$P_{\text{shadowed}} = \alpha_{\text{attn}} P_{\text{direct}} + P_{\text{scattered}}. \tag{14.1}$$

Here, α_{attn} represents the attenuation factor of the direct component of the signal power, and $P_{\text{scattered}}$ is the scattered signal from the vegetation. A typical shadowing attenuation of a building, bridge, or trees is on the order of 8–20 dB relative to the signal mean value.

The diffuse component results from various reflections from the surrounding terrain. This component varies randomly in amplitude and phase. Multipath propagation does not cause significant losses for land mobiles.

The shadowed diffuse component from vegetations is identical in form to the diffuse component for nonshadowed propagation. The diffuse component is assumed to be received randomly from all angular directions.

Hence, the total shadowed signal is the sum of the shadowed direct component and the diffuse component [6]:

$$P_{\text{shadowed}} = \alpha_{\text{attn}} P_{\text{direct}} + P_{\text{scattered}} + P_{\text{diffuse}}. \tag{14.2}$$

We can use this equation to calculate the total path loss within the LSC link. For this purpose, we need to use the corresponding models, pure statistical or physical–statistical, based on some special experiments and numerous measurements.

14.3. STATISTICAL MODELS

These models correspond to cases for which multipath and LOS are present simultaneously [6–13, 16, 19]. In this section, we will describe only two models; Loo's [7, 9] and Lutz [16], which have been used in References 2, 3, 14, and 15 for designing the unified algorithm for predicting fading phenomena in the land-satellite links. Therefore, we will describe briefly these models and compare their results with those obtained in References 2, 3, 14, and 15 and with those obtained on the basis of the multiparametric stochastic approach proposed for the land-satellite links in Reference 20 for the special series of experiments described in References 3, 17, and 18.

We also refer the reader to some other statistical models by Suzuki [10], Corazza-Vatalaro [11], Xie-Fang [12], Abdi [13], and the three-state model [19].

14.3.1. Loo's Model

Loo's model [7, 9] is a statistical model for a land mobile-satellite link with applications to rural environments. The model assumes that the amplitude of the LOS component under foliage attenuation is distributed according to the log-normal probability density function (PDF) and the received multipath component is described by a Rayleigh PDF. The model is statistically described in terms of its PDF or cumulative distributed function (CDF), which were obtained under the hypothesis that foliage not only attenuates but also scatters radio waves. In such assumptions, the total complex fading signal is the sum of a log-normally distributed random signal and a Rayleigh signal [7, 9]:

$$r\exp(j\theta) = z \cdot \exp(j\phi_0) + w \cdot \exp(j\phi), z > 0, w > 0, \tag{14.3}$$

where the phase ϕ_0 and ϕ are uniformly distributed between 0 and 2π, z is log-normal distributed amplitude, and w is a Rayleigh distributed amplitude. If z is temporally kept constant, it can be assumed that the PDF $p(z)$ is log-normal. The signal random envelope r is log-normal distributed for large values and Rayleigh distributed for small values [7, 9]:

$$p(r) \approx \begin{cases} \dfrac{1}{r\sqrt{2\pi d_0}}\exp\left[-\dfrac{(\ln r - \mu)^2}{2d_0}\right] & \text{for } r \gg \sqrt{d_0} \\ \dfrac{r}{b_0}\exp\left[-\dfrac{r^2}{2b_0}\right] & \text{for } r \ll \sqrt{b_0} \end{cases} \tag{14.4}$$

In this equation, $\sqrt{d_0}$ and μ are the standard deviation and mean for the log-normal distribution, and b_0 is the variance for the Rayleigh distribution, respectively. The parameter b_0 represents also the average scattered power caused by multipath effects.

Many calculations with different values for b_0, d_0 and μ where carried out by Loo with the objective of fitting the results of his model to those derived from measurements made on simulated satellite paths. The measurements site was a rural area with about 35% tree coverage. The model parameters were obtained by trial and error to fit measured values. The parameters in decibels (dB) are shown in Table 14.3.

TABLE 14.3. Statistical Model Parameters

Conditions	Standard Deviation $10\log_{10}(\sqrt{d_0})$	Mean $10\log_{10}(\mu)$	Multipath Power $10\log_{10}(b_0)$
Infrequent: light shadowing	0.5	0.5	-8
Frequent: heavy shadowing	3.5	-14	-12
Overall results	1.0	-3	-6

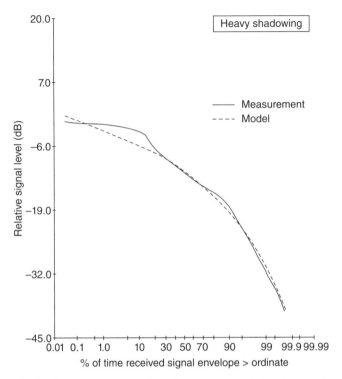

FIGURE 14.5. Loo's envelope model and measurement for heavy shadowing at $f = 18.925$ GHz.

Computational results for the signal envelope, based on Loo's model, are compared to measurements obtained in References 7 and 9 and presented in Figure 14.5.

As was mentioned in Chapter 11, the signal envelope PDF of the model facilitates the calculation of fade margins in the design of communications systems. As for the signal envelope phase distribution, Figure 14.6 shows a comparison of the complementary cumulative distribution function (CCDF) for the received signal phase calculated using the well-known equation (see Chapter 1):

$$\text{CCDF}(r) \equiv P(r > R) = 1 - \text{CCD}(r) = 1 - \int_{0}^{R} p(r)dr, \qquad (14.5)$$

where R is either maximum accepted path loss or noise floor figure of the system. We must note that for the case of infrequent light shadowing, the model shows the best fit around the median region and some deviation near the tails of the distribution (see Fig. 14.5). The results of the model showed a slightly higher shadowing effect than those from measured data. For the com-

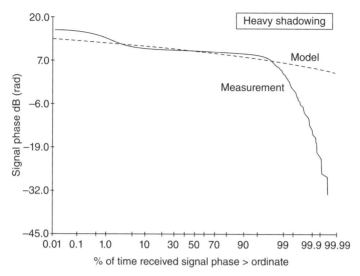

FIGURE 14.6. Loo's phase model and measurement for heavy shadowing at $f = 18.925$ GHz.

bined results (see Fig. 14.5 and Fig. 14.6), the fit was poor about the median but reasonably good in the weak signal range, which is most important for fade margin calculations. The model parameters were obtained by trial and error to fit the measured values.

The results indicate that the model shows a correlation between the rate of change of the envelope due to multipath and foliage attenuation both for heavy shadowing and for light shadowing. The disadvantage of this model is that the measurements were made up to 30°. Model parameters for higher elevation angles are not available.

14.3.2. Lutz Statistical Model

In this model, which can be considered as a generalization of Loo's model, the simple statistics of LOS and NLOS are modeled by two distinct states, *good* and *bad*, as shown in Figure 14.2. This is appropriate for describing the propagation situation in urban and suburban areas where there is a large difference between the shadowed and nonshadowed statistics. The parameters associated with each state and the transition probabilities for evolution between states are empirically derived. The LOS condition is represented by a good state, and the NLOS condition by a bad state. In the good state, the signal is assumed to be Rician K-factor distributed, which depends on the satellite elevation angle and the carrier frequency, so that the PDF of the signal amplitude is given by $P_{\text{good}} = P_{\text{Rice}}$. In the bad state, the fading statistics of the signal amplitude are assumed to be Rayleigh, with a mean power $S_0 = \sigma^2$, which varies with time.

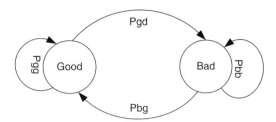

FIGURE 14.7. Markov's model of channel state.

So the PDF of amplitude is specified as the conditional distribution $P_{\text{Rayleigh}}(S|S_0)$, where S_0 varies slowly with a log-normal distribution $p_{\text{LN}}(S_0)$, representing the varying effects of shadowing with the NLOS component. For more details, we refer the reader to Reference 16.

Transitions between states are described by a first-order Markov chain. This is a state transition system, in which the transition from one state to another depends only on the current state rather than on any more distant history of the system. The transition probabilities, which summarize all models based on the Markov chain approach, are (see Fig. 14.7 [16]):

- probability of transition from good state to good state P_{gg}
- probability of transition from good state to bad state P_{gb}
- probability of transition from bad state to bad state P_{bb}
- probability of transition from bad state to good state P_{bg}.

For a digital communication system, each state transition is taken to represent the transition of one symbol. The transition probabilities can then be found in terms of the mean number of symbol duration spent in each state [16]:

$P_{gb} = \dfrac{1}{D_g}$ where D_g is the mean number of symbol duration in the good state;

$P_{bg} = \dfrac{1}{D_b}$ where D_b is the mean number of symbol duration in the bad state.

The sum of the probabilities leading from any state must equal to the sum of the unit, so

$$P_{gg} = 1 - P_{gb} \quad \text{and} \quad P_{bb} = 1 - P_{bg}. \tag{14.6}$$

The time share of shadowing (the proportion of a symbol in the bad state) is

$$A = \frac{D_b}{D_g + D_b}. \tag{14.7}$$

Later, we will use this equation to find the parameter A, denoted as *the time share of shadowing*, during comparison with physical–statistical and the stochastic multiparametric approaches, where this parameter is derived in another manner. In this comparison, we will use the Lutz model as a classical statistical approach.

14.4. PHYSICAL–STATISTICAL MODELS

On the one hand, in pure statistical models, the input data and computational effort are quite simple, as the model parameters are fitted to measured data. Such models only apply to hypothetical environments and lack the physical background of the realistic problem. On the other hand, pure deterministic physical models provide high accuracy, but they require actual analytical path profiles and time-consuming computations.

A combination of both approaches has been developed by the authors. The general method relates any channel simulation to the statistical distribution of physical parameters, such as building height, width and spacing, street width or elevation, and azimuth angles of the satellite link. This approach is henceforth referred to as the "*Physical–Statistical*" approach [2, 3, 14, 15, 20]. The main concept of such an approach is sketched in Figure 14.8.

As for the physical models, the input knowledge consists of electromagnetic theory and a full physical understanding of the propagation processes. However, this knowledge is then used to analyze a *statistical* input data set, yielding a *distribution* of the output predictions. The output predictions are not linked to specific locations. Physical–statistical models therefore require only simple input data such as distribution parameters (e.g., mean building height and building height variance, as was done in Chapter 5 for land communication links).

This modeling describes the geometry of mobile satellite propagation in built-up areas and proposes statistical distributions of building heights, which are used in subsequent analysis. We will consider only two of these models:

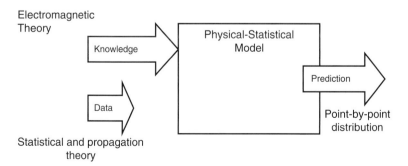

FIGURE 14.8. Algorithm of physical-statistical model.

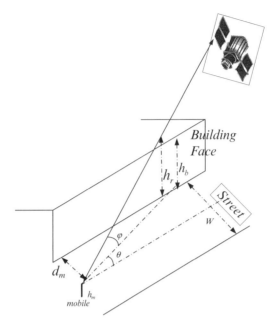

FIGURE 14.9. Geometry for mobile-satellite communication in built-up areas.

- a shadowing model based on the two-state channel Lutz model
- a multiparametric stochastic model.

14.4.1. The Model of Shadowing

The geometry of the situation, which was analyzed in References 2, 3, 14, and 15 by Saunders and Evans with their colleagues, is illustrated in Figure 14.9. It describes a situation where a mobile is situated along a straight street with the direct ray from the satellite impinging on the mobile from an arbitrary direction. The street has buildings lined up on both sides, with randomly varying heights. In the presented model, the statistics of building height in typical built-up areas is used as input data. A suitable form was sought by comparing with geographical data for the cities of Westminster and Guildford, UK [3, 14, 15]. The PDFs that were selected to fit the data are the log-normal and Rayleigh distributions with unknown parameters, the mean value, m, and standard deviation, σ_b. The PDF for the log-normal distribution is [2, 14, 15]

$$p_b(h_b) = \frac{1}{h_b\sqrt{2\pi}\sigma_b} \cdot e^{-\frac{\ln^2(h_b/m)}{2\sigma_b^2}}. \tag{14.8}$$

The PDF for the Rayleigh distribution is presented in Chapter 1. We will repeat it using notations made in References 2, 14, and 15:

TABLE 14.4. Best-fit Parameters for the Theoretical PDFs

| City | Log-Normal PDF | | Rayleigh PDF |
	Mean (m)	Standard Deviation	Standard Deviation
Westminster	20.6	0.44	17.6
Guildford	7.1	0.27	6.4

$$p_b(h_b) = \frac{h_b}{\sigma_b^2} \cdot e^{-\frac{h_b^2}{2\sigma_b^2}}. \tag{14.9}$$

To find the appropriate parameters for these functions in order to fit the data measurements as accurately as possible, the PDF was found by minimizing the maximum difference between the two CDFs. The parameters for each PDF are quoted in Table 14.4 from References 2, 14, and 15, where all parameters are in meters.

The direct ray is judged to be shadowed when the building height h_b exceeds some threshold height h_T relative to the direct ray height h_s (see Fig. 14.9). The shadowing probability (P_s) can then be expressed in terms of the PDF of the building height, $p_b(h_b)$ as [2, 14, 15]

$$P_s = \Pr(h_b > h_T) = \int_{h_T}^{\infty} p_b(h_b) dh_b. \tag{14.10}$$

The definition of h_T is obtained by considering shadowing to occur exactly when the direct ray is geometrically blocked by the building face. Using a simple geometry, the following expression is extracted for h_T [2, 14]:

$$h_T = h_r = \begin{cases} h_m + \dfrac{d_m \tan\phi}{\sin\theta} & \text{for } 0 < \theta < \pi \\ h_m + \dfrac{(w - d_m)\tan\phi}{\sin\theta} & \text{for } -\pi < \theta < 0 \end{cases}. \tag{14.11}$$

All notations and geometrical parameters in (14.11) are explained in Figure 14.9.

The shadowing model estimates the probability of shadowing for Lutz two-state model [16]. The same Markov chain shown in Figure 14.6 is used, but parameters A, P_{bad}, and P_{good} are obtained from actual random distribution of the obstructions above the terrain. Thus,

$$A = \int_{z_1}^{z_2} P_b(h) dh, \tag{14.12}$$

where h are different heights of obstacles, z_1 and z_2 are the minimum and maximum height of the built-up layer

$$P_b = \begin{cases} \log \text{normal} + \text{Ricean} \\ \log \text{normal} + \text{Rayleigh}, \end{cases} \tag{14.13}$$

where the log-normal PDF is for pure NLOS shadowing. The Rician PDF describes both the LOS and the multipath component, and the Rayleigh PDF describes the multipath component of the total signal, when the LOS component is absent (see Chapter 1).

Now, using theoretical results obtained by Lutz' pure statistical model and results from the Saunders and Evans model, we can combine them into one unified model. This unified model will be compared with the stochastic multi-parametric model described in the next section based on the theoretical framework analyzed in Reference 20. Thus, taking into account the Markov's chain (Fig. 14.7), we consider the bad status by using the Rayleigh PDF and the good status by using the Rician PDF, as well as shadowing by using the log-normal PDF. By combining all these PDFs in a Markov chain, we finally can obtain the total PDF that describes the effects of different kinds of fading occurring within the LSC link, caused by terrain obstructions; natural and manmade. As a result we get

$$p(S) = (1 - A) \cdot P_{good} + A \cdot P_{bad} = (1 - A) \cdot P_{Rice}(S)$$
$$+ A \int_0^\infty p_{Rayleigh}(S \mid S_0) \cdot p_{\log norm}(S_0) dS_0. \tag{14.14}$$

Then, we introduce the corresponding CCDF, which describes the signal stability, being the received signal with amplitude r that prevails upon the maximum accepted path loss (R) in the multipath channel, caused by fading phenomena. This can be presented in the following form:

$$\text{CCDF} = \text{Pr}(r > R) = \int_0^R p(S) dS. \tag{14.15}$$

All earlier equations allow us to present the unified algorithm for fading phenomena estimation in LSC links, both stationary and mobile.

14.4.2. Multiparametric Stochastic Model

As an example of a physical–statistical model, we present the same stochastic approach which was used successfully for land communication channels, rural, suburban, and urban in Chapters 6 and 11. The reason for that is based on the fact that the previous physical–statistical model, as was shown in References 2, 3, 14, and 15, predicts more strictly the fading effects in different LSC links compared to the pure statistical models [7–13].

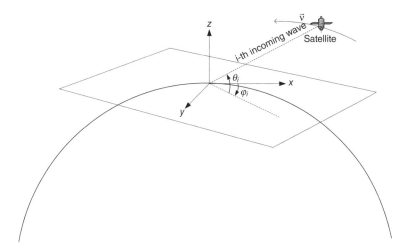

FIGURE 14.10. General geometry of the land-satellite link.

At the same time, as was mentioned in Reference 20, the physical–statistical model, which is based on a deterministic distribution of the local built-up geometry (see previous section), cannot strictly predict any situation when a satellite moving around the world has different elevation angles θ_i, with respect to a subscriber located at the ground surface, as shown in Figure 14.10. As the result, the radio path between the desired subscriber and the satellite crosses different overlay profiles of the buildings because of continuously changing elevation angle of the satellite during its rotation around the Earth, with respect to the ground-based subscriber antenna. To predict continuously the outage probability of shadowing in real time, a huge amount of data is needed regarding each building, geometry of each radio path between desired user and the satellite during its rotation around the Earth, and finally, high-speed powerful computer.

Most of these difficulties can be handled by using the multiparametric physical–statistical model [21–23]. Following Reference 20, we take into account both the buildings' distribution at the ground surface and their height profile changing in the vertical plane, that is, accounting for the 3-D stochastic model of multiple scattering, reflection, and diffraction rearranging the corresponding equations in the case of the LSC link. For convenience, we will repeat some equations presented in Chapter 6, which are needed to introduce the main features of this model.

Buildings' Overlay Profile. The LSC link is very sensitive to the overlay profile of the buildings as shown in Figure 14.11, because during its movement around the Earth, depending on the elevation angle ϕ, the buildings' profile will be continuously changed leading to different effects of shadowing in the current communication link (see Fig. 14.12).

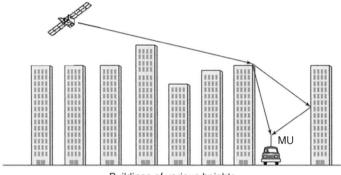

Buildings of various heights

FIGURE 14.11. Local effect of buildings' overlay profile.

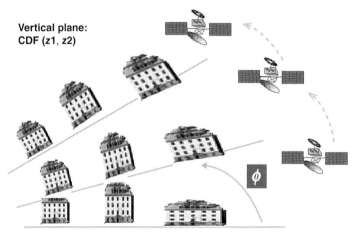

FIGURE 14.12. Change of the profile function $F(z_1, z_2)$, in the vertical plane during the movements of a satellite.

Taking into account the fact that real profiles of urban environment are randomly distributed, the probability function $P_h(z)$, which describes the overlay profile of the buildings, can be presented in the following form [20, 23] (see also Chapter 6):

$$P_b(z) = H(h_1 - z) + H(z - h_1) \cdot H(h_2 - z) \cdot \left| \frac{(h_2 - z)}{(h_2 - h_1)} \right|^n \quad n > 0, \quad 0 < z < h_2, \quad (14.16)$$

where the function $H(x)$ is the Heaviside step function, which equals to 1 if $x > 0$, and equals to 0 if $x < 0$. The graph of this function versus height z of a built-up overlay is presented in Figure 14.13.

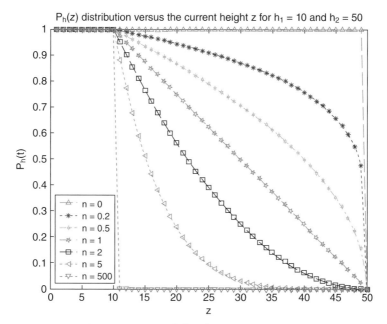

FIGURE 14.13. Buildings' overlay profile PDF.

For $n \gg 1$, $P_b(z)$ describes the case where buildings higher than h_1 (minimum level) very rarely exist. The case, where all buildings have heights close to h_2 (maximum level of the built-up layer), is given by $n \ll 1$. For n close to zero, or n approaching infinity, most buildings have approximately the same level that equals h_2 or h_1, respectively. For $n = 1$, we have the case of building height uniformly distributed in the range of h_1 to h_2. The average building height ($\bar{h}$) can be found as [20, 23] (see also Chapter 6):

$$\bar{h} = h_2 - n \cdot \frac{(h_2 - h_1)}{n+1}. \tag{14.17}$$

Then, analyzing the built-up layer profile $F(z_1, z_2)$, "illuminating" by the terminal antennas of the heights z_1 and z_2, we get

- for the case when the antenna height is above the rooftop level, that is,

$$z_2 > h_2 > h_1$$

$$\begin{aligned}
F(z_1, z_2) &= H(h_1 - z_1)\left[(h_1 - z_1) + \frac{(h_2 - h_1)}{(n+1)}\right] \\
&\quad + H(z_1 - h_1)H(h_2 - z_1)\frac{(h_2 - z_1)^{n+1}}{(n+1)(h_2 - h_1)^n}
\end{aligned} \tag{14.18a}$$

- for the case when the antenna height is below the rooftop level, that is, $z_2 < h_2$

$$
\begin{aligned}
F(z_1, z_2) &= H(h_1 - z_1)\left[(h_1 - z_1) + \frac{(h_2 - h_1)^{n+1} - (h_2 - z_2)^{n+1}}{(n+1)(h_2 - h_1)^n}\right] \\
&\quad + H(z_1 - h_1)H(h_2 - z_1)\frac{(h_2 - h_1)^{n+1} - (h_2 - z_2)^{n+1}}{(n+1)(h_2 - h_1)^n}.
\end{aligned}
\tag{14.18b}
$$

Then, the CDF of the event that any subscriber located in the built-up layer is affected by obstructions due to shadowing effect can be presented as [20, 23]

$$
CDF(z_1, z_2, n) = \frac{1}{z_2 - z_1}\int_{z_1}^{z_2} P_h(z)\,dz \equiv \frac{1}{z_2 - z_1}F(z_1, z_2). \tag{14.19}
$$

The built-up profile function presented by (14.19) allows us to account for continuous changes of the overlay building profile during the movement of the satellite around the Earth and the corresponding changes of its elevation angle with respect to the position of any subscriber located in areas of service. This CDF will be used in the next section to find the total outage probability of fading phenomena occurring in LSC links.

Multiray Pattern Distribution on the Ground Surface. As is shown in References 21–23 and mentioned in Chapter 5, at the horizontal plane, the array of buildings are randomly distributed according to Poisson's law. During movement around the Earth, the satellite antenna "illuminates" different land areas with various distributions of obstructions on the ground surface, that is, in the horizontal plane, as shown in Figure 14.14.

The proposed multiparametric model [21–24] allows us to consider the strength of the total field at the receiver as the additive summation of n-time independently scattered waves with independent strengths. The real situation with multipath phenomena occurring in the urban environment is shown in Figure 14.14. As was mentioned in Chapter 6, in microcell land communication links ($d < 1$–3 km; d is the range from the BS antenna), the single scattered

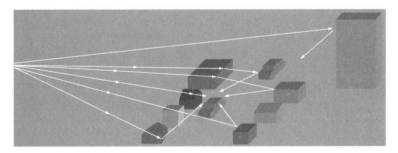

FIGURE 14.14. Horizontal map of multipath phenomenon occurring in the urban scene.

waves are dominant, whereas the two- and three-time scattered waves begin to prevail in macrocell scenarios for $d > 5$–10 km. Therefore, as was shown in Reference 20, for land-satellite megacell links, we must additionally consider the three-time reflected and scattered waves. As was shown in Reference 20, the strengths r_i of these waves are distributed according to the Gaussian law with the zero-mean value and dispersion σ_1^2 (for single scattered waves), σ_2^2 (for two-time scattered waves), and σ_3^2 (for three-time scattered waves), which depend strongly on the characteristic features of the terrain.

The average number of scattered waves involves also a dependence on the distance from the BS antenna d. Following Reference 20, we can obtain a new multiray distribution for the LSC link. Thus,

- for the average number of single scattered waves

$$\bar{N}_1 = \frac{\pi \nu d^2}{4} K_2(\gamma_0 d) \tag{14.20a}$$

- for the average number of two-time scattered waves

$$\bar{N}_2 = 9\left(\pi \nu d^2\right)^2 \left[\frac{K_4(\gamma_0 d)}{8!} + \sqrt{\frac{2}{\pi \gamma_0 d}} \frac{K_{7/2}(\gamma_0 d)}{7!} \right] \tag{14.20b}$$

- for the average number of three-time scattered waves

$$\bar{N}_3 = 8\left(\pi \nu d^2\right)^3 \left[\frac{1}{\gamma_0 d} \frac{K_5(\gamma_0 d)}{10!} + \sqrt{\frac{2}{\pi \gamma_0 d}} \frac{K_{11/2}(\gamma_0 d)}{11!} \right], \tag{14.20c}$$

where $K_n(\gamma_0 d)$ is the MacDonald's function of n-order and γ_0 [km^{-1}] is the density of building contours defined in Chapter 6.

The probability of receiving one- to three-time (i.e., for $n = 1, 2, 3$) scattered waves at the moving subscriber (MS) antenna can be computed according to the following equation [20]:

$$P_n = 1 - \exp[\bar{N}_n], n = 1, 2, 3. \tag{14.21}$$

In computations of Equation (14.19), we take into account the effects of independent single (the first term), double (the second term), and triple (the third term) scattering of rays with the random amplitude r, as well as their mutual influences on each other (the fourth term), that is,

$$CDF(r) = \frac{P_1(1 - P_2)(1 - P_3)}{P_0} \frac{r}{\sigma_1^2} e^{-\frac{r^2}{2\sigma_1^2}} + \frac{P_2(1 - P_1)(1 - P_3)}{P_0} \frac{r}{\sigma_2^2} e^{-\frac{r^2}{2\sigma_2^2}}$$

$$+ \frac{P_3(1 - P_1)(1 - P_2)}{P_0} \frac{r}{\sigma_3^2} e^{-\frac{r^2}{2\sigma_3^2}} + \frac{P_1 P_2 P_3}{P_0} \frac{r}{\sigma_1^2 + \sigma_2^2 + \sigma_3^2} e^{-\frac{r^2}{2(\sigma_1^2 + \sigma_2^2 + \sigma_3^2)}}, \tag{14.22}$$

where

$$P_0 = 1 - (1 - P_1)(1 - P_2)(1 - P_3) = 1 - e^{-(\bar{N}_1 + \bar{N}_2 + \bar{N}_3)} \tag{14.23}$$

is the probability of direct visibility (e.g., LOS component), P_1, P_2, and P_3 are defined by Equation (14.21) combined with (14.20a), (14.20b), and (14.20c), respectively.

Equation (14.22) is a more general form of the CDF of the multipath effects occurring in the land-satellite links with respect to that describing the same multiray effects in the land-to-land communication scenarios (see Chapter 6). This CDF will be used in the next section (together with CDF described by (14.19)) to estimate the total probability of fading for predicting successful communication and service by the satellite antenna of any ground-based subscriber.

14.5. FADING EFFECTS ESTIMATION IN LAND-SATELLITE LINKS VIA THE EXPERIMENTAL DATA

In this section, we will use the unified algorithm, based on the combination of the physical–statistical (Saunders and Evans) model and the pure statistical (Lutz) model of the mobile-satellite communication channel, using Equations (14.14) and (14.15), and compare it with the proposed stochastic multiparametric model based on the corresponding $CCDF(r) = 1 - CDF(r)$ of signal random strength envelope. This CCDF describes the total probability to achieve a successful communication between any ground-based subscriber and the satellite. It can be presented as [20]

$$CCDF(r) = 1 - CDF(r) \cdot CDF(z_1, z_2, n), \tag{14.24}$$

where both CDFs are described by (14.19) and (14.22), respectively.

In our comparative analysis, we used measured data from References 17 and 18, done in several cities in Europe. These tests were narrowband measurements at a single frequency, representing the channel within its coherence bandwidth. The test was transmitted from the European Space Agency (ESA) ground station in Villafranca, Spain, and relayed by the geostationary satellite, Maritime European Communications Satellite (MARECS) at L-band (1.54 GHz). The measurements were conducted in areas with different satellite elevations (see Table 14.5). Using these measurements and using a Rayleigh PDF for building distribution heights and the corresponding Equation (14.15), for CCDF, we constructed a corresponding numerical code to see if there is a good agreement with the measured data. It is clear that the CCDF gives us the knowledge of stability of received signal with respect to noise caused by fading phenomena.

The same was done for the multiparametric model by using Equation (14.24) with the corresponding CDF (14.19) and (14.22) for each city. The

TABLE 14.5. Parameters of Channel Model Measured by References 17 and 18

Satellite Elevation	A	10log(c) (dB)	μ (dB)	σ (dB)
13° Stockholm	0.24	10.2	−8.9	5.1
18° Copenhagen	0.8	6.4	−11.8	4.0
24° Munich	0.66	6.0	−10.8	2.8
34° Barcelona	0.58	6.0	−10.6	2.6
43° Cadiz	0.54	5.5	−13.6	3.8

results of *fading* estimations were compared with those obtained from the pure statistical Lutz model based on Markov's chain.

The main goal of such simulations was to define which one of the three models presented the best fit to the measured data and also the simplest one. In such a comparison we used for the Lutz statistical model simulation $P_{gg} = (0.8, 0.95, 0.85, 0.83, 0.7)$ and $P_{bb} = (0.08, 0.15, 0.25, 0.22, 0.5)$. These results are shown in Figure 14.15a,b,c,d,e for five cities presented in Table 14.5.

The standard deviation (σ), taken from our estimations for each built-up profile, was not more than 2.6 dB. Nevertheless, in Reference 16, the authors used $\sigma = 3$–4 dB, which is not a realistic case when NLOS regime is very small compared with the LOS component of the total filed strength. It is clearly seen from the results of the comparative analysis presented in Figure 14.15a, b,c,d,e that the physical–statistical models, Saunders and Evans, and the stochastic multiparametric, are closer to the experimental data compared to the pure statistical Lutz model. Therefore, in our further analysis, we will compare the proposed physical–statistical models with each other.

Before doing this comparative analysis, let us check the accuracy of the proposed stochastic model in the case when the standard deviation of the CCDF cannot be directly estimated via numerical evaluations. Thus, in Figure 14.16, the CCDF of received signal for the city of Stockholm [20] at satellite elevation angle 13° is presented. The computations were carried out taking the following parameters: The building contour density parameter varied from 8 km^{-1} to 12 km^{-1} with the mean value $\bar{\gamma}_0 = 10$ km^{-1}. The built-up profile parameter n was changed from 0.65 to 1.48. The obtained terrain data allow us to estimate the standard deviation of CCDF within the range of $\sigma \in (-4, 4)$dB.

The results of comparison between the experimental data (continuous curve) and those obtained numerically for the range in the estimated parameters σ shown in Figure 14.16 show that the experimental data lie between the two theoretical curves computed for σ changing from −4 to −2 dB, which is close to that parameter estimated by analyzing the topographic map of Stockholm [20].

Another land-satellite experiment was compared with the Saunders–Evans physical–statistical model [2, 3, 14, 15], and was carried out in England for two

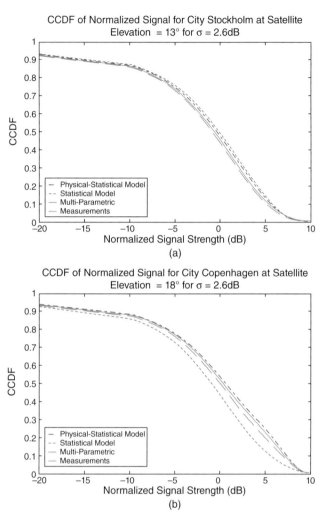

FIGURE 14.15. (a) CCDF of normalized signal for city Stockholm at satellite elevation angle 13° for three different models. (b) CCDF of normalized signal for city Copenhagen at satellite elevation angle 18° for three different models. (c) CCDF of normalized signal for city Munich at satellite elevation angle 24° for three different models. (d) CCDF of normalized signal for city Barcelona at satellite elevation angle 34° for three different models. (e) CCDF of normalized signal for city Cadiz at satellite elevation angle 43° for three different models.

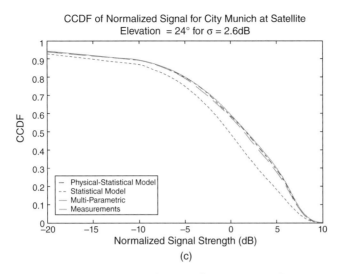

(c)

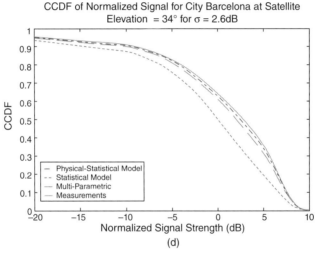

(d)

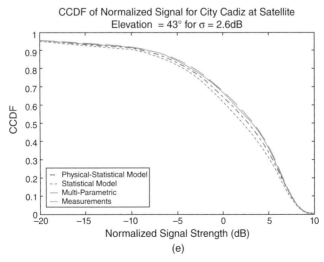

(e)

FIGURE 14.15. (*Continued*)

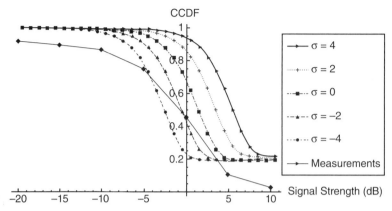

FIGURE 14.16. Comparison between the experimental data obtained in satellite observations over Stockholm (continuous line) and those obtained theoretically from the stochastic approach.

cities, Westminster and Guildford. Here, the CDF, as a total probability of shadowing, was investigated using both models. We do not go deeply into the description of these experiments that are fully described in References 2, 3, 14, and 15. Here, we will only mention that a comparison showed that the best fit was obtained when the CDF of the combined fast/slow fading was described by the corresponding Rayleigh law with the corresponding mean height of buildings $\bar{h} = 20.6$ m and standard deviation $\sigma_b = 17.6$ m (for Westminster) and $\bar{h} = 7.6$ m and $\sigma_b = 6.4$ m (for Guildford). Therefore, in our additional comparison with the Saunders–Evans, physical–statistical model, and with experimental data, we accounted for these estimations as well as for the average density of buildings' contours within 1 km^2, which was estimated as $\bar{\gamma}_0 \approx 10.6$ km^{-1} (for Westminster) and $\bar{\gamma}_0 \approx 7.5$ km^{-1} (for Guildford). We also estimated the corresponding mean height of buildings for Westminster, $\bar{h} = 18.2$ m with a standard deviation of $\sigma_b = 7.5$ m and for Guildford, $\bar{h} = 5.8$ m and $\sigma_b = 4.6$ m. These estimations have suggested that the two cities have fully different built-up terrain profiles: the parameter of the overlay building profile is $n = 2$ (the amount of small buildings exceeds that of tall buildings) for Westminster and $n = 10$ (most buildings are small) for Guilford.

We should note that these parameters were taken in a form of average values, but not exactly, as was done by Saunders and Evans with their coauthors in References 2, 3, 14, and 15 using the local parameters presented in Figure 14.9 for each position of the moving satellite. Despite this fact, our estimations are within the ranges of estimations obtained in References 2, 3, 14, and 15, and we can finally compare our computations of CDF = 1 − CCDF according to (14.19) and (14.22) with those obtained in References 2, 3, 14, and 16 using the Rayleigh CDF, as a best fit of measured data. This comparison is

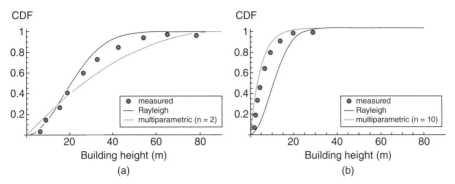

FIGURE 14.17. (a) Comparison between experimental data (dots) obtained in Westmister, physical statistical model (green line) and data obtained from the stochastic approach (blue line). (b) Comparison between experimental data (dots) obtained in Guildford, physical statistical model (blue line) and data obtained from the stochastic approach (green line).

shown in Figure 14.17a for Westminster and Figure 14.17b for Guildford, respectively, where the corresponding experimental data were plotted by dots.

Even with the data on terrain features using average parameters of the built-up terrain, not the local parameters, we obtained a satisfactory agreement between the theoretical prediction based on both the Saunders–Evans physical–statistical and multiparametric stochastic models, and experimental data. This means that the designers of satellite-land links do not need every time to have information on the local built-up terrain parameters as shown in Figure 14.9. It is enough to obtain average parameters of the terrain during satellite movements above the corresponding city, town, village, and so on, and we can use for prediction of fading effects the stochastic multiparametric approach.

Now, in order to show how the proposed stochastic multiparametric algorithm is working, we first varied the satellite elevation angle in a "virtual computer experiment." Figure 14.18a,b,c show the increase in the probability of getting less path loss with a decrease in elevation angle.

Then, we investigated the effect of the type of land usage on the probability of obtaining the maximum acceptable path loss. Figure 14.19a,b,c show a difference between the obtained results for various land environments: (a) urban, (b) suburban, and (c) open rural. The results obtained were for a low elevation satellite ($\varphi = 40°$).

As expected, and as seen from Figure 14.19a,b,c, following the knowledge of the kind of terrain topography from good (open) to bad (urban) scenarios, the main difference is noticeable in the urban environment (as the worst case of strong fading), due to its special propagation features such as multiple diffraction, scattering, and reflection from buildings surrounding the subscriber antenna.

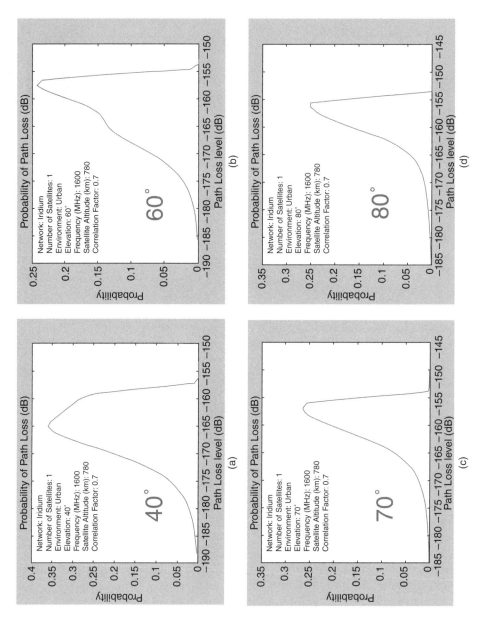

FIGURE 14.18. Simulation output of probability to obtain maximum loss for varying satellite elevation angles.

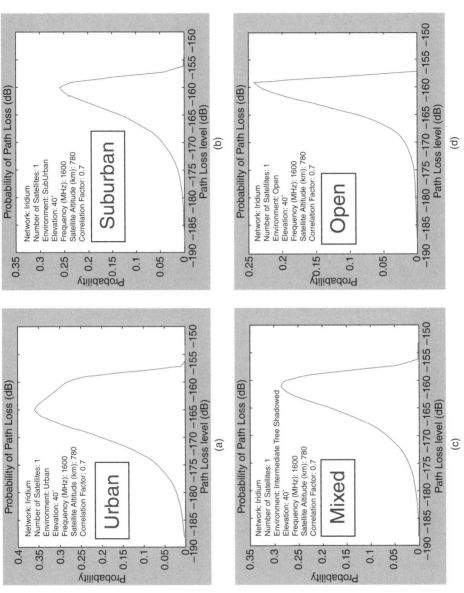

FIGURE 14.19. Simulation output of probability to obtain smaller acceptable path loss for varying land environments.

667

14.6. MEGACELL GLOBAL NETWORKS DESIGN

To give full radio coverage of the Earth and the corresponding megacell map, a net of satellites, assembled into global satellite networks with specific constellations, was developed. Modern global satellite systems were built both for personal and mobile communications. Presently, a strong effort is being made toward the assessment of main specifications for the IMT-2000 system for mobile and personal multimedia communications worldwide, which in Europe is named Universal Mobile Telecommunication System (UMTS). IMT-2000 is intended as a "family" of interoperable systems capable of assisting the roaming/desired user in any mobile network where it may be temporarily present. This system is called a third generation (3G) system, after the first generation (1G) incompatible analogue cellular systems and the second generation (2G) digital systems, the most successful one being Global System for Mobile (GSM) communication [1–4].

Prediction of radio and cellular maps for full coverage of the Earth's surface, using determined constellations of satellites for specific networks, GEO, MEO or LEO, is based on rigorous statistical and physical–statistical propagation models, which take into account the average path loss along all three "sublinks": land, atmospheric, and ionospheric. These effects can be postulated as worst (or bad) or convenient (or good) with intermediate variants "good-bad" and "bad-good," obtained from numerous measurements carried out in these three "sublinks." In Chapters 12 and 13, it was shown how to take into account fading phenomena for link budget designs and how to predict radio coverage for the atmospheric and ionospheric links. It was also shown that for the frequencies of interest operated in land-satellite links, the effects of these two "subchannels" on total path loss and fading are not so significant. More essential fading effects are observed in the land communication "subchannel."

To show the reader how to predict the land link total path loss and to obtain "megacell" radio coverage using determined satellite constellations corresponding to more applicable satellite networks, different mathematical tools were developed [2, 3, 13–20]. The proposed earlier stochastic multiparametric approach showed the efficiency of predicting fade margins and the probability of fading. At the same time, another planning tool was developed, which combined the two models: Loo's pure statistical model [7, 9] (which is close to Lutz [16] and Abdi [13] models) and three-stated Markov model [19]. For this purpose, the authors in References 25–29, analyzing different satellite networks, introduced the PDFs, $P_a(r)$, $P_b(r)$, and $P_c(r)$ for LOS, multipath, and shadow effects description, respectively. Here, a, b, and c, correspond to the transaction from bad to good situations within a wireless link, depending on various environment phenomena. Then, for each situation within the channel, the authors in References 25–29 took the corresponding PDFs, $P(a)$, $P(b)$, and $P(c)$, (the log-normal, Rayleigh, and Rician, respectively). Next, the following

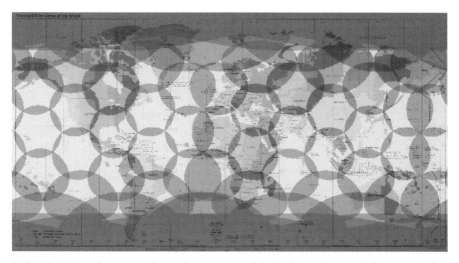

FIGURE 14.20. Computed foot print patterns of the "Iridium" network (Source: Reference 25. Reprinted with permission © 1991 IEEE.)

cumulative equation for determining the probability of fading phenomena within the satellite link was found as

$$P_{total}(r) = P(a)P_a(r) + P(b)P_b(r) + P(c)P_c(r). \tag{14.25}$$

Below, we present the different coverage maps based on the planning tool developed in References 25–29 on the basis of the cumulative Equation (14.25), in order to understand the differences between satellite networks and their effect on the network's *footprint*. Thus, in Figure 14.20, the cellular map of "Iridium" network [25, 26] with a uniform radio coverage of the Earth's surface is presented. Figure 14.21 illustrates the cellular map for "GlobalStar" network [27,28]. The corresponding constellation covers most of the populated area of the Earth. But the polar areas of the Earth cannot be covered by this system. Another example of how to create the coverage radio map is shown in Figure 14.22 for ICO network according to computations made in Reference 29. It is seen that a MEO constellation covers the Earth with only 12 satellites with great overlapping, which is caused by their relatively high altitude.

Now, we present some outputs from our calculations of the general physical–statistical model described by another cumulative Equation (14.24), proposed by the authors on the basis of results obtained in Reference 20, to demonstrate the actions of their own simulation tool and its ability to be adapted to different cellular networks. The output parameters that can be computed by the proposed tool are probability of fading, path loss, link budget, radio and cell

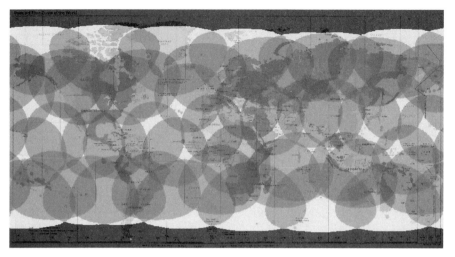

FIGURE 14.21. Computed footprint patterns of the GlobalStar network (Reprinted with permission from GlobalStar).

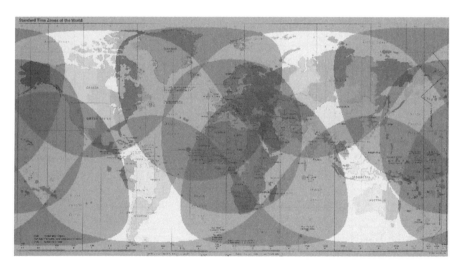

FIGURE 14.22. Computed footprint patterns of the "Intermediate Circular Orbit" (ICO) network according to Reference 29.

coverage, lever crossing rate (LCR), average fade duration (AFD), and bit error rate (BER).

As an example, we varied the types of the satellite networks and investigated the outage probability of shadowing (fading) versus the maximum accepted path loss for different satellite networks. Results are presented in Figure 14.23a,b,c,d, where one network is virtually created by the authors. The

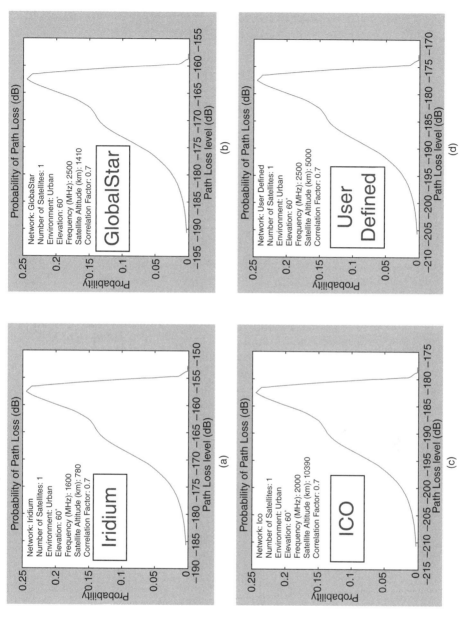

FIGURE 14.23. Simulation output of probability to obtain a smaller acceptable path loss for varying global networks.

671

difference between the networks is seen in the x-axis. It is clear that the differences in results are due to the varying satellite altitudes and the downlink operational frequency.

SUMMARY

- From the presented results, we can conclude that the pure statistical model is the weakest model, with respect to physical–statistical models, in predicting fading phenomena and link budget in LSC links.
- Also, we must state that in almost all simulations, the physical–statistical model based on built-up profiles described by the corresponding CDFs (14.19) and (14.22), as well as by (14.24), following the multiparametric model, are the best fit to measurements with respect to pure statistical models. Furthermore, the difference between two physical–statistical approaches, the Saunders and Evans [2, 3, 14, 15] and the multiparametric [20], on how to describe the overlay profile of the buildings is small, but the multiparametric model is much simpler to use and implement. The reason for this is the fact that for the physical–statistical model, we need exact knowledge of the distribution of the heights of the buildings in each city. This information is sometimes difficult to measure at every land site. On the contrary, for the multiparametric model, we need to know only the heights of the smallest and highest buildings in the city and also the average height of buildings in every land site. From these values, we could easily find the "relief" parameter n, using Equation (14.17).
- Both the Saunders and Evans physical–statistical and multiparametric stochastic approaches are more accurate models, which can be used in predicting fading phenomena and link budgets both for personal and mobile land satellite radio communication links.
- A simulation tool based on the combination of pure statistical and physical–statistical model was shown for the link budget performance and for the outage probability of path loss prediction for more applicable satellite networks based on radio propagation characteristics within each communication channel.
- The simulation tool is designed to be used both by Land Mobile Satellite (LMS) designers and by customers. The designers can determine, using the simulator, the best satellite constellation and channel characteristics for each satellite system, in terms of desired performance and cost. The customers can use the tool to determine which system is best suited for their needs in their specific location in the world.

We can outline some practical conclusions and remarks, which emphasize the advantages in LMS systems. They are

- A high elevation angle between satellite and LMS customer increases the link quality significantly. The main advantage in using the LMS system is emphasized dramatically in urban environments, where one satellite in a high elevation angle eliminates the use of dozens of base stations.
- LEO satellites provide better link quality than MEO satellites but demand more satellites to fully cover the Earth's surface. The main obstacle in designing an LMS system is determining the best-suited satellite constellation for potential customers
- A system designed for personal stationary subscriber will be very much different. from a system designed for MSs. The proposed tools, described by (14.24) or (14.25), give the designer the option to set its own user-defined system and check if it suits its customers.

REFERENCES

1 Farserotu, J. and R. Prasad, *IP/ATM Mobile Satellite Networks*, Artech House, Boston-London, 2002.

2 Saunders, S.R., *Antennas and Propagation for Wireless Communication Systems*, John Wiley& Sons, New York, 2001.

3 Evans, J.V., "Satellite systems for personal communications," *Proc. IEEE*, Vol. 86, No. 7, 1998, pp. 1325–1341.

4 Wu, W.W., "Satellite communication," *Proc. IEEE*, Vol. 85, No. 6, 1997, pp. 998–1010.

5 Fontan, F.P., J.P. Gonzalez, M.J.S. Ferreiro, M.A.V. Castro et al., "Complex envelope three-state Markov model based simulator for the narrow-band LMS channel," *Int. J. Satellite Communications*, Vol. 15, No. 1, 1997, pp.1–15.

6 Barts, R.M., W.L. Stutzman, "Modeling and simulation of mobile satellite propagation," *IEEE Trans. Antennas Propagat.*, Vol. 40, No. 4, 1992, pp.375–382.

7 Loo, C. and J.S. Butterworth, "Land mobile satellite channel measurements and modeling," *Proc. IEEE*, Vol. 86, No. 7, 1998, pp. 1442–1462.

8 Vatalaro, F. and F. Mazzenga, "Statistical channel modeling and performance evaluation in satellite personal communications," *Int. J. Satellite Communications*, Vol. 16, No. 2, 1998, pp. 249–255.

9 Loo, C., "A statistical model for land mobile satellite link," *IEEE Trans. Veh. Technol.*, Vol. VT-34, No. 3, 1985, pp.122–127.

10 Patzold, M., U. Killat, and F. Laue, "An extended Suzuki model for land mobile satellite channels and its statistical properties," *IEEE Trans. Veh. Technol.*, Vol. 47, No. 2, 1998, pp. 617–630.

11 Corazza, G.E. and F. Vatalaro, "A statistical model for land mobile satellite channels and its application on nongeostationary orbit systems," *IEEE Trans. Veh. Technol.*, Vol. 43, No. 3, 1994, pp.738–741.

12 Xie, Y. and Y. Fang, "A general statistical channel model for mobile satellite systems," *IEEE Trans. Veh. Technol.*, Vol. 49, No. 3, 2000, pp.744–752.

13 Abdi, A., W.C. Lau, M.-S. Alouini, and M. Kaveh, "A new simple model for land mobile satellite channels: first- and second-order statistics," *IEEE Trans. Wireless Communications.*, Vol. 2, No. 3, 2003, pp. 519–528.

14 Saunders, S.R. and B.G. Evans, "A physical model of shadowing probability for land mobile satellite propagation," *Electronics Letters*, Vol. 32, No. 17, 1996, pp. 1548–1549.

15 Tzaras, C., S.R. Saunders, and B.G. Evans, "A tap-gain process for wideband mobile satellite PCN channels," *Proc. COST 252/259 Joint Workshop*, Bradford, UK, April 21–22, 1998, pp. 156–161.

16 Lutz, E., D. Cygan, M. Dippold, F. Dolainsky, W. Papke, "The land mobile satellite communication channel-recording, statistics and channel model," *IEEE Trans. Veh. Technol.*, Vol. 40, No. 2, 1991, pp.375–385.

17 Butt, G., G. Evans, and M. Richharia, "Narrowband channel statistics from multi-band propagation measurements applicable to high elevation angle land-mobile satellite systems," *IEEE Trans. Select. Areas Communic.*, Vol. 10, No. 8, 1992, pp. 1219–1226.

18 Parks, M.A.N., B.G. Evans, G. Butt, and S. Buonomo, "Simultaneous wideband propagation measurements for mobile satellite communication systems at L- and S-bands," *Proc. 16th Int. Communic. Systems Conf.*, Washington DC, 1996, pp. 929–936.

19 Karasawa, Y., K. Kimura, and K. Minamisono, "Analysis of availability improvement in LMSS by means of satellite diversity based on three-state propagation state model," *IEEE Trans. Veh. Technol.*, Vol. 46, No. 4, 1997, pp. 1047–1056.

20 Blaunstein, N., Y. Cohen, and M. Hayakawa, "Prediction of fading phenomena in land-satellite communication links," *Radio Sci.*, Vol. 45, RS6005, doi:10.1029/2010RS004352, 2010, pp. 1–13.

21 Blaunstein, N. and M. Levin, "VHF/UHF wave attenuation in a city with regularly spaced buildings," *Radio Sci.*, Vol. 31, No. 2, pp. 313–323, 1996.

22 Blaunstein, N., "Distribution of angle-of-arrival and delay from array of building placed on rough terrain for various elevation of base station antenna," *Journal of Communic. and Networks*, Vol. 2, No. 4, 2000, pp. 305–316.

23 Blaunstein, N., D. Katz, D. Censor, A. Freedman, I. Matityahu, and I. Gur-Arie, "Prediction of loss characteristics in built-up areas with various buildings' overlay profiles," *IEEE Anten. Propagat. Magazine*, Vol. 43, No. 6, 2001, pp. 181–191.

24 Summers, R.A. and R.J. Lepkowski, "ARIES: Global communication through a constellation of low earth orbit satellites," *AIAA 14th Int. Communications Satellite Systems Conf., Collection of Technical Papers*, Washington, DC, 1992, pp. 628–638.

25 Sterling, D.E. and J.E. Harlelid, "The Iridium^{TM} System – A revolutionary satellite communications system developed with innovative applications of technology," *Proc. IEEE Military Satellite Communications Conf.*, McLean, VA, 1991, pp. 436–440.

26 Brunt, P., "IRIDIUM – Overview and status," *Space Communic.*, Vol. 14, No. 1, 1996, pp. 61–68.

27 Wiedeman, R.A., A.B. Salmasi, and D. Rouffet, "Globalstar: Mobile communications wherever you are," *AIAA 14th Int. Communications Satellite Systems Conf., Collection of Technical Papers*, Washington, DC, 1992, pp. 123–129.

28 Smith, D., "Operational innovations for the 48-satellite Globalstar constellation," *AIAA 15th Int. Communications Satellite Systems Conf., Collection of Technical Papers*, San Diego, CA, Feb./Mar. 1994, pp. 1107–1112.

29 Singh, J., "Project 21/Inmarsat-P: Putting reality into the handheld satphone vision," *Int. Mobile Satellite Conf.*, Paris, France, October 1993, pp. 834–839.

Index

Absorption, of radiowaves, 540, 602
Adaptive antennas, 216
    applications, 30, 219
Adaptive array antennas, 218, 219, 220
Adaptive beamforming, 219, 234–237
Additive white Gaussian noise
    (AWGN), 2, 4, 515
Aerosols, 537
Amplitude field pattern, 35
Analog beamforming, 226
Angle of arrival (AOA), 190, 252
Angle diversity, 245
Angular spread, 253, 254, 259
Antenna(s). *See also* Adaptive antennas
    array, 217, 220, 221
    directivity dependence, 391
    efficiency, 42
    gain, 42
    main loop azimuth dependence, 388
    tilt dependence, 387
AOA-distance joint signal power
    distribution, 343–344
AOA-EOA joint signal power
    distribution, 332–339
AOA-TD (TOA) joint signal power
    distribution, 329–331
Array factor, 219

Atmosphere, 536
    turbulence, 539
Autocorrelation function, 625
Average fade duration (AFD), 27, 29,
    531
Average refractive index, 625
AWGN channel, 526
Axial ratio, 44
Azimuth of arrival (AOA), 281, 286
    -distance joint signal power
        distribution, 343–344
    -EOA joint signal power
        distribution, 332–339
    -TD (TOA) joint signal power
        distribution, 329–331
Azimuth-distance field distribution,
    343

Back lobes, 36
Bandpath (RF) signal, 13
Base station, 55
Beam-forming matrix, 234
Beam-forming technique, 225
Beam tracing algorithm, 185
Bending, of radiowave trajectory, 605,
    506
Bertoni's model, 194–197

Radio Propagation and Adaptive Antennas for Wireless Communication Networks: Terrestrial, Atmospheric, and Ionospheric, Second Edition. Nathan Blaunstein and Christos G. Christodoulou.
© 2014 John Wiley & Sons, Inc. Published 2014 by John Wiley & Sons, Inc.

WILEY SERIES IN MICROWAVE AND OPTICAL ENGINEERING

KAI CHANG, Editor
Texas A&M University